Characterization, Applications and New Technologies of Civil Engineering Materials and Structures

Characterization, Applications and New Technologies of Civil Engineering Materials and Structures

Editors

Wensheng Wang
Qinglin Guo
Jue Li

Basel • Beijing • Wuhan • Barcelona • Belgrade • Novi Sad • Cluj • Manchester

Editors

Wensheng Wang	Qinglin Guo	Jue Li
College of Transportation	School of Civil Engineering	College of Traffic
Jilin University	Hebei University	& Transportation
Changchun	of Engineering	Chongqing Jiaotong
China	Handan	University
	China	Chongqing
		China

Editorial Office

MDPI

St. Alban-Anlage 66

4052 Basel, Switzerland

This is a reprint of articles from the Special Issue published online in the open access journal *Materials* (ISSN 1996-1944) (available at: www.mdpi.com/journal/materials/special_issues/7BI450OD53).

For citation purposes, cite each article independently as indicated on the article page online and as indicated below:

Lastname, A.A.; Lastname, B.B. Article Title. *Journal Name* **Year**, *Volume Number*, Page Range.

ISBN 978-3-7258-1102-1 (Hbk)

ISBN 978-3-7258-1101-4 (PDF)

doi.org/10.3390/books978-3-7258-1101-4

Contents

 *materials*

MDPI

Editorial

Characterization, Applications and New Technologies of Civil Engineering Materials and Structures

Wensheng Wang [1],* , Qinglin Guo [2],* and Jue Li [3],*

1   College of Transportation, Jilin University, Changchun 130022, China
2   School of Civil Engineering, Hebei University of Engineering, Handan 056038, China
3   College of Traffic & Transportation, Chongqing Jiaotong University, Chongqing 400074, China
*   Correspondence: wangws@jlu.edu.cn (W.W.); guoql@hebeu.edu.cn (Q.G.); lijue1207@cqjtu.edu.cn (J.L.)

1. Introduction

With the continuous development of large-scale maintenance of infrastructure, accurate, reasonable, and efficient mechanical behavior evaluation and performance prediction of civil materials and structures have become the keys to improving service durability and intelligent maintenance management for infrastructure. The multi-component composition, multi-scale characteristics, and multi-field dependence of civil materials lead to extremely complex mechanical behaviors. The phenomenological method based on empirical tests is an important means to understand and evaluate civil materials, but its low efficiency and high consumption cannot meet the design and application requirements of civil materials. Numerical simulation has gradually become an important tool to study and understand the mechanical behavior of civil materials and structures, including the finite element method, discrete element method, molecular dynamics simulation, etc. In addition, the rapid development of numerical simulations has greatly promoted the modeling and simulation of civil materials. Artificial intelligence is known for including powerful computational techniques and is now being used more frequently by civil engineers to solve real problems related to civil materials and structures. This field is under rapid development, and numerous novel technologies have been proposed to characterize and evaluate the performance of civil materials and structures. Considering the above, the aim of this Special Issue is to bring together cutting-edge research and application. To share, present, and discuss innovative materials, structures, and characterization methods may help us further develop the technology used in civil engineering.

2. Highlights in the Present Issue

A total of twenty research papers and one systematic review are presented in this Special Issue, covering the application of solid wastes in civil engineering materials, environmental impact, material properties, mechanical behavior of civil engineering materials, etc. Multiple studies explored the use of industrial solid waste and various materials in enhancing soil, concrete, asphalt, and tunnel grouting materials, aiming to optimize their properties and processing parameters. Qiu et al. used an all-solid-waste binder composed of general industrial solid waste calcium carbide residue, ground granulated blast furnace slag, and fly ash instead of cement and combined with polypropylene fiber to strengthen the silty soil [1]. Lv et al. investigated the influence of recycled aggregate on different cement mortar pretreatment methods, including natural aggregate concrete and concrete with recycled aggregate after the wetting pretreatment and cement mortar pretreatment [2]. Pan et al. investigated the storage stability of crumb rubber asphalt with epoxidized soybean oil and polyester fiber based on the response surface methodology [3]. Liu et al. explored the influence of material design parameters on the physical and mechanical properties of recycled asphalt to determine the optimal processing parameters using response surface methodology [4]. Jiang et al. presented a comprehensive review

Citation: Wang, W.; Guo, Q.; Li, J. Characterization, Applications and New Technologies of Civil Engineering Materials and Structures. *Materials* **2024**, *17*, 2058. https://doi.org/10.3390/ma17092058

Received: 17 April 2024
Accepted: 24 April 2024
Published: 27 April 2024

of the recent advancements in tunnel grouting materials using industrial solid waste [5]. Xu et al. used controlled low-strength materials for backfilling narrow spaces to achieve efficient and high-quality backfilling effects for pipeline trenches [6]. Liu et al. investigated the effects of foam densities, foaming agent types, and the presence of slurry on the pore structure and mechanical properties of foamed lightweight soils to evaluate their stability [7]. Zhao et al. examined the foam stability of different foaming agent dilution ratios by analyzing foam density, foaming ratio, settlement distance, and bleeding volume, developing and evaluating the effectiveness of foamed lightweight soil grouting material for goaf treatment [8]. Chong et al. investigated the influence of dry–wet cycles and sulfate attack on the performance of magnesium potassium phosphate cement as well as the effect of waterglass [9]. Xu et al. evaluated the effect of polyphosphoric acid on the ultraviolet aging resistance of styrene butadiene rubber-modified asphalt [10]. Klimczak et al. used the respective methodology to identify the parameters of the well-established Bodner–Partom material model [11]. Tan et al. investigated the applicability and reliability of ultrasonic testing technology in evaluating the performance (ultrasonic, indirect tensile, compression, and dynamic modulus tests) of asphalt mixtures at various temperatures [12]. Diao et al. selected hydrated lime and basalt fiber to prepare a modified asphalt mixture to enhance the fatigue resistance of aging asphalt pavement [13]. Yang et al. conducted monotonic tensile tests on three kinds of stone-mastic asphalt mixtures to determine the optimum dosage of each kind of fiber [14]. Civil materials and structures are essential to engineering, but are very vulnerable to harsh environments, freeze–thaw actions, loading, etc. These factors can cause damage to civil structure and infrastructure by exerting negative influences on the mechanical and functional properties of civil materials (e.g., asphalt and cement concrete). Liu et al. discussed a series of tensile and puncture tests of geotextiles under different low temperatures and at different moisture content levels [15]. Haimei et al. clarified the influence of moisture and aging on the healing ability of asphalt–aggregate bonding interfaces based on laboratory experiments and the meso-finite element method [16]. Yang et al. evaluated the effect of the number of drains per pile and the orientation of the drains relative to the direction of shaking [17]. Tang et al. studied the flexural performance of the combined structure of steel pipe and steel slag powder in ultra-high-performance concrete via a pure bending test and a finite element simulation [18]. Wang et al. investigated the strength and failure characteristics of silty mudstone using different stress paths through triaxial unloading tests [19]. Obara et al. carried out a parametric analysis of tensegrity structures subjected to periodic loads, determining the main region of dynamic instability [20]. Zhang et al. investigated the temperature effect, including hydration heat and the sunshine temperature effect, of the construction process of a rigid frame-tied steel box arch bridge [21].

Author Contributions: Conceptualization, W.W., Q.G. and J.L.; formal analysis, W.W., Q.G. and J.L.; writing—original draft preparation, W.W.; writing—review and editing, Q.G. and J.L.; project administration, W.W.; funding acquisition, W.W. All authors have read and agreed to the published version of the manuscript.

Funding: This research was funded by the Natural Science Foundation of Jilin Province (grant number: 20240101141JC), and the Scientific and Technological Research Project of the Education Department of Jilin Province (grant number: JJKH20241300KJ).

Acknowledgments: Thanks to all the authors and peer reviewers for their valuable contributions to this Special Issue titled 'Characterization, Applications, and New Technologies of Civil Engineering Materials and Structures'. I would also like to express my gratitude to all the staff and people involved in this Special Issue.

Conflicts of Interest: The authors declare no conflicts of interest.

References

1. Qiu, X.; Yang, J.; Wu, Y.; Yan, L.; Liu, Q. Effect of fiber content on mechanical properties of fiber-reinforced cgf all-solid-waste binder-solidified soil. *Materials* **2024**, *17*, 388. [CrossRef] [PubMed]
2. Lv, D.B.; Huang, K.N.; Wang, W.S. Influence of pretreatment methods on compressive performance improvement and failure mechanism analysis of recycled aggregate concrete. *Materials* **2023**, *16*, 3807. [CrossRef] [PubMed]
3. Pan, J.; Jin, J.; Liu, S.; Xiao, M.; Qian, G.; Wang, Z. Synergistic effects of epoxidized soybean oil and polyester fiber on crumb rubber modified asphalt using response surface methodology. *Materials* **2023**, *16*, 3469. [CrossRef] [PubMed]
4. Liu, H.; Wang, J.; Lu, W.; Zhang, N. Optimization design and mechanical performances of plant-mix hot recycled asphalt using response surface methodology. *Materials* **2023**, *16*, 5863. [CrossRef] [PubMed]
5. Jiang, B.; Wu, M.; Wu, S.; Zheng, A.; He, S. A review on development of industrial solid waste in tunnel grouting materials: Feasibility, performance, and prospects. *Materials* **2023**, *16*, 6848. [CrossRef]
6. Xu, J.; Luo, Q.; Tang, Y.; Zeng, Z.; Liao, J. Experimental study and application of controlled low-strength materials in trench backfilling in suqian city, china. *Materials* **2024**, *17*, 775. [CrossRef] [PubMed]
7. Liu, H.; Shen, C.; Li, J.; Zhang, G.; Wang, Y.; Wan, H. Study on the effect of foam stability on the properties of foamed lightweight soils. *Materials* **2023**, *16*, 6225. [CrossRef] [PubMed]
8. Zhao, Z.Z.; Chen, J.; Zhang, Y.P.; Jiang, T.H.; Wang, W.W. Study on preparation and performance of foamed lightweight soil grouting material for goaf treatment. *Materials* **2023**, *16*, 4325. [CrossRef]
9. Chong, L.; Yang, J.; Chang, J.; Aierken, A.; Liu, H.; Liang, C.; Tan, D. Corrosion behavior of magnesium potassium phosphate cement under wet-dry cycle and sulfate attack. *Materials* **2023**, *16*, 1101. [CrossRef]
10. Xu, Y.; Niu, K.; Zhu, H.; Chen, R.; Ou, L. Evaluating the effects of polyphosphoric acid (ppa) on the anti-ultraviolet aging properties of sbr-modified asphalt. *Materials* **2023**, *16*, 2784. [CrossRef]
11. Klimczak, M.; Tekieli, M.; Zielinski, P.; Strzepek, M. Dic-enhanced identification of bodner-partom model parameters for bitumen binder. *Materials* **2023**, *16*, 1856. [CrossRef] [PubMed]
12. Tan, X.; Wu, C.; Li, L.; Li, H.; Liang, C.; Zhao, Y.; Li, H.; Zhao, J.; Wang, F. Sensitivity and reliability analysis of ultrasonic pulse parameters in evaluating the laboratory properties of asphalt mixtures. *Materials* **2023**, *16*, 6852. [CrossRef] [PubMed]
13. Diao, H.; Ling, T.; Zhang, Z.; Peng, B.; Huang, Q. Multiscale fatigue performance evaluation of hydrated lime and basalt fiber modified asphalt mixture. *Materials* **2023**, *16*, 3608. [CrossRef] [PubMed]
14. Yang, S.; Zhou, Z.; Li, K. Influence of fiber type and dosage on tensile property of asphalt mixture using direct tensile test. *Materials* **2023**, *16*, 822. [CrossRef]
15. Liu, L.; Zhang, H.; Zhu, J.; Lv, S.; Liu, L. Evaluation of the tensile and puncture properties of geotextiles influenced by soil moisture under freezing conditions. *Materials* **2024**, *17*, 376. [CrossRef] [PubMed]
16. Haimei; Li, L.; Guo, Q.; Zhao, T.; Zuo, P.; E, F. Effects of aging and immersion on the healing property of asphalt-aggregate interface and relationship to the healing potential of asphalt mixture. *Materials* **2023**, *16*, 3574. [CrossRef]
17. Yang, Y.; Xin, G.; Chen, Y.; Stuedlein, A.W.; Wang, C. Seismic performance of drained piles in layered soils. *Materials* **2023**, *16*, 5868. [CrossRef]
18. Tang, X.; Feng, C.; Chang, J.; Ma, J.; Hu, X. Research on the flexural performance of steel pipe steel slag powder ultra-high-performance concrete components. *Materials* **2023**, *16*, 5960. [CrossRef] [PubMed]
19. Wang, J.; Zhang, H.; Qi, S.; Bian, H.; Long, B.; Duan, X. Study on the strength and failure characteristics of silty mudstone using different unloading paths. *Materials* **2023**, *16*, 5155. [CrossRef] [PubMed]
20. Obara, P.; Tomasik, J. Dynamic stability of tensegrity structures—Part II: The periodic external load. *Materials* **2023**, *16*, 4564. [CrossRef]
21. Zhang, T.; Wang, H.; Luo, Y.J.; Yuan, Y.; Wang, W.S. Hydration heat control of mass concrete by pipe cooling method and on-site monitoring-based influence analysis of temperature for a steel box arch bridge construction. *Materials* **2023**, *16*, 2925. [CrossRef]

Article

Effect of Fiber Content on Mechanical Properties of Fiber-Reinforced CGF All-Solid-Waste Binder-Solidified Soil

Xinyi Qiu [1,2], Junjie Yang [1,2], Yalei Wu [1,2,*], Lijun Yan [1,2] and Qiang Liu [1,2]

1   College of Environmental Science and Engineering, Ocean University of China, Qingdao 266100, China;
     qxy839540904@163.com (X.Q.); jjyang@ouc.edu.cn (J.Y.); yanlijun313@126.com (L.Y.); oucliuq@163.com (Q.L.)
2   The Key Laboratory of Marine Environment and Ecology of the Ministry of Education, Ocean University of
     China, Qingdao 266100, China
*   Correspondence: wuyalei@ouc.edu.cn; Tel.: +86-130-7378-7368

Abstract: In order to realize the resource utilization of solid waste and improve the tensile strength and toughness of soil, CCR-GGBS-FA all-solid-waste binder (CGF) composed of general industrial solid waste calcium carbide residue (CCR), ground granulated blast furnace slag (GGBS) and fly ash (FA) was used instead of cement and combined with polypropylene fiber to strengthen the silty soil taken from Dongying City, China. An unconfined compressive strength test (UCS test) and a uniaxial tensile test (UT test) were carried out on 10 groups of samples with five different fiber contents to uncover the effect of fiber content on tensile and compressive properties, and the reinforcement mechanism was studied using a scanning electron microscopy (SEM) test. The test results show that the unconfined compressive strength, the uniaxial tensile strength, the deformation modulus, the tensile modulus, the fracture energy and the residual strength of fiber-reinforced CGF-solidified soil are significantly improved compared with nonfiber-solidified soil. The compressive strength and the tensile strength of polypropylene-fiber-reinforced CGF-solidified soil reach the maximum value when the fiber content is 0.25%, as the unconfined compressive strength and the tensile strength are 3985.7 kPa and 905.9 kPa, respectively, which are 116.60% and 186.16% higher than those of nonfiber-solidified soil, respectively. The macro–micro tests identify that the hydration products generated by CGF improve the compactness through gelling and filling in solidified soil, and the fiber enhances the resistance to deformation by bridging and forming a three-dimensional network structure. The addition of fiber effectively improves the toughness and stiffness of solidified soil and makes the failure mode of CGF-solidified soil transition from typical brittle failure to plastic failure. The research results can provide a theoretical basis for the application of fiber-reinforced CGF-solidified soil in practical engineering.

Keywords: fiber content; alkali-activated binder; tensile strength; fracture energy; reinforcement mechanism; tensile–compression ratio

Citation: Qiu, X.; Yang, J.; Wu, Y.; Yan, L.; Liu, Q. Effect of Fiber Content on Mechanical Properties of Fiber-Reinforced CGF All-Solid-Waste Binder-Solidified Soil. *Materials* **2024**, *17*, 388. https://doi.org/10.3390/ma17020388

Academic Editor: René de Borst

Received: 30 November 2023
Revised: 27 December 2023
Accepted: 29 December 2023
Published: 12 January 2024

1. Introduction

Tensile failure is one of the basic states of soil [1], as it is difficult for soil to resist tensile forces; under the action of external forces, soil structure is easily destroyed, leading to various engineering and geological problems. For instance, problems have been posed in the following areas: electric towers, due to the tension of the wire caused by tower foundation soil pile cracking; water conservation projects, with embankment slopes occurring under mixing piles due to bending or stretching and cracking, leading to destruction; and tunnels, with soil around the holes experiencing tensile stress, leading to tension cracks, etc. [2–5]. Therefore, it is of significant engineering importance to properly improve soil to effectively enhance its tensile strength.

Currently, soil improvement methods can be divided into three categories: chemical reinforcement, physical reinforcement and biological reinforcement [6]. Cement–soil is the commonly used chemical reinforcement method; its compressive strength has been greatly improved in comparison with soil, but it remains a relatively brittle material. Under brittle damage or tension cracking, the strength of cement–soil is reduced abruptly, and it is generally difficult to meet the deformation and stability requirements of engineering under complex stress conditions [7–9]. In order to increase the tensile strength and improve the plasticity of cement–soil, a composite reinforcement method combining chemical reinforcement and physical reinforcement by incorporating fibers has emerged in recent years [10,11]. In general, fibers have the advantages of high tensile strength and good dispersion, and they form an isotropic, high-tensile-strength composite with cement–soil [12–16]. The strength of fiber-reinforced, solidified soil is much greater than that of pure fiber–soil and solidified soil without fibers [17–19]. For example, the incorporation of PET fiber can effectively improve the tensile strength and durability of cement-based composite materials by providing corrosion resistance [20].

The soil type, binder type, curing condition, fiber type and content affect the strength of fiber-reinforced, solidified soil [21]. The combination of polypropylene fiber and fly ash can effectively increase the peak strain of cement–soil and maintain a certain residual stress after failure. Polypropylene fiber enhances the deformation resistance of fiber–fly ash cement–soil through anchorage and bridging, and improves its failure mode [3,22]. The addition of the plant fiber can effectively improve the tensile properties of cement-based materials, including the fracture energy and the tensile modulus, as well as increase the ductility of cement-based materials [23]. Simultaneously, the fiber enhances the tensile strength of solidified soil significantly more than the compressive strength [24]. As such, there is a good linear relationship between the tensile strength and compressive strength of fiber-reinforced solidified soil, i.e., the ratio of tensile and compressive strength is 0.105 [25]. In addition, the fiber content determines the strength variation in the fiber-reinforced, solidified soil to some extent. There is an optimum fiber content, where the strength increases with a fiber content below the optimum fiber content but decreases above the optimum fiber content [26]. The optimum fiber content is related to the material to be solidified and the fiber type [17,27–30]. On the other hand, the interaction between fibers and soil has a significant effect on the reinforcement mechanism of fiber-reinforced, solidified soil. The reinforcing effect of fibers mainly depends on the friction and adhesion of the fiber–soil, and the effect is influenced by soil conditions and interfacial contact conditions [25]. In cement–soil, the reinforcing effect of fibers and the cementing effect of cement hydration products on strength enhancement are not just a simple superposition, but enhance the strength of cement–soil through a complex coupling effect [31].

At present, most of the research on fiber-reinforced, solidified soil selects traditional binders such as cement and lime, and the use of all-solid-waste binders to replace traditional binders can alleviate the resource and energy consumption, ecological damage and environmental pollution brought about by the production process of traditional binders. Studies on fiber-reinforced, solidified soil mostly focus on sand and clay, and there are limited studies on silty soil. The tensile modulus, fracture energy and tensile–compression ratio of fiber-reinforced, solidified soil are also important indicators for characterizing tensile strength, and there remains a lack of research in this area.

In this paper, polypropylene fibers were selected as the reinforcing material, and an all-solid-waste binder (abbreviated as CGF) with calcium carbide residue (CCR), ground granulated blast furnace slag (GGBS) and fly ash (FA) as components was used to solidify the silt. The unconfined compressive strength test and the uniaxial tensile test were used to investigate the effect of fiber content on the mechanical properties of fiber-reinforced, solidified soil and the relationship between the tensile modulus and the deformation modulus, as well as between the tensile strength and the compressive strength. The results of this research can provide a theoretical basis for the application of fiber-reinforced, solid waste-based binder-solidified soil in practical engineering.

2. Materials and Methods

2.1. Materials

2.1.1. Test Soil

The soil used in this study was taken from the subsurface soil (30–50 cm) of the first-class terrace near the estuary of the Yellow River in Dongying City, Shandong Province, China. The soil was dried for 12 h, crushed and sieved through a 0.075-mm sieve as the test soil. The basic physical properties of the test soil were tested according to the Standard for Geotechnical Test Methods [32]. The basic physical properties of the test soil are shown in Table 1; the cumulative curve of particle size gradation is shown in Figure 1, and the contents of clay and silt particles are 9.2% and 82.8%, respectively.

Table 1. Basic physical properties of test soil.

Particle Density (G_s)	Plastic Limit (%)	Liquid Limit (%)	Plasticity Index (I_P)
2.68	26.5	29.7	3.2

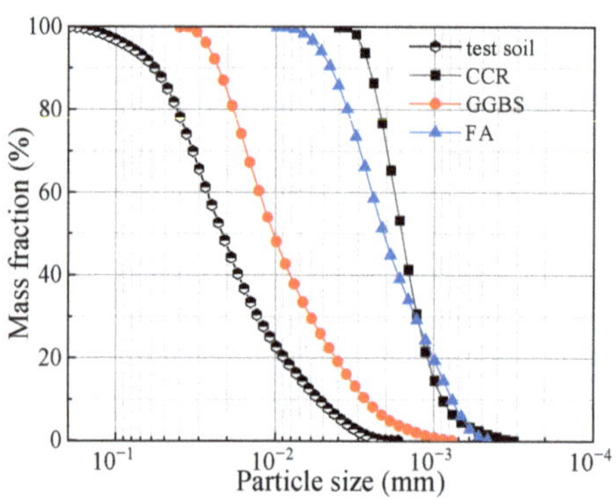

Figure 1. Particle size distribution curves of each component of binder and test soil.

2.1.2. CGF Binder and Polypropylene Fiber

The CGF binder used in the test is an all-solid-waste, alkali-activated binder, which was prepared using raw calcium carbide residue (CCR), ground granulated blast furnace slag (GGBS) and fly ash (FA), according to a mass ratio of 4:4:2 (Figure 2) [33,34]. CCR is a by-product of acetylene obtained from calcium carbide hydrolysis, and its pH value exceeds 12.0, providing an alkaline environment as an alkali activator. GGBS is the waste residue produced in the iron-making process, and FA is fly ash collected by bag dust collectors in coal-fired power plants. The strength of CGF-solidified soils is 1.38–2.30 times higher than that of cement-solidified soil; however, the carbon emissions per unit strength of cement-solidified soils are 71.43–125.02 times higher than those of CGF-solidified soils, and the costs per unit strength are 3.53–5.88 times higher than those of CGF-solidified soils [32]. The chemical compositions of the components of the CGF binder are shown in Table 2, and the particle size distribution curves are shown in Figure 1.

Figure 2. CGF binder's components.

Table 2. Chemical compositions of components of CGF binder.

Component \ Chemical Component	CaO	SiO$_2$	Al$_2$O$_3$	MgO	Fe$_2$O$_3$	Na$_2$O	K$_2$O	SO$_3$	P$_2$O$_5$
CCR	68.8	3.59	1.56	1.21	0.09	/	0.028	0.75	/
GGBS	41.17	29.47	13.61	8.04	0.425	0.676	0.354	4.90	0.03
FA	6.60	61.29	12.66	0.02	4.48	3.75	1.32	0.66	0.01

The reinforcing material selected for the test was polypropylene fiber with a length of 12 mm (Figure 3), and its physical and mechanical properties are shown in Table 3.

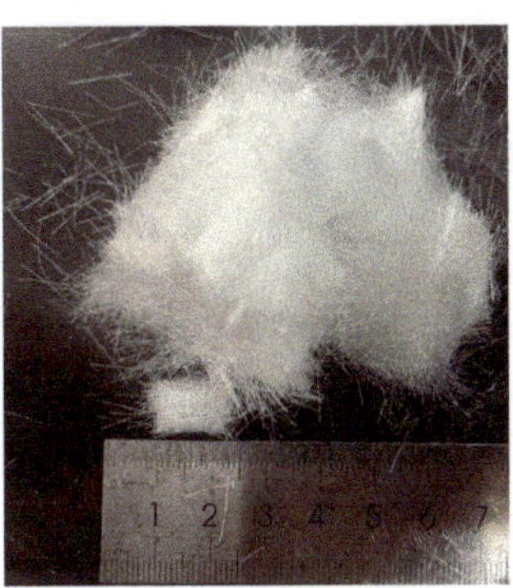

Figure 3. Polypropylene fiber for testing.

Table 3. Physical and mechanical parameters of polypropylene fiber.

Type	Tensile Strength (MPa)	Ultimate Elongation in Percent (%)	Elasticity Modulus (MPa)	Diameter (μm)	Length (mm)
Fasciculate monofilament	469	28.4	4236	32.7	12

2.2. Testing Equipment

2.2.1. Uniaxial Tensile Test Equipment

The uniaxial tensile test equipment is shown in Figure 4, including the uniaxial tensile platform, the data collector and the computer (data output). The loading mode was equal to the strain control, and the loading rate was set to 0.5 mm/min. The sample is an "hourglass" sample with a thickness of 20 mm. In order to limit the tensile failure location to a small range, the tensile section was set at 5 mm to alleviate the influence of stress concentration, and the guide was smooth with low friction [1].

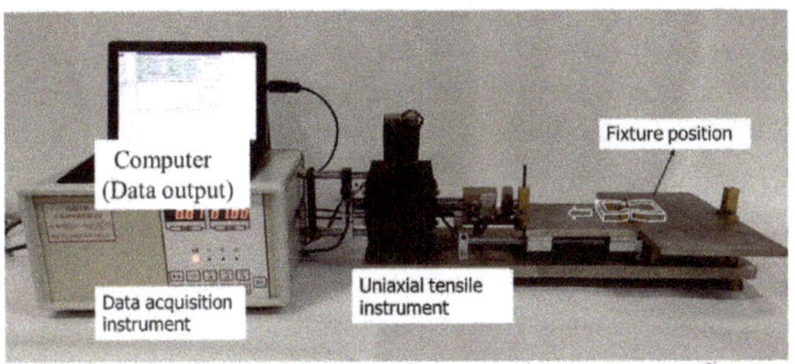

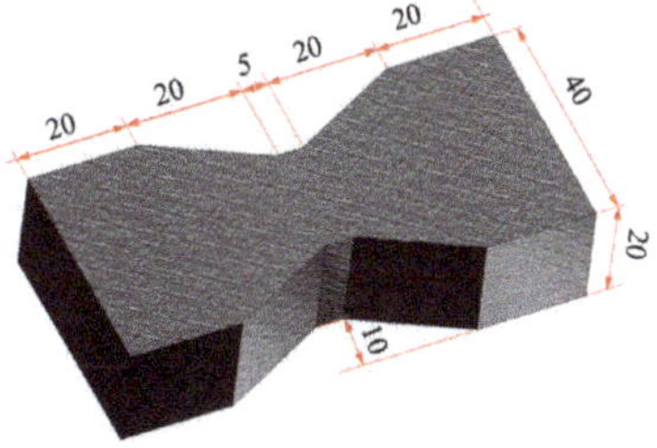

Figure 4. Uniaxial tensile test equipment and specimen size (unit: mm).

2.2.2. Unconfined Compressive Strength Test Equipment

The WCY-1 unconfined compressive strength test equipment is shown in Figure 5, used according to the Standard for Geotechnical Test Methods [32]. This device can apply a load to the sample in a controlled manner. The loading rate was set to 1 mm/min, and the maximum measuring range is 30 kN. The specimens are cylindrical, with a diameter of 50 mm and a height of 100 mm.

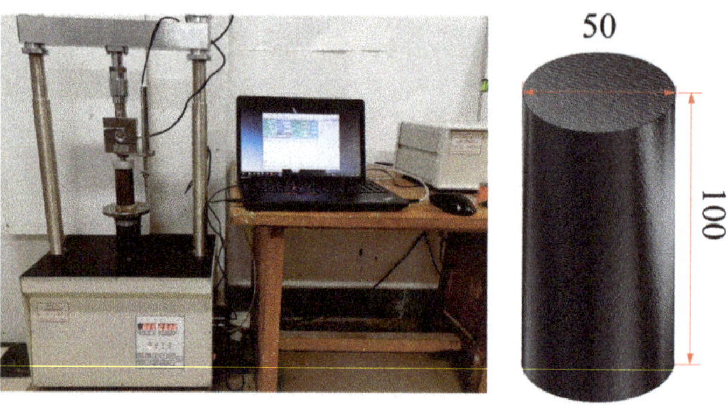

Figure 5. Unconfined compressive strength test equipment and specimen size (unit: mm).

2.3. Testing Program

This paper focuses on the effect of fiber content on the tensile strength properties of fiber-reinforced CGF-solidified soil. For this purpose, the experimental program was designed as shown in Table 4. Fiber content is the mass ratio of fiber to dry soil and binder content is the mass ratio of binder to dry soil.

Table 4. Testing program.

Test Soil	Water Content	Binder Type	Binder Content (%)	Fiber Content (%)	Curing Age (d)	Type of Test
Dongying silt	$1.2w_L$	CGF	15	0, 0.1, 0.25, 0.4, 0.5	28	Uniaxial tensile test, inconfined compression test, SEM

Note: CGF is prepared using CCR, GGBS and FA according to the mass ratio of 4:4:2.

2.4. Sample Preparation Method

The binder was mixed with fiber and test soil to form a dry mixture, and water was added to the dry mixture to form a fiber-reinforced CGF-solidified soil mixture after mixing. The uniaxial tensile specimens were filled in a single layer, and the unconfined compressive specimens were filled in three layers. After filling the sample, the sample and the mold were placed into a curing chamber for standard curing (temperature 20 ± 2 °C, humidity $\geq 95\%$), and the mold was removed after 24 h. The demolded sample was placed in a sealed bag, and standard maintenance was continued until the set curing age, and then a uniaxial tensile test and an unconfined compressive test were carried out. After the strength test, the sample was broken, and a fresh section was selected for SEM testing. The sample preparation and testing process are shown in Figure 6.

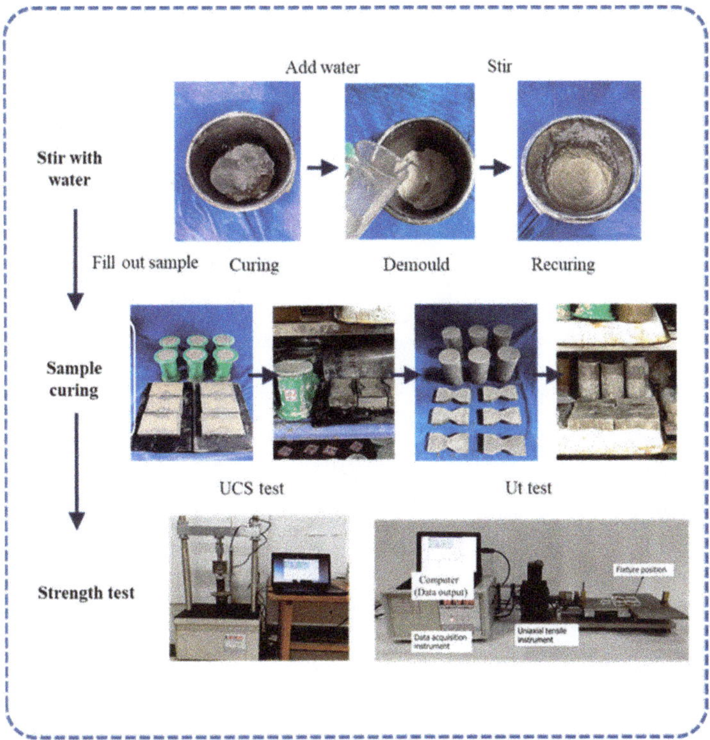

Figure 6. Sample preparation and testing process.

3. Results and Discussion

3.1. Deformation Characteristics

3.1.1. Stress–Strain Relationship and Failure Mode

Figure 7 shows the stress–strain curves of the fiber-reinforced, CGF-solidified soils with different fiber contents obtained from the unconfined compressive strength tests. The stress–strain curves of the solidified soils with and without fibers both showed peak stress. The peak stresses of the fiber-reinforced, CGF-solidified soils were all higher than those of the solidified soils without fibers, and the peak stresses of the solidified soils were the highest at a fiber content of 0.25%.

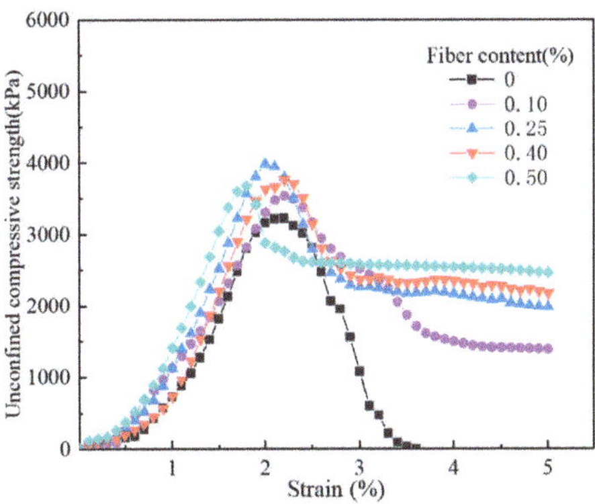

Figure 7. Unconfined compressive strength curve of fiber-reinforced, solidified soil.

As can be seen in Figure 7, the stress–strain of the solidified soil increases linearly before reaching its peak stress. After the peak stress, the stress of the solidified soil without fiber dropped abruptly, the cracks induced by the damage extended in the load direction (Figure 8a), and the stress curve declined rapidly with the strain; at the same time, the specimens were seriously damaged and presented typical brittle failure characteristics. However, the stress of the fiber-reinforced CGF-solidified soil decreased slowly and maintained a certain residual stress, and the stress–strain curves show a strain-softening trend. Meanwhile, cracks appeared on the surface of the specimens and gradually lengthened and developed with the increase in stress, and the integrity of the specimen was still high under the adhesive effect of the fibers (Figure 8b); this represents plastic failure damage. In addition, the residual stress increased with the increase in fiber content. The above results indicate that fibers can effectively improve the toughness of solidified soil, and the failure mode of specimens gradually transitions from brittle damage to plastic failure [35]. The addition of fiber can increase the peak stress of solidified soil and retain some residual stress after failure, which is in line with fiber-reinforced cement soils [13,22,36].

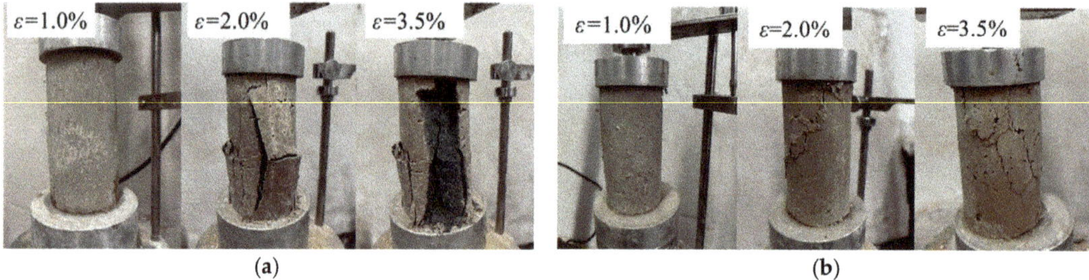

Figure 8. Failure morphology of unconfined compressive specimens at each stage: (**a**) fiber content: 0; (**b**) fiber content: 0.5%.

Figure 9 shows the tensile stress–strain curves of fiber-reinforced CGF-solidified soils with different fiber contents obtained from uniaxial tensile tests. The stress–strain curves of solidified soil with and without fibers both show peak stress. The peak tensile stresses of the fiber-reinforced CGF-solidified soils were all greater than those of the solidified soil without fibers, and the peak tensile stresses were the greatest at 0.25% fiber content.

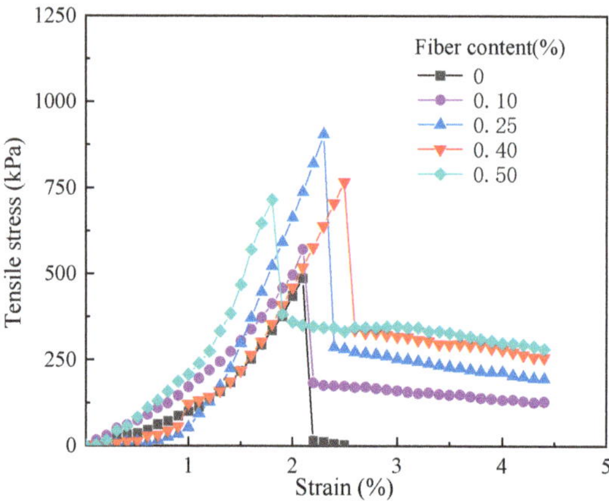

Figure 9. Tensile stress–strain curve of fiber-reinforced, solidified soil.

As can be seen in Figure 9, the stress of the solidified soils with and without fibers showed a linear growth trend with the increase in strain before reaching the peak stress. The tensile stress of the solidified soil without fibers reaches its peak stress and then decreases sharply to zero, presenting a typical brittle failure. Therefore, the tensile stress of the fiber-reinforced CGF-solidified soil decreases slowly with the increase in strain, and there exists a longer strain-softening stage after dropping to approximately 20~40% of the peak stress, which is obviously different from the stress trend of the solidified soil without fibers. This indicates that the incorporation of fiber not only effectively enhanced the peak tensile stress of the solidified soil, but also improved its toughness.

As the strain increases, the fiber-reinforced, CGF-solidified soil specimen produces cracks, and the tensile stress exceeds the maximum tensile stress produced by the solidified fiber-CGF hydration product-soil particle interface. As the tensile stress exceeds the pullout resistance of the fiber, tiny cracks appear perpendicular to the tensile direction in the tensile section, resulting in tensile damage. The cracks continue to expand and extend with the increase in strain, and the fibers play the role of a "Bridge", which is responsible for transferring and dispersing tensile stress; therefore, the specimen began to retain a certain residual stress [18]. The residual stress increases with increasing fiber content and decreases slowly with increasing strain, and the fibers break and the specimen is completely destroyed when the tensile stress exceeds the tensile strength of the fibers (Figure 10). The strength of the solidified soil without fibers was completely destroyed after the peak tensile stress, with no residual stress.

Figure 10. Failure morphology of uniaxial tensile specimen at each stage (fiber content 0.5%).

The tensile damage on the surface of the solidified soil with and without fibers (0.25% fiber content) is shown in Figure 11. The damage locations in the solidified soil were confined to a 5-mm-long neck stretch in the middle of the specimen, with the damage surface nearly perpendicular to the direction of stretching. The damage surface of the solidified soil without fibers is basically flat, which is typical brittle damage. Meanwhile, on the tensile damage surface of the fiber-reinforced, solidified soil at a fiber content of 0.25%, the fibers were partially pulled out and broken under the action of tensile stress, and some solidified soil particles adhered to the surface of the exposed fibers. The fibers play the role of a "bridge" across the fracture surface in the pulling process [16] to hinder further development of cracks and then improve the toughness of the specimens. The bridging effect of fiber in the process of CGF-solidified soil and the phenomenon of "cracking and ceasing" after fracture are consistent with the phenomenon of fiber acting on cement-solidified soil [13].

Figure 11. Tensile failure surface of fiber-reinforced solidified soil.

3.1.2. Determination Method of Strength, Residual Strength and Modulus

As shown in Figure 12, the peak stress is taken as the strength, unconfined compressive strength is obtained from the unconfined compressive strength test (referred to as compressive strength, q_u) and uniaxial tensile strength is obtained from the uniaxial tensile test (referred to as tensile strength, q_t) [1]. The strain corresponding to the strength is the destructive strain (ε_{max}). Take the stress corresponding to the point where the degree of reduction starts to slow down after the peak stress as the residual strength, i.e., set the intersection point of straight line 1, the section of sudden drop in stress after the peak, and straight line 2, the section of the stress–strain-stabilizing section, to be A; the transverse coordinate of the vertical line passing through point A and point B of the intersection point on the stress–strain curve to be the residual strain; and the longitudinal coordinate to be the residual strength.

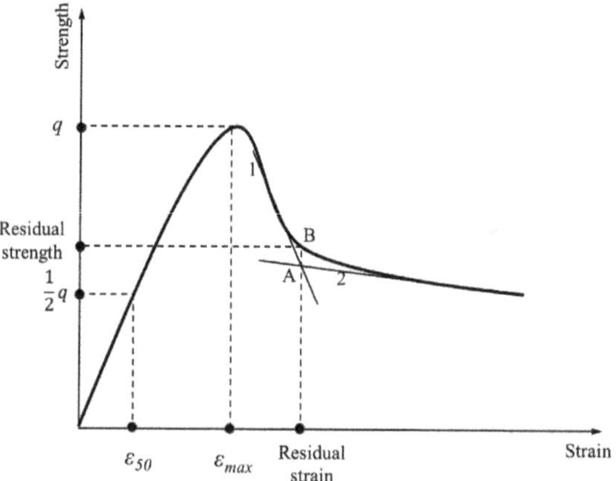

Figure 12. Definition diagram of indicators of strength and deformation characteristics.

The ratio of one-half strength to the corresponding strain is defined as the modulus. The deformation modulus (Equation (1)) [33] and the tensile modulus (Equation (2)) [1] are calculated as follows:

$$E_{50u} = \frac{\frac{1}{2}q_u}{\varepsilon_{50u}} \tag{1}$$

where E_{50u} is the deformation modulus, MPa; q_u is the compressive strength, MPa; and ε_{50u} is the strain corresponding to one half of the compressive strength, %.

$$E_{50t} = \frac{\frac{1}{2}q_t}{\varepsilon_{50t}} \tag{2}$$

where E_{50t} is the tensile modulus, MPa; q_t is the tensile strength, MPa; and ε_{50t} is the strain corresponding to one half of the tensile strength, %.

3.1.3. Tensile Modulus and Deformation Modulus

The modulus indicates the stiffness of the soil and is an important physical quantity to describe the deformation characteristics of the soil [1]. The larger the tensile modulus E_{50t}, the stronger the soil resistance to tensile deformation. Similarly, the larger the deformation modulus E_{50u}, the stronger the soil resistance to compressive deformation [33,34].

Figures 13 and 14 show the relationship between the deformation modulus, tensile modulus and fiber content of fiber-reinforced CGF-solidified soil, respectively. The addition of fiber increases the deformation modulus of solidified soil by 2.1–30.1% and the tensile modulus by 12.9–58.7%. The deformation modulus and tensile modulus both increased and then decreased with the increase in fiber content, reaching their peak value when the fiber content was 0.25%. When the fiber content was greater than 0.25%, the deformation modulus and tensile modulus showed a decreasing trend. The above phenomenon demonstrated that the low fiber content (less than 0.25%) is less distributed in the solidified soil, which is related to the limited contact area with the soil particles and the small interface friction. In this case, the fiber mainly plays the role of pillar reinforcement in the soil, which has poor resistance to deformation. When the fiber content is too high (greater than 0.25%), the excessive fiber congregates in the solidified soil, which tends to form a network of overhead fibers and does not come into contact with the soil particles, forming a weak stress zone and reducing friction. Macroscopically, the strength is reduced, and the resistance to deformation is reduced. However, the strengths of the fiber-reinforced CGF-solidified soils were still larger than those of the solidified soils without fibers. It is further shown that the incorporation of fiber can effectively improve the ability of solidified soil to resist deformation, and the effect of fiber on the resistance to tensile deformation is better than that of compressive deformation.

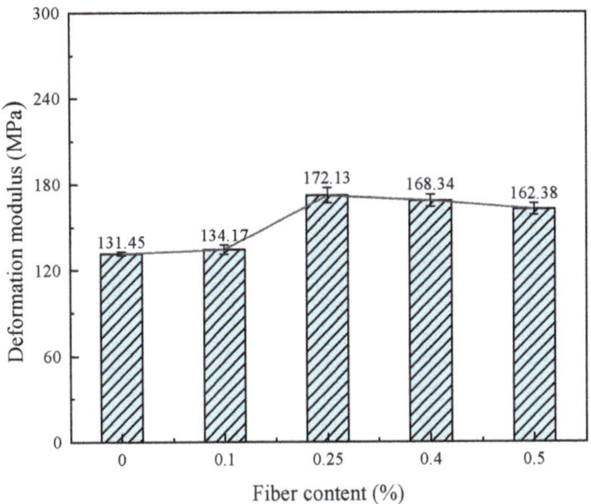

Figure 13. Relationship between deformation modulus and fiber content of fiber-reinforced, solidified soil.

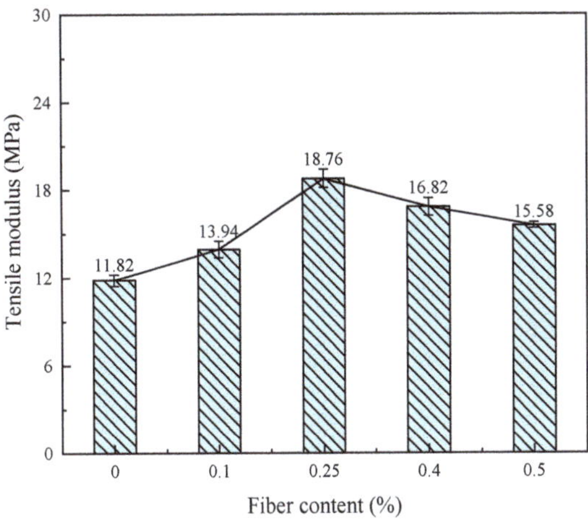

Figure 14. Relationship between tensile modulus and fiber content of fiber-reinforced, solidified soil.

The ratio of tensile modulus to compressive modulus is called the modulus tensile–compression ratio, or tensile–compression ratio for short, which can reflect the ability of fiber-reinforced solidified soil to resist deformation under external forces. The higher the tensile–compression ratio, the stronger the resistance to deformation. The tensile–compression ratio shows a trend of increasing and then slowly decreasing with the increase in fiber content (Figure 15). This indicates that there exists an optimal fiber content (0.25%), which gives the fiber-reinforced CGF-solidified soil the strongest ability to resist deformation under tensile stress or external load. Nevertheless, the ability of fiber-reinforced CGF-solidified soils to resist deformation becomes weaker beyond the optimal fiber content.

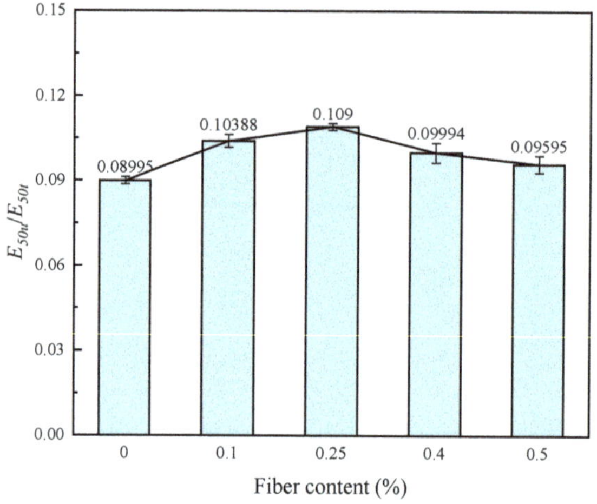

Figure 15. Relationship between E_{50}/E_{50u} and fiber content of fiber-reinforced, solidified soil.

3.1.4. Fracture Energy

Fracture energy is the energy consumed on the fracture surface during the whole process, from the beginning of stretching to destruction, expressed in terms of the work exerted by the tension per unit of tension on the surface. Generally speaking, the larger the fracture energy, the stronger the material toughness. Fracture energy is an indicator for evaluating the toughness effect of fiber-reinforced, solidified soil [18,37] and is calculated using Equation (3).

$$G_F = l \int_0^{\varepsilon_{tmax}} \sigma d\varepsilon \tag{3}$$

where G_F is the fracture energy, kJ/m^2; l is the specimen tensile section length, and the test $l = 0.005$ m; ε_{tmax} is the destructive strain; σ is the stress, kPa; and ε is the strain.

Figure 16 shows the relationship between fracture energy and fiber content. The fracture energy of the fiber-reinforced, CGF-solidified soil increased by 130.1~171.7% compared with the solidified soil without fibers. With the increase in fiber content, the tensile fracture energy of the sample first increases and then decreases, and the tensile fracture energy reaches its peak at 0.25% fiber content. This is due to the fact that fiber can increase the failure strain and peak tensile stress of the sample through friction and bonding, thus improving the tensile fracture of the sample; this indicates that fiber incorporation is conducive to enhancing the toughness of the sample. When the fiber content is 0.25%, the failure strain and failure stress are greater than those of the other four amounts of fiber content, and the energy consumed on the corresponding fracture plane is also the highest.

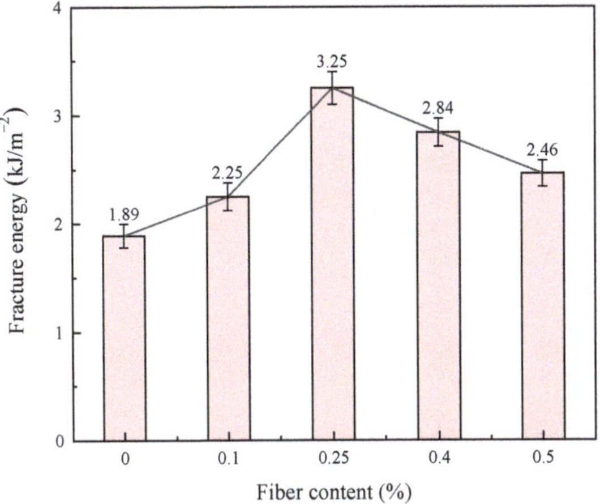

Figure 16. Relationship between fracture energy and fiber content of fiber-reinforced, CGF-solidified soil.

3.2. Strength Characteristics

3.2.1. Influence of Fiber Content on Unconfined Compressive Strength

Figure 17 shows the relationship between compressive strength and fiber content. The compressive strength increases first and then decreases with the increase in fiber content. When the fiber content is 0.25%, the compressive strength reaches 3985.7 kPa. Compared with solidified soil without fibers, the compressive strength of solidified soil with fiber content of 0.1%, 0.25%, 0.4% and 0.5% increased by 109.8%, 123.6%, 116.6% and 114.0%, respectively. It can be seen that adding the appropriate amount of fiber to solidified soil can improve the compressive strength of the soil, and there is an optimal fiber content (0.25%). This is attributed to the fact that, as the fiber content is too low, the fiber is scattered in the solidified soil, and the reinforcement effect on the soil is only one-dimensional single fiber reinforcement. As the fiber content increased, the fiber formed a fiber network in the soil, and the structural strength and unconfined compressive strength of the soil increased significantly. When the fiber content is too high, there will be an uneven distribution phenomenon, which affects the overall durability of the soil and fiber combined force [28,38].

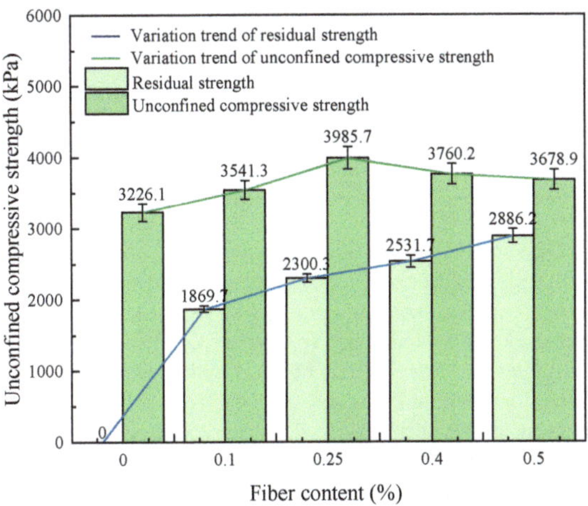

Figure 17. Relationship between UCS, residual strength and fiber content.

Fiber-reinforced, CGF-solidified soil retains a certain amount of residual strength after damage occurs, and the residual strength increases with increasing fiber content. The residual stresses of 0.1%, 0.25%, 0.4% and 0.5% were 184.1 kPa, 288 kPa, 337.1 kPa and 382.2 kPa, respectively. This is mainly due to the fact that the fibers on the fracture surface have the effect of bearing and dispersing certain stresses after damage to the specimen, and this effect is related to the number of fibers on the fracture surface. Therefore, unlike the unconfined compressive strength, the residual strength of fiber-reinforced, CGF-solidified soil did not reach its maximum at 0.25% of the fiber content, but was positively correlated with the fiber content.

3.2.2. Influence of Fiber Content on Uniaxial Tensile Strength

Figure 18 shows the relationship between the tensile strength of the fiber-reinforced CGF-solidified soil and the fiber content. The tensile strength of fiber-reinforced CGF-solidified soil increases and then decreases with the increase in fiber content. At fiber contents between 0 and 0.25%, the tensile strength increases with the increase in fiber content. Compared to the solidified soil without fibers, the tensile strength reaches its peak at 0.25% fiber content, 905.9 kPa, and increases by 186.16%. After exceeding 0.25% fiber content, the tensile strength remains higher than that of the solidified soil without fibers. Similar to the optimum fiber content for unconfined compressive strength, the fiber content also has an optimal fiber content to reach the peak uniaxial tensile strength, and this pattern is the same as the findings of other scholars [38,39]. The tensile residual strength of the fiber-reinforced CGF-solidified soil showed a positive correlation with fiber content and the same pattern as the compressive residual strength.

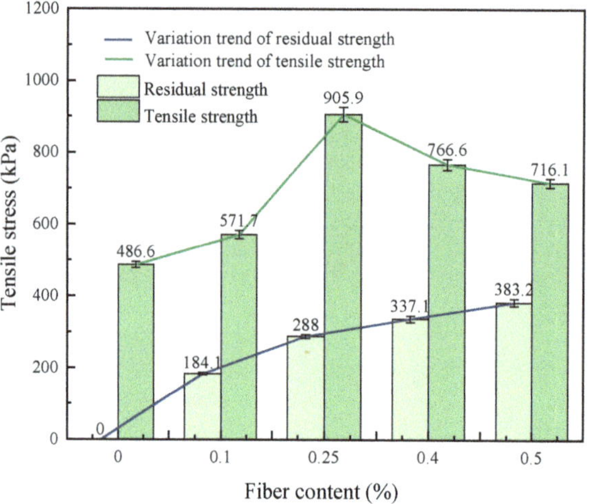

Figure 18. The relationship between tensile strength, residual strength and fiber content.

The above result is due to the low fiber content (0–0.10%), where the fiber distribution in the soil is relatively dispersed, the contact area with the solidified soil particles is limited and the interface friction generated by the two is also small, making it difficult to form a network space structure. The main role of the fiber in the solidified soil is to provide a one-dimensional reinforcement effect with low-strength macroscopic mechanical properties. As the fiber content increases (0.10–0.25%), the fiber forms a three-dimensional spatial grid structure in the solidified soil, which can transfer and disperse tensile stress [40]. The hydration products formed by CGF have high bonding strength, which can effectively improve the bonding force between the fiber and the soil; at the same time, these hydration products will wrap the fiber to improve the stiffness of the fiber. When the soil is subjected to tensile stress, the fibers play one-dimensional and three-dimensional roles at the same time, limiting the relative sliding of soil particles through friction and bonding forces, and the macroperformance is significantly improved in terms of integrity and strength. When the fiber content is too high (0.25–0.50%), excessive fiber is easily wound into a group in the stirring process, and the distribution is uneven, forming a weak force area; thus, the number of effective fibers bearing the tension is reduced, which affects the durability of the fiber and the solidified soil under stress, and the macro performance is that the strength of the sample is reduced.

3.3. Relationship between Tensile–Compression Ratio of Strength and Fiber Content

The tensile–compression strength ratio is the ratio of tensile strength to unconfined compressive strength, or simply the tensile–compression ratio, which can be used to assess a material's brittle characteristics [41]. The relationship between the tensile–compression ratio and fiber content of the fiber-reinforced CGF-solidified soil is shown in Figure 19. With the increase in fiber content, the tensile–compression ratio first increases and then decreases. The tensile–compression ratio of the solidified soil without fibers is the smallest, which is 0.151. The maximum tensile–compression ratio of 0.227 occurs in the fiber-reinforced CGF-solidified soil at 0.25% fiber content. This indicates that the fiber content can significantly improve the toughness of solidified soil, and the optimal fiber content is 0.25%; when exceeding this fiber content, the toughness of solidified soil will grow slowly, but it is still greater than the toughness of solidified soil without fibers.

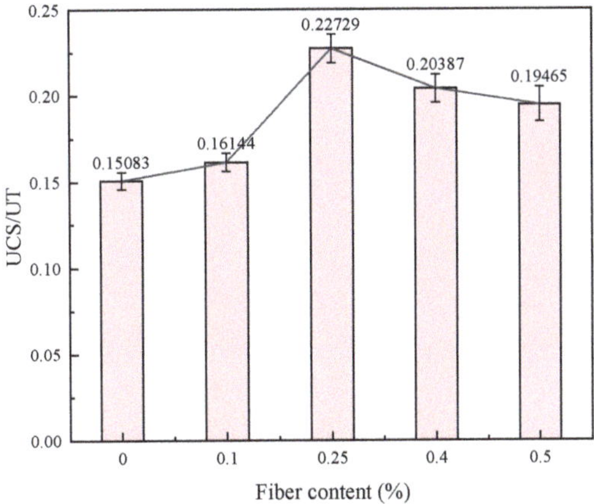

Figure 19. The relationship between UCS/UT and fiber content.

3.4. Micromechanisms

3.4.1. Microscopic Morphology

The results of the SEM test are shown in Figure 20. As seen in Figure 20a, the spaces between the soil particles are filled with needle-like, flocculent and blocky CGF hydration products. These hydration products cemented the soil particles and fibers into a whole, partly filled in the pores (Figure 20b,c) and partly attached to the fiber surface (Figure 20f).

After magnifying 50 times, it can be seen that the fibers are randomly distributed in the solidified soil in a three-dimensional mesh structure, and some of the fibers are bent under the action of external force (Figure 20d); after magnifying 200 times, it can be observed that there exists an anchoring effect at the interface of the fibers and the CGF-solidified soil, so the fibers cannot be easily pulled and pulled out (Figure 20e); after magnifying 1000 times, it can be observed that the fibers' surface is adhered to the CGF hydration products, making the originally smooth fibers rough and uneven (Figure 20f).

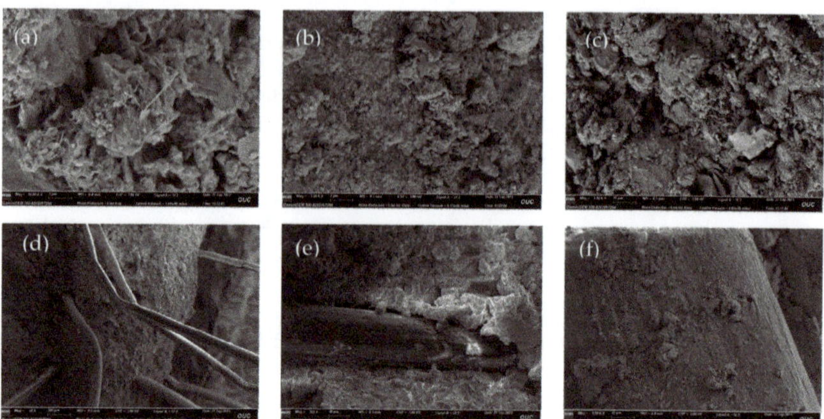

Figure 20. SEM diagram of fiber CGF-solidified soil: (**a**) ×10,000, (**b**) ×5000, (**c**) ×1000, (**d**) ×50, (**e**) ×200, (**f**) ×1000.

3.4.2. Fiber–Binder Coupling Reinforcement Mechanism

Combined with the results of the unconfined compressive strength, uniaxial tensile and microscopic tests, the fiber–binder coupling reinforcement mechanism was analyzed: the cementing and filling effect of CGF (solidification effect), the embedding and bridging effect of fibers (reinforcing effect) and the spatial reticulation structure are the main factors controlling the reinforcement effect. A mechanism diagram is shown in Figure 21.

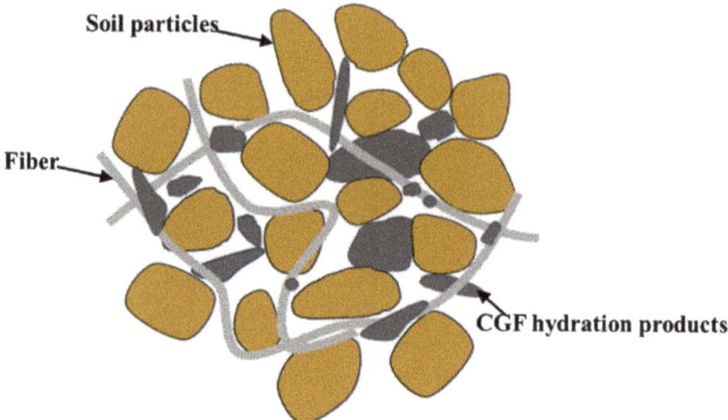

Figure 21. Interface diagram of fiber/CGF-solidified soil.

1. Gelling and filling effect of CGF: In the alkaline environment formed by the dissolution of a large amount of OH^- from CCR in contact with water, GGBS and FA undergo the pozzolanic reaction to generate hydration products such as C-A-H, C-S-H and C-A-S-H [33,34], which can gel the soil particles and fibers, fill the pore spaces in the soil and increase the density of the soil, increasing the strength of solidified soil.
2. Embedding and bridging effect of fibers: The enhancement effect of fibers on the strength of solidified soil depends on the mechanical interaction between fibers and the interface of the solidified soil matrix. As the hydration products of CGF have strong gelling properties, they attached to the fiber surface to form a layer of hard shell, which effectively enhanced the stiffness of the fiber. At the same time, the crystals attached to the fiber surface and the hydration products inside the solidified soil combined with each other, resulting in the overall firm embedding of the fiber in the solidified soil. Due to the large difference in the modulus of elasticity between the fiber

and the solidified soil, the difference in deformation between the two produces relative displacement; thus, the originally smooth fiber surface attaches to the solidified soil particles, resulting in its surface roughness and unevenness, which increases the occlusal force of the fiber–solidified soil interface and improves the ability of the fiber to bear tensile stress. Under the action of external forces, the interface forces of adhesion and friction can inhibit the fiber's slip in the solidified soil, preventing the fiber from being pulled out. When the external force continues to increase, the soil body becomes damaged, and fiber plays a "bridging" role through the fracture surface to inhibit the expansion of the damaged surface; when the external force exceeds the force in the fiber-solidified soil interface or the fiber's own strength, then the fiber will be pulled out of the soil. When the external force exceeds the force at the fiber-solidified soil interface or the strength of the fiber itself, the fiber on the fracture surface will be pulled out or broken, and the reinforcing effect of the fiber will be completely lost [42–44].

3. Spatial network structure: When the optimal fiber content (0.25%) is used in the CGF-solidified soil, the fibers cross and overlap each other inside the soil, presenting a randomly distributed mesh structure. When one or more bundles of fibers in the mesh structure are subjected to relative movement with the solidified soil, the fibers overlapping with them will transfer and disperse the stresses in different directions and improve the overall strength of the solidified soil [45]. However, when the fiber content is too high, the excess fiber will be unevenly distributed in the solidified soil and knotted into a ball, resulting in the existence of "weak areas" in the soil but also damage to the structure of the soil, thus reducing its strength [46].

4. Conclusions

This paper proposes the use of polypropylene-fiber-reinforced CGF all-solid-waste binder (abbreviated as CGF)-solidified soil modified by carrying out unconfined compressive strength and uniaxial tensile tests, and the effects of different fiber contents on the deformation characteristics and strength characteristics of the CGF-solidified soil were comparatively analyzed. The micro-mechanism and reinforcement mechanism under fiber-CGF coupling were also analyzed with SEM results. The following main conclusions were obtained:

- Compared with the nonfiber-solidified soil, the unconfined compressive strength, uniaxial tensile strength, deformation modulus, tensile modulus, fracture energy and residual strength of the fiber-reinforced CGF-solidified soil are significantly improved. Compared with the solidified soil without fiber, the fracture energy of the fiber-reinforced CGF-solidified soil can be increased by 130.1–171.7%, the deformation modulus can be increased by 102.1–130.1% and the tensile modulus can be increased by 112.9–158.7%. The increased amplitudes of the tensile strength and tensile modulus are greater than those of the compressive strength and deformation modulus, respectively. The addition of fiber makes the failure mode of the CGF-solidified soil transition from typical brittle failure to plastic failure.
- The compressive strength and tensile strength of the polypropylene-fiber-reinforced CGF-solidified soil increased and then decreased with the increase in fiber content, and both reached their maximum values when the fiber content was 0.25%. The tensile strength is 905.9 kPa, which is 186.6% higher than that of the solidified soil without fiber, and the compressive strength is 3985.7 kPa, which is 116.6% higher than that of the solidified soil without fiber. However, both the compressive residual strength and tensile residual strength increased continuously with the increase in fiber content, and there was no optimal content.
- The modulus tensile–compression ratio and strength tensile–compression ratio of the fiber-reinforced CGF-solidified soil both increased and then decreased with the increase in fiber content, reaching the maximum at 0.25% fiber content. The modulus tensile–compression ratio is 0.109 and the strength tensile–compression ratio is 0.227.

The incorporation of fiber effectively improved the toughness and stiffness of the solid soil.

- The fiber-CGF coupling effect is mainly embodied by two processes: first, the gelling and filling effect of the hydration products generated by CGF, enhancing the compactness of the soil body and then enhancing the strength of the soil body; second, the one-dimensional embedding and bridging effect of the fibers and the three-dimensional spatial mesh structure, resulting in a combination of friction and adhesion, preventing the development of cracks and enhancing the ability of the soil body to resist deformation.

Author Contributions: Conceptualization, X.Q.; methodology, L.Y.; validation, Y.W.; formal analysis, Q.L.; resources, J.Y.; data curation, X.Q.; writing—original draft preparation, X.Q.; writing—review and editing, J.Y. and Y.W.; visualization, L.Y. and Q.L.; supervision, J.Y.; funding acquisition, J.Y. All authors have read and agreed to the published version of the manuscript.

Funding: This research was funded by the National Natural Science Foundation of China (Nos. 52378380, 52078474, and 51779235).

Institutional Review Board Statement: Not applicable.

Informed Consent Statement: Not applicable.

Data Availability Statement: Some or all of the data, models or codes that support the findings of this study are available from the corresponding author upon reasonable request.

Acknowledgments: We gratefully acknowledge the financial support of the above funds and the researchers of all reports cited in our paper.

Conflicts of Interest: The authors declare no conflicts of interest.

References

1. Li, S.C.; Yang, J.J.; Wu, Y.L.; Wang, M. Research on tensile characteristics of cement-treated soft soil. *J. Cent. South Univ. Sci. Technol.* **2022**, *53*, 2619–2632.
2. Wang, Y.S.; Liu, Z.Q.; Wan, W.L.; Nie, A.H.; Zhang, Y.W.; Han, C.P. Toughening effect and mechanism of rice straw Fiber-reinforced lime soil. *Constr. Build. Mater.* **2023**, *393*, 132133. [CrossRef]
3. Lu, L.F.; Ma, Q.; Hu, J.; Li, Q.F. Mechanical Properties, Curing Mechanism, and Microscopic Experimental Study of Polypropylene Fiber Coordinated Fly Ash Modified Cement-Silty Soil. *Materials* **2021**, *14*, 5441. [CrossRef] [PubMed]
4. Akbari, H.R.; Sharafi, H.; Goodarzi, A.R. Effect of polypropylene fiber and nano-zeolite on stabilized soft soil under wet-dry cycles. *Geotext. Geomembr.* **2021**, *49*, 1470–1482. [CrossRef]
5. Pradhan, P.K.; Kar, R.K.; Naik, A. Effect of Random Inclusion of Polypropylene Fibers on Strength Characteristics of Cohesive Soil. *Geotech. Geol. Eng.* **2012**, *30*, 15–25. [CrossRef]
6. Yuan, T.; Lei, X.W.; Ai, D.; An, R.; Chen, C.; Chen, X. Strength characteristics of expansive soil improved by coir fiber-MICP composite. *J. Water Resour. Archit. Eng.* **2023**, *21*, 105–111.
7. Sindhu Nachiar, S.; Anandh, S.; Ahamed Sarjune, N.; Kumaresan, M. Effect of steel fiber in SIFCON using GGBS as binder replacement. *Mater. Today Proc.* **2023**, *93*, 1–8. [CrossRef]
8. Divya, P.V.; Viswanadham, B.V.S.; Gourc, J.P. Evaluation of Tensile Strength-Strain Characteristics of Fiber-Reinforced Soil through Laboratory Tests. *J. Mater. Civ. Eng.* **2014**, *26*, 14–23. [CrossRef]
9. Coviello, C.G.; Lassandro, P.; Sabbà, M.F.; Foti, D. Mechanical and Thermal Effects of Using Fine Recycled PET Aggregates in Common Screeds. *Sustainability* **2023**, *15*, 16692. [CrossRef]
10. Cai, Y.; Shi, B.; Gao, W.; Chen, F.J.; Tang, Z.S. Experimental study on engineering properties of fiber-lime treated soils. *Chin. J. Geotech. Eng.* **2006**, *28*, 1283–1287.
11. Shukla, S.; Sivakugan, N.; Das, B. Fundamental concepts of soil reinforcement—An overview. *Int. J. Geotech. Eng.* **2009**, *3*, 329–342. [CrossRef]
12. Prabakar, J.; Sridhar, R.S. Effect of random inclusion of sisal fibre on strength behaviour of soil. *Constr. Build. Mater.* **2002**, *16*, 123–131. [CrossRef]
13. Wang, D.X.; Wang, H.W.; Larsson, S.; Benzerzour, M.; Maherzi, W.; Amar, M. Effect of basalt fiber inclusion on the mechanical properties and microstructure of cement-solidified kaolinite. *Constr. Build. Mater.* **2020**, *241*, 118085. [CrossRef]
14. Hejazi, S.M.; Sheikhzadeh, M.; Abtahi, S.M.; Zadhoush, A. A simple review of soil reinforcement by using natural and synthetic fibers. *Constr. Build. Mater.* **2012**, *30*, 100–116. [CrossRef]
15. Ghavami, K.; Toledo Filho, R.D.; Barbosa, N.P. Behaviour of composite soil reinforced with natural fibres. *Cem. Concr. Compos.* **1999**, *21*, 39–48. [CrossRef]

16. Mahattanatawee, K.; Manthey, J.A.; Luzio, G.; Talcott, S.T.; Goodner, K.; Baldwin, E.A. Total antioxidant activity and fiber content of select Florida-grown tropical fruits. *J. Agric. Food Chem.* **2006**, *54*, 7355–7363. [CrossRef]
17. Bouaricha, L.; Henni, A.D.; Lancelot, L. A Laboratory Investigation On Shear Strength Behavior Of Sandy Soil: Effect Of Glass Fiber And Clinker Residue Content. *Stud. Geotech. Mech.* **2017**, *39*, 3–15. [CrossRef]
18. Tang, C.S.; Li, H.; Pan, X.H.; Yin, L.Y.; Cheng, L.; Cheng, Q.; Liu, B.; Shi, B. Coupling effect of biocementation-fiber reinforcement on mechanical behavior of calcareous sand for ocean engineering. *Bull. Eng. Geol. Environ.* **2022**, *81*, 163. [CrossRef]
19. Consoli, N.C.; Bassani, M.A.A.; Festugato, L. Effect of fiber-reinforcement on the strength of cemented soils. *Geotext. Geomembr.* **2010**, *28*, 344–351. [CrossRef]
20. Rostami, R.; Zarrebini, M.; Mandegari, M.; Mostofinejad, D.; Abtahi, S.M. A review on performance of polyester fibers in alkaline and cementitious composites environments. *Constr. Build. Mater.* **2020**, *241*, 117998. [CrossRef]
21. Sabbà, M.F.; Tesoro, M.; Falcicchio, C.; Foti, D. Rammed Earth with Straw Fibers and Earth Mortar: Mix Design and Mechanical Characteristics Determination. *Fibers* **2021**, *9*, 30. [CrossRef]
22. Duan, X.L.; Zhang, J.S. Mechanical Properties, Failure Mode, and Microstructure of Soil-Cement Modified with Fly Ash and Polypropylene Fiber. *Adv. Mater. Sci. Eng.* **2019**, *2019*, 9561794. [CrossRef]
23. Marvila, M.T.; Rocha, H.A.; de Azevedo, A.R.G.; Colorado, H.A.; Zapata, J.F.; Vieira, C.M.F. Use of natural vegetable fibers in cementitious composites: Concepts and applications. *Innov. Infrastruct. Solut.* **2021**, *6*, 180. [CrossRef]
24. Anggraini, V.; Asadi, A.; Huat, B.B.K.; Nahazanan, H. Effects of coir fibers on tensile and compressive strength of lime treated soft soil. *Measurement* **2015**, *59*, 372–381. [CrossRef]
25. Duong, N.T.; Tran, K.Q.; Satomi, T.; Takahashi, H. Effects of agricultural by-product on mechanical properties of cemented waste soil. *J. Clean. Prod.* **2022**, *365*, 132814. [CrossRef]
26. Miller, C.J.; Rifai, S. Fiber reinforcement for waste containment soil liners. *J. Environ. Eng.-Asce* **2004**, *130*, 891–895. [CrossRef]
27. Kakooei, S.; Akil, H.M.; Jamshidi, M.; Rouhi, J. The effects of polypropylene fibers on the properties of reinforced concrete structures. *Constr. Build. Mater.* **2012**, *27*, 73–77. [CrossRef]
28. Li, Q.; Chen, J.H.; Hu, H.X. The tensile and swelling behavior of cement-stabilized marine clay reinforced with short waste fibers. *Mar. Georesour. Geotechnol.* **2019**, *37*, 1236–1246. [CrossRef]
29. Kumar, A.; Walia, B.S.; Bajaj, A. Influence of fly ash, lime, and polyester fibers on compaction and strength properties of expansive soil. *J. Mater. Civ. Eng.* **2007**, *19*, 242–248. [CrossRef]
30. Afroughsabet, V.; Biolzi, L.; Ozbakkaloglu, T. High-performance fiber-reinforced concrete: A review. *J. Mater. Sci.* **2016**, *51*, 6517–6551. [CrossRef]
31. Tang, C.S.; Shi, B.; Gu, K. Microstructural study on interfacial interactions between fiber reinforcement and soil. *J. Eng. Geol.* **2011**, *19*, 610–614.
32. *GB/T50123*; Standard of Geotechnical Test Method. Ministry of Housing and Urban-Rural Development: Beijing, China, 2019. (In Chinese)
33. Wu, Y.L.; Yang, J.J.; Li, S.C.; Wang, M. Experimental Study on Mechanical Properties and Micro-Mechanism of All-Solid-Waste Alkali Activated Binders Solidified Marine Soft Soil. *Mater. Sci. Forum.* **2021**, *1036*, 327–336. [CrossRef]
34. Wu, Y.L.; Yang, J.J.; Chang, R.Q. The design of ternary all-solid-waste binder for solidified soil and the mechanical properties, mechanism and environmental benefits of CGF solidified soil. *J. Clean. Prod.* **2023**, *429*, 139439. [CrossRef]
35. Tang, C.-S.; Wang, D.-Y.; Cui, Y.-J.; Shi, B.; Li, J. Tensile Strength of Fiber-Reinforced Soil. *J. Mater. Civ. Eng.* **2016**, *28*, 04016031. [CrossRef]
36. Pakravan, H.R.; Jamshidi, M.; Latifi, M.; Pacheco-Torgal, F. Evaluation of adhesion in polymeric fibre reinforced cementitious composites. *Int. J. Adhes. Adhes.* **2012**, *32*, 53–60. [CrossRef]
37. Golewski, G.L. Fracture Performance of Cementitious Composites Based on Quaternary Blended Cements. *Materials* **2022**, *15*, 6023. [CrossRef] [PubMed]
38. Tran, K.Q.; Satomi, T.; Takahashi, H. Tensile behaviors of natural fiber and cement reinforced soil subjected to direct tensile test. *J. Build. Eng.* **2019**, *24*, 100748. [CrossRef]
39. Fang, X.W.; Yang, Y.; Chen, Z.; Liu, H.L.; Xiao, Y.; Shen, C.N. Influence of Fiber Content and Length on Engineering Properties of MICP-Treated Coral Sand. *Geomicrobiol. J.* **2020**, *37*, 582–594. [CrossRef]
40. Tran, K.Q.; Satomi, T.; Takahashi, H. Improvement of mechanical behavior of cemented soil reinforced with waste cornsilk fibers. *Constr. Build. Mater.* **2018**, *178*, 204–210. [CrossRef]
41. Zhang, G.L.; Ren, M.H.; Zhang, D.W.; Xu, H.M.; Song, T. Experimental study on tensile properties enhancement of cement-treated soil by basalt fiber. *Hydro-Sci. Eng.* **2022**, *02*, 109–116.
42. Li, Z.G.; Ning, J.G.; Huang, X. Causes for different stabilization effects of cementitious-expansible compound soft soil stabilizer. *Chin. J. Geotech. Eng.* **2011**, *33*, 420–426.
43. Chen, R.M.; Jian, W.B.; Zhang, X.F.; Fang, Z.H. Experimental study on performance of sludge stabilized by CSFG-FR synergy. *Rock Soil Mech.* **2022**, *43*, 1020–1030. [CrossRef]
44. Shen, C.; Zhang, D.W.; Zhang, G.L.; Song, T.; Xu, H.M. Analysis on tensile properties of cemented soil compounded with unequal length mixed basalt fiber. *J. Archit. Civ. Eng.* **2021**, *38*, 33–39. [CrossRef]

45.	Xiao, Y.; Tong, L.Y.; Che, H.B.; Guo, Q.W.; Pan, H.S. Experimental studies on compressive and tensile strength of cement-stabilized soil reinforced with rice husks and polypropylene fibers. *Constr. Build. Mater.* **2022**, *344*, 128242. [CrossRef]
46.	Li, P.D. Experimental study on California Bearing Ratio of Fiber Reinforced Soil and Analysis of Road Working Condition. Master's Thesis, Northwest A & F University, Xianyang, China, 2022.

materials

Article

Influence of Pretreatment Methods on Compressive Performance Improvement and Failure Mechanism Analysis of Recycled Aggregate Concrete

Dongbin Lv [1,2], Kainan Huang [3,*] and Wensheng Wang [2,*]

1 Guangxi Fuhe Expressway Co., Ltd., Hezhou 542800, China
2 College of Transportation, Jilin University, Changchun 130025, China
3 Guangxi Transportation Science and Technology Group Co., Ltd., Nanning 530007, China
* Correspondence: 201620105179@mail.scut.edu.cn (K.H.); wangws@jlu.edu.cn (W.W.)

Abstract: The utilization of recycled aggregate can avert the squandering of resources and the destruction of the environment. Nevertheless, there exists a slew of old cement mortar and microcracks on the surface of recycled aggregate, which give rise to the poor performance of aggregates in concrete. In this study, for the sake of ameliorating this property of recycled aggregates, the surface of the recycled aggregates is covered with a layer of cement mortar to compensate for the microcracks on the surface and reinforce the bond between old cement mortar and aggregates. In order to demonstrate the influence of recycled aggregate by different cement mortar pretreatment methods, this study prepared natural aggregate concrete (NAC) and concretes with recycled aggregate after the wetting pretreatment (RAC-W) and cement mortar pretreatment (RAC-C), and conducted uniaxial compressive strength tests on different types of concrete at different curing ages. The test results indicated that the compressive strength of RAC-C at a 7 d curing age was higher than that of RAC-W and NAC, and the compressive strength of RAC-C at a 28 d curing age was higher than RAC-W but lower than NAC. The compressive strength of NAC and RAC-W at a 7 d curing age was about 70% of that at a 28 d curing age, and the compressive strength of RAC-C at a 7 d curing age was about 85–90% of that at a 28 d curing age. The compressive strength of RAC-C increased dramatically at the early stage, while the post-strength of the NAC and RAC-W groups increased rapidly. The fracture surface of RAC-W mainly occurred in the transition zone between the recycled aggregates and old cement mortar under the pressure of the uniaxial compressive load. However, the main failure of RAC-C was the crushing destruction of cement mortar. With changes in the amount of cement added beforehand, the proportion of aggregate damage and A-P interface damage of RAC-C also changed accordingly. Therefore, the recycled aggregate pretreated with cement mortar can significantly improve the compressive strength of recycled aggregate concrete. The optimal amount of pre-added cement was 25%, which is recommended for practical engineering.

Keywords: recycled aggregate; pretreatment method; compressive performance improvement; failure mechanism

Citation: Lv, D.; Huang, K.; Wang, W. Influence of Pretreatment Methods on Compressive Performance Improvement and Failure Mechanism Analysis of Recycled Aggregate Concrete. *Materials* **2023**, *16*, 3807. https://doi.org/10.3390/ma16103807

Academic Editor: Francesco Fabbrocino

Received: 14 April 2023
Revised: 7 May 2023
Accepted: 15 May 2023
Published: 18 May 2023

1. Introduction

With the innovation of technology, plenty of concrete buildings need to be updated [1]. According to the relevant literature, about 40% of the construction garbage generated by the demolition of old buildings is concrete [2]. The recycling of concrete garbage not only satisfies the requirement of sustainable development, but also makes a huge contribution to the protection of the environment and the conservation of land resources [3]. At present, various types of recycled aggregates can replace natural aggregates to varying degrees for use in concrete materials, such as waste powder, fine and coarse marble aggregates [4,5], waste glass [6–8], recycled coal bottom ash [9], plastic waste [10], etc. With the increasing demand for sustainable development, recycled aggregate concrete (RAC), as a sustainable material, has received much attention in recent years due to its potential benefits in reducing environmental impact and conserving natural resources [11]. However, RAC has some drawbacks, such as low mechanical properties and poor durability than those of conventional concrete, which limit its widespread use in construction [12,13].

To address this issue, various pretreatment methods have been developed to enhance the mechanical properties of RAC [14–17]. These methods include surface modification, alkali activation, thermal treatment, and chemical treatment. Furthermore, the bond behaviour of passively confined RAC has been investigated based on centre pull-out tests [18]. Research about the preprocessing of recycled aggregate has already evolved. The effectiveness of these pretreatment methods on RAC has been studied by many researchers, and it has been found that they can significantly enhance the compressive strength and other mechanical properties of RAC [19–21]. Some researchers tried to remove the old cement mortar on the recycled aggregate with the preprocessing of HCI and Na_2SO_4 and found that the pretreatment of HCI can conspicuously enhance the compressive strength of RAC [22]. Furthermore, the silica fume solution was also used to improve the bond of the transition interface and fill the microcracks of recycled aggregate. Subsequently, the compactness and strength of the recycled aggregate were significantly improved [23]. In addition, removing the adhesion mortar was an effective method to improve the performance of recycled aggregates [15]. Meanwhile, there were also a few researchers who found that the concrete prepared by the pre-wetted recycled aggregate had the highest compressive strength [24] because the water plays an internal curing role in the concrete, which improved the compressive strength and freeze–thaw performance of the recycled concrete [25]. However, others found that recycled concrete prepared with dry aggregate had the best compressive strength; the concrete comprised of pre-wetted recycled aggregate had the best workability [26]. Through the uniaxial compression test of recycled aggregate concrete, researchers found that the compressive failure of recycled aggregate concrete mainly occurred in the interface transition zone of mortar and aggregate. Consequently, the performance of recycled aggregate concrete was significantly improved by increasing the strength of the interface transition zone [27]. Hou et al. found that the microcracks on the surface of the recycled aggregate would absorb cement particles. This process can enhance the transition zone between the new cement mortar and the old cement mortar, which results in a faster increase in the early compressive strength of the recycled concrete [28].

On the basis of previous research, the major approach to reusing concrete garbage is to convert it to recycled aggregate [24,29–31]. However, recycled aggregate is denser and different from natural aggregate [32]. Moreover, the surface of the recycled aggregate is covered with a layer of old cement mortar, and in the process of mechanical crushing, some microcracks are produced. These microcracks give rise to more water absorption and porosity. Subsequently, the workability of RAC is lower than that of natural aggregate concrete [33]. Hence, microcracks and old mortar on the surface of aggregate have a negative effect on the properties of recycled aggregate. Aggregate cracking notably reduces the performance of the entire structure, which is often investigated by numerical models in order to propose real strategies to prevent cracking phenomena. The discrete and smeared crack approaches provide reliable results in terms of crack patterns and tensile and compressive strength [34,35]. Hence, preprocessing the recycled aggregates would be a valid method to ameliorate the performance of the concrete. Some researchers introduced numerous methods to preprocess the aggregate, such as pre-wetting the aggregate and removing the old mortar adhered to the aggregate. Whereas, there remained relatively few applications of the method that blended the cement mortar and the aggregate preprocessed by cement paste directly. Furthermore, the paste wrapped on the aggregate would fill the cracks on the surface of the recycled aggregate and improve the bonding effect between the recycled aggregate and the old mortar, as well as the old mortar and the new mortar [36]. Compared with traditional methods, this way not only improves the performance of the concrete, but also saves time in the actual project.

This paper aims to investigate the influence of different pretreatment methods on the compressive performance improvement of RAC and analyze the failure mechanism of RAC under compression. The research will focus on two commonly used pretreatment methods, including wetting pretreatment and cement mortar pretreatment. Therefore, the paper's innovation lies in exploring the effects of different treatments on recycled aggregate concrete's properties and analyzing its failure mechanisms. The results of this study will provide valuable insights into the optimization of pretreatment methods for RAC, which can provide insights into the sustainable use of construction materials and lead to the development of more sustainable and high-performance construction materials.

2. Experimental Investigation

2.1. Raw Materials

In this study, the cement used in the experiment is ordinary Portland cement with a strength grade of 42.5 MPa, an initial setting time of 86 min and a final setting time of 142 min, which has 62.3% of CaO, 22.8% of SiO_2, 5.6% of Al_2O_3, 4.4% of Fe_2O_3, 1.7% of MgO, etc. For the used coarse aggregates, the recycled aggregate was produced by crushing concrete waste from construction waste demolished in a project in Changchun, and the natural aggregate was basalt. Both recycled coarse aggregate and natural coarse aggregate were from Jilin Province, and the water absorption test results at different immersion times are summarized in Table 1. The crush index of recycled coarse aggregate and natural coarse aggregate was 10.96% and 4.08%, respectively. The sieve analysis results for both recycled coarse aggregate and natural coarse aggregate are shown in Figure 1. The fine aggregate used was natural sand, with an apparent density of 2541.0 kg/m^3, a water absorption rate of 0.7%, and a fineness modulus of 2.38. Table 2 shows the particle size distribution of natural sand. In addition, the water used in the concrete was drinking water, and the additional agent was a polycarboxylic acid superplasticizer with a 25% water reduction rate.

Table 1. Water absorption test results of recycled coarse aggregate at different immersion times.

Immersion Time (h)	1/6	1/3	0.5	1	2	3
Water absorption (%)	6.52	6.58	6.77	6.82	6.85	7.03

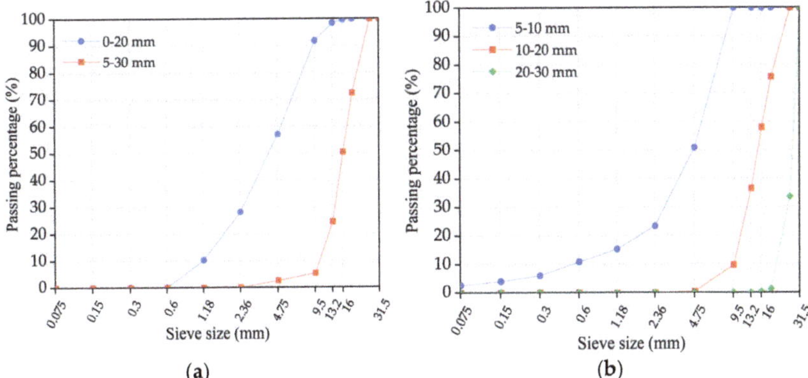

Figure 1. The sieve analysis results: (**a**) recycled coarse aggregate; (**b**) natural coarse aggregate.

Table 2. Particle size distribution of natural sand.

Mesh Size (mm)	4.75	2.36	1.18	0.6	0.3	0.15
Cumulative residual percentage (%)	5.1	19.3	27.9	4.26	81.8	92.5

2.2. Specimen Preparation and Experimental Method

In this study, two pretreatment methods were used to improve the performance of recycled aggregate, preparing the concrete specimens with recycled aggregate after the wetting pretreatment (i.e., RAC-W) and after the cement mortar pretreatment (i.e., RAC-C). At the same time, natural aggregate concrete (NAC) was also prepared as a control group for the following comparative analysis.

For the RAC-W specimen preparation, based on the water absorption test results of recycled coarse aggregate at different immersion times in Table 1, the recycled coarse aggregate immersed for 1 h was selected to carry out the wetting pretreatment, and then natural sand, cement, and water were poured into the mixer one by one. After mixing thoroughly, the RAC-W specimens were prepared.

The RAC-C specimen was divided into two groups, including the RAC-CA group and the RAC-CB group. Firstly, the recycled coarse aggregates with different particle sizes were poured into a mixer and mixed until the differently-sized recycled coarse aggregates were evenly distributed. Then, the cement and water were added to the mixer, with 20%, 25%, and 30% of their total amount, respectively. Next, the remaining cement, water, and natural sand were poured into the mixer to prepare cement mortar, mixing with the recycled aggregate after cement mortar pretreatment to prepare the RAC-CA specimens. On the other hand, the cement mortar pretreatment of recycled coarse aggregate for the RAC-CB group was the same as the RAC-CA group. Whereas the normal amount of cement, water, and natural sand—rather than the remaining or partial materials—were poured into the mixer to prepare the cement mortar. Then, the recycled coarse aggregates after cement mortar pretreatment were blended with the normal amount of cement mortar to prepare the RAC-CB specimens. The RAC-CB group used more cement and natural sand compared with the RAC-CA group; hence, the cost of the RAC-CB group increased by contrast with the RAC-CA group. The purpose of pouring the recycled aggregate after cement mortar pretreatment directly into the cement mortar is that the cement paste wrapped on the surface of the recycled aggregate could be prevented from washing away by adding water.

The flow diagram of RAC-C preparation is shown in Figure 2. Based on the considerations, including the availability of materials, sustainability goals, and desired performance characteristics, the different types of concrete proportions are shown in Table 3. Specimens of 100 mm × 100 mm × 100 mm were produced and then placed in a standard curing room with a temperature of (20 ± 2) °C and humidity greater than 95% for curing. The uniaxial compression loading tests of different types of concrete (i.e., NAC, RAC-W, RAC-CA, and RAC-CB) were conducted using a mechanical testing machine, with a loading rate maintained at 0.05 MPa/s [37].

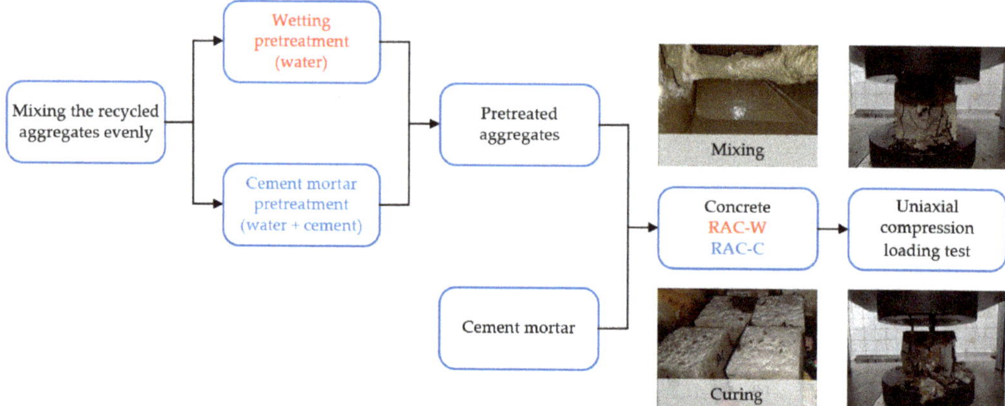

Figure 2. The flow diagram of RAC preparation and test.

Table 3. The proportion of different types of concrete.

Type	Superplasticizer (%)	Ingredients (kg/m^3)					
		Recycled Coarse Aggregate	Natural Coarse Aggregate	Total Cement	Total Water	Pre-Add Cement	Pre-Add Water
NAC	0.5	0	1888.4	400	160	0	0
RAC-W	0.5	1191.9	0	400	160	0	0
RAC-CA20	1	1191.9	0	400	160	80	32
RAC-CA25	1	1191.9	0	400	160	100	40
RAC-CA30	1	1191.9	0	400	160	120	48
RAC-CB20	1	1191.9	0	480	192	80	32
RAC-CB25	1	1191.9	0	500	200	100	40
RAC-CB30	1	1191.9	0	520	208	120	48

3. Results and Discussions

3.1. Analysis of Compressive Strength of Concrete at a 7-d Curing Age

In order to more intuitively display the changes in compressive strength of four different types of concrete, the bar charts of compressive strength results for these four types of concrete at a 7 d curing age are drawn in this study, as shown in Figure 3.

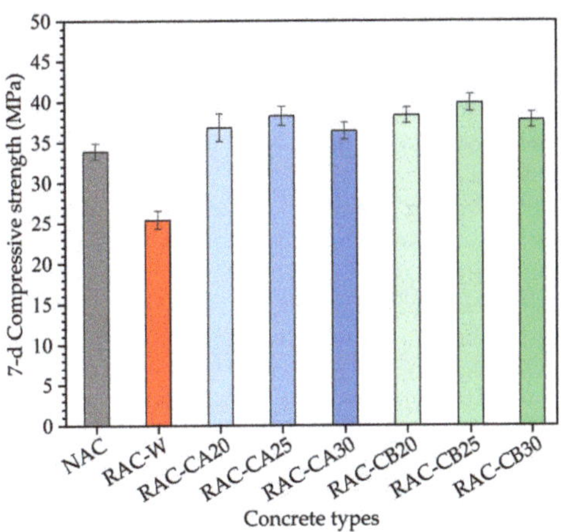

Figure 3. Bar chart of compressive strength for three types of concrete at a 7 d curing age.

From the compressive strength results in Figure 3, it can be clearly seen that the compressive strengths of two RAC-C groups after two kinds of cement mortar pretreatment methods were generally higher than the NAC and RAC-W groups. For these two RAC-C groups, when the amount of cement and water added to the mixer in the pretreatment step was 20%, 25%, and 30%, respectively, the compressive strength of RAC-CA and RAC-CB showed a trend of first increasing and then decreasing. When the amount of pre-added cement and water reached 30%, after the 7 d curing age, the corresponding compressive strength of recycled aggregate concrete (i.e., RAC-CA30 and RAC-CB30) was the lowest among RAC-C20, RAC-C25, and RAC-C30. This may be because the cement mortar wrapped on the surface of recycled aggregate was thicker, and a higher cement content results in insufficient hydration reaction during a shorter curing age, resulting in a decrease in the compressive strength of concrete. When the amount of pre-added cement and water was 25%, the compressive strength of recycled aggregate concrete (i.e., RAC-CA25 and RAC-CB25) reached its maximum value in each RAC-C group. Therefore, based on the RAC-C compressive strength results at a 7-day curing age, it can be preliminarily predicted that the optimal amount of pre-added cement was 25%. Furthermore, it can be clearly seen that the compressive strength of the RAC-CB group was higher than that of the RAC-CA group as a whole. When the amount of cement added in the pretreatment step was 20% of the total, the compressive strength of the RAC-CB group (i.e., RAC-CB20) was about 4.1% higher than that of the RAC-CA20. For the 25% pre-added cement, the compressive strength of RAC-CB25 was about 4.2% higher than that of the RAC-CA25; for the 30% pre-added cement content, the compressive strength of the RAC-CB30 was about 3.8% higher than the RAC-CA30. In Table 3, the cement content of the RAC-CB group was higher than the RAC-CA group. Hence, the overall compressive strength of the RAC-CB group was higher than the RAC-CA group.

According to the compressive strength test results of the RAC-W and RAC-C groups, at a 7-day curing age, the compressive strength of the RAC-C group was higher than the RAC-W group. The average compressive strength for the 7-day curing RAC-CA group was about 46.0% higher than that of the RAC-W group, and the average compressive strength of the 7-day curing age RAC-CB group was about 51.9% higher than that of the RAC-W group. The reason may be that part of the water for the wetting pretreatment would become the water for mixing [38], which resulted in an increase in the water–cement ratio, thus reducing the compressive strength. In addition, part of the water absorbed by the recycled aggregate in the wetting pretreatment step would return to the cement mortar ascribed to the hydrostatic pressure during the process of mixing [39]. The water–cement ratio of concrete would also be increased, and therefore, the compressive strength was reduced. Moreover, the recycled aggregate of the RAC-CA and RAC-CB groups is pretreated with cement mortar, and thus the microcracks on the surface of the recycled aggregate would be filled by the cement mortar. The cement mortar pretreatment would strengthen the combination of the old cement mortar and the recycled aggregate and improve the integrity of the recycled aggregate, sequentially increasing the compressive strength of the recycled aggregate concrete.

At a 7-day curing age, the compressive strength of the RAC-C group was higher than that of NAC based on the comparison of the experimental results. The average compressive strength at a 7-day curing age of the RAC-CA and RAC-CB groups was 9.8% and 14.3% higher than that of the NAC group, respectively. The reasons are that the overall performances of the recycled aggregate of the RAC-CA and RAC-CB groups were improved with the effect of pre-added cement mortar, and part of the water for mixing was still absorbed by the recycled aggregate during the blending process. However, the recycled aggregate is pretreated with cement mortar, which would result in a decrease in the water–cement ratio. Because the initial strength of concrete is mainly determined by the water–cement ratio, the compressive strength of the 7-day curing age RAC-CA and RAC-CB groups was higher than that of the NAC group.

3.2. Analysis of Compressive Strength of Concrete at a 28 d Curing Age

Similarly, in order to clearly display the changes in compressive strength of different types of concrete at a 28 d curing age, a bar chart of compressive strength results for these four types of concrete is shown in Figure 4.

According to the compressive strength results in Figure 4, compared to the compressive strength at a 7 d curing age (shown in Figure 3), the compressive strength of different types of concrete at a 28 d curing age was improved to varying degrees. The compressive strength of the 28 d curing age RAC-CB group was higher than that of the RAC-CA group in general. When the amount of cement added in the pretreatment step reached 20%, 25%, and 30%, the compressive strength of the RAC-CB group was 3.2%, 4.9%, and 2.6% higher than that of the RAC-CA group, respectively. This is because the sand ratio of the RAC-CB group was lower, which gives rise to the stronger adhesion of cement mortar and recycled aggregate. Thus, the strength of the RAC-CB group was higher than that of the RAC-CA group. In addition, the compressive strength gap between the RAC-CA and RAC-CB groups decreased, compared with the compressive strength of the RAC-C type at a 7 d curing age. The reason might be that the cement amount of the RAC-CB group was relatively larger, which resulted in a relative weakness of the skeleton effect of the aggregate [40]. Consequently, the compressive strength gap between these two groups of RAC-CA and RAC-CB was relatively reduced. When the amount of pre-added cement was 30%, the compressive strength at a 28 d curing age of these two groups had a significant increase compared with the 7 d curing age. The reason is that the cement mortar on the surface of the recycled aggregate was cured for a longer time, which has already formed the hard cement stone giving rise to greater strength. When the amount of pre-added cement was 25%, the compressive strength of concrete at a 28 d curing age for RAC-C was the largest. Taking into account the compressive strengths at the 7 d and 28 d curing ages, it can be inferred that the optimal amount of pre-added cement is 25%.

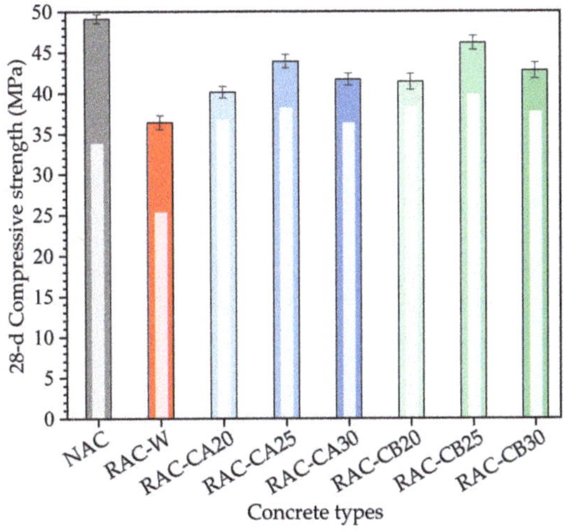

Figure 4. Bar chart of compressive strength for three types of concrete at a 28 d curing age.

In Figure 4, it can be seen from the compressive test results that the compressive strength at a 28 d curing age of the RAC-CA and RAC-CB groups is 44.8% and 57.4%, respectively, higher than that of the RAC-W group on average. The reason is that the microcracks of the recycled aggregate in the RAC-C group have been filled with cement mortar, and the surface of the recycled aggregate is covered with a layer of hard cement, which improves the performance of the recycled aggregate. Therefore, the compressive strength of the RAC-C group is higher than that of the RAC-W group. The average compressive strength at a 28 d curing age of the RAC-CA and RAC-CB groups is 14.5% and 11.4% lower than that of the NAC group, respectively. The reason may be that the compressive strength at the later period of concrete mainly depends on the quality of the aggregate. Although the quality of the recycled aggregate in the pretreatment step has been improved with cement mortar, its mechanical properties are still inferior compared with the natural aggregate. Thus, the compressive strength of the RAC-CA and RAC-CB groups was lower than that of the NAC group at the later period.

The compressive strength ratio of different concrete types at the 7 d curing age to 28 d curing age was calculated and is listed in Table 4. From the compressive strength ratio results shown in Table 4, it can be seen that the compressive strength of the NAC and RAC-W groups at a 7 d curing age was about 70% of the compressive strength at a 28 d curing age. The compressive strength of the RAC-CA and RAC-CB groups at a 7 d curing age was about 85–90% of the compressive strength at a 28 d curing age. Figure 5 shows the compressive strength increases amplitude at the curing age of 28 d vs. 7 d. This shows that the posted strength of the NAC and RAC-W groups increases rapidly; however, the compressive strength amplitude increase of the RAC-CA and RAC-CB groups at a 28 d curing age was relatively smaller. Among the RAC-C group, 25% of the pre-added cement would lead to the obvious compressive strength increase at a 28 d curing age.

Table 4. The compressive strength ratio of 7 d curing age concrete to 28 d curing age concrete.

Type	NAC	RAC-W	RAC-CA20	RAC-CA25	RAC-CA30	RAC-CB20	RAC-CB25	RAC-CB30
Ratio (%)	68.9	69.7	91.6	86.8	87.3	92.3	86.2	88.3

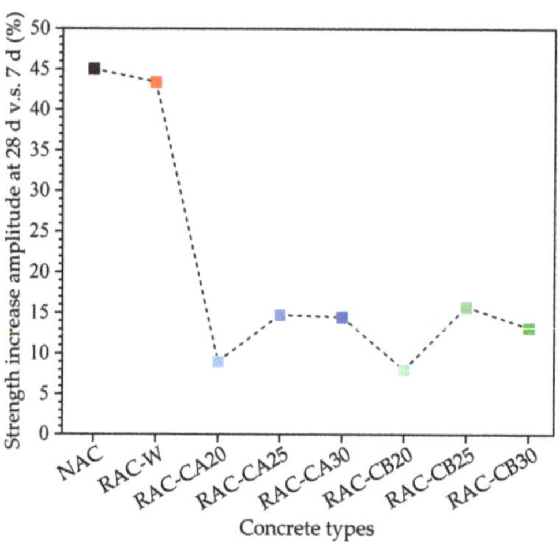

Figure 5. The compressive strength amplitude increase at the curing age of 28 d vs. 7 d.

3.3. Failure Mechanism of Recycled Aggregate Concrete under the Compressive Effect

The fracture of the NAC group under the pressure of a uniaxial compressive load would be mainly located in the transition zone between the aggregates and cement mortar. This is because aggregate and cement mortar are considered two kinds of media. When different kinds of media are blended together, their connection interface is considered to be the weakest position in the mixture of the two media. As a result, when the NAC group was damaged under pressure, the fracture mostly appeared at the connection between aggregates and cement mortar.

For the RAC-W group, the compressive test results show that the fracture surface mainly occurs in the transition zone between the recycled aggregates and old cement mortar under the pressure of the uniaxial compressive load, as shown in Figure 6a. The recycled aggregate in Figure 6b is usually attached to a layer of old cement mortar on the surface. Once it is used to form the concrete specimen, the recycled aggregates and old cement mortar would be wrapped in new cement mortar. In general, on account of mechanical crushing, the old cement mortar of the recycled aggregates on the surface gets rougher, which contributes to a close combination between the new cement mortar and the old cement mortar. However, the old cement mortar has a lower performance attribute to ageing and vibration intensity during the mechanical crushing process. Hence, the strength of the transition zone between the old mortar and the new mortar is greater than the strength of the transition zone between the recycled aggregates and the old mortar, which would lead to the recycled aggregates being mostly exposed on the surface when the RAC-W group is damaged under pressure.

Figure 6. The fracture of the RAC-W group under the pressure of uniaxial compressive load: (**a**) broken diagram of RAC-W; (**b**) surface of recycled aggregate.

From the above analysis, it can be seen that the transition zone between the recycled aggregates and old cement mortar is the weakest position of the recycled aggregate concrete. Consequently, it would be a valid way to strengthen the mechanical properties of recycled concrete by strengthening the performance of the transition zone. In this study, the surface of the recycled aggregates is covered with a layer of cement mortar, which can form the shell structure. The structure of the recycled aggregate after the cement mortar pretreatment and the real pretreated aggregate are shown in Figure 7, in which the microcracks on the surface of the recycled aggregate are fully filled by the cement mortar. Part of the cement mortar is absorbed by the old cement mortar on the surface of the aggregate, which improves the properties of the old cement mortar. Moreover, the cement mortar would also strengthen the bond between the recycled aggregate and the old cement mortar. Therefore, the recycled aggregates pretreated with cement mortar have an overall increase in positive properties. Meanwhile, it can be found through the compression test that the main failure of the RAC-C group is the crushing failure of cement mortar, accompanied by different amounts of destructed aggregates and the aggregate–cement paste interface (i.e., A-P interface). The failure mainly occurs at the transition zone between old cement mortar and cement paste and the transition zone between aggregate and cement paste.

Figure 7. The RAC-C group after the cement mortar pretreatment: (**a**) structure of the recycled aggregate; (**b**) the real pretreated aggregates.

In this study, based on the software ImagePro Plus 6.0, the ratio of aggregate damage and A-P interface damage in the RAC-C group could be calculated. To distinguish the aggregate damage and A-P interface damage from the broken RAC specimens, ImagePro

Plus software was used for image analysis as follows: First, the acquired RAC cross-sectional image was imported into the ImagePro Plus software. Then, the built-in image processing functions of the software were used, such as binarization and edge detection, to extract the damaged area of the aggregate damage and A-P interface damage. Next, using the software's area statistics function, the area of the aggregate damage and A-P interface damage within the entire statistical area could be calculated. Finally, the ratio of the area of the aggregate damage and A-P interface damage could be used to calculate their respective proportions in the entire statistical area. The area ratio results of aggregate damage and A-P interface damage in the RAC-C group are listed as shown in Table 5. From the results, we see that as the proportion of aggregate damage increases, the compressive strength of the RAC-C group has a growing tendency and as the amount of pre-added cement increases, the proportion of A-P interface damage decreases.

Table 5. The ratio of aggregate damage and A-P interface damage in the RAC-C group.

Damage	RAC-CA20	RAC-CB20	RAC-CA25	RAC-CB25	RAC-CA30	RAC-CB30
Aggregate damage (%)	1.01	1.18	3.32	3.95	1.36	1.42
A-P interface damage (%)	6.37	8.15	5.84	2.91	3.04	1.72

The comparison of aggregate and A-P interface damages in the RAC-CA and RAC-CB groups is shown in Figure 7. According to Table 5, the concrete of the RAC-CA20 group had relatively more damage to the A-P interface compared with the aggregate damage. The reasons are that the RAC-CA20 group has a small amount of pre-added cement, and the shell structure formed on the surface of the aggregate is relatively thin, which results in the inferior strength of the shell structure. Thus, the fracture of the RAC-CA20 concrete mainly appeared at the A-P interface, as shown in Figure 8a. The failure mode of the RAC-CB group was similar to that of the RAC-CA group; because there was comparatively less cement added in the pretreatment step, the shell structure formed on the surface of the aggregates was thinner, so the strength of the RAC-CA group was insufficient, easily resulting in damage, as shown in Figure 8b.

From Table 5, the ratio of aggregate damage in the RAC-CA25 group relatively increases, which also has more damage at the A-P interface. This is ascribed to the relatively smaller amount of cement mortar in the RAC-CA25 group compared with the RAC-CA20 group, and hence the main strength of concrete in this group is served by the embedding effect between the aggregates. As a result, during the compression test, most of the compressive load is resisted by the aggregates, which gives rise to a partial crushing failure of aggregates. Once the aggregate is broken, the A-P interface would also be destroyed, as shown in Figure 8c. Compared with the damage of the A-P interface, there were more crushing failures of the aggregate that occurred in the RAC-CB25 group. The reasons are that the quantity of the cement mortar was larger compared to the RAC-CA25 group, which contributes to better adhesion to the aggregates, enhancing their embedding effect. Consequently, when a small portion of cement mortar and the A-P interface are destroyed under the compressive load, the aggregate becomes the main bearer of the load, which eventually gives rise to the crushing failure of the aggregate, as shown in Figure 8d.

By comparing the RAC-CA30 and RAC-CB30 groups, relatively less damage of the A-P interface and aggregate occurred in the RAC-CA30 group. The crushing failure of cement mortar is the main failure form. The reasons are that the RAC-CA30 group has a greater amount of cement added in the pretreatment step, and hence the shell structure is continuous and complete and formed on the surface of the aggregates. In addition, the sand ratio of the RAC-CA30 group was larger than the RAC-CB30 group, which resulted in the relatively lower strength of the cement mortar. Consequently, almost all of the destruction of the RAC-CA30 group was caused by the crushing failure of the cement mortar. There were fewer exposed aggregates on the surface after the compressive failure, as shown in Figure 8e. The dominant destruction of the RAC-CB30 group was the crushing failure of cement mortar with little damage to the A-P interface and aggregates. This is because there was an enormous amount of cement added in the pretreatment step of the RAC-CB30 group. Thus, it is thicker for the shell structure wrapped on the surface of the aggregate. In addition, the amount of cement mortar was huge, which contributed to a relatively weakened embedding effect of the aggregate. Therefore, when the concrete was damaged under compression, the cement mortar bore most of the load, so the RAC-CB30 group was mainly damaged by cement mortar, as shown in Figure 8f.

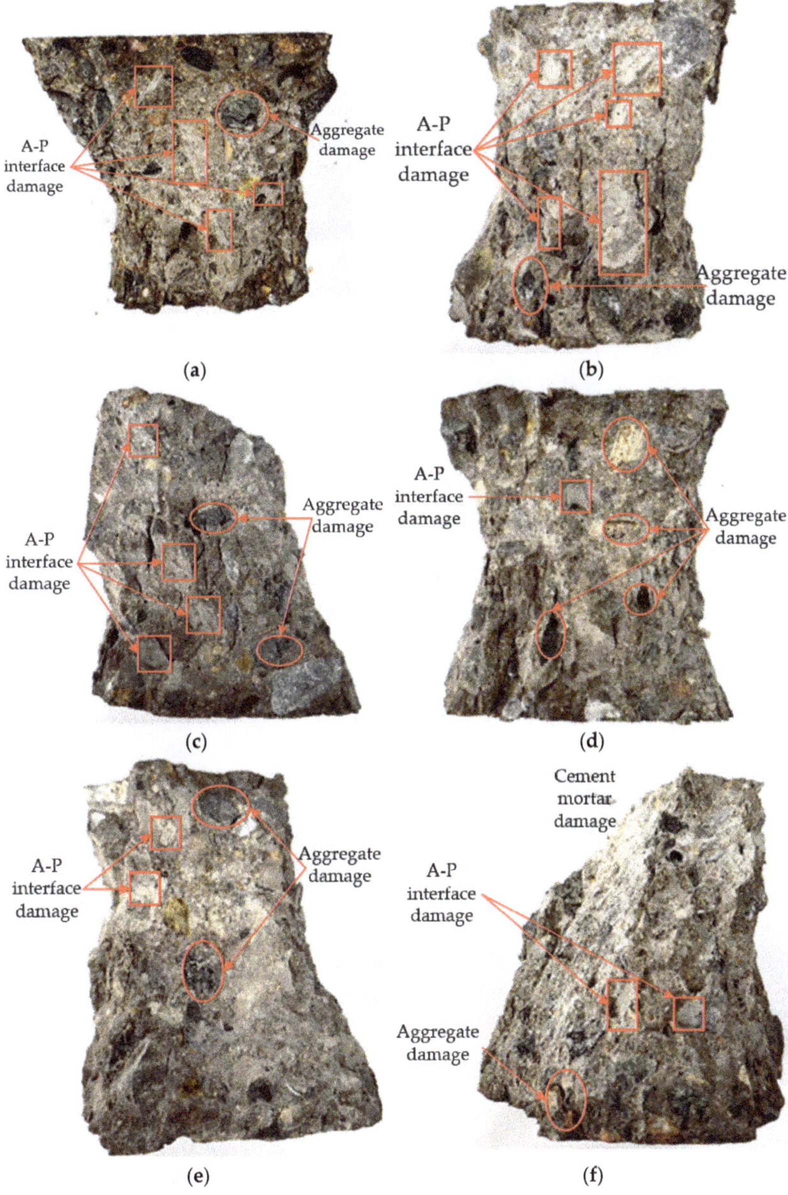

Figure 8. The damage comparison for the aggregate and A-P interface in the RAC-C group: (**a**) RAC-CA20; (**b**) RAC-CB20; (**c**) RAC-CA25; (**d**) RAC-CB25; (**e**) RAC-CA30; (**f**) RAC-CB30.

4. Conclusions

This study prepared natural aggregate concrete (NAC) and concretes with recycled aggregate after the wetting pretreatment (RAC-W) and after the cement mortar pretreatment (RAC-C), and conducted uniaxial compressive strength tests on different types of concrete at different curing ages. Based on the compression test results and failure interfaces, the failure mechanisms of different types of concrete were analyzed. The following conclusions are drawn:

(1) The compressive strength of the RAC-C group, including RAC-CA and RAC-CB, after two kinds of cement mortar pretreatment methods at a 7 d curing age is higher than that of RAC-W and NAC, and the compressive strength of the RAC-C group at a 28 d curing age is higher than RAC-W but lower than NAC. Therefore, it may be a suitable option for projects that require quick strength development.

(2) At a 7 d curing age, the compressive strength of RAC-CB is about 3.8–4.2% higher than that of the RAC-CA, the compressive strength of RAC-CA is about 46.0–51.9% higher than that of RAC-W, the compressive strength of RAC-C is about 9.8–14.3% higher than that of NAC.

(3) At a 28 d curing age, the compressive strength of RAC-CB is about 2.6–4.9% higher than that of the RAC-CA, the compressive strength of RAC-CA is about 44.8–57.4% higher than that of RAC-W, the compressive strength of RAC-C is about 11.4–14.5% lower than that of NAC.

(4) The compressive strength of NAC and RAC-W at a 7 d curing age is about 70% of that at a 28 d curing age, and the compressive strength of RAC-C at a 7 d curing age is about 85–90% of that at a 28 d curing age. The compressive strength of RAC-C increased dramatically at the early stage, while the post-strength of the NAC and RAC-W groups increased rapidly.

(5) The fracture surface of RAC-W mainly occurs in the transition zone between the recycled aggregates and old cement mortar under the pressure of the uniaxial compressive load. The main failure of RAC-C is the crushing destruction of cement mortar. With the amount of cement added beforehand changed, the proportion of aggregate damage and A-P interface damage of RAC-C also changed accordingly.

Therefore, the appropriate type of concrete should be selected based on the project's specific requirements. Concretes with recycled aggregate after the cement mortar pretreatment are suggested to improve the utilization rate of recycled aggregate and the compressive strength of concrete. The curing age and failure mechanisms should be considered while selecting the appropriate type of concrete.

Author Contributions: Conceptualization, D.L. and W.W.; methodology, D.L. and W.W.; validation, K.H.; formal analysis, D.L. and K.H.; investigation, D.L. and K.H.; writing—original draft preparation, D.L.; writing—review and editing, K.H. and W.W.; project administration, W.W.; funding acquisition, W.W. All authors have read and agreed to the published version of the manuscript.

Funding: This research was funded by the Scientific and Technological Project of the Science and Technology Department of Jilin Province (grant number: 20210508028RQ), Scientific Research Project of the Department of Education of Jilin Province (grant number: JJKH20221019KJ), and Postdoctoral Researcher Selection Funding Project of Jilin Province.

Institutional Review Board Statement: Not applicable.

Informed Consent Statement: Not applicable.

Data Availability Statement: Not applicable.

Conflicts of Interest: The authors declare no conflict of interest.

References

1. Kumar, G.; Gupta, R.C.; Shrivastava, S. Sustainable Zero-Slump Concrete Containing Recycled Aggregates from Construction and Demolition Waste of a 63-Year-Old Demolished Building. *J. Mater. Civ. Eng.* **2022**, *34*, 04022115. [CrossRef]
2. Patra, I.; Al-Awsi, G.R.L.; Hasan, Y.M.; Almotlaq, S.S.K. Mechanical properties of concrete containing recycled aggregate from construction waste. *Sustain. Energy Technol. Assess.* **2022**, *53*, 102722. [CrossRef]
3. Alani, A.A.; Lesovik, R.; Lesovik, V.; Fediuk, R.; Klyuev, S.; Amran, M.; Ali, M.; de Azevedo, A.R.G.; Vatin, N.I. Demolition Waste Potential for Completely Cement-Free Binders. *Materials* **2022**, *15*, 6018. [CrossRef] [PubMed]
4. Basaran, B.; Kalkan, I.; Aksoylu, C.; Ozkilic, Y.O.; Sabri, M.M.S. Effects of Waste Powder, Fine and Coarse Marble Aggregates on Concrete Compressive Strength. *Sustainability* **2022**, *14*, 4388. [CrossRef]
5. Karalar, M.; Özkılıç, Y.O.; Aksoylu, C.; Sabri Sabri, M.M.; Beskopylny, A.N.; Stel'makh, S.A.; Shcherban', E.M. Flexural behavior of reinforced concrete beams using waste marble powder towards application of sustainable concrete. *Front. Mater.* **2022**, *9*, 1068791. [CrossRef]
6. Zeybek, O.; Ozkilic, Y.O.; Karalar, M.; Celik, A.I.; Qaidi, S.; Ahmad, J.; Burduhos-Nergis, D.D.; Burduhos-Nergis, D.P. Influence of Replacing Cement with Waste Glass on Mechanical Properties of Concrete. *Materials* **2022**, *15*, 7513. [CrossRef]
7. Qaidi, S.; Najm, H.M.; Abed, S.M.; Ozlahc, Y.O.; Al Dughaishi, H.; Alosta, M.; Sabri, M.M.S.; Alkhatib, F.; Milad, A. Concrete Containing Waste Glass as an Environmentally Friendly Aggregate: A Review on Fresh and Mechanical Characteristics. *Materials* **2022**, *15*, 6222. [CrossRef]
8. Celik, A.I.; Ozkilic, Y.O.; Zeybek, O.; Karalar, M.; Qaidi, S.; Ahmad, J.; Burduhos-Nergis, D.D.; Bejinariu, C. Mechanical Behavior of Crushed Waste Glass as Replacement of Aggregates. *Materials* **2022**, *15*, 8093. [CrossRef]
9. Karalar, M.; Bilir, T.; Cavuslu, M.; Ozkilic, Y.O.; Sabri, M.M.S. Use of recycled coal bottom ash in reinforced concrete beams as replacement for aggregate. *Front. Mater.* **2022**, *9*, 1064604. [CrossRef]
10. Qaidi, S.; Al-Kamaki, Y.; Hakeem, I.; Dulaimi, A.F.; Özkılıç, Y.; Sabri, M.; Sergeev, V. Investigation of the physical-mechanical properties and durability of high-strength concrete with recycled PET as a partial replacement for fine aggregates. *Front. Mater.* **2023**, *10*, 1101146. [CrossRef]
11. Liu, Z.H.; Takasu, K.; Suyama, H.; Koyamada, H.; Liu, S.L.; Hao, Q. The Effect of Cementitious Materials on the Engineering Properties and Pore Structure of Concrete with Recycled Fine Aggregate. *Materials* **2023**, *16*, 305. [CrossRef]
12. Kim, J. Influence of quality of recycled aggregates on the mechanical properties of recycled aggregate concretes: An overview. *Constr. Build. Mater.* **2022**, *328*, 127071. [CrossRef]
13. Bu, C.M.; Liu, L.; Lu, X.Y.; Zhu, D.X.; Sun, Y.; Yu, L.W.; OuYang, Y.H.; Cao, X.M.; Wei, Q.K. The Durability of Recycled Fine Aggregate Concrete: A Review. *Materials* **2022**, *15*, 1110. [CrossRef]
14. Sasanipour, H.; Aslani, F.; Taherinezhad, J. Chloride ion permeability improvement of recycled aggregate concrete using pretreated recycled aggregates by silica fume slurry. *Constr. Build. Mater.* **2021**, *270*, 121498. [CrossRef]
15. Ouyang, K.; Shi, C.J.; Chu, H.Q.; Guo, H.; Song, B.X.; Ding, Y.H.; Guan, X.M.; Zhu, J.P.; Zhang, H.B.; Wang, Y.L.; et al. An overview on the efficiency of different pretreatment techniques for recycled concrete aggregate. *J. Clean. Prod.* **2020**, *263*, 121264. [CrossRef]
16. Li, Y.; Yang, X.B.; Lou, P.; Wang, R.J.; Li, Y.L.; Si, Z. Sulfate attack resistance of recycled aggregate concrete with NaOH-solution-treated crumb rubber. *Constr. Build. Mater.* **2021**, *287*, 123044. [CrossRef]
17. Al-Baghdadi, H.M. Experimental study on sulfate resistance of concrete with recycled aggregate modified with polyvinyl alcohol (PVA). *Case Stud. Constr. Mater.* **2021**, *14*, e00527. [CrossRef]
18. Fayed, S.; Madenci, E.; Ozkilic, Y.O.; Mansour, W. Improving bond performance of ribbed steel bars embedded in recycled aggregate concrete using steel mesh fabric confinement. *Constr. Build. Mater.* **2023**, *369*, 130452. [CrossRef]
19. Zhong, C.H.; Tian, P.; Long, Y.H.; Zhou, J.Z.; Peng, K.; Yuan, C.X. Effect of Composite Impregnation on Properties of Recycled Coarse Aggregate and Recycled Aggregate Concrete. *Buildings* **2022**, *12*, 1035. [CrossRef]
20. Reddy, R.R.K.; Yaragal, S.C. A novel approach for optimizing the processing of recycled coarse aggregates. *Constr. Build. Mater.* **2023**, *368*, 130480. [CrossRef]
21. Xue, H.J.; Zhu, H.Y.; Guo, M.Z.; Shao, S.W.; Zhang, S.Y.; Zhang, Y. Modeling for predicting triaxial mechanical properties of recycled aggregate concrete considering the recycled aggregate replacement. *Constr. Build. Mater.* **2023**, *368*, 130447. [CrossRef]
22. Kim, Y.; Hanif, A.; Kazmi, S.M.S.; Munir, M.J.; Park, C. Properties enhancement of recycled aggregate concrete through pretreatment of coarse aggregates—Comparative assessment of assorted techniques. *J. Clean. Prod.* **2018**, *191*, 339–349. [CrossRef]
23. Katz, A. Treatments for the Improvement of Recycled Aggregate. *J. Mater. Civ. Eng.* **2004**, *16*, 597–603. [CrossRef]
24. Tian, Y.; Guo, Y.; Guo, L.; Wang, L.; Cao, Y.; Liu, J.; Luo, Y. Study on the Effect of Pretreatment on Property of Recycled Concrete. *Yellow River* **2019**, *41*, 131–134, 143.
25. Zhang, M.Q.; Liu, B.J.; Xie, Y.L.; Luo, Y.Q. Influence of Pre-wetted Recycled Fine Aggregate on Drying Shrinkage and Freeze-thaw Resistance of Concrete. *Build. Sci.* **2016**, *32*, 89–95.
26. Mefteh, H.; Kebaïli, O.; Oucief, H.; Berredjem, L.; Arabi, N. Influence of moisture conditioning of recycled aggregates on the properties of fresh and hardened concrete. *J. Clean. Prod.* **2013**, *54*, 282–288. [CrossRef]
27. Wengui, L.; Chu, L.; Luo, Z.Y.; Zhengyu, H. Investigation on Failure Mechanism of Nanomodified Recycled Aggregate Concrete. *J. Build. Mater.* **2017**, *20*, 685–691, 786.
28. Hou, Y.-L.; Zheng, G. Mechanical Properties of Recycled Aggregate Concrete in Different Age. *J. Build. Mater.* **2013**, *16*, 683–687.

29. Fernandes, B.; Carre, H.; Mindeguia, J.C.; Perlot, C.; La Borderie, C. Spalling behaviour of concrete made with recycled concrete aggregates. *Constr. Build. Mater.* **2022**, *344*, 128124. [CrossRef]
30. da Franca, A.P.M.; da Costa, F.B.P. Evaluating the effect of recycled concrete aggregate and sand in pervious concrete paving blocks. *Road Mater. Pavement* **2023**, *24*, 560–577. [CrossRef]
31. Liu, J.; Ma, K.L.; Shen, J.T.; Zhu, J.B.; Long, G.C.; Xie, Y.J.; Liu, B.J. Influence of recycled concrete aggregate enhancement methods on the change of microstructure of ITZs in recycled aggregate concrete. *Constr. Build. Mater.* **2023**, *371*, 130772. [CrossRef]
32. Mehrabi, P.; Shariati, M.; Kabirifar, K.; Jarrah, M.; Rasekh, H.; Trung, N.T.; Shariati, A.; Jahandari, S. Effect of pumice powder and nano-clay on the strength and permeability of fiber-reinforced pervious concrete incorporating recycled concrete aggregate. *Constr. Build. Mater.* **2021**, *287*, 122652. [CrossRef]
33. Medina, C.; Zhu, W.; Howind, T.; Sánchez de Rojas, M.I.; Frías, M. Influence of mixed recycled aggregate on the physical—Mechanical properties of recycled concrete. *J. Clean. Prod.* **2014**, *68*, 216–225. [CrossRef]
34. De Maio, U.; Gaetano, D.; Greco, F.; Lonetti, P.; Pranno, A. The damage effect on the dynamic characteristics of FRP-strengthened reinforced concrete structures. *Compos. Struct.* **2023**, *309*, 116731. [CrossRef]
35. Rimkus, A.; Cervenka, V.; Gribniak, V.; Cervenka, J. Uncertainty of the smeared crack model applied to RC beams. *Eng. Fract. Mech.* **2020**, *233*, 107088. [CrossRef]
36. Chen, Y.W.; Pan, W.H.; Sun, X.W. Performance study of pretreated concrete with recycled aggregate. *Concrete* **2012**, *1*, 81–83.
37. Shang, X.; Yang, J.; Wang, S.; Zhang, M. Fractal analysis of 2D and 3D mesocracks in recycled aggregate concrete using X-ray computed tomography images. *J. Clean. Prod.* **2021**, *304*, 127083. [CrossRef]
38. Guo, Y.D.; Liu, Y.Z.; Chen, Y.; Wang, W.J.; Li, K. Effect of pre-wetting of recycled coarse aggregate on compressive strength of concrete. *J. Guangxi Univ.* **2017**, *42*, 1548–1553.
39. Zhang, S.J.; Li, X.; Wang, F.J.; Li, F.; Zhou, L.A. Research on the water absorption and desorption properties of recycled brick aggregate. *Concrete* **2017**, 185–188.
40. Xiao, W.; Ding, C.P.; Xie, X.J. Experimental Investigation on Influence of Cement Content on Concrete Performance. *Bull. Chin. Ceram. Soc.* **2018**, *37*, 2083–2087.

materials

Article

Synergistic Effects of Epoxidized Soybean Oil and Polyester Fiber on Crumb Rubber Modified Asphalt Using Response Surface Methodology

Jie Pan [1,2], Jiao Jin [1,*], Shuai Liu [1], Mengcheng Xiao [1], Guoping Qian [1] and Zhuo Wang [2]

[1] School of Traffic and Transportation Engineering, Changsha University of Science & Technology, Changsha 410114, China; 202027010307@stu.csust.edu.cn (J.P.); shuailiu@stu.csust.edu.cn (S.L.); mengc0739@163.com (M.X.); guopingqian@csust.edu.cn (G.Q.)
[2] International College of Engineering, Changsha University of Science & Technology, Changsha 410114, China; 202027010124@stu.csust.edu.cn
* Correspondence: jinjiao@csust.edu.cn; Tel.: +86-731-85258255

Abstract: The incorporation of crumb rubber (CR) into asphalt pavement materials can improve the performance of asphalt pavement and generate environmental benefits. However, the storage stability of the crumb rubber asphalt (CRA) remains an issue that needs to be resolved. This study explores the interaction laws among various modified materials based on the response surface methodology. Optimal preparation dosages of each material are determined, and performance predictions and validations are conducted. The storage stability of the CRA compounded with epoxidized soybean oil (ESO) and polyester fiber (PF) is investigated by combining traditional compatibility testing methods with refined characterization methods. The results indicate that the modification of CRA exhibits better rheological properties when the percentages of CR, PF, and ESO are 22%, 0.34%, and 3.21%, respectively. The addition of ESO effectively complements the light components of CRA to improve asphalt compatibility, and the addition of PF alleviates the adverse effects of ESO's softening effect on rheological properties through stabilization and three-dimensional strengthening. The scientifically compounded additions of ESO and PF can effectively enhance the storage stability and rheological properties of CRA, promoting the development of sustainable and durable roads.

Keywords: crumb rubber; epoxidized soybean oil; polyester fiber; response surface methodology; storage stability

Citation: Pan, J.; Jin, J.; Liu, S.; Xiao, M.; Qian, G.; Wang, Z. Synergistic Effects of Epoxidized Soybean Oil and Polyester Fiber on Crumb Rubber Modified Asphalt Using Response Surface Methodology. *Materials* **2023**, *16*, 3469. https://doi.org/10.3390/ma16093469

Academic Editor: Giovanni Polacco

Received: 9 April 2023
Revised: 27 April 2023
Accepted: 27 April 2023
Published: 29 April 2023

1. Introduction

A substantial quantity of waste rubber tires and waste rubber materials are produced due to the fast growth of industry. This causes serious harm to the ecological environment and must be effectively treated [1]. Crumb rubber is a byproduct of the tire recycling process, made by grinding used tires into small particles [2]. Crumb rubber has good elasticity, resilience, and energy absorption, making it an ideal additive for modified asphalt [3]. The crumb rubber swells in the asphalt, and the cross-linking reaction between the effective components can form a mesh structure to improve the high- and low-temperature performances of asphalt [4]. Crumb rubber can also enhance asphalt's elasticity and fatigue resistance, resulting in enhanced performance and a longer service life [5]. A high content (about 62% of the volume of modified asphalt) of crumb rubber can double the fatigue resistance of modified asphalt [6]. However, crumb rubber and asphalt are made from different materials and have different properties, which can make it difficult to achieve a homogeneous mixture [7]. A high content of crumb rubber can result in asphalt segregation and poor storage stability [8]. Therefore, it is necessary to improve the segregation and storage stability of crumb rubber asphalt while guaranteeing the dosage of crumb rubber in order to enhance the performance of asphalt pavement.

Different additives, such as compatibilizers, stabilizers, and viscosity reducers, have been used to solve this problem [9]. Chemical compatibilizers can improve the compatibility of crumb rubber with asphalt through the grafting reaction [10]. However, chemical compatibilizers have fewer practical applications owing to their toxicity. Viscosity reducers can improve the thermal storage stability of crumb rubber asphalt by breaking the long chains of rubber to form soluble radicals, or via grafting reaction [11]. Stabilizers exhibit a high dissolution with asphalt and form a stable cross-linked structure with enhanced compatibility with asphalt components [12]. Epoxidized soybean oil has been widely used as a vegetable oil-based plasticizer due to its excellent permeability, fluidity, non-toxicity, and non-hazardous nature [13]. The addition of epoxidized soybean oil supplements the asphalt's light components and improves compatibility [14]. The epoxidized bonds in epoxidized soybean oil can react with the unsaturated bonds in asphalt via ring-opening to form compounds with a three-dimensional network structure [15], which improves the storage stability of the crumb rubber asphalt. The combination of ESO and Sasobit enhances the low-temperature performance of styrene–butadiene rubber modified asphalt (SBRMA), and mitigates the negative effects of short-term aging [16]. Although epoxidized soybean oil improves the thermal storage stability of crumb rubber asphalt, its softening effect deteriorates the asphalt's high-temperature stability [17]. The stabilizing and reinforcing effect of fibers can compensate for this shortcoming.

Polyester fibers are synthetic fibers made from polyethylene terephthalate (PET) [18], with high strength, low weight, and a good resistance to UV radiation and chemical degradation [19]. Polyester fibers can improve the tensile strength, fatigue resistance, and crack resistance of asphalt [20]. The addition of polyester fiber leads to the formation of a fiber network structure within the asphalt mixture, enhancing its viscosity and high-temperature stability [21]. The mechanical properties of the asphalt concrete are optimal when the polyester fiber content is 1% [22]. The addition of polyester fiber can enhance the storage stability of the asphalt mixture while meeting the performance requirements of the pavement [23]. Therefore, the stabilizing, adsorbing, and reinforcing effects of polyester fiber have the potential to improve the compatibility of asphalt and it works in synergy with epoxidized soybean oil to improve the thermal stability of crumb rubber asphalt storage.

This study uses crumb rubber, polyester fiber, and epoxidized soybean oil as modified asphalt materials, which were modified by melting and mixing. BBR and MSCR experimental data were used as response indicators to establish response surface models to obtain scientific ratios of the three modified materials and to fully utilize the synergistic effects of the three materials. The rheological properties of modified asphalt at high and low temperatures were analyzed at the same time. Additionally, the softening point difference, fluorescence microscopy (FM), and scanning electron microscopy (SEM) were used to characterize the storage stability of the composite modified asphalt. This study contributes to the practical use of crumb rubber in asphalt binder to develop sustainable and durable pavements.

2. Materials and Methods

2.1. Materials

Polyester fiber: The basic technical indicators are shown in Table 1, purchased from Changzhou Tianyi Fiber Co., Ltd. (Changzhou, China). Epoxidized soybean oil: The basic technical indicators are shown in Table 2, purchased from Guangzhou Xinjinlong Plastic Additives Co., Ltd. (Guangzhou, China). Tire crumb rubber: 80 mesh (0.178 mm), purchased from Lantian Hongrui Industrial rubber dealership. Matrix asphalt: Basic technical performance indicators are shown in Table 3, purchased from Sinopec (Beijing, China).

Table 1. Technical indexes of polyester fiber.

Material Behavior		Modified Polyesters	
Diameter (μm)	20 ± 5	Extension at break (%)	30 ± 9
Length (mm)	6	Density (g·cm^{-3})	1.36
Acid–alkali resistance	Strong	Melting point (°C)	258
Tensile strength (Mpa)	≥ 700	Ignition point (°C)	556

Table 2. Technical indexes of epoxidized soybean oil.

Testing Item	Enterprise Standard E-10	Test Value
Appearance	Light yellow transparent liquid	Coincidence
Tincture (Platinum–cobalt colorimetric)	≤ 170	130
Epoxidized value (Hydrochloric acid–acetone)%	≥ 6.0	6.27
Acid value (mg KOH·g^{-1})	≤ 0.5	0.4
Iodine number (gI2·(100 g)$^{-1}$)	≤ 5.0	2.03
Density (20 °C)	0.993 ± 0.005	0.992
Heating loss (125 °C $\times$ 2 h)	≤ 0.2	0.05
Thermal stability (177 °C $\times$ 3 h)	≥ 5.7	6.20
Flash point (Opening cup)	≥ 280	306

Table 3. Technical indexes of matrix asphalt.

Index	Technical Requirement	Test Value
Penetration (25 °C, 100 g, 5 s), 0.1 mm	60~80	71
Penetration index (PI)	-1.5~$+1.0$	-1.16
Softening point, °C	≥ 46	47.1
Ductility (10 °C, 5 cm·min^{-1}), cm	≥ 15	76.3
Dynamic viscosity (60 °C), Pas	≥ 180	197
Flash point (COC), °C	≥ 260	314
Solubility, %	≥ 99.5	99.89
Wax content, %	≤ 2.2	1.5
Density (15 °C), g·cm^{-3}	—	1.026

2.2. Preparation of Composite Modified Asphalt

A preliminary investigation was carried out using a test study methodology [24,25] based on prior research, with the content of polyester fiber set between 0.2% and 0.4%, tire crumb rubber at 18% to 22%, and epoxidized soybean oil at 2% to 6%, all relative to the mass of the matrix asphalt. The preparation process parameters for 17 groups of composite modified asphalt were determined using the Box-Behnken design, as shown in Table 4. The matrix asphalt was heated to the liquid state and subjected to high-speed shearing at a rotational speed of 5000 r·min^{-1}. The process involved mechanically stirring the epoxidized soybean oil at 160 °C to 180 °C for 5 min, followed by crumb rubber stirring at 180 ± 5 °C for 1 h, and polyester fiber stirring at 160 ± 5 °C for 1 h. The composite modified asphalt was finally prepared by swelling and developing in an oven at 140 °C for 1 h. The preparation process is illustrated in Figure 1. The modified asphalt is prepared via commercial melting and mixing. During the production period in the premix plant, it is sufficient to heat the asphalt binder to the appropriate temperature range and then to sequentially add the modified materials for adequate mixing so that the modified materials are uniformly dispersed in the asphalt. The production process is relatively simple and easy to implement.

Table 4. Dosage of modified materials in each group.

Test Group Number	Epoxidized Soybean Oil	Crumb Rubber	Polyester Fiber
1	4%	18%	0.2%
2	4%	22%	0.2%
3	4%	18%	0.4%
4	4%	22%	0.4%
5	2%	18%	0.3%
6	2%	22%	0.3%
7	6%	18%	0.3%
8	6%	22%	0.3%
9	2%	20%	0.2%
10	2%	20%	0.4%
11	6%	20%	0.2%
12	6%	20%	0.4%
13	4%	20%	0.3%
14	4%	20%	0.3%
15	4%	20%	0.3%
16	4%	20%	0.3%
17	4%	20%	0.3%

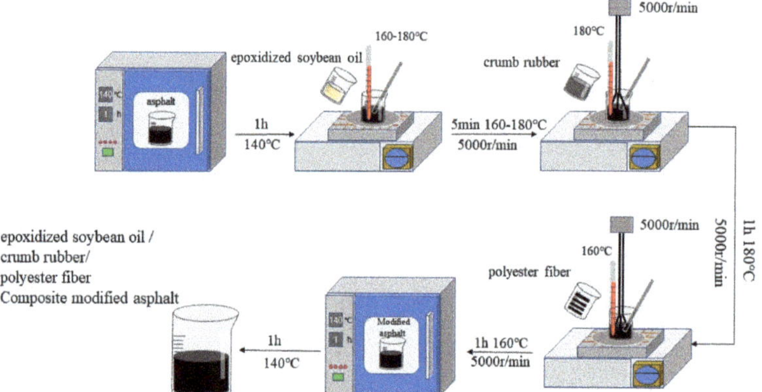

Figure 1. Preparation process of composite modified asphalt.

2.3. Tests

The relationship between the contents of modified materials and the experimental data of BBR and MSCR are established based on the response surface method. The effects of different modifiers on the rheological properties of asphalt are compared using temperature scanning. Finally, softening point difference, fluorescence microscopy (Leika DM4, Germany), and scanning electron microscopy (EVO10, Germany) are combined to characterize the storage stability of the composite modified asphalt. The flow chart of the experimental plan is shown in Figure 2.

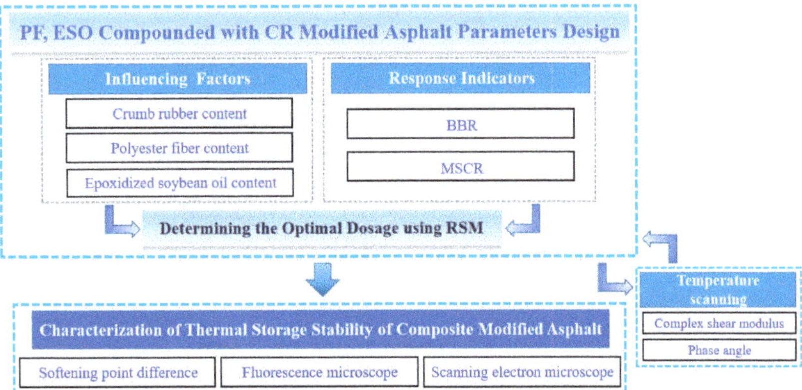

Figure 2. Flow chart of the experimental plan.

2.3.1. Rheological Properties of Asphalt

A dynamic shear rheometer (DSR) was utilized on the asphalt samples according to ASTM D7405-2010a [26] by using an Anton Paar MCR 301 device from Austria. A multiple stress creep recovery test (MSCR) was conducted, and its non-recoverable creep compliances at 0.1 kPa and 3.2 kPa were employed as evaluation indices of the high-temperature rheological properties of the asphalt [27]. The bending beam rheometer (BBR) test was used to evaluate the low-temperature rheological properties of composite modified asphalt in accordance with ASTM D6648-08(2016) [28] by using a TE-BBR device from America, with the creep rate (m) and stiffness modulus (S) at $-18\,^\circ$C for 60 s serving as evaluation indices. The asphalt samples were tested via temperature scanning (scanning range: 40–80 $^\circ$C), and the phase angle (δ) and complex shear modulus (G^*) were used as the evaluation indexes of high-temperature rheological properties.

2.3.2. Storage Stability Tests

The procedure for evaluating the storage stability of the composite modified asphalt involved pouring the sample into a 2.5 cm segregation tube, incubating it in a $163 \pm 5\,^\circ$C oven for 48 h, and then transferring it to $-12\,^\circ$C for 4 h, as specified in JTG E20-2011 [29]. The softening point difference was then measured at the upper and lower sections of the segregation tube to assess the stability of the asphalt [30]. Fluorescence microscopy (FM) was utilized to observe the upper and lower sections of the asphalt in the segregation tube after segregation with a magnification of 50 μm, to assess the degree of asphalt segregation based on the dispersion concentration of the modified substances [31]. Two samples were taken from each of the upper and lower sections of the separation tube, and three points were taken for each sample. This made a total of six images generated for each asphalt section to avoid statistically significant errors [32]. The quantification of the asphalt microstructure was carried out using Fiji ImageJ (NIH, Bethesda, Maryland, USA) by enabling the standard threshold selection to mark the green parts, as shown in Figure 3. The microscopic distribution characteristics of the three crumb rubber modified asphalts were further investigated using the stability coefficients. Scanning electron microscopy (SEM) was also used to observe the morphology of modified asphalt and to characterize its storage stability.

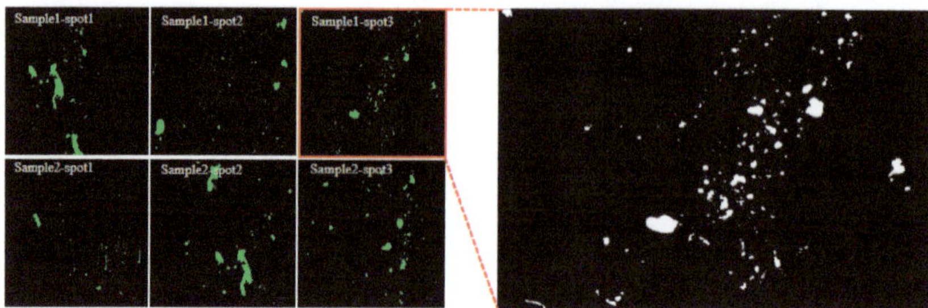

Figure 3. Six images of each part of modified asphalt (**left**) and quantitative analysis results (**right**).

2.4. Response Surface Methodology

The study aims to examine the effect of the quantity of epoxidized soybean oil, crumb rubber, and polyester fiber on the high- and low-temperature performances of composite modified asphalt. Seventeen groups of composite modified asphalt were prepared using the Box-Behnken design with varying quantities of the three modifiers as a percentage of the matrix asphalt mass. The influences of the dosages of these modifiers on the performance of the composite modified asphalt were analyzed using regression analysis. It was necessary to adopt a systematic and appropriate method of data processing and analysis, in order to assure the validity of the results and to meet the desired performance requirements.

The Response Surface Methodology (RSM) is a statistical technique that combines the design of experiments and mathematical modeling to establish a second-order polynomial relationship between the design variables and the response variables, thereby determining the optimal combination of design variables and the corresponding optimal response value [33]. RSM has the advantage of high precision and reduced experimentation, which makes it suitable for the analysis of multi-factor interactions and the prediction of target values based on different levels of design variables. RSM has been extensively employed in the optimization of processes and enhancement of product parameters in pavement technology [34].

3. Results and Discussion

3.1. Response Surface Methodology

In this section, the optimization and analysis of the properties of composite modified asphalt composed of polyester fiber, crumb rubber, and epoxidized soybean oil were performed. The study utilized 17 groups of composite modified asphalt prepared based on the previously designed test scheme. The influencing factors considered in the study included the content of polyester fiber, crumb rubber, and epoxidized soybean oil. The creep rate, stiffness modulus, and non-recoverable creep compliance at 0.1 kPa and 3.2 kPa were used as response indicators to characterize the high- and low-temperature properties of the composite modified asphalt. Results from the experiments were used to establish the response surface model that predicts the optimal dosage of modified materials.

The degree of approximation of the response surface was evaluated using the complex correlation coefficient (R^2) and correction determination coefficient ($Adj.R^2$), as indicated by the calculation formulas in Equations (1) and (2). The results of the model's variance analysis are shown in Table 5.

$$R^2 = \frac{SS_{residual}}{SS_{model} + SS_{residual}} \tag{1}$$

$$Adj.R^2 = 1 - \frac{\frac{SS_{residual}}{DF_{residual}}}{\frac{SS_{model} - SS_{residual}}{DF_{model} + DF_{residual}}} \tag{2}$$

Table 5. Results of variance analyses of models.

Index	m	S	$J_{nr0.1}$	$J_{nr3.2}$
R^2	0.9643	0.9592	0.9695	0.9875
$Adj.R^2$	0.9185	0.9067	0.9303	0.9715
F-value	21.03	18.28	24.71	61.53
p-value	0.0003	0.0005	0.0002	<0.0001
Lack-of-Fit p-value	0.8851	0.6421	0.5066	0.1464
Std. Dev.	0.0049	3.07	1.23	1.47
C.V.%	1.42	2.23	1.63	4.76

In the equation above: $SS_{residual}$ is the residual sum of squares; SS_{model} is the model sum of squares; $DF_{residual}$ is the residual degree of freedom; DF_{model} is the model degree of freedom.

3.1.1. Low-Temperature Rheological Property

The low-temperature creep characteristics of the composite modified asphalt were evaluated using the bending beam rheometer test (BBR). An increase in m and a decrease in S are indicative of an enhancement in the anti-cracking performance of the asphalt pavement at low temperatures. The response surface diagram was created using Design Expert software from Stat-Ease, Inc. (Minneapolis, Minnesota, USA), and the results show that when the polyester fiber content is kept constant, the increase in crumb rubber content results in an increase in m and a decrease in S, as depicted in Figure 4a,d. It indicates that the addition of crumb rubber improves the rheological properties of asphalt at low temperatures due to the cross-linking reaction of crumb rubber in asphalt, which increases the asphalt's toughness at low temperatures. Figure 4a,d also shows that as the amount of polyester fiber increases, the m and S values initially decrease and then increase, while the amount of crumb rubber remains constant. Additionally, an increase in epoxidized soybean oil leads to a rise in m and a drop in S when the crumb rubber content is constant, as shown in Figure 4b,e, thereby enhancing the low-temperature crack resistance of the modified asphalt. This is due to the fact that the epoxidized soybean oil complements the lighter components of the asphalt. A comprehensive analysis of the results demonstrates that a suitable amount of polyester fiber can improve the low-temperature performance of the modified asphalt by forming a three-dimensional network structure within it [35]. The addition of epoxidized soybean oil can also enhance the rheological properties of asphalt at low temperatures.

The following coded multiple quadratic regression equation can be derived from the experimental results of the m and S values:

$$m = 0.4566 - 0.033625X + 0.85775Y + 0.013112Z \\ -0.0775XY + 0.00025XZ + 0.00875YZ + 0.001494X^2 + 1.2225Y^2 - 0.001694Z^2 \tag{3}$$

$$S = 711.075 - 35.8125X - 1173.25Y - 17.525Z + 22.5XY + 0.9375XZ - 23.75YZ + 0.6125X^2 + 1395Y^2 + 0.3Z^2 \tag{4}$$

The response surface model was subjected to variance and significance analysis, with the results presented in Table 5. The p and F values indicate the significance of the correlation coefficient [36]. As seen in Table 5, the p values for the creep rate (m) and stiffness modulus (S) are 0.0003 and 0.0005, while the corresponding F values are 21.03 and 18.28, respectively. The p values are below the significance threshold of 0.05, indicating that the model exhibits significant correlation and statistical significance. The complex correlation coefficients (R^2) of the creep rate and stiffness modulus are 0.96 and 0.95, and the adjusted determination coefficients ($Adj.R^2$) are 0.91 and 0.90, respectively. The closer the values of R^2 and $Adj.R^2$ are to 1, the better the fit of the regression equation and the closer the correlation between the predicted and actual values. This model can be employed to analyze and to predict related indexes.

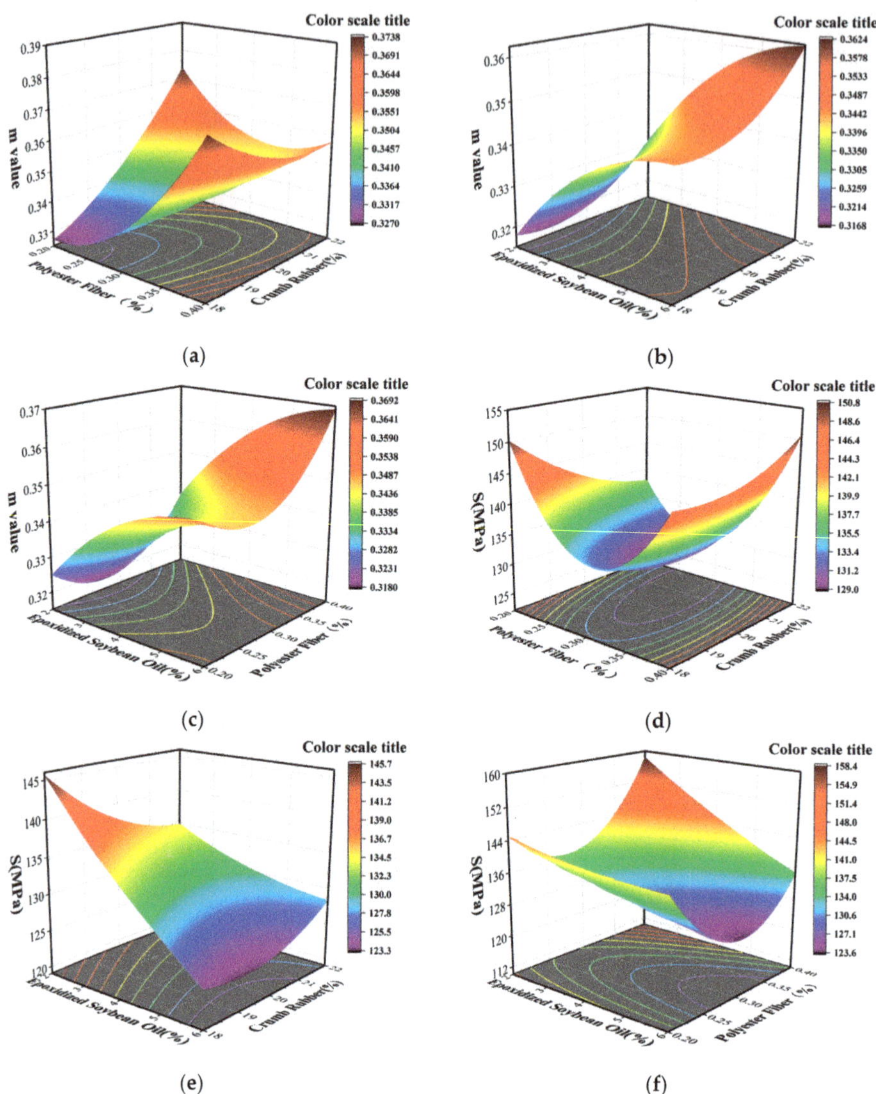

Figure 4. Interaction between various test factors and low-temperature index of composite modified asphalt. (**a**) Effect of interaction between CR and PF; (**b**) Effect of interaction between CR and ESO; (**c**) Effect of interaction between ESO and PF; (**d**) Effect of interaction between CR and PF; (**e**) Effect of interaction between CR and ESO; (**f**) Effect of interaction between ESO and PF.

3.1.2. High-Temperature Rheological Property

A multiple stress creep recovery test (MSCR) was conducted on 17 groups of composite modified asphalt to determine their non-recoverable creep compliance at 0.1 kPa and 3.2 kPa. The utilization of Design Expert software enabled the creation of three-dimensional response surface diagrams. The analysis of the results depicted in Figure 5 suggests that an increase in the content of crumb rubber leads to a rise in the non-recoverable creep compliance and a decrease in the permanent deformation resistance of asphalt under varying levels of stress when the content of polyester fiber remains constant. On the other hand, the resistance to the permanent deformation of asphalt increases and then slows down with the increase in the amount of epoxidized soybean oil, as shown in Figure 5b,e, when the amount of crumb rubber is kept constant. Figure 5c,f show that the addition of polyester fiber mitigates the detrimental impact of the softening effect of epoxidized soybean oil on the high-temperature rheological properties of asphalt. The results further indicate that a suitable content of polyester fiber can boost the permanent deformation resistance of asphalt and improve the high-temperature properties of asphalt.

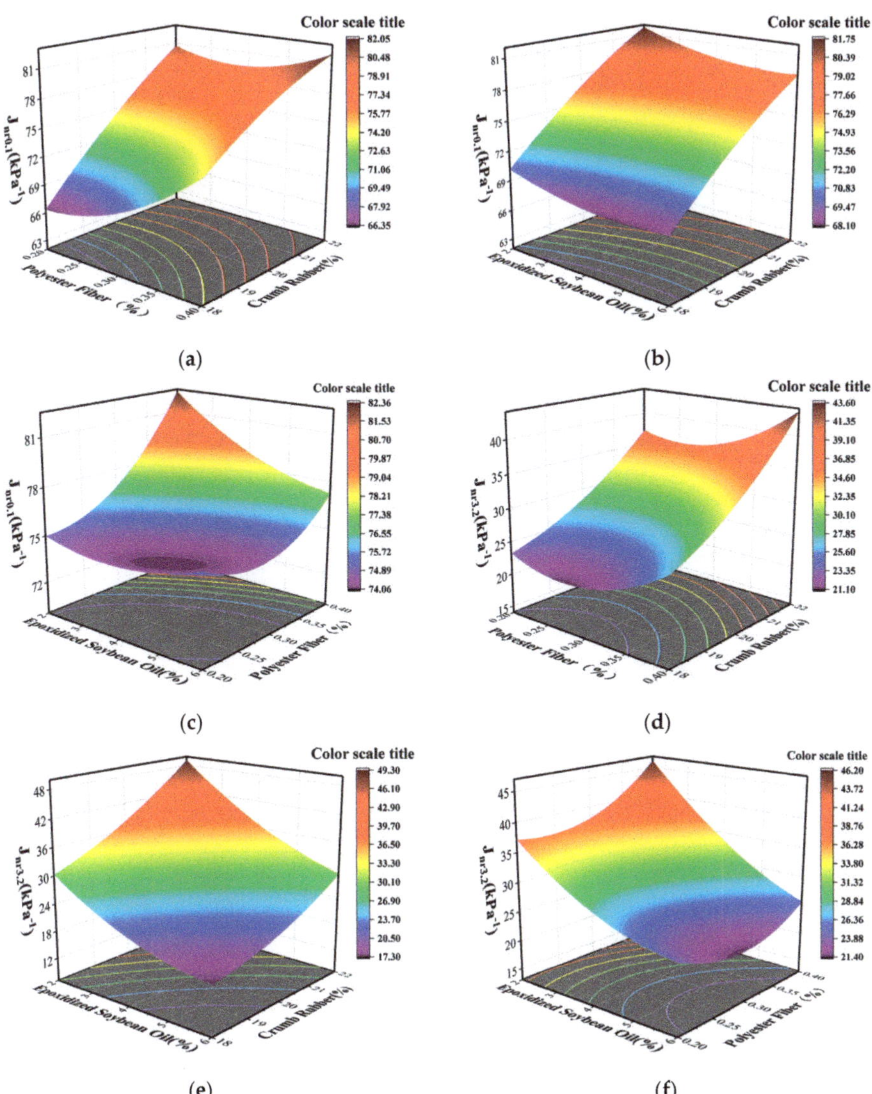

Figure 5. Interaction between various test factors and high-temperature index of composite modified asphalt. (**a**) Effect of interaction between CR and PF; (**b**) Effect of interaction between CR and ESO; (**c**) Effect of interaction between ESO and PF; (**d**) Effect of interaction between CR and PF; (**e**) Effect of interaction between CR and ESO; (**f**) Effect of interaction between ESO and PF.

The following coding multiple quadratic regression equation can be derived from the experimental results:

$$J_{nr0.1} = -111.657 \quad + 14.02687X + 96.9325Y + 0.79275Z - 7.8375XY - 0.05375XZ - 7YZ \\ -0.217062X^2 + 184.175Y^2 + 0.227312Z^2 \tag{5}$$

$$J_{nr3.2} = 151.8325 \quad -13.02375X - 225.0123Y + 2.78062Z + 3.575XY - 0.445XZ \\ -9.8375YZ + 0.439063X^2 + 363.875Y^2 + 0.617188Z^2 \tag{6}$$

Variance analysis was performed on the multiple quadratic regression equation established based on the results of the multiple stress creep recovery test (MSCR) conducted on 17 groups of composite modified asphalt. The results of the analysis are presented in Table 5. It can be observed from Table 5 that the complex correlation coefficients (R^2) for $J_{nr0.1}$ and $J_{nr3.2}$ are 0.96 and 0.98, respectively, and that the adjusted determination coefficients ($Adj.R^2$) are 0.93 and 0.97. These values, which are close to 1, indicate a high degree of fitting and accuracy for the model, and this model can be utilized to predict the change behavior of $J_{nr0.1}$ and $J_{nr3.2}$.

3.1.3. Design Optimization Comparison

A response surface model for each index was established based on the results above, and its predicted value was compared with the experimentally measured value, as illustrated in Figure 6a,b. A strong correlation is observed between the predicted and measured values, indicating that the model can be used effectively to analyze and to predict the relevant indicators.

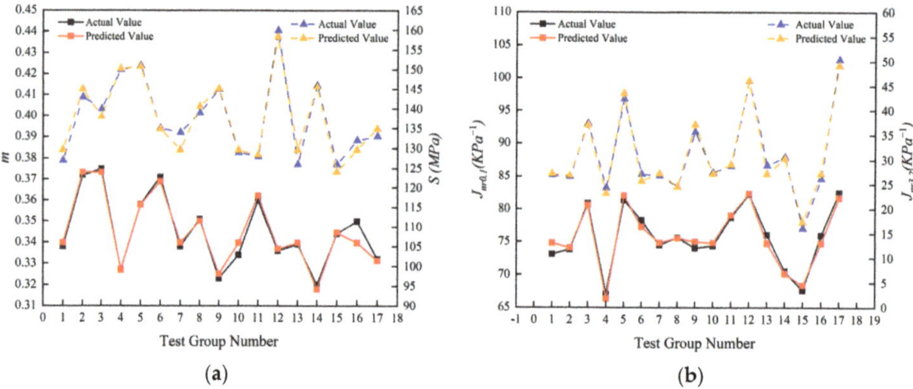

Figure 6. Comparison of predicted and actual values; (a) Comparison of measured creep rate and stiffness modulus with the predicted values of the model; (b) Comparison of measured non-recoverable creep compliance with the predicted values of the model.

Figure 6a,b demonstrates a high degree of concordance between the predicted values and the corresponding actual values. These results provide the basis for optimizing design variables to specific requirements and expectations in practical engineering applications. In this study, an optimization of the response surface model was performed to determine the optimal design parameters for the composite modified asphalt. The optimal composition was found to be 22% crumb rubber, 0.34% polyester fiber, and 3.21% epoxidized soybean oil, relative to the mass of the matrix asphalt. The composite modified asphalt exhibits the best overall performance under these conditions. Compared to a commercial asphalt binder such as SBS, the percentage of modified materials added in the above ratio is comparatively low, and the average market prices of crumb rubber, polyester fiber, and epoxidized soybean oil are not high. The current prices of polyester fiber, epoxidized soybean oil, and crumb rubber are relatively cheap. The price of SBS modified asphalt is USD 880 per ton, while the cost of composite modified asphalt is about USD 710 per ton according to the best addition ratio for SBS modified asphalt. Modified materials are less costly and more cost-effective. From a long-term perspective, this composite modified asphalt can extend the service life of asphalt pavement, reduce the number of asphalt pavement repairs, and lower pavement maintenance costs, with certain economic benefits and investment value.

3.2. Experimental Analysis of Temperature Scanning

Four modified asphalt phase angles (δ) and complex shear moduli (G*) of CR, ESO + CR, ESO + PF + CR, and SBS are compared and analyzed. SBS modified asphalt is chosen as the reference. Figure 7a shows that the phase angle of different modified asphalts changed positively with temperature, except for SBS modified asphalt. This symbolizes the asphalt's increase in viscosity and decrease in elasticity. The particular variation in the phase angle of SBS modified asphalt may be attributed to the inherent nature of the SBS modifier itself. The phase angles of CR, ESO + CR, and ESO + PF + CR modified asphalt all increase with increasing temperature. The curve of CR modified asphalt is located below the curve of SBS modified asphalt at temperatures below 64 °C, indicating that its ability to resist deformation is higher than that of SBS modified asphalt at lower temperatures. It is worth noting that the curve of ESO + CR intersects with the curve of ESO + PF + CR at around 70 °C, and then the curve of ESO + PF + CR modified asphalt is lower than that of ESO + CR modified asphalt. This indicates that at higher temperatures, the reinforcing and stabilizing effect of PF mitigates the adverse effects of ESO softening on the high-temperature deformation resistance of asphalt, and the addition of PF improves the high-temperature rheological properties of modified asphalt. The complex shear modulus of the four modified asphalts falls monotonically as the temperature rises, as shown in Figure 7b, and the change trend is essentially the same. The complex shear moduli of CR, ESO + CR, and ESO + PF + CR modified asphalt are always higher than that of SBS modified asphalt, indicating that the addition of modifiers effectively improves the high-temperature rheological properties of asphalt.

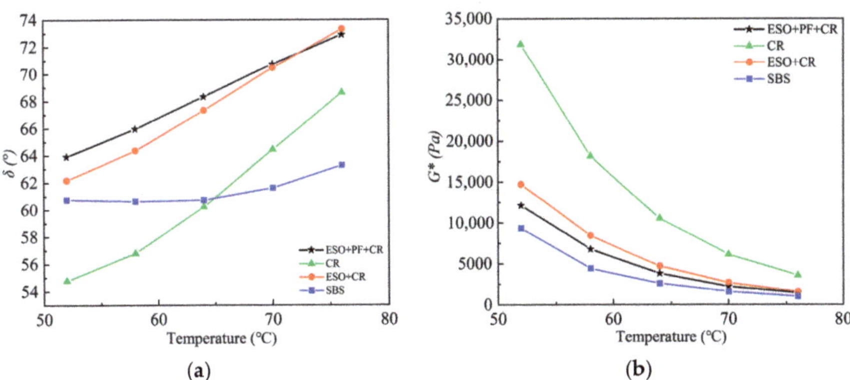

Figure 7. Temperature scan experimental data graph. (**a**) Phase angles of different modified asphalts; (**b**) Complex shear moduli of different modified asphalts.

3.3. Storage Stability

A partial three-dimensional network structure forms between crumb rubber and asphalt as the content of crumb rubber increases, leading to heightened segregation [25]. Despite the fact that crumb rubber asphalt improves the performance of asphalt, the compatibility issues between the two materials results in segregation during construction and transportation. Thus, it is important to find measures to stabilize rubber modified asphalt and to increase its usage. This section utilized the comparative analysis of asphalt storage stability via the softening point difference, fluorescence microscopy, and scanning electron microscopy to investigate the influence of polyester fiber (PF) and epoxidized soybean oil (ESO) on the storage stability of crumb rubber modified asphalt.

3.3.1. Softening Point Difference after Segregation

The softening point difference, a commonly employed index for evaluating the storage stability of modified asphalt, was utilized in this study to examine the influence of PF and ESO on the storage stability of crumb rubber modified asphalt. The experiment involved storing various modified asphalt samples at 163 °C for 48 h and then dividing the sealed aluminum tubes into three sections. The upper and lower sections were selected for softening point analysis, with the difference in softening point between the sections being used to assess the degree of segregation and storage stability. The results are presented in Table 6.

Table 6. Different modified asphalt upper and lower softening points and differences.

Asphalt Type	Upper Sections	Lower Sections	Differentials
CR	54.5	52.45	1.75
ESO + CR	50.95	49.35	1.6
ESO + PF + CR	54.7	53.85	0.85

The analysis of Table 6 reveals that the difference in the softening point between the upper and lower sections of the modified asphalts ESO + PF + CR, ESO + CR, and CR after segregation are less than 2.5 °C, fulfilling the specification requirements. The CR modified asphalt has the biggest softening point difference among the three modified asphalts, at 1.75 °C. The addition of ESO results in a reduction in the softening point difference of ESO + CR to 1.6 °C, likely due to the ring-opening reaction between the epoxidized bond in ESO and the unsaturated bond in asphalt, which forms a network structure compound and improves the compatibility between crumb rubber and asphalt [15]. The ESO + PF + CR modified asphalt has the lowest softening point difference at 0.85 °C, indicating minimum segregation and the best storage stability. The stability of ESO + PF + CR modified asphalt is greatly improved through the combination of the multi-directional reinforcement of polyester fiber [37] and the improved compatibility between crumb rubber and asphalt due to ESO. The softening point difference of ESO + PF + CR modified asphalt is reduced by 46.9% relative to ESO + CR, and by 51.9% relative to CR, demonstrating a considerable improvement in storage stability and a decrease in crumb rubber segregation within asphalt under the synergistic effects of ESO and PF.

3.3.2. Fluorescence Microscope Test after Segregation

The use of fluorescence microscopy has been shown to be an effective tool for observing the phase structures of polymers in the process of blending polymers and asphalt and for providing insight into the morphological structure of modified asphalt. This technique works by exposing the polymer in the modified asphalt to light of a specific wavelength, which causes the energy absorbed by the polymer to jump to an excited state and emit fluorescence [38]. The fluorescence microscope was utilized to examine the modified materials dispersions of ESO + PF + CR, ESO + CR, and CR in the upper and lower sections of a segregation tube after being stored at 163 °C for 48 h. This investigation was performed to assess the thermal storage stability of asphalt, and the results are presented in Figure 8.

Figure 8 shows that in the comparison between the upper and lower sections of ESO + PF + CR, ESO + CR, and CR, the fluorescent substance difference in ESO + PF + CR is minimal, indicating a more uniform dispersion. This can be attributed to the stability provided by PF adsorption and three-dimensional reinforcement, as well as improved compatibility due to ESO. In contrast, the difference in fluorescent substances between the upper and lower sections of CR modified asphalt in Figure 8e,f is the largest, with an obvious accumulation of fluorescent substances in the lower section. ESO may have enhanced asphalt compatibility, promoting gradual fusion with the fluorescent material, which results in uniform dispersion and reduced segregation, according to Figure 8c,d. CR modified asphalt exhibits the highest amount of fluorescent substances based on the analysis of fluorescent substances in ESO + PF + CR, ESO + CR, and CR in Figure 8. This can be attributed to the incomplete decomposition and dispersion of rubber particles, which are then exposed using fluorescence microscopy. On the other hand, the fluorescent substance in ESO + CR is relatively lower than that of CR alone, due to the promoting effect of ESO on the compatibility between CR and asphalt. Moreover, ESO + PF + CR displays the best storage stability and the least amount of fluorescent substance due to the multi-directional reinforcement and stabilization provided by PF, and the improved compatibility of ESO with asphalt.

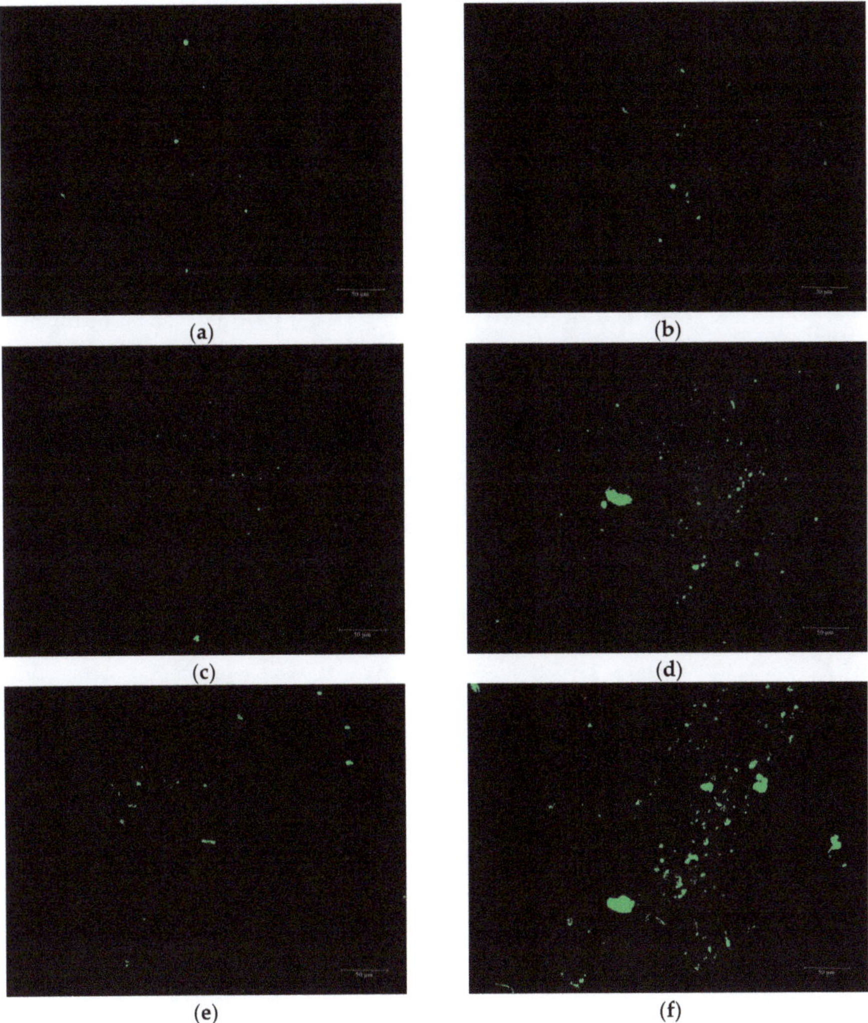

Figure 8. Fluorescence microscopy images of three modified asphalts after 48 h storage. (**a**) ESO + PF + CR upper sections; (**b**) ESO + PF + CR lower sections; (**c**) ESO + CR upper sections; (**d**) ESO + CR lower sections; (**e**) CR upper sections; (**f**) CR lower sections.

Fiji ImageJ was utilized to quantitatively assess the fluorescent area of the added substance in the asphalt binder. Additionally, the stability coefficient (I_{se}) of the modified asphalt was computed using Formula (7). The closer the value of I_{se} is to 1, the better the storage stability of the modified asphalt is. The results of this analysis are presented in Table 7.

$$I_{se} = \frac{S_t}{S_b} \tag{7}$$

Table 7. Fluorescence partial area and stability index of different modified asphalts (I_{se}).

Species	ESO + PF + CR	ESO + CR	CR
Upper sections (S_t)	185.361	401.814	729.18
Lower sections (S_b)	234.442	1118.366	2954.095
Coefficient of stability (I_{se})	0.79	0.36	0.25

In the equation above: S_t and S_b represent the fluorescence areas of the upper and lower sections of the segregation tube, respectively.

Table 7 shows that the fluorescent area of ESO + PF + CR composite modified asphalt is smaller compared to the other two types of modified asphalt. Therefore, the segregation of ESO + PF + CR composite modified asphalt is not obvious. The stability coefficient of ESO + PF + CR composite modified asphalt has an I_{se} value that is approximately 2.1 times higher than that of CR asphalt and 1.2 times higher than that of ESO + CR asphalt, suggesting that it has a high level of stability. Furthermore, its stability coefficient's I_{se} value is closest to 1, indicating that the degradation degree of modified substances is high, with uniform dispersion and forming a stable system within the matrix asphalt, which is consistent with the previous analysis.

3.3.3. Scanning Electron Microscopy Analysis

Scanning electron microscopy (SEM) was used to observe the microscopic morphology of the modified asphalt and to compare the compatibilities of different modified materials with asphalt. Then, the binding mechanism and storage stability of the modified material with asphalt were analyzed.

Figure 9 presents the results of the SEM experiments at $1000\times$ for three different asphalts. Figure 9a indicates the matrix asphalt, which is nearly homogeneous and exhibits a relatively smooth and flat surface. The irregular lumps in Figure 9b are crumb rubber particles with rough surfaces and more obvious boundaries with the asphalt. This is due to the rough surface of the crumb rubber particles and their poor adhesion to the asphalt, which makes it easy to segregate. It can be seen in Figure 9c that the crumb rubber and asphalt binder adhere to the surface of the polyester fiber, and that the polyester fiber reinforces the asphalt binder as a whole via cross-lap. PF is in contact with the crumb rubber and asphalt at an angle, and no agglomeration occurs. The interfacial behavior in the ESO + PF + CR composite modified asphalt is flat, the boundary is blurred, and the modified materials are uniformly distributed. A comprehensive comparison of Figure 9 shows that the polyester fibers in ESO + PF + CR modified asphalt adsorb the crumb rubber and asphalt and provide reinforcement to the asphalt binder through the cross-lap. The interface is blurred, and the problem of poor storage stability of crumb rubber is improved. The modified material is uniformly distributed in the asphalt binder, forming a solid physical co-mingling.

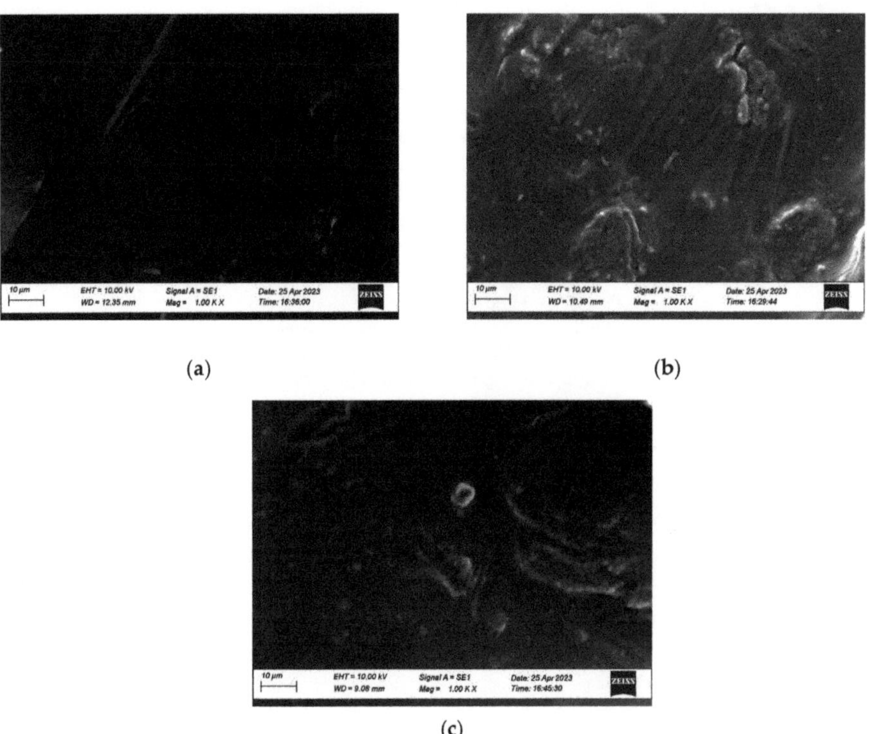

Figure 9. Scanning electron micrograph. (**a**) Matrix asphalt; (**b**) ESO + CR; (**c**) ESO + PF + CR.

4. Conclusions

This study proved that epoxidized soybean oil and polyester fiber could effectively improve the thermal storage stability of crumb rubber modified asphalt. The synergistic effect and scientific ratio of the three modified materials could improve the high- and low-temperature rheological properties of asphalt. The following conclusions can be obtained:

(1) Polyester fiber's stabilizing and reinforcing effects mitigate the negative effect of epoxidized soybean oil's softening effect on the high-temperature rheological properties of modified asphalt. Epoxidized soybean oil can effectively supplement the light components of crumb rubber asphalt to improve the compatibility between crumb rubber and asphalt. The synergistic effects of the CR, PF, and ESO modifiers improve the high- and low-temperature performances of asphalt.

(2) The optimal relative percentage dosages of crumb rubber, polyester fiber, and epoxidized soybean oil with respect to the matrix asphalt quality are determined to be 22%, 0.34%, and 3.21%, respectively, through the response surface method.

(3) Variance analysis and significance analysis of the RSM show that the p-values of the model are all less than the significance threshold of 0.05, and that the correlation coefficient and the adjusted determination coefficient are close to 1. The fitting degree is high, and the correlation is significant.

(4) The softening point difference of ESO + PF + CR modified asphalt is 46.9% lower than that of ESO + CR and 51.4% lower than CR. Quantification of the fluorescence area reveals that the stability coefficient of ESO + PF + CR is closest to 1. Polyester fibers have good adhesion with crumb rubber and asphalt, and the overlap between fibers forms a three-dimensional network structure, which plays the role of reinforcement

and consolidation in asphalt binder. All these indicate that PF and ESO can improve the storage stability of crumb rubber modified asphalt.

Author Contributions: Conceptualization, J.J., G.Q. and J.P.; methodology, J.J. and G.Q.; software, J.P.; validation, J.P. and S.L.; formal analysis, J.P. and M.X.; investigation, J.P., M.X. and Z.W.; resources, J.J. and S.L.; data curation, J.P.; writing—original draft preparation, J.P. and Z.W.; writing—review and editing, J.J.; visualization, S.L.; supervision, J.J.; project administration, J.P.; funding acquisition, J.J. and J.P. All authors have read and agreed to the published version of the manuscript.

Funding: This work was supported by the National Natural Science Foundation of China (52174237, 51704040), the Science Foundation for Outstanding Youth of Hunan Province (2022JJ10051), the Science and Technology Youth Talent Support Project of China Association, the Innovative Project of Hunan Province—Huxiang Young Talents Program (2020RC3039), the Science and Technology Project of Changsha—Outstanding Innovative Youth (kq2206031), the Guangdong Provincial Key Laboratory of Modern Civil Engineering Technology (2021B1212040003), and the Innovation and Entrepreneurship Training Program for college students of Changsha University of Science & Technology (202210536049).

Institutional Review Board Statement: Not applicable.

Informed Consent Statement: Not applicable.

Data Availability Statement: Not applicable.

Acknowledgments: The authors acknowledge the support provided by the National Engineering Research Center of Highway Maintenance Technology.

Conflicts of Interest: The authors declare no conflict of interest.

References

1. Gao, Y.; Xie, Y.; Liao, M.; Li, Y.; Zhu, J.; Tian, W. Study on the Mechanism of the Effect of Graphene on the Rheological Properties of Rubber-Modified Asphalt Based on Size Effect. *Constr. Build. Mater.* **2023**, *364*, 129815. [CrossRef]
2. Sun, D.; Kandare, E.; Maniam, S.; Zhou, A.; Robert, D.; Buddhacosa, N.; Giustozzi, F. Thermal-Based Experimental Method and Kinetic Model for Predicting the Composition of Crumb Rubber Derived from End-of-Life Vehicle Tyres. *J. Clean. Prod.* **2022**, *357*, 132002. [CrossRef]
3. Mohamed, A.S.; Cao, Z.; Xu, X.; Xiao, F.; Abdel-Wahed, T. Bonding, Rheological, and Physiochemical Characteristics of Reclaimed Asphalt Rejuvenated by Crumb Rubber Modified Binder. *J. Clean. Prod.* **2022**, *373*, 133896. [CrossRef]
4. Zhu, Y.; Xu, G.; Ma, T.; Fan, J.; Li, S. Performances of Rubber Asphalt with Middle/High Content of Waste Tire Crumb Rubber. *Constr. Build. Mater.* **2022**, *335*, 127488. [CrossRef]
5. Poovaneshvaran, S.; Mohd Hasan, M.R.; Putra Jaya, R. Impacts of Recycled Crumb Rubber Powder and Natural Rubber Latex on the Modified Asphalt Rheological Behaviour, Bonding, and Resistance to Shear. *Constr. Build. Mater.* **2020**, *234*, 117357. [CrossRef]
6. Nanjegowda, V.H.; Biligiri, K.P. Utilization of High Contents of Recycled Tire Crumb Rubber in Developing a Modified-Asphalt-Rubber Binder for Road Applications. *Resour. Conserv. Recycl.* **2023**, *192*, 106909. [CrossRef]
7. Mousavi, M.; Hosseinnezhad, S.; Kabir, S.F.; Burnett, D.J.; Fini, E.H. Reaction Pathways for Surface Activated Rubber Particles. *Resour. Conserv. Recycl.* **2019**, *149*, 292–300. [CrossRef]
8. Zheng, W.; Wang, H.; Chen, Y.; Ji, J.; You, Z.; Zhang, Y. A Review on Compatibility Between Crumb Rubber and Asphalt Binder. *Constr. Build. Mater.* **2021**, *297*, 123820. [CrossRef]
9. Duan, K.; Wang, C.; Liu, J.; Song, L.; Chen, Q.; Chen, Y. Research Progress and Performance Evaluation of Crumb-Rubber-Modified Asphalts and their Mixtures. *Constr. Build. Mater.* **2022**, *361*, 129687. [CrossRef]
10. Wang, X.; Hong, L.; Wu, H.; Liu, H.; Jia, D. Grafting Waste Rubber Powder and its Application in Asphalt. *Constr. Build. Mater.* **2021**, *271*, 121881. [CrossRef]
11. Liu, Q.; Liu, J.; Yu, B.; Zhang, J.; Pei, J. Evaluation and Optimization of Asphalt Binder and Mixture Modified with High Activated Crumb Rubber Content. *Constr. Build. Mater.* **2022**, *314*, 125676. [CrossRef]
12. Feng, X.; Liang, H.; Dai, Z. Rheological Properties and Microscopic Mechanism of Waste Cooking Oil Activated Waste Crumb Rubber Modified Asphalt. *J. Road Eng.* **2022**, *2*, 357–368. [CrossRef]
13. Kuang, D.; Jiao, Y.; Ye, Z.; Lu, Z.; Chen, H.; Yu, J.; Liu, N. Diffusibility Enhancement of Rejuvenator by Epoxidized Soybean Oil and its Influence on the Performance of Recycled Hot Mix Asphalt Mixtures. *Materials* **2018**, *11*, 833. [CrossRef]
14. Wei, C.; Zhang, H.; Duan, H. Effect of Catalytic-Reactive Rejuvenator on Structure and Properties of Aged Sbs Modified Asphalt Binders. *Constr. Build. Mater.* **2020**, *246*, 118531. [CrossRef]
15. Si, J.; Li, Y.; Wang, J.; Niyigena, A.R.; Yu, X.; Jiang, R. Improving the Compatibility of Cold-Mixed Epoxy Asphalt Based on the Epoxidized Soybean Oil. *Constr. Build. Mater.* **2020**, *243*, 118235. [CrossRef]

16. Li, Q.; Zhang, H.; Chen, Z. Improvement of Short-Term Aging Resistance of Styrene-Butadiene Rubber Modified Asphalt by Sasobit and Epoxidized Soybean Oil. *Constr. Build. Mater.* **2021**, *271*, 121870. [CrossRef]
17. Lei, Y.; Wang, H.; Fini, E.H.; You, Z.; Yang, X.; Gao, J.; Dong, S.; Jiang, G. Evaluation of the Effect of Bio-Oil on the High-Temperature Performance of Rubber Modified Asphalt. *Constr. Build. Mater.* **2018**, *191*, 692–701. [CrossRef]
18. Ma, J.; Hesp, S.A.M. Effect of Recycled Polyethylene Terephthalate (Pet) Fiber on the Fracture Resistance of Asphalt Mixtures. *Constr. Build. Mater.* **2022**, *342*, 127944. [CrossRef]
19. Slebi-Acevedo, C.J.; Lastra-González, P.; Pascual-Muñoz, P.; Castro-Fresno, D. Mechanical Performance of Fibers in Hot Mix Asphalt: A Review. *Constr. Build. Mater.* **2019**, *200*, 756–769. [CrossRef]
20. Zarei, M.; Abdi Kordani, A.; Ghamarimajd, Z.; Khajehzadeh, M.; Khanjari, M.; Zahedi, M. Evaluation of Fracture Resistance of Asphalt Concrete Involving Calcium Lignosulfonate and Polyester Fiber Under Freeze–Thaw Damage. *Theor. Appl. Fract. Mech.* **2022**, *117*, 103168. [CrossRef]
21. Kou, C.; Chen, Z.; Kang, A.; Zhang, M.; Wang, R. Rheological Behaviors of Asphalt Binders Reinforced by Various Fibers. *Constr. Build. Mater.* **2022**, *323*, 126626. [CrossRef]
22. Kim, M.; Kim, S.; Yoo, D.; Shin, H. Enhancing Mechanical Properties of Asphalt Concrete Using Synthetic Fibers. *Constr. Build. Mater.* **2018**, *178*, 233–243. [CrossRef]
23. Sun, Z.; Ma, Y.; Liu, S.; Li, Y.; Qiu, X.; Luo, Z. Evaluation of Engineering Properties of Fiber-Reinforced Usual-Temperature Synthetic Pitch (Usp) Modified Cold Mix Patching Asphalt. *Case Stud. Constr. Mater.* **2022**, *16*, e00997. [CrossRef]
24. Lu, Q.; Xin, C.; Alamri, M.; Alharthai, M. Development of Porous Asphalt Mixture with Bio-Based Epoxy Asphalt. *J. Clean. Prod.* **2021**, *317*, 128404. [CrossRef]
25. Zhou, Y.; Xu, G.; Leng, Z.; Kong, P.; Wang, H.; Yang, J.; Qiu, W.; Xu, Z.; Chen, X. Investigation on Storage Stability and Rheological Properties of Environment-Friendly Devulcanized Rubber Modified Asphalt. *Phys. Chem. Earth Parts A/B/C* **2023**, *129*, 103332. [CrossRef]
26. *ASTM D7405-2010a*; Standard Test Method for Multiple Stress Creep and Recovery (MSCR) of Asphalt Binder Using a Dynamic Shear Rheometer. ASTM International: West Conshohocken, PA, USA, 2010.
27. Shi, K.; Ma, F.; Liu, J.; Song, R.; Fu, Z.; Dai, J.; Li, C.; Wen, Y. Development of a New Rejuvenator for Aged Sbs Modified Asphalt Binder. *J. Clean. Prod.* **2022**, *380*, 134986. [CrossRef]
28. *ASTM D6648-08(2016)*; Standard Test Method for Determining the Flexural Creep Stiffness of Asphalt Binder Using the Bending Beam Rheometer (BBR). ASTM International: West Conshohocken, PA, USA, 2016.
29. Ministry of Transport of the People's Republic of China. *Standard Test Methods for Bitumen and Bituminous Mixtures for Highway Engineering (JTG E20-2011)*; China Communications Press: Beijing, China, 2011.
30. Zhao, Y.; Chen, M.; Wu, S.; Jiang, Q.; Xu, H.; Zhao, Z.; Lv, Y. Effects of Waterborne Polyurethane on Storage Stability, Rheological Properties, and Vocs Emission of Crumb Rubber Modified Asphalt. *J. Clean. Prod.* **2022**, *340*, 130682. [CrossRef]
31. Ke, Y.; Cao, J.; Xu, S.; Bian, C.; Zhang, C.; Jia, X. Storage Stability and Anti-Aging Performance of Sebs/Organ-Montmorillonite Modified Asphalt. *Constr. Build. Mater.* **2022**, *341*, 127875. [CrossRef]
32. Mirwald, J.; Hofko, B.; Pipintakos, G.; Blom, J.; Soenen, H. Comparison of Microscopic Techniques to Study the Diversity of the Bitumen Microstructure. *Micron* **2022**, *159*, 103294. [CrossRef]
33. Zhang, Z.; Luo, S.; Wang, J.; Sun, M.; Yan, S.; Wang, Q.; Zhang, Y.; Liu, X.; Lei, X. Optimization of Preparation of Lignite-Based Activated Carbon for High-Performance Supercapacitors with Response Surface Methodology. *J. Energy Storage* **2022**, *56*, 105913. [CrossRef]
34. Reza Omranian, S.; Geluykens, M.; Van Hal, M.; Hasheminejad, N.; Rocha Segundo, I.; Pipintakos, G.; Denys, S.; Tytgat, T.; Fraga Freitas, E.; Carneiro, J.; et al. Assessing the Potential of Application of Titanium Dioxide for Photocatalytic Degradation of Deposited Soot on Asphalt Pavement Surfaces. *Constr. Build. Mater.* **2022**, *350*, 128859. [CrossRef]
35. Zhang, J.; Huang, W.; Zhang, Y.; Lv, Q.; Yan, C. Evaluating Four Typical Fibers Used for Ogfc Mixture Modification Regarding Drainage, Raveling, Rutting and Fatigue Resistance. *Constr. Build. Mater.* **2020**, *253*, 119131. [CrossRef]
36. Xie, Y.C.; Kang, K.; Zheng, C.; Lan, L.; Song, H.; Li, H.L.; Kang, J.; Bai, S.P. Optimised Synthesis of Stainless Steel Fibre-Entrapped Activated Carbon Composites Using Response Surface Methodology. *Chem. Phys. Lett.* **2023**, *815*, 140355. [CrossRef]
37. Zhang, T.; Wu, J.; Hong, R.; Ye, S.; Jin, A. Research on Low-Temperature Performance of Steel Slag/Polyester Fiber Permeable Asphalt Mixture. *Constr. Build. Mater.* **2022**, *334*, 127214. [CrossRef]
38. Padhan, R.K.; Sreeram, A. Enhancement of Storage Stability and Rheological Properties of Polyethylene (Pe) Modified Asphalt Using Cross Linking and Reactive Polymer Based Additives. *Constr. Build. Mater.* **2018**, *188*, 772–780. [CrossRef]

Article

Optimization Design and Mechanical Performances of Plant-Mix Hot Recycled Asphalt Using Response Surface Methodology

Honglin Liu, Jinping Wang, Weiwei Lu * and Naitian Zhang

National Engineering Research Center of Highway Maintenance Technology, Changsha University of Science & Technology, Changsha 410114, China; hlliu@huuc.edu.cn (H.L.); wangjp@stu.csust.edu.cn (J.W.); zntcs@stu.csust.edu.cn (N.Z.)
* Correspondence: lww_cs@csust.edu.cn

Abstract: This study aimed to explore the influence of material design parameters on the physical and mechanical properties of recycled asphalt. A Box–Behnken design was employed to determine the optimal preparation scheme for 17 groups of recycled asphalt. The effects of styreneic methyl copolymer (SMC) regenerant content, styrene–butadiene–styrene (SBS)-modified asphalt content, and shear temperature on the mechanical properties of recycled asphalt were analyzed using conventional and high/low-temperature rheological tests. The optimal processing parameters were determined by a response surface model based on multiple response indexes. The results revealed that the SBS-modified asphalt content had the most significant effect on the penetration of recycled asphalt. An increase in SMC regenerant content led to a gradual decrease in the rutting factor, while SBS-modified asphalt content had the opposite effect. The usage of SMC regenerant helped to reduce non-recoverable creep compliance by adjusting the proportion of viscoelastic–plastic components in recycled asphalt. Furthermore, the stiffness modulus results indicated that the addition of SMC regenerant improved the recovery performance of recycled asphalt at a low temperature. The recommended contents of SMC regenerant and SBS-modified asphalt are 7.88% and 150%, respectively, with a shear temperature of 157.7 °C.

Keywords: recycled asphalt; materials design; mechanical performance; response surface methodology; regression model

Citation: Liu, H.; Wang, J.; Lu, W.; Zhang, N. Optimization Design and Mechanical Performances of Plant-Mix Hot Recycled Asphalt Using Response Surface Methodology. *Materials* **2023**, *16*, 5863. https://doi.org/10.3390/ ma16175863

Academic Editor: Simon Hesp

Received: 7 July 2023
Revised: 15 August 2023
Accepted: 19 August 2023
Published: 27 August 2023

1. Introduction

Due to intricate traffic conditions and heavy traffic loads, the performance of asphalt pavement gradually deteriorates, failing to meet the demands of traffic. Consequently, rehabilitation projects of asphalt pavement generate approximately 100 million tons of reclaimed asphalt pavement (RAP) per year [1]. To address this issue, the utilization of RAP has gained significant attention [2]. The recycling of RAP offers benefits such as reduced energy consumption, preserved landfill space, minimized air pollution, conservation of non-renewable resources, and decreased production costs [3]. Previous research has indicated that while a high RAP content in asphalt mixtures improves resistance to rutting, it may negatively affect workability and promote fatigue cracking [4–6]. Moreover, the mechanical properties of recycled asphalt can be enhanced by incorporating a specific amount of regenerant [7]. Therefore, it is worth noting that an optimal value for both RAP content and regenerant amount may exist, enabling the enhancement of the mechanical properties of the asphalt mixture [8].

The plant-mix hot recycled technology of RAP encompasses a multitude of advantages, including the reutilization of waste materials; the conservation of non-renewable resources, such as mineral aggregate and asphalt; and the fulfillment of the requirements for low carbon development. Moreover, its construction process is straightforward, leading to its increasing popularity [9,10]. Numerous scholars have delved into the hot recycled technology of RAP. For example, Ferreira et al. [11] examined the influence of RAP content and aging heterogeneity on the properties of asphalt mixtures. Chen et al. [12] discovered a continuous improvement in the high-temperature performance of a plant-mix hot recycled asphalt mixture as the aged asphalt content increased. Lee et al. [13] examined the self-healing ability of RAP and evaluated the influence of different regenerants with a three-point bending test. Wan et al. [14] found that the use of hot recycled asphalt mixtures for the upper surface layer could improve the ability to resist the permanent strain of the pavement. Lee et al. [15] scrutinized the absolute viscosity of aged asphalt in RAP and the nominal maximum aggregate size of recycled aggregate, studying their influence on RAP performance. Hettiarachchi et al. [16] extracted aged asphalt from RAP and investigated the factors affecting the blending degree of recycled asphalt mixtures using Fourier transform infrared spectroscopy. Blanc et al. [17] explored the use of biomaterials as regenerants to achieve secondary utilization of RAP. Zhang et al. [18] evaluated the performance of four recycled asphalt mixtures with RAP contents of 0%, 15%, 30%, and 50%. Zaumanis M et al. [19] determined that the performance of a recycled asphalt mixture with 100% RAP approached the requirements of high modulus designs, but fell short of fully satisfying the fatigue, modulus, and rutting requirements. Meroni et al. [20] conducted a comprehensive evaluation of four recycled asphalt mixtures produced and laid in Virginia. Several studies have concluded that RAP enhances the rutting resistance and tensile strength of recycled asphalt mixtures while reducing moisture sensitivity. However, it has been observed that RAP inclusions can have a negative influence on the fatigue resistance of recycled asphalt mixtures [21]. Consequently, recycled asphalt technology has become a topic of great interest among scholars [22]. Recently, researchers have focused on studying the aging of asphalt in RAP to promote its regeneration and restore its fundamental properties [23]. It has been observed that utilizing less aged asphalt from RAP in recycled asphalt results in a brittle and hard state. This characteristic can be fully considered to optimize the performance of recycled asphalt.

Response surface methodology (RSM) represents a mathematical and statistical approach employed in the design of experiments, the mathematical modeling of variables (both univariate and multivariate), the evaluation of independent variables, and the optimization of procedures [24]. This method has been successfully applied in several fields, including civil engineering, mechanical and materials engineering, biology, and earth science [25–27]. In recent years, RSM has been found to have applications in the design and optimization of asphalt materials. For example, Rafiq et al. [28] optimized the content of binders and additives in stone mastic asphalt mixtures using RSM on the basis of volumetric properties and Marshall tests. Bala et al. [29] developed a performance-based statistical model utilizing RSM to optimize the content of asphalt binder and fine aggregates in hot mix asphalt. Guo et al. [30] employed RSM to assess the effects of waste engine oil content and gradation on indirect tensile strength (ITS) under dry and saturated conditions. Vatanparast et al. [31] optimized the asphalt content of a warm mix asphalt (WMA) mixture with RSM, considering volume characteristics and strength as responses. Relying on RSM, Taherkhani and Noorian [32] investigated the influence of preparation parameters, including gradation and aggregate types, on the ITS of WMA mixtures. Additionally, Khairuddin et al. [33] examined the influences of asphalt and polyurethane content on penetration, softening, and viscosity values using RSM. Previous research has demonstrated the successful application of RSM in establishing relationships between the parameters and performance characterization of asphalt materials. Hence, it is indeed feasible to optimize the preparation process of recycled asphalt utilizing RSM.

This study aimed to investigate the effects of preparation process parameters on the mechanical properties of recycled asphalt through RSM. The experimental data obtained in this study were utilized to establish a comprehensive multivariate quadratic regression equation and construct a three-dimensional response surface. These tools facilitated the identification of optimal values for the styrene–butadiene–styrene (SBS)-modified asphalt content, the styreneic methyl copolymer (SMC) regenerate content, and the shear temperature. Subsequently, the recycled asphalt was prepared according to the optimal design parameters. Both its conventional performance and rheological properties at high and low temperatures were assessed. Finally, the test results were compared with the predicted values generated from the regression equation, thereby further validating its reliability.

2. Materials and Methods

2.1. Materials

2.1.1. Base Asphalt

In this study, a SBS (I-D) modified asphalt was selected as the base asphalt in recycled asphalt. Its main performance indexes are shown in Table 1.

Table 1. General physical properties of SBS (I-D) modified asphalt.

Indicators		Test Results	Specification Requirement	Test Method
Penetration at 25 °C, 100 g, and 5 s (0.1 mm)		71	60–80	ASTM D5
Penetration index		0.15	−1.5–1.0	ASTM D5
Ductility at 5 cm/min and 5 °C (cm)		>150	≥100	ASTM D113
Softening point (°C)		48	≥46	ASTM D36
Kinematic viscosity at 135 °C (Pa·s)		2.45	≤3	ASTM D2171
Solubility (%)		99.78	>99.5	ASTM D2042
Flash point (°C)		310	≥260	ASTM D92
Residue from the film oven test *	Mass loss (%)	−0.04	≤±1.0	ASTM D402
	Penetration at 25 °C (%)	78	≥65	ASTM D5
	Ductility at 5 °C (cm)	22	≥20	ASTM D113

* Notes: The film oven test was conducted by situating a vessel containing asphalt in a film heating oven. The test temperature was set at 163 °C, and the duration was 5 h. The asphalt film after the test was defined as the residue from the film oven test. This test aimed to determine the mass changes of asphalt films after heating. Additionally, the changes in the properties of asphalt residue, including penetration and ductility, were measured to assess the aging resistance performance of original asphalt.

2.1.2. Regenerant

To recycle the asphalt from RAP, the aged asphalt, the regenerant, and the original asphalt were mixed in appropriate proportions. The recycled asphalt after this process can achieve the desired viscosity and the optimal performance for its intended applications. The regenerant plays a crucial role in adjusting the viscosity of the aged asphalt, modifying its colloidal structure, and improving its rheological properties. In this study, the SMC-III modifier was chosen as the regenerant, and its key performance indicators are outlined in Table 2.

Table 2. Basic properties of SMC regenerant.

Performance	Test Results
Appearance	Yellow-brown viscous liquid
Density (g/cm^3)	0.88
Flash point by open cup method (°C)	70
hydrocarbon content of rub (%)	84
Viscosity at 25 °C (Pa·s)	0.71

2.2. Preparation of Recycled Asphalt

2.2.1. Aging Simulation of Asphalt Binders

Generally, the asphalt pavement has suffered the impact of various environmental factors including light, oxygen, and humidity for a long time before undergoing the recycled process. Consequently, the asphalt binder obtained from RAP is aging badly. To replicate the long-term aging of asphalt in a laboratory, several accelerated aging methods have been developed in the last few decades. In this study, an accelerated aging test was performed using a pressure aging vessel.

Firstly, the original asphalt was subjected to softening in a rotary thin film oven (Shanghai Changji Geological Instrument Co., Ltd., Shanghai, China). Subsequently, the softened asphalt was transferred into aging pans (Shanghai Changji Geological Instrument Co., Ltd., Shanghai, China), with each pan containing a sample size of 50.0 g $\pm$ 0.5 g. These pans were placed on a burn-in tray. Once the temperature reached the tested value, the pressure vessel (Prentex Alloy Fabricators Inc, Dallas, TX, USA) was swiftly opened and the burn-in tray was inserted into the vessel. The vessel was sealed to initiate the aging process. The experiment was conducted under a pressure of 2.1 MPa $\pm$ 0.1 MPa and an aging duration of 20 h $\pm$ 10 min at a temperature of 100 °C. Upon completion of the predetermined aging time, the pressure inside the aging container was released to equalize the internal and external pressures. The long-term aged asphalt was obtained by removing the asphalt from the aging tray.

In response to the challenge of accurately predicting the long-term aging of asphalt binders, numerous researchers have undertaken the laboratory testing procedures to prepare aged asphalt specimens, serving as a viable substitute for the field asphalt derived from RAP. Qian et al. [34] suggested that although current laboratory aging methods might not completely replicate field conditions, the variability in performance of aged asphalt obtained through laboratory tests was comparatively lower. Moreover, Chen et al. [12] discovered that the influence of the aging asphalt content on the properties of the base asphalt aligned with the influence of the RAP content on the performance of the recycled asphalt mixture. Building upon these findings, this study employed the laboratory-prepared long-term aged asphalt as a substitute for the aged asphalt within RAP materials.

2.2.2. Preparation of Recycled Asphalt

In this study, the procedure involved heating aged asphalt and SBS-modified asphalt to a molten state at 165 °C using an oven (Shanghai Changji Geological Instrument Co., Ltd., Shanghai, China). The SMC regenerative agent was gradually introduced to the aged asphalt while stirring to ensure thorough blending. After the addition was complete, the mixture was further stirred with a glass rod for an additional 10 min to enhance compatibility. Subsequently, the SBS-modified asphalt was weighed and added to the recycled asphalt, and the components were mixed using the glass rod. Finally, the mixture underwent shearing for 30 min using a high-speed shear apparatus at test temperature and a rotation rate of approximately 4000 r/min. The content of the SMC regenerative agent content, the SBS-modified asphalt, and the shear temperature were determined by the following test design.

2.3. Test Methods

2.3.1. Penetration Test

Penetration serves as an indicator of the hardness, consistency, shear resistance, and viscosity of asphalt under specific circumstances. The test was operated by adjusting the thermostatic water (Shanghai Changji Geological Instrument Co., Ltd., Shanghai, China) bath to 25 °C and maintaining its stability as per specifications. Then, the sample was placed in the water bath for 1.5 h. After five seconds, the vertical penetration depth was measured on the displacement meter (Shanghai Changji Geological Instrument Co., Ltd., Shanghai, China) using a standard needle of 100 g, with an accuracy of 0.1 mm. The reading of depth represents the penetration of asphalt at 25 °C.

2.3.2. Softening Point Test

The softening point plays a significant role in evaluating the temperature sensitivity of asphalt in road engineering. To determine this crucial parameter, a systematic procedure was followed. Firstly, the asphalt sample was allowed to cool naturally at room temperature for 30 min. Next, a steel ball was delicately placed on the surface of the sample, which was then carefully introduced into the softening point tester. The apparatus was heated gradually in a water bath, with the rate of increase set at 5 °C per minute. During this process, the asphalt gradually softened, leading to the sagging of the steel ball. The softening point was identified as the temperature at which the steel ball made contact with the surface of the bottom plate. To ensure accuracy, the softening point was meticulously measured and recorded with an accuracy of 0.5 °C.

2.3.3. Temperature Sweep Test

To assess the high-temperature rheological properties of recycled asphalt, a temperature sweep analysis was conducted using a dynamic shear rheometer (Anton Paar, Shanghai, China). This analysis aimed to acquire essential parameters such as the shear modulus G^* and the phase angle δ of the regenerated asphalt composite, as well as the rutting factor $G^*/\sin \delta$. The rutting factor can be defined as the ratio of the complex shear modulus to the sine of the phase angle. In this study, the strain control mode was selected for loading, employing a loading speed of 10 rad/s $\pm$ 0.1 rad/s. The temperature range for the test was set from 46 °C to 82 °C, with the temperature increasing at a controlled rate of 2 °C/min.

2.3.4. Multiple Stress Creep Recovery Test

The determination of the non-recoverable creep compliance (J_{nr}) of asphalt involved conducting a multi-stress creep recovery test in the stress control mode. This test was performed in two stress stages. The first stage employed a controlled stress level of 0.1 kPa, while the second stage utilized a controlled stress level of 3.2 kPa. During the test, repetitive loading was applied over a total of 30 cycles. Among these cycles, the initial 10 cycles served as a conditioning phase for the specimen under a stress level of 0.1 kPa; however, no data recording took place during this phase. The subsequent 10 cycles were conducted to record data while maintaining a stress level of 3.2 kPa. Finally, the remaining 10 cycles were loaded, and data recording was carried out. The recorded data were then utilized to calculate the irrecoverable creep compliance and creep recovery rate. Both the data recording and calculation processes were facilitated by the data acquisition system. With regard to the loading process for each cycle, the asphalt was subjected to constant stress loading for a duration of 1 s, followed by a recovery phase at zero stress for nine seconds. During the constant stress loading phase, stress and strain measurements were recorded at intervals of 0.1 s, while during the zero stress recovery phase, recordings were made at intervals of at least 0.45 s.

2.3.5. Bending Creep Stiffness Test of Asphalt

To assess the toughness and resistance to cracking of a recycled asphalt binder at low temperature, a flexural creep stiffness test was conducted on the asphalt. The low temperature performance of the recycled asphalt was evaluated using the stiffness modulus (S) and creep rate (m) at a temperature of 60 °C. The stiffness modulus of the asphalt was kept below 300 MPa, and the creep rate was greater than 0.3. These criteria indicated that the asphalt exhibited good performance in low temperature conditions, showcasing its ability to withstand cracking.

2.4. Experimental Design of Response Surface Methodology

The RSM technique has been utilized to optimize the process parameters, minimize the number of required experiments, and evaluate the relationships between various influencing factors [35–37]. Unlike traditional single-factor and orthogonal experiments,

RSM establishes a multivariate quadratic regression equation that relates the influencing factors to the response values [35,36]. RSM offers two methods in terms of experimental designs, including the Box–Behnken and the central composite design. In this study, the Box–Behnken method was employed with the SBS-modified asphalt content, the SMC regenerant content, and the shear test temperature designated as independent variables, which were denoted as X_1, X_2, and X_3, respectively. The range of SMC regenerant content was set at 4% to 12%. The range of SBS-modified asphalt content was 66% to 150%, while the range of shear temperature was 150 °C to 170 °C. The coded levels of the independent variables are presented in Table 3.

Table 3. Factor and level coding table for Box–Behnken design.

Factor	Symbol	Unit	Code Level		
			−1	0	1
SMC regenerant content	X_1	%	4	8	12
SBS-modified asphalt content	X_2	%	66	108	150
Shearing temperature	X_3	°C	150	160	170

Based on the Box–Behnken design, 17 groups of preparation parameters of recycled asphalt were determined. The SMC regenerant content and SBS-modified asphalt content were determined as a percentage of the long-term aged asphalt by mass. The preparing scheme for recycled asphalt samples is outlined in Table 4. Following modeling with RSM, an analysis of variance (ANOVA) was conducted to quantitatively examine the influence of the preparation process parameters on the properties of recycled asphalt. The suitability of the model was assessed through residuals, F-values, and p-values obtained from the ANOVA.

Table 4. Preparation scheme of recycled asphalt samples for each group.

Serial Number	Recycled Content (%)	Ratio of SBS-Modified Asphalt (%)	The Temperature at Shear Test (°C)
1	8.00	66.00	150.00
2	12.00	150.00	160.00
3	8.00	150.00	170.00
4	8.00	108.00	160.00
5	12.00	108.00	170.00
6	8.00	108.00	160.00
7	8.00	108.00	160.00
8	4.00	66.00	160.00
9	8.00	108.00	160.00
10	12.00	108.00	150.00
11	8.00	66.00	170.00
12	8.00	108.00	160.00
13	8.00	150.00	150.00
14	4.00	108.00	170.00
15	4.00	108.00	150.00
16	4.00	150.00	160.00
17	12.00	66.00	160.00

3. Results

The evaluation of asphalt recycled technology mainly depends on the physical and mechanical properties of recycled asphalt. The performance test results of recycled asphalt with different material design parameters are presented in Table 5.

Table 5. Performance test results of recycled asphalt.

Serial Number	Penetration (0.1 mm)	Softening Point (°C)	Rutting Factor ($G^*/\sin\delta$)	Non-Recoverable Creep Compliance (kPa^{-1})	Creep Rate (%)	Stiffness Modulus (Pa)
1	97.63	69.1	2.03	0.85	0.3	87.65
2	63.41	77.1	2.46	0.79	0.43	34.46
3	67.67	72.3	2.7	0.87	0.34	65.75
4	61.4	69.1	2.47	0.73	0.32	53.4
5	79.58	70.4	1.47	0.78	0.39	49.8
6	61.4	70.3	2.39	0.73	0.32	53.4
7	61.4	69.1	2.35	0.73	0.32	53.4
8	68.53	68.5	2.54	0.58	0.2	108
9	61.4	69.1	2.26	0.73	0.32	53.4
10	89.2	69.3	1.79	1	0.42	54.9
11	79.35	58.5	1.72	0.66	0.2	70.8
12	61.4	69.1	2.37	0.73	0.32	53.4
13	56.36	69.6	2.7	0.82	0.34	67.9
14	64.36	61.3	2.19	0.81	0.26	83.5
15	61.38	69	2.49	0.75	0.31	97.8
16	47.5	74.8	3.24	0.92	0.34	44.9
17	95.33	72.7	1.65	0.97	0.4	48.4

3.1. Characterization of Recycled Asphalt Performance

To investigate the differences between the regression equations using different methods, an ANOVA test was conducted on the same dataset. The null and alternative hypotheses for each characteristic are presented in Table 6, while the results of the ANOVA for each characteristic are shown in Table 7. The null hypotheses can be rejected since the computed p-value is less than 0.05, which suggests that the quadratic model adequately represents the penetration at a significance level of less than 0.05.

Table 6. Summary of the null and alternative hypotheses.

Asphalt Performance Index	Test Hypotheses	
	Null	Alternative
Penetration Softening point Rutting factor Non-recoverable creep compliance Creep rate Stiffness modulus	Quadratic model has no influence on asphalt performance index.	Quadratic model has influence on asphalt performance index.

Table 7. Summary of the one-way ANOVA results.

Asphalt Performance Index	Null Hypothesis	
	F	p-Value
Penetration	45,444.21	<0.0001
Softening point	50.95	<0.0001
Rutting factor	27.65	0.0003
Non-recoverable creep compliance	225.37	<0.0001
Creep rate	40.75	<0.0001
Stiffness modulus	8.16	0.0110

3.1.1. Penetration Test Results and Analysis

Using the penetration test data, the least square method was utilized to establish a multiple quadratic regression equation, as shown in Equation (1). To assess the significance of each factor within the fitted equation, a variance test was executed to eliminate elements that are not significant. The null hypothesis assumes that the parameters exhibit significance, while the alternative hypothesis suggests the parameters are not significant.

If the *p*-value is less than 0.05, the null hypothesis is embraced; otherwise, the alternative hypothesis is favored. As shown in Table 8, the ANOVA results reveal a high significance level for the model ($p < 0.0001$). The predicted values derived from the regression equation exhibited a strong correlation with the experimental values ($R^2 = 1.0000$), thereby confirming the excellence of the model in accurately reflecting the actual outcomes. The lack of fit attributed to pure error is not significant.

$$Pen = 61.40 + 10.72X_1 - 13.24X_2 - 1.70X_3 - 2.72X_1X_2 \\ - 3.15X_1X_3 + 7.40X_2X_3 + 2.84X_1^2 + 4.46X_2^2 + 9.39X_3^2 \tag{1}$$

where, *Pen* is 25 °C penetration (0.1 mm); X_1 is the content of the SMC regenerant; X_2 is the content of the SBS-modified asphalt; and X_3 is the shear temperature.

Table 8. ANOVA of regression model of penetration.

Variance Source	Sum of Squares	Degree of Freedom	Mean Square	F Value	*p*-Value	Significance
Model	3162.58	9	351.40	90,350.25	<0.0001	significant
Residual error	0.0272	7	0.0039	–	–	–
Total	3162.61	16	–	–	–	–
X_1	919.13	1	919.13	2.363×10^5	<0.0001	significant
X_2	1401.85	1	1401.85	3.604×10^5	<0.0001	significant
X_3	23.15	1	23.15	5953.28	<0.0001	significant
X_1X_2	29.65	1	29.65	7623.00	<0.0001	significant
X_1X_3	39.69	1	39.69	10,204.96	<0.0001	significant
X_2X_3	218.89	1	218.89	56,280.78	<0.0001	significant
X_1^2	33.84	1	33.84	8701.07	<0.0001	significant
X_2^2	83.66	1	83.66	21,510.44	<0.0001	significant
X_3^2	371.65	1	371.65	95,556.47	<0.0001	significant
			Adjust $R^2 = 1.0000$			

The *p*-value represents the likelihood of an event occurring and is important in determining the significance of a sample. If the *p*-value exceeds 0.05, it implies that the observed difference may be attributed to factors beyond mere sampling errors, rendering the item not significant. Conversely, a *p*-value below 0.05 indicates statistical significance. In this specific scenario, the content of SMC regenerant, SBS-modified asphalt, and shear temperature significantly influence the penetration. Notably, the regression equation for penetration remains unaltered, as all the *p*-values hold a value of 0.0001.

3.1.2. Softening Point Test Results and Analysis

The softening point test results were analyzed using the least square method to fit the multiple quadratic regression equation (as shown in Equation (2)), with *Spt* representing the softening point. An ANOVA was performed on the model, and the results are shown in Table 9. The null hypothesis states that the parameters are significant, while the alternative hypothesis suggests that the parameters are not significant. The null hypothesis is accepted when the *p*-value is less than 0.05. The statistical analysis demonstrated that the model holds immense significance ($p < 0.0001$). Specifically, the content of SMC regenerant, SBS-modified asphalt, and the shear temperature are identified as significantly influencing the softening point. The strong correlation between predicted and experimental values ($R^2 = 0.9860$) confirm the accuracy of the model in reflecting the actual outcomes.

$$Spt = 69.29 + 1.99X_1 + 3.14X_2 - 1.82X_3 - 0.4875X_1X_2 \\ + 2.21X_1X_3 + 3.33X_2X_3 + 2.04X_1^2 + 1.91X_2^2 - 3.84X_3^2 \tag{2}$$

Table 9. Variance analysis of softening point regression model.

Variance Source	Sum of Squares	Degree of Freedom	Mean Square	F Value	p-Value	Significance
Model	292.23	9	32.47	54.93	<0.0001	significant
Residual error	4.14	7	0.5911	–	–	–
Total	296.36	16	–	–	–	–
X_1	31.60	1	31.60	53.46	0.0002	significant
X_2	79.07	1	79.07	133.76	<0.0001	significant
X_3	26.46	1	26.46	44.77	0.0003	significant
X_1X_2	0.9506	1	0.9506	1.61	0.2453	not significant
X_1X_3	19.58	1	19.58	33.13	0.0007	significant
X_2X_3	44.22	1	44.22	74.82	<0.0001	significant
X^2_1	17.57	1	17.57	29.72	0.0010	significant
X^2_2	15.28	1	15.28	25.85	0.0014	significant
X^2_3	62.25	1	62.25	105.31	<0.0001	significant
			Adjust R^2 = 0.9860			

Based on the results in Table 9, the fitting function of the softening point after eliminating the terms that are not significant can be reconstructed in Equation (3).

$$Spt = 69.29 + 1.99X_1 + 3.14X_2 - 1.82X_3 + 2.21X_1X_3 \\ + 3.33X_2X_3 + 2.04X_1^2 + 1.91X_2^2 - 3.84X_3^2 \tag{3}$$

3.1.3. Temperature Scanning Test Results and Analysis

Utilizing the test results in Table 5, the multiple quadratic regression equation was fitted using the least square method. This equation is denoted as Equation (4).

$$G^*/\sin\delta = 2.37 - 0.3863X_1 + 0.3950X_2 - 0.1162X_3 + 0.0275X_1X_2 \\ - 0.0050X_1X_3 + 0.0775X_2X_3 - 0.099X_1^2 + 0.2035X_2^2 - 0.284X_3^2 \tag{4}$$

An ANOVA was performed on the model, and the results are shown in Table 10. The null hypothesis suggests that the parameters hold significance, while the alternative hypothesis posits their insignificance. If the p-value is less than 0.05, the null hypothesis is accepted; otherwise, the alternative hypothesis is accepted. Notably, as observed from Table 10, a p-value of less than 0.0001 for the model signifies its utmost significance. The constituents of SMC regenerant, SBS-modified asphalt, and the shear temperature notably influence the rutting factor. Moreover, the predicted value strongly correlated with the experimental value (R^2 = 0.9857), effectively reflecting the actual outcomes.

Table 10. Variance analysis of rutting factor regression model.

Variance Source	Sum of Squares	Degree of Freedom	Mean Square	F Value	p-Value	Significance
Model	3.11	9	0.3458	53.60	<0.0001	significant
Residual error	0.0452	7	0.0065	–	–	–
Total	3.16	16	–	–	–	–
X_1	1.19	1	1.19	185.02	<0.0001	significant
X_2	1.25	1	1.25	193.50	<0.0001	significant
X_3	0.1081	1	0.1081	16.76	0.0046	significant
X_1X_2	0.0030	1	0.0030	0.4689	0.5155	not significant
X_1X_3	0.0001	1	0.0001	0.0155	0.9044	not significant
X_2X_3	0.0240	1	0.0240	3.72	0.0949	not significant
X^2_1	0.0413	1	0.0413	6.40	0.0393	significant
X^2_2	0.1744	1	0.1744	27.03	0.0013	significant
X^2_3	0.3396	1	0.3396	52.65	0.0002	significant
			Adjust R^2 = 0.9857			

The term, which is not significant, is removed, and the function of the rutting factor obtained anew is shown in Equation (5).

$$G^* / \sin \delta = 2.37 - 0.3863X_1 + 0.395X_2 - 0.1162X_3 - 0.099X_1^2 + 0.2035X_2^2 - 0.284X_3^2 \quad (5)$$

3.1.4. Multiple Stress Creep Recovery Test Results and Analysis

Using the least squares method, a multiple quadratic regression equation, depicted as Equation (6), was fitted for the multiple stress creep recovery tests.

$$\begin{aligned} J_{nr} = {}& 0.7300 + 0.0600X_1 + 0.0425X_2 - 0.0375X_3 - 0.1300X_1X_2 \\ & - 0.0700X_1X_3 + 0.0600X_2X_3 + 0.0600X_1^2 + 0.0250X_2^2 + 0.0450X_3^2 \end{aligned} \quad (6)$$

An ANOVA analysis was conducted on the model, and the results were detailed in Table 11. The null hypothesis posits the significance of the parameters, while the alternative hypothesis assumes their insignificance. Should the p-value fall below 0.05, the null hypothesis is accepted; otherwise, the alternative hypothesis is favored. Notably, the model yielded a p-value of less than 0.0001, indicating its significance. The content of SMC regenerant, SBS-modified asphalt, and the shear temperature were observed to wield significant influence over non-recoverable creep compliance. The predicted values strongly correlated with the experimental values ($R^2 = 0.9984$), and the lack of fit, attributed to pure error, was insignificant. Hence, the model fit is deemed satisfactory to accurately reflect the real-world outcomes. Given that all the p-values are below 0.05, the regression equation for non-recoverable creep compliance remains unchanged.

Table 11. Variance analysis of non-recoverable creep compliance regression model.

Variance Source	Sum of Squares	Degree of Freedom	Mean Square	F Value	p-Value	Significance
Model	0.1851	9	0.0206	479.83	<0.0001	significant
Residual error	0.0003	7	0.0000	–	–	–
Total	0.1854	16	–	–	–	–
X_1	0.0288	1	0.0288	672.00	<0.0001	significant
X_2	0.0144	1	0.0144	337.17	<0.0001	significant
X_3	0.0112	1	0.0112	262.50	<0.0001	significant
X_1X_2	0.0676	1	0.0676	1577.33	<0.0001	significant
X_1X_3	0.0196	1	0.0196	457.33	<0.0001	significant
X_2X_3	0.0144	1	0.0144	336.00	<0.0001	significant
X^2_1	0.0152	1	0.0152	353.68	<0.0001	significant
X^2_2	0.0026	1	0.0026	61.40	0.0001	significant
X^2_3	0.0085	1	0.0085	198.95	<0.0001	significant
			Adjust R^2 = 0.9984			

3.1.5. Test Results and Analysis of Flexural Creep Stiffness of Asphalt

Based on the flexural creep stiffness test results, a multiple quadratic regression equation was fitted using the least squares method, as presented in Equations (7) and (8).

$$\begin{aligned} m = {}& 0.3200 + 0.0662X_1 + 0.0438X_2 - 0.0225X_3 - 0.0275X_1X_2 \\ & + 0.0050X_1X_3 + 0.0250X_2X_3 + 0.0363X_1^2 - 0.0137X_2^2 - 0.0113X_3^2 \end{aligned} \quad (7)$$

$$\begin{aligned} S = {}& 53.40 - 18.33X_1 - 12.73X_2 - 4.80X_3 + 12.29X_1X_2 \\ & + 2.330X_1X_3 + 3.68X_2X_3 + 2.01X_1^2 + 3.53X_2^2 + 16.09X_3^2 \end{aligned} \quad (8)$$

where m is the creep rate and S is the stiffness modulus.

ANOVA analyses were performed on both the creep rate and stiffness modulus models, and the corresponding results are displayed in Tables 12 and 13, respectively. The null hypothesis asserts the significance of the parameters, while the alternative hypothesis suggests their insignificance. Acceptance of the null hypothesis occurs when the p-value is below 0.05; otherwise, the alternative hypothesis is favored. The statistical analysis unveiled that the model exhibits remarkable significance, as evidenced by the p-value of less than 0.0001. Notably, the content of SMC regenerant, SBS-modified asphalt, and the shear temperature significantly influence the creep rate, while the content of SMC regenerant and SBS-modified asphalt exert a noteworthy influence on the stiffness modulus. Furthermore, a strong correlation between the predicted and experimental values is observed, with R^2 values of 0.9944 and 0.9459 for the creep rate and stiffness modulus, respectively. The result signifies that the model accurately reflects the actual results with high precision.

Table 12. Variance analysis of creep rate regression model.

Variance Source	Sum of Squares	Degree of Freedom	Mean Square	F Value	p Value	Significance
Model	0.0666	9	0.0074	138.23	<0.0001	significant
Residual error	0.0004	7	0.0001	–	–	–
Total	0.0670	16	–	–	–	–
X_1	0.0351	1	0.0351	655.43	<0.0001	significant
X_2	0.0153	1	0.0153	285.83	<0.0001	significant
X_3	0.0040	1	0.0040	75.60	<0.0001	significant
X_1X_2	0.0030	1	0.0030	56.47	0.0001	significant
X_1X_3	0.0001	1	0.0001	1.87	0.2141	not significant
X_2X_3	0.0025	1	0.0025	46.67	0.0002	significant
X^2_1	0.0055	1	0.0055	103.28	<0.0001	significant
X^2_2	0.0008	1	0.0008	14.86	0.0063	significant
X^2_3	0.0005	1	0.0005	9.95	0.0161	significant
			Adjust R^2 = 0.9944			

Table 13. Variance analysis of stiffness modulus regression model.

Variance Source	Sum of Squares	Degree of Freedom	Mean Square	F Value	p Value	Significance
Model	6060.03	9	673.34	13.60	<0.0001	significant
Residual error	346.53	7	49.50	–	–	–
Total	6406.56	16	–	–	–	–
X_1	2687.91	1	2687.91	54.30	0.0002	significant
X_2	1296.42	1	1296.42	26.19	0.0014	significant
X_3	184.32	1	184.32	3.72	0.0950	not significant
X_1X_2	604.18	1	604.18	12.20	0.0101	significant
X_1X_3	21.16	1	21.16	0.4274	0.5341	not significant
X_2X_3	54.02	1	54.02	1.09	0.3309	not significant
X^2_1	16.97	1	16.97	0.3428	0.5766	not significant
X^2_2	52.54	1	52.54	1.06	0.3372	not significant
X^2_3	1090.39	1	1090.39	22.03	0.0022	significant
			Adjust R^2 = 0.9459			

Removing the non-significant terms, the functions of creep rate and stiffness modulus are presented in Equations (9) and (10), respectively.

$$m = 0.3200 + 0.0662X_1 + 0.0438X_2 - 0.0225X_3 - 0.0275X_1X_2 \\ +0.0250X_2X_3 + 0.0363X_1^2 - 0.0137X_2^2 - 0.0113X_3^2 \tag{9}$$

$$S = 53.4 - 18.33X_1 - 12.73X_2 + 12.29X_1X_2 + 16.09X_3^2 \tag{10}$$

3.2. Response Surface Interaction Analysis

3.2.1. Penetration

Figure 1 presents a response surface plot showcasing the outcomes of the penetration test. The results illustrate that with an increase in the temperature during the shear test, the permeability shows an initial rise followed by a subsequent decline. In contrast, the levels of SMC regenerant and SBS-modified asphalt remain constant throughout. Moreover, elevating the quantity of SBS-modified asphalt decreases penetration, while the addition of SMC regenerant produces the opposite effect.

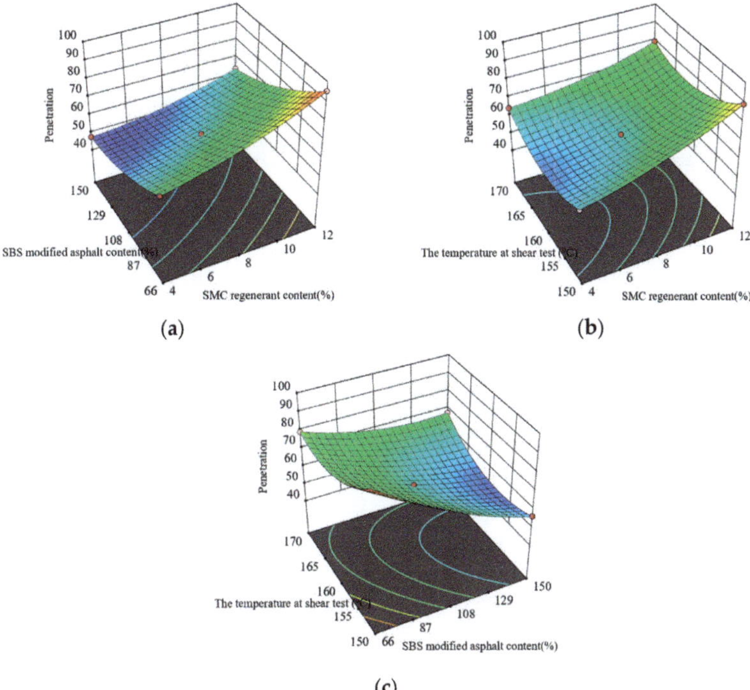

Figure 1. Interaction between test factors and penetration: (**a**) regenerate agent and original SBS-modified asphalt content; (**b**) regenerant content and the temperature at shear test; and (**c**) the temperature at shear test and SBS-modified asphalt content.

3.2.2. Softening Point

Figure 2 shows a response surface plot representing the outcomes of the softening point analysis, indicating that while SBS-modified asphalt is kept at a constant level, the softening point increases as the SMC regenerant content rises. Likewise, when the SMC regenerant content remains consistent, the softening point experiences an increase with higher quantities of SBS-modified asphalt. Specifically, when the content of SMC regenerant is fixed at a specific value and the content of SBS-modified asphalt is set at 108%, a decrease in the softening point is observed as the shearing temperature increases for SBS-modified asphalt contents below 108%. However, when the content of SBS-modified asphalt surpasses 108%, the softening point initially rises and then declines with an increase in shearing temperature.

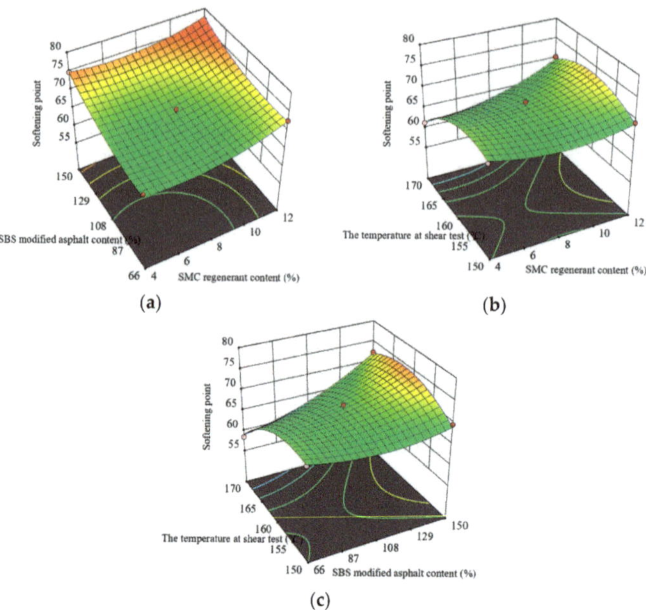

Figure 2. Interaction between experimental factors and softening point: (**a**) regenerate agent and original SBS-modified asphalt content; (**b**) regenerant content and the temperature at shear test; and (**c**) the temperature at shear test and SBS-modified asphalt content.

3.2.3. Temperature Sweep Test

Figure 3 indicates a decrease in the rutting factor as the content of regenerant and SBS-modified asphalt increases. Additionally, the rutting factor shows a decrease with higher shear temperatures.

3.2.4. Multiple Stress Creep Recovery Test

Based on the results of the multiple stress creep recovery test, Figure 4 illustrates a response surface plot. It is apparent that when the content of SBS-modified asphalt is lower than 111.41%, and the shear temperature remains constant, the non-recoverable creep compliance demonstrates an increase as the SMC regenerator content increases. However, when the content of SBS-modified asphalt exceeds 111.41%, the non-recoverable creep compliance shows decreases with an increase in the content of SMC regenerant. Additionally, as the shear temperature increases, and the content of SMC regenerant is high while the SBS-modified asphalt content is low, the irrecoverable creep compliance experiences a decrease.

3.2.5. Multiple Stress Creep Recovery Test

Figures 5 and 6 illustrate the response surface plots derived from the creep test results. The results reveal a gradual elevation in the creep rate of asphalt as the content of SMC regenerant increases, while maintaining a constant content of SBS-modified asphalt. Conversely, when the SMC regenerant content remains constant, the creep velocity initially decreases and then increases as the shear temperature rises. At a specific shear temperature and SMC regenerant content, the stiffness modulus of SBS-modified asphalt diminishes with an increase in the content of SBS-modified asphalt. Furthermore, when the shear temperature and SBS-modified asphalt content remain constant, the stiffness modulus decreases as the content of SMC regenerant increases. When the content of SMC regenerant and SBS-modified asphalt remains constant, the stiffness modulus encounters an initial decline followed by an increase with an elevated shear temperature.

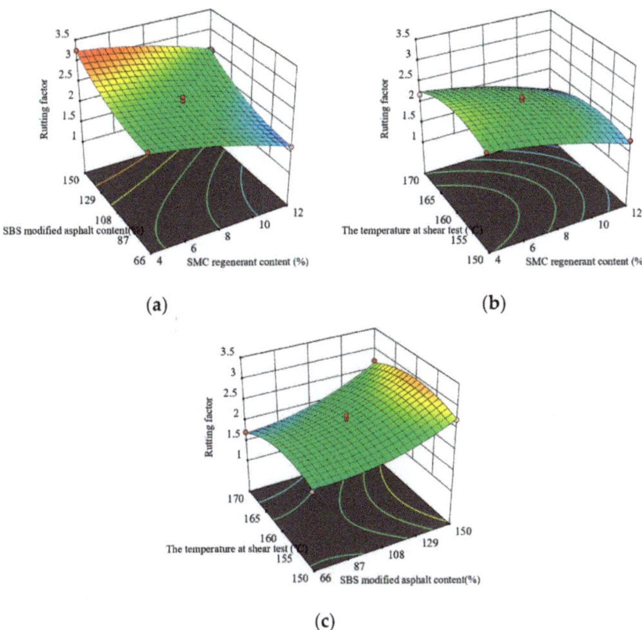

Figure 3. Interaction between test factors and rutting factor: (**a**) regenerate agent and original SBS-modified asphalt content; (**b**) regenerant content and the temperature at shear test; and (**c**) the temperature at shear test and SBS-modified asphalt content.

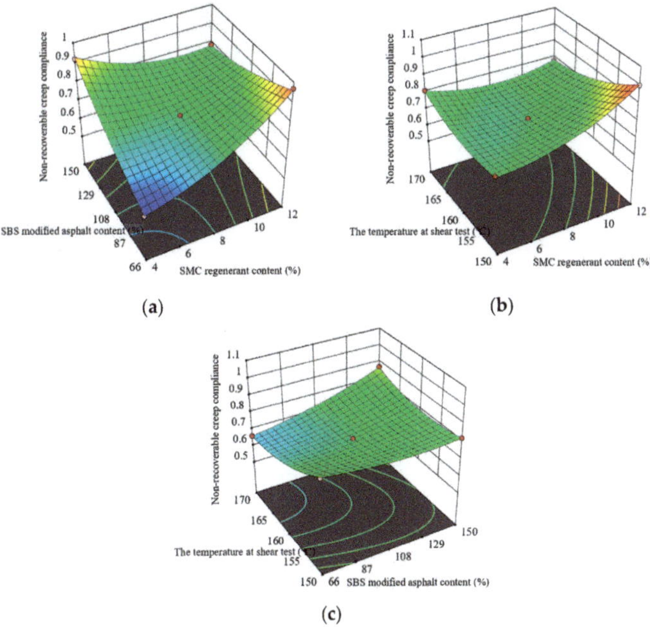

Figure 4. Interaction between test factors and non-recoverable creep compliance: (**a**) regenerate agent and original SBS-modified asphalt content; (**b**) regenerant content and the temperature at shear test; and (**c**) the temperature at shear test and SBS-modified asphalt content.

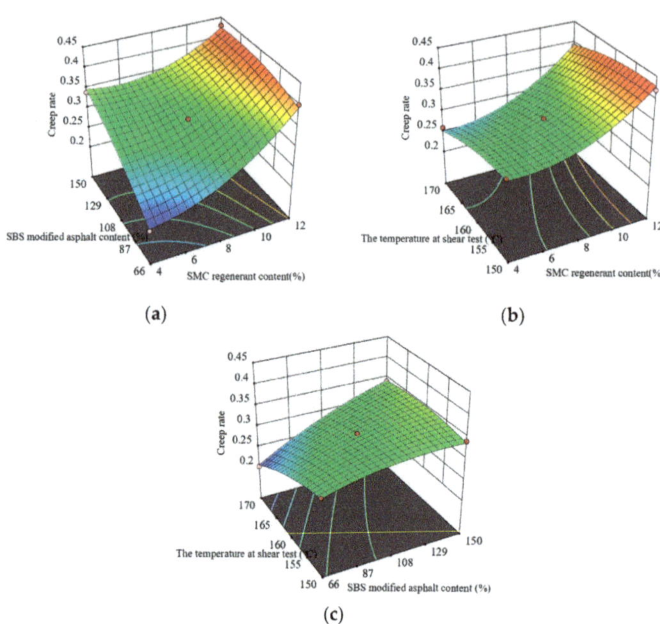

Figure 5. Interaction between test factors and creep rate of recycled asphalt: (**a**) regenerate agent and original SBS-modified asphalt content; (**b**) regenerant content and the temperature at shear test; and (**c**) the temperature at shear test and original SBS-modified asphalt content.

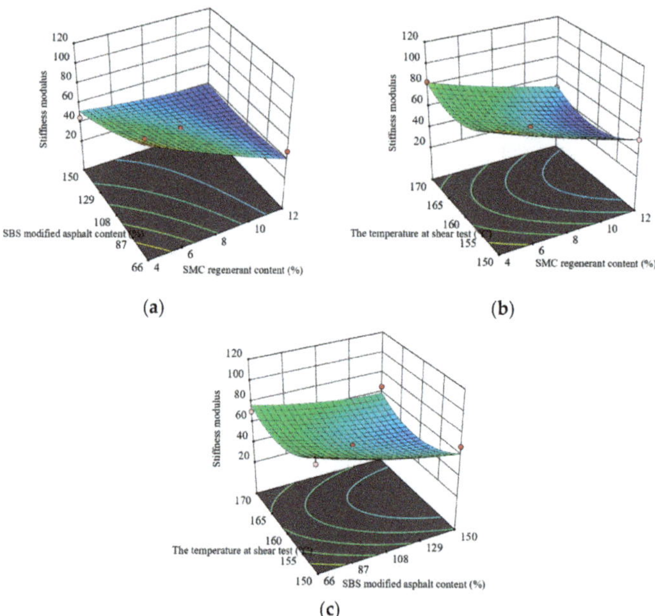

Figure 6. Interaction between test factors and stiffness modulus of recycled asphalt: (**a**) regenerate agent and original SBS-modified asphalt content; (**b**) regenerant content and the temperature at shear test; and (**c**) the temperature at shear test and original SBS-modified asphalt content.

3.3. Performance Optimization and Model Verification of Recycled Asphalt

Based on the above test analysis, this study has successfully determined the optimal design parameters of recycled asphalt. Specifically, the SMC regenerant content is 7.88% (8% in subsequent investigations), the content of SBS-modified asphalt has been increased to 150%, and the shear temperature has been set at 157.68 °C (160 °C in subsequent analysis). By employing these carefully chosen parameters, the projected values for key performance indicators of recycled asphalt under stress levels of 0.1 kPa and 3.2 kPa are as follows: 52.1 (0.1 mm) for penetration, 73.88 °C for softening point, 2.94 kPa for rutting factor, 0.347 for creep rate, 44.95 MPa for stiffness modulus, 0.796 kPa^{-1} for average non-recoverable creep compliance at 0.1 kPa, and 1.185 kPa^{-1} for average non-recoverable creep compliance at 3.2 kPa.

To validate the optimal preparation process parameters obtained through response surface methodology (RSM), recycled asphalt was prepared using the specified conditions: 8% SMC regenerant content, 150% SBS-modified asphalt, and a shear test temperature of 160 °C. Subsequently, several performance tests were conducted, including conventional performance evaluation and high and low-temperature rheological performance assessment. The outcomes of these tests are presented in Table 14. In the table, m denotes the creep rate and S denotes the stiffness modulus. The comparison between the predicted and actual values of the key recycled asphalt indexes, along with the maximum error of 5.04%, falls within an acceptable range.

Table 14. Validation test results of RSM.

Performance	Regenerant Dosage	New Asphalt Content	Shearing Temperature	Penetration	Softening Point	$G^*/\sin\delta$	m	S	J_{nr}
Unit	%	%	°C	0.1 mm	°C	kPa	—	MPa	kPa^{-1}
test value	8	150	160	49.6	72.5	2.87	0.364	43.17	0.773
predicted value				52.1	73.9	2.94	0.347	44.95	0.796
Relative error (%)	—	—	—	−5.04	—	2.44	+4.67	−4.12	−2.97

4. Conclusions

This study employed the response surface method to investigate the effects of SMC regenerant content, SBS-modified asphalt content, and shear temperature on the mechanical properties of recycled asphalt. The application of RSM in optimizing recycled asphalt design was demonstrated through constructing regression equations, significance testing, prediction of optimal mix proportions, and experimental verification. The main findings are as follows.

(1) When the content of SMC regenerant or SBS-modified asphalt is fixed, the penetration exhibits a trend of initially increasing and then decreasing with an increase in shear temperature. With the increase in SBS-modified asphalt content, the penetration decreases, and with the increase in shear temperature, the penetration first increases and then decreases. Among the three influencing factors, SBS-modified asphalt content has the greatest impact on penetration. The increasing content of SBS-modified asphalt and SMC regenerant decreased the softening point of recycled asphalt. The softening point tends to decrease with an increase in shear temperature.

(2) When the content of SMC regenerant and SBS-modified asphalt is fixed, an increase in shear temperature initially increases, and then the rutting factor decreases. The increasing content of SMC regenerant leads to a gradual decrease in the rutting factor, while SBS-modified asphalt content has the opposite effect. In a comprehensive comparison, SBS-modified asphalt content has the most significant effect on the rutting factor. With a constant content of SMC regenerant and SBS-modified asphalt,

an increase in shear temperature results in a first decrease and then an increase in non-recoverable creep compliance. The addition of SMC regenerant reduces the non-recoverable creep compliance by regulating the proportion of viscoelastic and plastic components in recycled asphalt.

(3) The stiffness modulus of recycled asphalt gradually decreases with an increase in the content of SBS-modified asphalt, while keeping the shear temperature and SMC regenerant content constant. Similarly, with constant shear temperature and SMC regenerant content, the stiffness modulus gradually decreases with an increase in SMC regenerant content. Additionally, the creep rate increases gradually with a constant content of SMC regenerant. Notably, when the content of SBS-modified asphalt is 66%, the creep rate increases by 50% as the SMC regenerant content increases from 4% to 12%. These findings highlight the significant enhancement of low-temperature properties through the beneficial use of SMC regenerant.

In summary, this study contributes to the field of road engineering by demonstrating the feasibility of RSM in optimizing recycled asphalt design. The study findings emphasize the effects of the shear temperature, the SBS-modified asphalt content, and the SMC regenerant content on the performance of recycled asphalt, including penetration, softening point, rut factor, and stiffness modulus. These insights provide valuable guidance for the development of sustainable and high-performance asphalt pavements.

Author Contributions: Conceptualization, H.L. and W.L.; methodology, H.L.; software, J.W.; validation, J.W., W.L. and N.Z.; formal analysis, N.Z.; investigation, H.L. and J.W.; resources, W.L.; data curation, J.W.; writing—original draft preparation, H.L.; writing—review and editing, W.L. and N.Z.; visualization, H.L.; supervision, W.L.; project administration, N.Z.; funding acquisition, W.L. All authors have read and agreed to the published version of the manuscript.

Funding: This research was funded by Open Fund of the Key Laboratory of Highway Engineering of Ministry of Education (Changsha University of Science & Technology), grant number kfj150204; Postgraduate Scientific Research Innovation Project of Hunan Province, grant number CX2015B340; Key R&D and Promotion Special Project of Henan Province, grant number 232102211012 and 232102241009.

Institutional Review Board Statement: Not applicable.

Informed Consent Statement: Not applicable.

Data Availability Statement: Not applicable.

Conflicts of Interest: The authors declare no conflict of interest.

References

1. Guo, M.; Ren, X.; Jiao, Y.-B.; Liang, M.-C. Review of aging and antiaging of asphalt and asphalt mixtures. *China J. Highw. Transp.* **2022**, *35*, 41.
2. Surehali, S.; Singh, A.; Biligiri, K.P. A state-of-the-art review on recycling rubber in concrete: Sustainability aspects, specialty mixtures, and treatment methods. *Dev. Built Environ.* **2023**, *14*, 100171. [CrossRef]
3. Hong, F.; Prozzi, J.A. Evaluation of recycled asphalt pavement using economic, environmental, and energy metrics based on long-term pavement performance sections. *Road Mater. Pavement Des.* **2018**, *19*, 1816–1831. [CrossRef]
4. Ali, H.; Sobhan, K. On the road to sustainability: Properties of hot in-place recycled Superpave mix. *Transp. Res. Rec.* **2012**, *2292*, 88–93. [CrossRef]
5. Anthonissen, J.; Braet, J. Review and environmental impact assessment of green technologies for base courses in bituminous pavements. *Environ. Impact Assess. Rev.* **2016**, *60*, 139–147. [CrossRef]
6. Behnood, A. Application of rejuvenators to improve the rheological and mechanical properties of asphalt binders and mixtures: A review. *J. Clean. Prod.* **2019**, *231*, 171–182. [CrossRef]
7. Sojobi, A.O.; Xuan, D.; Li, L.; Liu, S.; Poon, C.S. Optimization of gas-solid carbonation conditions of recycled aggregates using a linear weighted sum method. *Dev. Built Environ.* **2021**, *7*, 100053. [CrossRef]
8. Yu, B.; Gu, X.; Zhang, L.; Ni, F. Performance evaluation of aged asphalt mix for hot in-place recycling. *J. Test. Eval.* **2016**, *44*, 770–780. [CrossRef]
9. Xu, T.; Huang, X. Investigation into causes of in-place rutting in asphalt pavement. *Constr. Build. Mater.* **2012**, *28*, 525–530. [CrossRef]

10. Li, N.; Tang, W.; Yu, X.; Zhan, H.; Wang, X.; Wang, Z. Laboratory investigation on blending process of reclaimed asphalt mixture. *Constr. Build. Mater.* **2022**, *325*, 126793. [CrossRef]

11. Ferreira, W.L.; Branco, V.T.C.; Vasconcelos, K.; Bhasin, A.; Sreeram, A. The impact of aging heterogeneities within RAP binder on recycled asphalt mixture design. *Constr. Build. Mater.* **2021**, *300*, 124260. [CrossRef]

12. Chen, Y.; Chen, Z.; Xiang, Q.; Qin, W.; Yi, J. Research on the influence of RAP and aged asphalt on the performance of plant-mixed hot recycled asphalt mixture and blended asphalt. *Case Stud. Constr. Mater.* **2021**, *15*, e00722. [CrossRef]

13. Lee, S.-H.; Tam, A.B.; Kim, J.; Park, D.-W. Evaluation of rejuvenators based on the healing and mechanistic performance of recycled asphalt mixture. *Constr. Build. Mater.* **2019**, *220*, 628–636. [CrossRef]

14. Wang, W.; Cheng, H.; Sun, L.; Sun, Y.; Liu, N. Multi-performance evaluation of recycled warm-mix asphalt mixtures with high reclaimed asphalt pavement contents. *J. Clean. Prod.* **2022**, *377*, 134209. [CrossRef]

15. Lee, S.-Y.; Le, T.H.M.; Kim, Y.-M. Full-scale and laboratory investigations on the performance of asphalt mixture containing recycled aggregate with low viscosity binder. *Constr. Build. Mater.* **2023**, *367*, 130283. [CrossRef]

16. Hettiarachchi, C.; Hou, X.; Xiang, Q.; Yong, D.; Xiao, F. A blending efficiency model for virgin and aged binders in recycled asphalt mixtures based on blending temperature and duration. *Resour. Conserv. Recycl.* **2020**, *161*, 104957. [CrossRef]

17. Blanc, J.; Hornych, P.; Sotoodeh-Nia, Z.; Williams, C.; Porot, L.; Pouget, S.; Boysen, R.; Planche, J.-P.; Presti, D.L.; Jimenez, A. Full-scale validation of bio-recycled asphalt mixtures for road pavements. *J. Clean. Prod.* **2019**, *227*, 1068–1078. [CrossRef]

18. Zhang, J.; Zhang, X.; Liang, M.; Jiang, H.; Wei, J.; Yao, Z. Influence of different rejuvenating agents on rheological behavior and dynamic response of recycled asphalt mixtures incorporating 60% RAP dosage. *Constr. Build. Mater.* **2020**, *238*, 117778. [CrossRef]

19. Zaumanis, M.; Arraigada, M.; Poulikakos, L. 100% recycled high-modulus asphalt concrete mixture design and validation using vehicle simulator. *Constr. Build. Mater.* **2020**, *260*, 119891. [CrossRef]

20. Meroni, F.; Flintsch, G.W.; Habbouche, J.; Diefenderfer, B.K.; Giustozzi, F. Three-level performance evaluation of high RAP asphalt surface mixes. *Constr. Build. Mater.* **2021**, *309*, 125164. [CrossRef]

21. Bowers, B.F.; Huang, B.; Shu, X.; Miller, B.C. Investigation of reclaimed asphalt pavement blending efficiency through GPC and FTIR. *Constr. Build. Mater.* **2014**, *50*, 517–523. [CrossRef]

22. Wei, H.; Zhang, H.; Li, J.; Zheng, J.; Ren, J. Effect of loading rate on failure characteristics of asphalt mixtures using acoustic emission technique. *Constr. Build. Mater.* **2023**, *364*, 129835. [CrossRef]

23. Xing, C.; Li, M.; Liu, L.; Lu, R.; Liu, N.; Wu, W.; Yuan, D. A comprehensive review on the blending condition between virgin and RAP asphalt binders in hot recycled asphalt mixtures: Mechanisms, evaluation methods, and influencing factors. *J. Clean. Prod.* **2023**, *398*, 136515. [CrossRef]

24. Chelladurai, S.J.S.; Murugan, K.; Ray, A.P.; Upadhyaya, M.; Narasimharaj, V.; Gnanasekaran, S. Optimization of process parameters using response surface methodology: A review. *Mater. Today: Proc.* **2021**, *37*, 1301–1304. [CrossRef]

25. Wei, H.; Wang, Y.; Li, J.; Zhang, Y.; Qian, G. Optimization design of cold patching asphalt liquid based on performance experiments and statistical methods. *Constr. Build. Mater.* **2023**, *391*, 131881. [CrossRef]

26. Li, J.; Zhang, J.; Yang, X.; Zhang, A.; Yu, M. Monte Carlo simulations of deformation behaviour of unbound granular materials based on a real aggregate library. *Int. J. Pavement Eng.* **2023**, *24*, 2165650. [CrossRef]

27. Isam, M.; Baloo, L.; Kutty, S.R.M.; Yavari, S. Optimisation and modelling of Pb (II) and Cu (II) biosorption onto red algae (*Gracilaria changii*) by using response surface methodology. *Water* **2019**, *11*, 2325. [CrossRef]

28. Rafiq, W.; Napiah, M.; Habib, N.Z.; Sutanto, M.H.; Alaloul, W.S.; Khan, M.I.; Musarat, M.A.; Memon, A.M. Modeling and design optimization of reclaimed asphalt pavement containing crude palm oil using response surface methodology. *Constr. Build. Mater.* **2021**, *291*, 123288. [CrossRef]

29. Bala, N.; Napiah, M.; Kamaruddin, I. Nanosilica composite asphalt mixtures performance-based design and optimisation using response surface methodology. *Int. J. Pavement Eng.* **2020**, *21*, 29–40. [CrossRef]

30. Guo, P.; Cao, Z.; Chen, S.; Chen, C.; Liu, J.; Meng, J. Application of design-expert response surface methodology for the optimization of recycled asphalt mixture with waste engine oil. *J. Mater. Civ. Eng.* **2021**, *33*, 04021075. [CrossRef]

31. Vatanparast, M.; Sarkar, A.; Sahaf, S.A. Optimization of asphalt mixture design using response surface method for stone matrix warm mix asphalt incorporating crumb rubber modified binder. *Constr. Build. Mater.* **2023**, *369*, 130401. [CrossRef]

32. Taherkhani, H.; Noorian, F. Investigating permanent deformation of recycled asphalt concrete containing waste oils as rejuvenator using response surface methodology (RSM). *Iran. J. Sci. Technol. Trans. Civ. Eng.* **2021**, *45*, 1989–2001. [CrossRef]

33. Khairuddin, F.H.; Alamawi, M.Y.; Yusoff, N.I.M.; Badri, K.H.; Ceylan, H.; Tawil, S.N.M. Physicochemical and thermal analyses of polyurethane modified bitumen incorporated with Cecabase and Rediset: Optimization using response surface methodology. *Fuel* **2019**, *254*, 115662. [CrossRef]

34. Qian, Y.; Guo, F.; Leng, Z.; Zhang, Y.; Yu, H. Simulation of the field aging of asphalt binders in different reclaimed asphalt pavement (RAP) materials in Hong Kong through laboratory tests. *Constr. Build. Mater.* **2020**, *265*, 120651. [CrossRef]

35. Lee, E.Y.; Wong, S.Y.; Phang, S.J.; Wong, V.-L.; Cheah, K.H. Additively manufactured photoreactor with immobilized thermoset acrylic-graphitic carbon nitride nanosheets for water remediation: Response surface methods and adsorption modelling studies. *Chem. Eng. J.* **2023**, *455*, 140633.

36. Toor, U.A.; Duong, T.T.; Ko, S.-Y.; Hussain, F.; Oh, S.-E. Optimization of Fenton process for removing TOC and color from swine wastewater using response surface method (RSM). *J. Environ. Manag.* **2021**, *279*, 111625. [CrossRef] [PubMed]
37. Zhu, X.; Tu, X.; Mei, D.; Zheng, C.; Zhou, J.; Gao, X.; Luo, Z.; Ni, M.; Cen, K. Investigation of hybrid plasma-catalytic removal of acetone over CuO/γ-Al2O3 catalysts using response surface method. *Chemosphere* **2016**, *155*, 9–17. [CrossRef]

 *materials*

Review

A Review on Development of Industrial Solid Waste in Tunnel Grouting Materials: Feasibility, Performance, and Prospects

Bolin Jiang [1,2,3], Mengjun Wu [2,3], Shanshan Wu [3,4], Aichen Zheng [1,3] and Shiyong He [3,*]

1   Chongqing Vocational Institute of Engineering, Chongqing 402260, China; bolinjiang@cqvie.edu.cn (B.J.); aichenzheng@cqvie.edu.cn (A.Z.)
2   China Merchants Chongqing Communications Technology Research & Design Institute Co., Ltd., Chongqing 400067, China; mengjuwu@163.com
3   School of Civil Engineering, Chongqing Jiaotong University, Chongqing 400074, China; shanshanwu84@163.com
4   Chongqing Vocational College of Public Transportation, Chongqing 402247, China
*   Correspondence: heshiyong@cqjtu.edu.cn

Abstract: With rapid infrastructure development worldwide, the generation of industrial solid waste (ISW) has substantially increased, causing resource wastage and environmental pollution. Meanwhile, tunnel engineering requires large quantities of grouting material for ground treatment and consolidation. Using ISW as a component in tunnel grouts provides a sustainable solution to both issues. This paper presented a comprehensive review of the recent advancements in tunnel grouting materials using ISW, focusing on their feasibility, mechanical characteristics, and future development directions. Initially, the concept and classification of ISW were introduced, examining its feasibility and advantages as grouting materials in tunnels. Subsequently, various performances of ISW in tunnel grouting materials were summarized to explore the factors influencing mechanical strength, fluidity, durability, and microstructure characteristics. Simultaneously, this review analyzed current research trends and outlines future development directions. Major challenges, including quality assurance, environmental risks, and lack of standardized specifications, are discussed. Future research directions, including multifunctional grouts, integrated waste utilization, and advanced characterization techniques, are suggested to further advance this field. These findings provided useful insights for the continued development of high-performance and environmentally friendly ISW-based grouting materials.

Keywords: literature review; industrial solid waste; tunnel grouting materials; performance; applications

 **check for updates**

Citation: Jiang, B.; Wu, M.; Wu, S.; Zheng, A.; He, S. A Review on Development of Industrial Solid Waste in Tunnel Grouting Materials: Feasibility, Performance, and Prospects. *Materials* **2023**, *16*, 6848. https://doi.org/10.3390/ma16216848

Academic Editor: Francisco Agrela

Received: 15 September 2023
Revised: 13 October 2023
Accepted: 18 October 2023
Published: 25 October 2023

1. Introduction

In recent years, infrastructure investment around the world has accelerated, resulting in a surge of tunnel construction projects for transportation, hydropower, and urban development [1–4]. Tunneling often takes place in soft ground or fractured rock, requiring large quantities of grouting materials for ground improvement and water inflow control [5–8]. Traditional cement and sodium silicate-based grouts used in tunnels incur high costs and substantial carbon emissions. Meanwhile, rapid industrialization and urbanization have led to a massive generation of industrial solid waste (ISW), including slag, fly ash, steel slag, red mud, etc. [9–12]. Annual global ISW production continues to increase, as shown in Figure 1. The annual production of ISW worldwide continues to increase, and China produces more than 10 billion tons of ISW annually [13–15]. Stockpiling these ISWs wastes resources and poses contamination risks to soil and groundwater. The requirement of reasonable disposal and the recycling of ISW is urgent in the world. Some studies reveal that ISW has the potential and feasibility to be used as grouting material (double-liquid grouts and single-liquid grouts) in tunnel engineering [16–21]. Therefore, increasing research efforts have focused on recycling ISW as substitute constituents in cementitious grouting materials; the details are shown in Figure 2. This emerging approach provides a promising solution to cut tunnel costs while reducing environmental pollution from ISW.

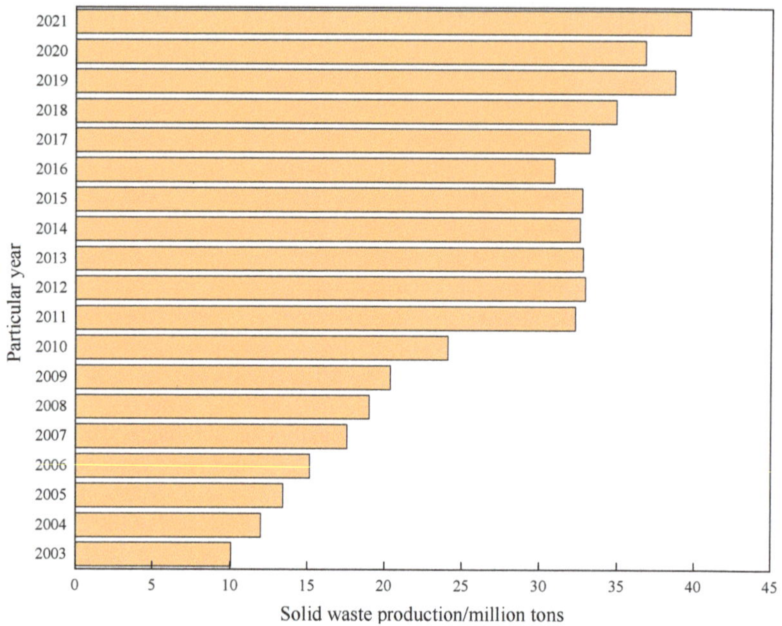

Figure 1. Annual output of solid waste in China.

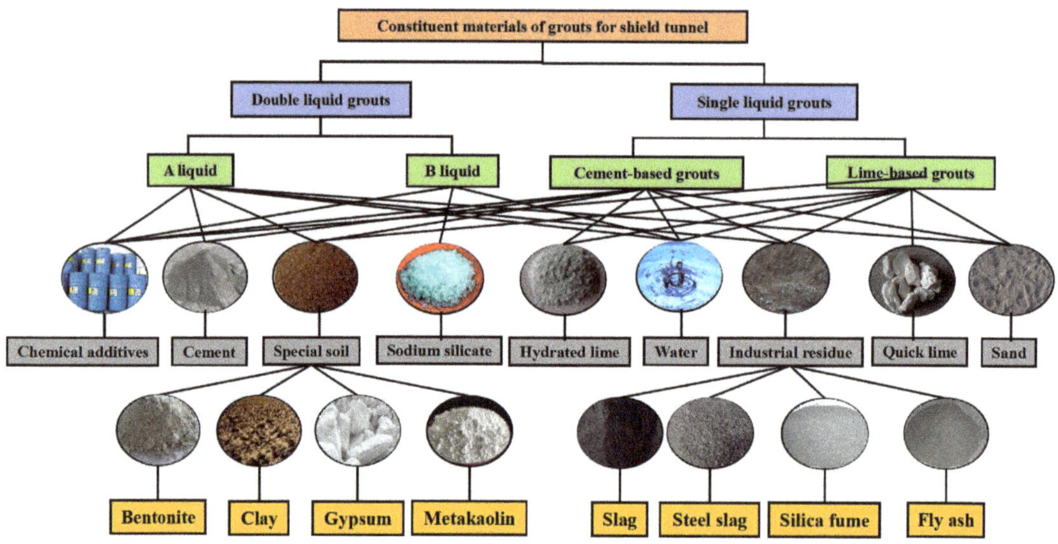

Figure 2. Composition of grouting material.

Compared to traditional grouting materials, ISW grouts exhibit certain advantages, such as high early strength, good impermeability, and improved durability [22–25]. Specific project examples are shown in Table 1. The complex chemical and mineral components in ISW can participate in pozzolanic reactions and generate cementitious gels. When rationally designed, ISW grouts can achieve equivalent or superior performance to conventional grouts [26]. However, utilizing ISW in grouting materials also faces some technical challenges. ISW comprises waste and by-products generated during industrial production processes, which contain various chemical components, resulting in different physical characteristics; the details are shown in Table 2. Due to the variability in ISW properties, obtaining a consistent grout quality that meets engineering requirements is difficult to guarantee; the details are shown in Figure 3 [27–29]. Furthermore, standardized codes and specifications for ISW grout production and application are still lacking. Zhang et al. [30] conducted a review study to examine the research progress of ISW as cementitious materials, highlighting various utilization methods and their application effects. As technology advanced and environmental concerns grew, research on ISW as grouting material gradually increased, leading to its practical application in engineering projects.

Table 1. The summary of case studies on the application of grouting materials.

Application Area	Engineering Environment	Material Composition	Reasons For Selection	Ref.
Changsha, Hunan Province, China	Karst ground	Sodium bentonite, lignin, meta-aluminate, cement	Better stability	[31]
Qingdao, Shandong Province, China	Marine environment	OPC, sodium silicate	Higher erosion resistance and better strength	[32]
Changsha, Hunan Province, China	Loose ground	Cement, clay, sodium bentonite, gypsum	Better stability and permeability	[33]
Wuhan, Hubei Province, China	High-pressure, water-rich strata	OPC, sand, fly ash, silica fume, steel slag, sodium bentonite	Better stability	[34]
Wuhan, Hubei Province, China	Under river	OPC, steel slag, fly ash, slag, metakaolin, sodium silicate,	Better corrosion resistance	[35]
Nanjing, Jiangsu Province, China	Soft soil strata	OPC, fly ash, hydrated lime, sand, clay, sodium bentonite,	Good workability and corrosion resistance	[36]
Wuhan, Hubei Province, China	Water-rich strata	OPC, sand, fly ash, silica fume, SPs, HEC	Good flow and strength properties	[37]
Xi'an, Shaanxi Province, China	Loess ground	OPC, sand, fly ash, sodium bentonite, micro-expansive agents	Higher strength and good workability	[38]
Beijing, China	Water-rich strata	OPC, fly ash, slag, sodium silicate, sodium bentonite, retarder agents	Better liquidity performance	[39]
Beijing, China	Sulfate-rich environment	Cement, fly ash, slag, sodium silicate, sodium bentonite	Good flow and strength properties	[40]
Chengdu, Sichuan Province, China	Complex ground	OPC, CSA, silica fume, early strength agent	High early strength and better durability	[41]

Table 2. Specific chemical composition of solid waste.

Species	Fe_2O_3 /%	FeO /%	Al_2O_3 /%	CaO /%	MgO /%	SiO_2 /%	MnO /%
Cyclone ash	66.45	10.42	3.21	1.74	0.32	4.91	0.12
Sintered dust	64.89	2.8	2.9	14.2	3.05	6.35	0.23
Converter Dephosphrized steel slag	3.83	12.52	3.55	39.66	3.02	15.25	6.08

Species	TiO_2 /%	Na_2O /%	K_2O /%	ZnO /%	Cl /%	P_2O_5 /%	Other /%
Cyclone ash	0.08	0.16	0.15	0.31	0.59	0.05	11.49
Sintered dust	0.09	0.06	0.3	0.03	0.28	0.06	4.76
Converter Dephosphrized steel slag	2.03	0.3	0.36	0.014	-	6.58	6.8

The potential and suitability of different ISW types in the production of grouting materials were explored by investigating their physical, chemical, and mechanical properties. For instance, waste slag, fly ash, and steel slag have been extensively studied and applied in grouting material preparation; the details are shown in Figure 4 [42–47]. Furthermore, the development of chemical grouting technology has opened up more possibilities for ISW utilization. Sun et al. [48] investigated the use of chemical grouting technology to convert waste steel slag into high-performance grouting materials. By adjusting the chemical formula and process parameters, they successfully transformed waste steel slag into grouting materials with exceptional properties; a range of 20–50% ISW content is proposed, which seems to be optimal for balancing the strength gain with the loss of workability and demonstrating the application potential of ISW in the field of grouting materials. Therefore, the research on utilizing ISW as a tunnel grouting material has undergone a significant and meaningful development process.

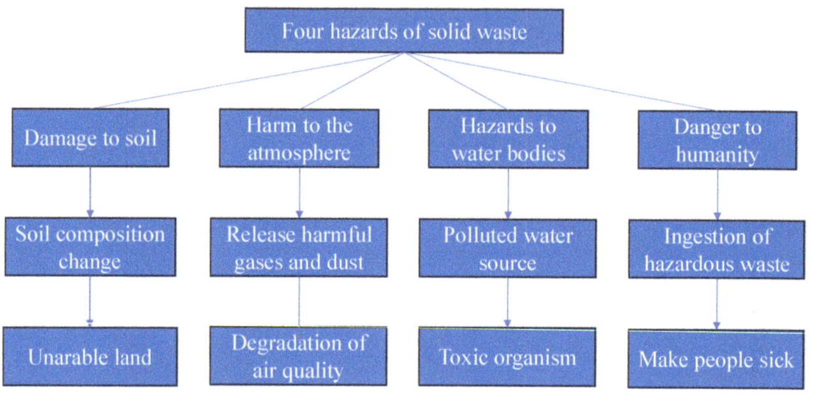

Figure 3. The four hazards of solid waste.

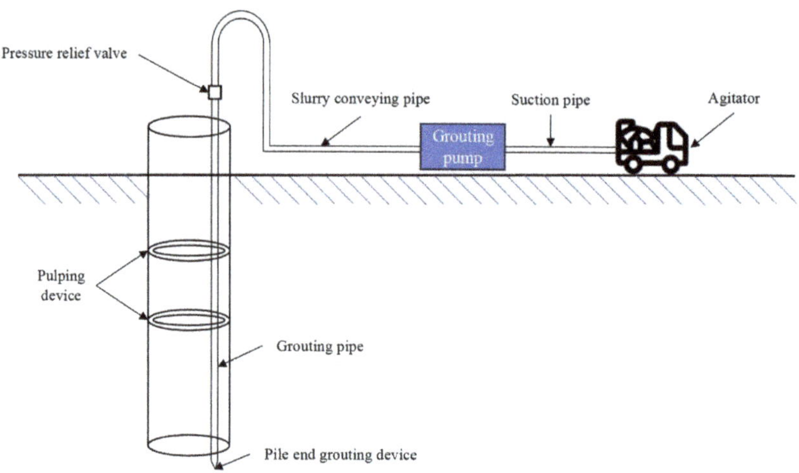

Figure 4. Schematic diagram of grouting process.

There is an increasing focus on the impact of ISW on the mechanical properties of grouting materials, prompting researchers to delve deeper by experimental and microscopic methods. Wang et al. [49] conducted experiments and simulation studies to investigate the influence of various waste components on grouting material performance, providing valuable insights for engineering applications. Ren et al. [50] examined the effects of different ISW components on the mechanical properties of grouting materials. They selected diverse ISWs, including waste slag, fly ash, and steel slag, and mixed them with cement to prepare a series of grouting materials. In a separate study, Sha et al. [51] focused on the influence of steel slag particle-size distribution on the mechanical properties of grouting materials. They prepared a range of grouting materials by sieving and classifying steel slag with varying particle size distributions. Additionally, Mirza et al. [52] investigated the effect of curing conditions on the mechanical properties of fly ash-containing grouting materials. They examined the impact of curing conditions on the strength, toughness, and deformation properties of these materials through compression and tensile tests. Therefore, a better understanding of the feasibility and optimization methods for utilizing ISW as a tunnel grouting material can be achieved, by studying the effects of different waste compositions, particle size distributions, and curing conditions on the mechanical properties of grouting materials.

The authors of this study concentrated on publications containing keywords such as "industrial solid waste", "tunnel grouting materials", "performance enhancement", "process optimization", "durability", and "fluidity". Various scientific databases, including "Web of Science", "Google Scholar", "Science Direct", "Science Citation Index", and "Scopus", were utilized for the literature search. Based on this approach, a literature review of over 130 articles from the past decade was conducted.

This paper aims to provide a comprehensive overview of the recent studies on ISW grouting materials for tunnel engineering, analyze key factors influencing their properties, summarize the major issues and challenges, and discuss future research directions. It is divided into four main sections. Section 2 examines the feasibility of using different ISWs in tunnel grouting. Section 3 discusses the effects of ISW on key grout properties, including strength, fluidity, and durability. Section 4 identifies major challenges and suggests future work to realize the potential of ISW grouts. Section 5 provides the overall conclusions.

2. Feasibility of Industrial Solid Waste as Grouting Material

ISW primarily consists of waste slag, fly ash, and other ISW. The usage of ISW presents significant challenges due to its large output, varied types, and complex composition. A wide range of industrial wastes have been investigated for their viability as substitute materials in tunnel grouting formulations [53]. This section summarizes recent studies on the feasibility of incorporating different ISWs as partial replacements for cement or aggregates in grout mixtures. The effects of ISW composition and proportions on the properties of fresh and hardened grouts are also discussed.

2.1. Slag

Slag is a by-product of iron and steel smelting processes. Research shows that slag powder can be used to partly replace cement in tunnel grouts. Lin et al. [54] found that slag undergoes hydration reactions when activated by cement and lime; the details are shown in Figure 5. The reaction products include calcium silicate hydrates (C-S-H) and calcium aluminate hydrates (C-A-H), which enhance strength. Li et al. [55] discovered that the compressive strength of the slag reached 20 MPa after the treatment, and the slag was suitable as a cementing material to replace a part of Portland cement. Prentice et al. [56] addressed the issues of a prolonged setting time and low early strength of (low clinker) slag cement by incorporating admixtures and implementing a compound preparation scheme. Through X-ray diffraction (XRD) testing and analysis, Yuyou, Zengdi, Xiangqian, and Haijun [40] explored the mechanism of tricalcium silicate in cement minerals to enhance the setting time and early compressive strength. Li et al. [57] optimized the particle size distribution of slag powder to achieve a 7-day grout strength exceeding 6 MPa with 30% slag incorporation.

Steel slag is a by-product of steel manufacturing. Its hydraulic properties allow it to be used in grouting. Ghouleh et al. [58] showed that hardened and graded steel slag aggregates improve grout stability. Up to 70% of natural aggregates can be substituted by steel slag. However, a higher steel slag content increases shrinkage. Li et al. [59] developed a cement-based composite material using ISW (steel slag) and sulphoaluminate cement clinker. This material leverages the cementitious activity of the silicon-aluminum component of steel slag, resulting in improved fluidity, stability, and resistance to ion erosion of the grouting material. André et al. [60] prepared grouts using ground granulated blast furnace slag and found that finely ground steel slag accelerates C-S-H gel formation, enabling early strength development.

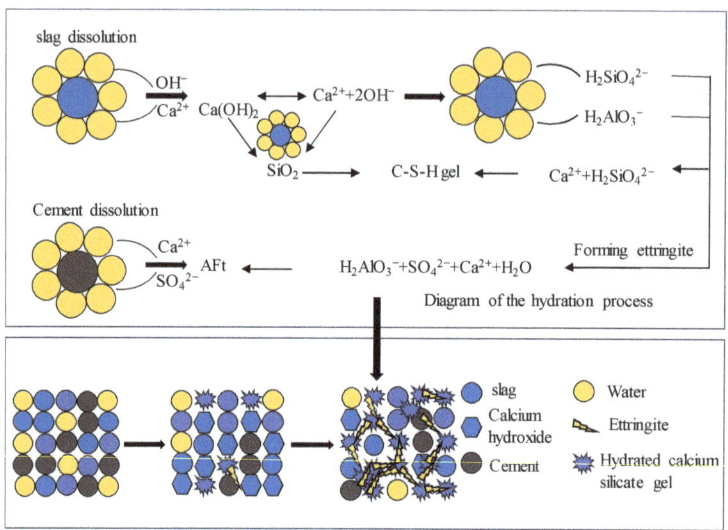

Figure 5. Schematic diagram of hydration reaction.

Gencel et al. [61] investigated the use of waste steel slag as a cementitious material in cement mortar. They analyzed the composition and structural characteristics of ISW and evaluated the active characteristics, curing degree, and performance of the activated cementitious body. Sun and Chen [62] studied the effects of steel slag fineness and the silicate modulus of sodium silicate solution on grout performance. A slag fineness of 7000 cm^2/g and a silicate modulus of 2.5 were optimal. Yu et al. [63] suggested that a steel slag content between 30–50% in grouts provided favorable rheology, setting time, and strength. Grade S95 GGBS (ground granulated blast slag) and class F fly ash were used as solid precursors to produce the AAISW-based two-component grout in this study. GGBS particles are irregular and angular, while FA particles are spherical in different sizes, as shown in Figure 6.

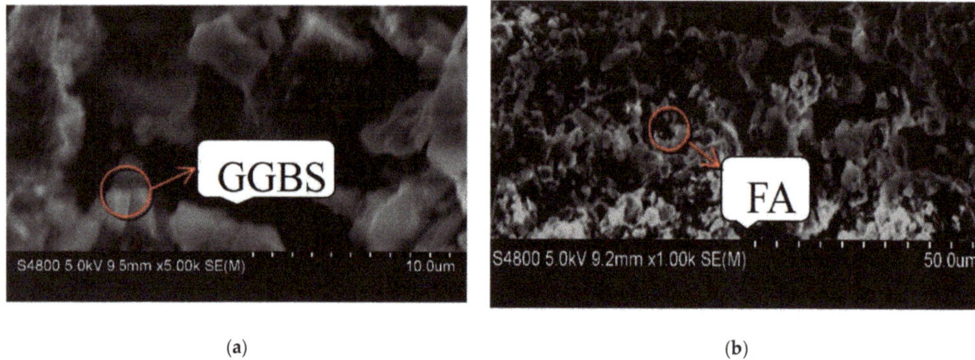

(a)   (b)

Figure 6. SEM images of solid precursors: (**a**) GGBS and (**b**) FA.

2.2. Fly Ash

Fly ash is a fine powder produced from the combustion of coal. Fly ash exhibits strong pozzolanic reactivity and has been widely studied for grout applications. Pei et al. [64]

found that class F fly ash can serve as a supplementary cementing material in grouting materials. The results revealed that incorporating 5% fly ash into the grouting material increased its compressive strength by more than 22% compared to the original grouting material. This enhancement is attributed to the filling of pores between aggregates by the calcium carbonate produced during the carbonation process of fly ash, facilitating the formation of a dense structure; the details are shown in Figure 7. Trimurtiningrum and Ekaputri [65] developed a grout mixture with a 70% class C fly ash replacement of cement. With a 10% gypsum addition, the grout achieved a 28-day strength of over 100 MPa. Li et al. [66] found that class C fly ash refinement is key to improving grout properties. Fly ash participation in pozzolanic reactions also enhances long-term strength. The optimal fly ash content is around 30% to 50% for strength and fluidity considerations.

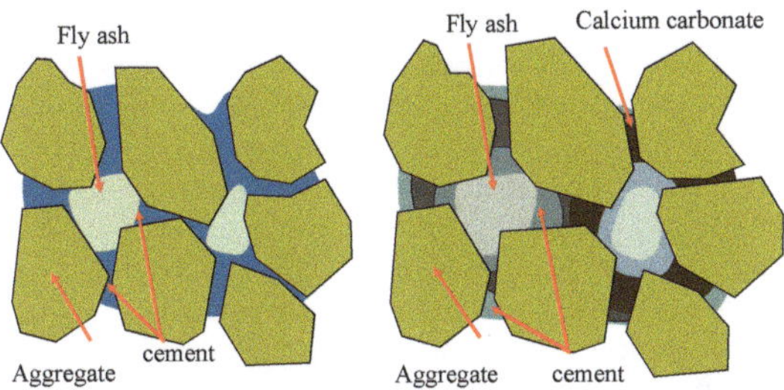

Figure 7. Schematic diagram of acidification reaction.

2.3. Other ISWs

Apart from the aforementioned categories, other ISWs, including red mud, carbide slag, and water treatment residuals, have also been studied for grout production. Composite blends using two or more ISWs can further improve the grout properties compared to single admixtures. Huang et al. [67] investigated the influence of the slag content, the steel slag content and fineness on the autogenous shrinkage of Portland cement. The results indicated that the hydration activity of steel slag plays a crucial role in determining the autogenous shrinkage of the cement. Higher hydration activity leads to greater autogenous shrinkage. Cui et al. [68] employed grouting reinforcement technology and used sulphoaluminate cement as the primary material mixed with prepared talc-like substances. Deng and Zheng [69] utilized zinc slag as a raw material for grouting material derived from ISW. Zhang et al. [70] investigated the chemical structure, surface characteristics, and physical and mechanical properties of sulphoaluminate cement-based grouting materials. They enhanced the mechanical properties of these grouting materials by incorporating an appropriate amount of ISW admixture and water-soluble polyurethane.

Guo et al. [71] employed a significant amount of fly ash, coal gangue, a suitable quantity of desulfurization gypsum, and an alkali activator, along with a small amount of admixture, to prepare coal-based ISW grouting material. Additionally, they utilized response surface analysis to determine the optimal dosage for achieving the desirable fluidity, gel time, and other properties. Overall, an ISW content of around 40% is appropriate for balancing strength enhancement versus workability reduction. The effectiveness also depends on the ISW characteristics and activation methods.

Table 3 shows the feasibility results of different ISWs as grouting materials. It can be seen that using solid waste as grouting material can effectively reduce construction costs while facilitating the safe and efficient recycling of solid waste in engineering applications. Furthermore, treating ISW requires careful consideration of grouting performances in tunnel engineering, as well as the potential environmental pollution resulting from solid waste. The usage of ISW holds significant practical value and contributes to environmental protection, human health preservation, and sustainable development promotion.

Table 3. Feasibility summary of various ISW materials as grouting materials.

ISW Type	Procedure	Improved Performance	Feasibility
Slag	Composite addition with admixture	Enhanced strength performance	Feasible
Steel slag	Optimize particle size distribution	Fluidity, stability, and resistance to ion erosion	Feasible
Fly ash	The amount of incorporation in the appropriate range	Increase compressive strength	Feasible
Other ISWs (including red mud, carbide slag, water treatment residuals, etc.)	Two or more than two types of composite incorporation	Improve fluidity, gelation time, and other working properties.	Feasible

3. Effect of Industrial Solid Waste on Performance of Grouting Material

With the rapid development of society, a generation of various industrial solid wastes (ISWs), including steel slag, slag, and fly ash, has surged. The large-scale stockpiling of these ISWs not only occupies land resources but also causes environmental pollution around the storage sites. Therefore, the comprehensive utilization of these resources is imperative. Currently, the incorporation of ISWs into building materials has become a research hotspot. Numerous studies have proven that the addition of appropriate amounts of ISWs to cement-based grouting materials can significantly reduce raw material costs while delivering enhanced mechanical performance [72]. This section examines how the incorporation of ISW influences the various grout properties essential for engineering performance, including strength, fluidity, impermeability, volume stability, and durability. The effects of the ISW type and dosage are analyzed based on recent studies.

3.1. Mechanical Strength of Grouting Material

Based on previous research, it summarizes the effects of different ISWs on the mechanical strength of cementitious grouting materials and discusses the mechanisms behind ISW incorporation to provide a theoretical basis for the resource utilization of ISWs.

3.1.1. Industrial Solid Waste Incorporation and Process Optimization

The strength development in ISW grouts is linked to the pozzolanic reactivity of ISW components [73]. Si, Al, and Fe elements in ISW source materials can react with the $Ca(OH)_2$ released during cement hydration to form additional strength-enhancing products. The porous structure of ISW particles also serves as nucleation sites, refining pore structure and increasing density.

Common ISWs, such as slag, fly ash, and steel slag, are widely used in cement-based grouting materials. For instance, the combination of slag powder and cement can produce materials with high early strength for full tailings filling, while the mixture of steel slag slurry and cement exhibits significantly improved compressive strength after carbonation [74]. The proportional incorporation of Bayer red mud, carbide slag, and desulfurized gypsum continuously enhances material strength during the carbonation process due to the gradual formation of calcium carbonate. At the same time, the Bayer process will generate the following types of industrial waste (including arsenic waste, silica ferrous slag, slurry wastewater, etc.), and to minimize the environmental impact and waste of resources, the relevant industrial waste needs to be treated and managed appropriately.

Cao et al. [75] combined slag powder and P.O. 42.5 cement to develop a cementitious material with specific early strength for full tailings filling. The composition consisted of 60% slag, 15% cement, 10% hemihydrate gypsum and desulfurization gypsum, and 5% lime. Xue et al. [76] utilized slag powder, along with clinker and gypsum, to develop a cementitious material for the purpose of whole tailings filling. This material exhibited a remarkable strength that was 2.7 times higher than that of cement-based filler in similar conditions. Furthermore, it proved to be a cost-effective alternative, with a significantly lower cost ranging from 40% to 45% of traditional cement. Sha et al. [77] combined pure steel slag slurry and Portland cement to form a steel slag mixture. After carbonization, the compressive strength of the mixture significantly improved. Deng et al. [78] developed a graded tailings filling cemented body using lead-zinc smelting slag, clinker, sodium silicate, gypsum, and other activating materials, with a cement–sand ratio of 1:6. The strength of the cured samples after 3 days reached 2.68 MPa, while after 28 days, it reached 3.97 MPa, representing a significant 21% increase compared to traditional P.O. 42.5 cement. Li et al. [79] developed a dual-liquid grouting material, incorporating industrial waste residues such as steel slag, slag, fly ash, and water glass. The material exhibited a gel time of 3–300 s and a strength of 15–25 MPa. This material significantly enhances the utilization of industrial waste residues and reduces cement grouting consumption. Lahalle et al. [80] developed an alkali-activated polymer double-liquid grouting material using sodium silicate-activated slag and metakaolin. They proposed that when the mass ratio of slag to metakaolin was 1:2, and the volume content of glass was 20%, the compressive strength at 3 days and 28 days reached 7.23 MPa and 17.26 MPa, respectively. Bai et al. [81] prepared a fly ash-based grouting material with an initial setting time of 13 min, a final setting time of 20 min, and a compressive strength of 104.5 MPa. Wan et al. [82] addressed the slow hydration rate of a steel slag–slag–barium slag composite grouting material by utilizing desulfurization gypsum as an activator. The results showed a 21% increase in the 28-day strength. Sha et al. [83] prepared a new type of cement-fly ash-modified sodium silicate grouting material. The results showed that the initial setting time increased, and the final setting time decreased with an increase in the volume fraction of modified sodium silicate. The gel time was proportional to the mass fraction of fly ash in the powder, while the slurry's fluidity was proportional to the volume fraction of modified sodium silicate and the mass fraction of fly ash in the powder.

Further enhancement of grouting material performance can be achieved by optimizing the addition amounts, combinational types, and process parameters of ISWs. Zhu et al. [84] prepared cementitious materials using slag and fly ash as raw materials under alkali activation. The findings revealed that the compressive strength initially increased and then decreased with a decreasing fly ash content when 3% sodium hydroxide (by mass) was used as an alkali activator and the water-to-solid ratio was 0.4. Juan-hong et al. [85] prepared a grouting material by mixing red mud, carbide slag, and desulfurized gypsum at a mass ratio of 74:11:15. The results demonstrated that the content of calcium carbonate in the material increased with carbonization time, leading to a gradual increase in compressive strength. Zhang et al. [86] developed a novel double-liquid grouting material with an adjustable gel time, high consolidation strength, and a stone rate of up to 100%. This material was produced by combining industrial waste residue with Portland cement clinker, optimizing particle size distribution, and alkaline-activating cementitious activity.

Carvalho et al. [87] investigated the impact of steel slag replacement on the performance of composite filling cementitious materials. The findings revealed a pattern of initial shrinkage followed by expansion as the steel slag replacement amount increased, with shrinkage reaching its lowest point at 83%. Chao et al. [88] conducted a comprehensive tailings filling test using a cost-effective grouting material they developed, primarily composed of fly ash, slag powder, carbide slag, and a composite activator. The results demonstrated that with a cementing agent-to-tailings ratio of 1:4–1:8 and a slurry concentration of 67–70%, the 28-day compressive strength ranged from 0.68–2.63 MPa. Ehsani et al. [89] examined the water absorption, fluidity, and compressive strength of a novel fly ash-cement grouting material and determined that optimal performance was achieved with a fly ash ratio of 17–20% and a water–cement ratio of 0.65–0.7. Rehman et al. [90] examined the performance of a steel slag-modified silicate-sodium silicate double-liquid grouting composite material. Their research determined that the optimal water–binder ratio ranged from 0.6–0.8, the ideal volume ratio of sodium silicate to the slurry was 1:4–1:6, and the steel slag content constituted 50–80% of the cement mass. After 3 days of curing, the material with the ratios achieved an early strength exceeding 40 MPa. To enhance the comprehensive utilization of solid waste, Xue et al. [91] conducted an optimization test on subsection iron removal from steel slag and investigated the influence of steel slag powder on the compressive strength of the grouting material. The findings revealed that through sectional magnetic separation, high-performance steel slag powder with less than a 0.5% metal iron content could be obtained. Additionally, when the specific surface area of the steel slag powder was 1:2.5, the material achieved improved 3-day and 28-day strength simultaneously.

Gopinathan and Anand [92] investigated the mechanical properties of grouting materials with a high content of slag powder. The results demonstrated that as the slag incorporation rate increased, the slurry density gradually increased, porosity decreased, and strength increased. Additionally, under low-temperature conditions, the high-content slag powder grouting material exhibited an improvement in compressive strength by 18–27% compared to conventional grouting materials.

Based on a comprehensive analysis of steel slag powder characteristics, Liu and Guo [93] incorporated it into a steel slag-cement and steel slag-mineral powder-fly ash composite system. The research indicated that the optimal replacement rate of steel slag powder in cement was 15%, and the optimum ratio of steel slag powder to mineral powder-fly ash was 1:2:1. Furthermore, the use of Na_2SO_4 as an activator in the steel slag powder-cement composite system effectively improved the early strength of cementitious materials.

3.1.2. Mechanisms of Performance Enhancement

The incorporation of ISWs promotes cement hydration reactions and generates more hydration products, such as ettringite and C-S-H gel. For example, the addition of steel slag and slag induces hydration reactions when activated by cement [94]. The initial hydration of steel slag promotes the depolymerization of steel slag, which subsequently reacts with cement to generate ettringite network structures that are subsequently filled by C-S-H gel, thereby increasing the density of the hardened paste system.

Li et al. [95], guided by grouting engineering performance, primarily utilizes red mud from ISW as the main raw material to prepare a red mud-based grouting material in collaboration with various solid waste materials. The occurrence of free water, the hydration exothermic law, hydration kinetics characteristics, alkaline components, and the heavy metal content of the slurry are analyzed. The hydration mechanics characteristics of the red mud-based grouting material are determined, and its hydration mechanism is revealed. The hydration dynamics characteristics are analyzed using the Krstulovic–Dabic model.

Zhang et al. [96] found that different reaction degrees and sizes of embedded microstructures were observed in fly ash particles at different ages, adversely affecting the mechanical properties of the materials. The activation degree of fly ash gradually increased with age, contributing continuously to the strength development in the later stages. This is due to the fact that with the increase of age, the active substances in the fly ash (the activated substances refer to the components in the fly ash that can react with water to generate cementitious substances, which play an important role in promoting the development of the later strength of the material) will be gradually depleted, and the degree of activation will be gradually reduced, which will adversely affect the development of the later strength.

Liu et al. [97] investigated the hydration reaction process of a clinker-free cementitious system comprising steel slag, slag, and gypsum. The results revealed that an increase in the specific surface area of steel slag powder enhanced the compressive strength of non-clinker concrete but reduced its slump. The network structure formed by ettringite during the early stages played a vital role in the material's early compressive strength, while the hydration products and C-S-H gel continuously filled the gaps in the hardened paste, enhancing the overall compressive strength of the material.

From the above studies, it is easy to see that the use of ISW as grouting material can not only realize the comprehensive utilization of resources but also have significant economic benefits. However, although the existing studies on various ISW-modification mechanisms can provide some theoretical references for practical projects, the systematic selection of ISW materials and optimal dosage to produce low-cost and high-performance grouting materials requires further exploration.

3.2. Fluidity of Grouting Material

Grouting applications in tunnels, foundations, and structural repairs demand controllable fluidity and strength [98]. High fluidity is required during grout injection to achieve good penetration and filling capacity. ISW particles can improve particle packing and lubrication between cement grains, enhancing grout fluidity [99]. However, an excessive ISW dosage can sharply increase viscosity due to a higher specific surface. This section comprehensively examines the interconnected factors influencing the fluidity of grouting material.

The effects of alkali metal cations (K^+, Na^+) on slurry rheology and setting behaviors were evaluated. Li et al. [100] investigated the slurry viscosity and setting time of grouting materials mixed with different alkali metal cations using Fourier transform infrared spectroscopy (FTIR) and viscosity tests. The results indicate that when the ratio of alkali metal cation (K^+) to silicon ion is less than 1.5, the initial setting time of alkali slag grouting material can be effectively prolonged. Additionally, the addition of an appropriate amount of K^+ can effectively reduce the viscosity of the solution and improve the material's fluidity.

The alkali type/content regulated the viscosity and setting, enhancing workability below a threshold. Fly ash and slag modulated the workability and accelerated the strength gain in optimal dosages. Moghadam et al. [101] activated coal gangue, steel slag, and fly ash using $NaOH$ and Na_2SO_4 to prepare cement-free, controllable, low-strength grouting material. When the mass fraction of fly ash is below 40%, the workability (fluidity) of the grouting material meets the ACI (American Concrete Institute) specifications. The phase composition of reaction products in the routing material system is influenced by the mass ratio of steel slag to fly ash. Xu et al. [102] studied the influence of fly ash on the working performance of grouting materials. The results demonstrate that the addition of fly ash significantly improves the fluidity of the materials, and the fluidity increases gradually with the increasing fly ash content up to 40%, resulting in good working performance. ElKhatib et al. [103] successfully prepared various compaction grouting materials for tunnel grouting by utilizing steel slag, fly ash, and other raw materials based on the activity characteristics of steel slag. The results show that the addition of steel slag improves the fluidity of the grouting material, reduces shrinkage, accelerates setting time, and significantly enhances the 7-day and 28-day strength.

Xiang et al. [104] investigated the effect of large amounts of mineral admixtures, such as fly ash and mineral powder, on the performance of grouting materials. The results indicate that a material with a composition of 30% fly ash and 10% mineral powder achieves optimal fluidity. Wu et al. [105] conducted an experimental study on the basic performance of solid waste grouting filling material using rubber particles, fly ash, and clay as the main materials. The results show that a grouting slurry with a fluidity of 293 mm, a 28-day compressive strength of 11.7 MPa, and an impermeability pressure of 0.8 MPa can be achieved when the content of rubber particles is 20%, fly ash is 65%, and clay is 15%.

In summary, the above studies have elucidated the rheological properties, microstructure-property linkages, and composition-property linkages of ISW grouting materials, providing theoretical guidance for the design of sustainable materials. However, due to the limitations of microscopic equipment and numerical simulation computational capabilities, the mechanistic explanations of the above phenomena have not yet gained a unified perception.

3.3. Durability of Grouting Material

Grouting materials are widely used in tunnel engineering to consolidate surrounding rock and soil, control groundwater infiltration, and prevent corrosion. However, grouting materials often face harsh and complex service environments that can impact their durability and performance. This section categorizes and summarizes the recent studies on various factors influencing the durability of grouting materials, including volume stability, water corrosion resistance, impermeability, sulfate attack resistance, and freeze–thaw resistance.

3.3.1. Volume Stability

Volume stability is critical for grouting materials to avoid excessive shrinkage and cracking. Given that the service environment of grouting material is dynamic and intricate, the volume stability of the grout body is closely associated with the engineering's service life. Excessive shrinkage of the grout can lead to crack formation, which in turn can cause leakage and intensify the erosion caused by corrosive ions. He et al. [106] found that high early strength cement grouts exhibit micro-expansion without compromising strength. This expansibility tends to stabilize within the initial 7 days without compromising the subsequent strength of the grout body. Furthermore, micro-expansion also contributes to improving the compactness of hardened slurry. Li and Hao [107] showed that the early strength agent increased the early chemical shrinkage of the double-liquid sulphoaluminate grouting material. Under standard oxidation conditions, the grout experienced volume expansion; however, under dry air curing, it exhibited volume shrinkage. Borçato et al. [108] determined that the shrinkage rate decreases with increasing metakaolin content in geopolymer grouts. The results demonstrated that the shrinkage rate of the grout body decreased with increasing high terrestrial content. However, when the content exceeded 30%, the effect began to weaken. The application of industrial solid waste flue gas desulfurization gypsum (FGDG) in magnesium oxy-sulfate cement (MOSC) was investigated by Li et al. [109]. The results showed that the volume stability of the mixture with the optimum admixture was better at the curing temperature of 40 °C, and its 28-day compressive strength could reach 56.6 MPa, which was 12.5% higher than that of the curing condition at 20 °C, and the softening coefficient could reach 0.99. Meanwhile, it was also confirmed that more hydration products ($5Mg(OH)_2 \cdot MgSO_4 \cdot 7H_2O$) were produced in the mixture under the curing condition of 40 °C through SEM and XRD, and its structure is denser.

3.3.2. Water Corrosion Resistance

Groundwater can cause the gel product in the hardened grout stone body to decompose and dissolve through physical and chemical actions, resulting in a decrease in the mechanical properties of the stone body [110]. This phenomenon is known as water corrosion. Therefore, flowing water can cause dissolution and deterioration of grout gel products and strength loss. Bentz and Garboczi [111] identified calcium hydroxide as vulnerable to leaching and dissolution under the influence of flowing water. This is especially true under conditions of high permeability and high-water pressure in hardened cement paste, where water can infiltrate the stone body, dissolve calcium hydroxide, and then permeate out, leading to a decrease in the compactness and strength of the hardened paste. Furthermore, the dissolution of calcium hydroxide leads to a decrease in the alkalinity of the matrix, which can further cause the decomposition of other hydration products.

Several comparative studies evaluated the water corrosion resistance of various grouts using strength loss, ion leaching, and microscopic analysis. Sun et al. [112] have analyzed the water corrosion resistance of different types of grouting materials, including single-fluid cement slurry, cement-water glass slurry, and industrial waste residue-based slurry. Zhao et al. [113] evaluated the water corrosion resistance characteristics of clay-cement paste slurry, clay-cement stabilized slurry, and single-fluid cement slurry by measuring the amount of calcium oxide dissolved in immersion water; they evaluated the water corrosion resistance characteristics of these three types of slurries. It was found that the viscosity of cement paste slurry exhibited the best water corrosion resistance. Weldes and Lange [114] conducted experimental research and evaluation using indexes such as the strength loss rate, sodium ion solidification rate, silicate dissolution amount, electrical conductivity, and total mass of dissolved matter. They compared these results with those of cement-based grout and discussed the reasons for the improved water corrosion resistance of industrial waste residue-based grout through microscopic tests. Gebauer and Coughlin [115] performed water corrosion resistance tests on hardened samples of different types of grouts, including pure cement grout, cement-water glass grout, cement mortar, and a newly developed grout.

The above results suggested that industrial waste-based grouts and viscosity-modified grouts exhibited improved resistance.

3.3.3. Impermeability

Impermeability is an important performance index of the hardened stone body of grouting materials. It reflects how difficult it is for liquids or ions to infiltrate, diffuse, or migrate within the hardened stone body under pressure, chemical potential, or electric field. A grout with good impermeability usually has a relatively dense microstructure, making it resistant to penetration by external liquids and corrosive ions, which can lead to the deterioration of grouting material performance. Current studies analyzed the effects of density, mineral admixtures, mix proportions, and additives on the impermeability of specialized grout mixes, identifying the optimal parameters. Wang et al. [116] conducted research on the relationship between impermeability and density of ISW grouting materials. The results indicate that the density of polymer materials (non-aqueous reactive polymer grouting materials represented by polyurethane foam) is generally in the range of 0.1–0.3 g/cm^3, and the initial seepage pressure is in the range of 0.3–0.7 MPa, which can bear 30 to 70 m's head pressure when seepage grouting is carried out in hydraulic engineering. In practical applications, the amount of polymer grouting should be controlled, and the density of the polymer material should be adjusted according to the actual needs of different projects to meet the needs of seepage control in different projects. Mangat and Khatib [117] studied the influence of fly ash, silica fume, and slag powder on the impermeability of cement paste using a mortar permeameter. Their findings indicated that each material had different optimal replacement rates, which could optimize the impermeability of the hardened slurry. Shana and Jeffrey [118] analyzed the influencing factors of the permeability coefficient of slag cement slurry. They concluded that slag cement slurry prepared under suitable ratio conditions exhibited good impermeability and had broad application prospects. Zhang et al. [119] highlighted that alkali-activated cementitious materials prepared from dolomite, sodium carbonate, and calcined bentonite possess excellent water resistance. Zhao et al. [120] used the softening coefficient as a parameter to characterize the water resistance of the materials. The study found that incorporating 10–20% rice husk ash improved the water resistance of the prepared cementitious materials significantly. Yang et al. [121] discovered that after undergoing the same high-temperature test, the damage in terms of compressive strength and bond strength was lower in alkali-activated mortar compared to cement mortar, indicating better high-temperature resistance. Li et al. [122] investigated the influence of the water-binder ratio and the ratio of liquid A to liquid B on the impermeability of hardened slurry samples during the preparation of cement-sodium silicate double slurry. The maximum impermeability pressure of the prepared samples reached 1.6 MPa. Porbaha et al. [123] developed a testing device for measuring the permeability coefficient of hardened grout to study the impermeability of clay-solidified grout. Song et al. [25] used a new additive, called Z201, to prepare a grouting material with excellent impermeability for shield tunnel walls. When the content of additive Z201 was 10%, the impermeability pressure reached 0.8 MPa.

3.3.4. Sulfate Attack Resistance

Groundwater sulfates can react with grout hydration products, causing expansion, strength loss, and cracks [124]. Groundwater used in grouting often contains sulfate substances such as sodium sulfate and magnesium sulfate. When the content of these sulfate substances is high, it can lead to the erosion and deterioration of the hardened paste over time, resulting in a decrease in its strength. Sulfate erosion can be categorized into internal erosion and external erosion based on the source of sulfate ions [125]. Internal erosion occurs when sulfate ions are present within the hardened paste itself, while external erosion is caused by sulfate ions from the surrounding environment. The process of external sulfate erosion can be divided into two stages: infiltration of sulfate ions into the matrix of the hardened paste from groundwater, which is influenced by the impermeability of the hardened paste, and the reaction between sulfate ions and substances in the hardened paste matrix. The corrosion mechanism of sulfate on cement-sodium silicate grout was studied [32]. It was found that the gel products of traditional cement-based grout, mainly $Ca(OH)_2$ and C-S-H with a high Ca/Si ratio, react with sulfate ions to form expansive crystals, leading to the destruction of the matrix structure. Additionally, the stability of the C-S-H gel with a high Ca/Si ratio is poor, and the reaction consumption of $Ca(OH)_2$ reduces the alkalinity in the matrix, causing decomposition and destruction of the C-S-H gel. Carde et al. [126] identified calcium hydroxide and C-S-H as vulnerable and described degradation mechanisms. Yoshikawa and Toda [127] suggested that low concentrations of sulfate solutions can promote the early-stage strength development of hardened paste, while high concentrations of sulfate solutions can cause a decline in strength. Kourounis et al. [128] found that incorporating high-viscosity materials and steel slag into traditional cement-sodium silicate grout as partial replacements for cement can enhance the sulfate corrosion resistance of the grout. Therefore, it is important to conduct the appropriate research to develop grouting materials with improved sulfate corrosion resistance.

3.3.5. Freeze–Thaw Resistance

Karakurt and Bayazıt [129] compared the freeze–thaw resistance of alkali-activated concrete (with fly ash partially replacing kaolin) with ordinary Portland cement concrete. They found that ordinary Portland cement concrete reached a mass loss rate of 5% after 45 freeze–thaw cycles, while alkali-activated concrete required 98 freeze–thaw cycles to reach the same mass loss rate. Liu et al. [130] studied the effect of silica fume on the freeze–thaw resistance of alkali-activated fly ash/slag cementitious materials through experiments. The results showed that the addition of silica fume led to more damage and deterioration of the hardened samples under freeze–thaw cycles.

Aiken et al. [131] investigated the effects of freeze–thaw cycles on the mechanical properties of a grouting material composed of basalt fiber, slag powder, and fly ash. The results indicated that the material with a basalt fiber content of 0.18 vol% exhibited superior frost resistance, as well as the highest compressive and tensile strengths compared to other fiber contents under the same freeze–thaw cycles.

Hou et al. [132] emphasized that increasing the slag content in alkali-activated fly ash/slag cementitious materials can reduce the porosity and pore diameter of the hardened sample matrix, thereby enhancing the freeze–thaw resistance of the materials.

Vegas et al. [133] conducted 300 freeze–thaw cycle tests on alkali-activated fly ash mortar and traditional Portland cement-based mortar. It was concluded that alkali-activated materials exhibit better freeze–thaw resistance than traditional Portland cement.

Slavik et al. [134] investigated the freeze–thaw resistance of alkali-activated fluidized bed combustion bottom ash cementitious material. The results showed that the compressive strength of hardened samples retained over 80%, even after 50 freeze–thaw cycles, indicating good freeze–thaw resistance. These studies collectively suggest that alkali-activated ISW cement materials generally exhibit superior freeze–thaw resistance compared to ordinary Portland cement-based materials. The slag content, silica fume addition, and other compositions affect the performance of the grouting materials.

3.3.6. Corrosion Resistance

Several research studies suggested that alkali-activated materials, after hardening, exhibit high mechanical strength and corrosion resistance compared to traditional cement-based materials. Wu et al. [135] studied the high-temperature resistance of alkali-activated fly ash/slag-based cementitious materials. They noted that the critical temperature for the degradation of physical and mechanical properties of the alkali-activated materials prepared from fly ash/slag is between 400–600 °C. Kanaan [136] compared the sulfate attack resistance of alkali-activated high territory/slag-based paste materials with pure cement-based paste materials. The results showed that the paste samples containing slag exhibited the best sulfate attack resistance, followed by pure kaolin samples, while the cement-based samples had the least resistance to high temperatures. These findings suggest that alkali-activated materials, represented by ISW-based grouting materials, offer improved performance in terms of sulfate attack resistance, making them advantageous for various applications in construction and engineering. Fernández-Jiménez et al. [137] investigated the corrosion resistance of alkali-activated cementitious materials compared to cement-based cementitious materials. The results demonstrated that the corrosion resistance coefficient of alkali-activated cementitious materials was significantly higher than that of cement-based materials. Meanwhile, Xin et al. [138] found that under corrosive conditions, the appearance damage of cement-based samples was more severe compared to alkali-activated fly ash-based cementitious material samples. This can be attributed to the gel products in cement-based samples, primarily composed of $Ca(OH)_2$, C-S-H gel, and C-A-H gel, with a high Ca/Si ratio. These gels are stable in alkaline environments but easily decompose and deteriorate in acidic environments. These findings suggest that alkali-activated cementitious materials possess superior corrosion resistance compared to cement-based materials, making them more suitable for applications where exposure to corrosive environments is a concern. From the above studies, it is easy to see that ISW grouting materials possess superior durability properties under specific conditions compared to ordinary materials. However, conventional modification methods can only improve a certain property, and it is necessary to find a convenient and efficient method to improve the overall durability of ISW grouting materials.

4. Challenges and Future Prospects

The rapid expansion of infrastructure has resulted in a global increase in tunnel construction. Concurrently, significant industrialization has generated substantial amounts of industrial solid waste (ISW), leading to resource wastage and environmental pollution. Utilizing ISW as grouting materials in tunnels offers an eco-friendly and cost-effective solution to address these interconnected issues. However, several key challenges must be addressed before the widespread application of ISW grouting materials can be achieved.

One primary challenge is ensuring a consistent high performance of ISW grouting materials. Due to the variability in ISW composition and properties, the mechanical strength, fluidity, impermeability, durability, and other engineering properties of ISW grouts can be unstable. Thorough characterization and optimal mix design are necessary to meet the demanding performance criteria. Research should systematically investigate the complex effects of different ISW types, proportions, chemical activators, and preparation processes on grout properties.

Another major concern is the environmental impact and ecological safety associated with the use of ISW grouting materials. While intended to mitigate pollution, the improper use of ISW with high levels of heavy metals or other hazardous substances can introduce new environmental risks. Comprehensive assessment frameworks must be established to evaluate the leaching potential and ecological effects of ISW grouts. Standardized procedures are required to treat ISW and purify grout mixes to comply with health and safety regulations. Further research should focus on developing high-quality grouting materials from industrial waste.

Furthermore, there is currently a lack of specific international codes, standards, and guidelines regarding ISW grout production and quality control. Extensive mechanical and durability testing needs to be conducted to determine reasonable technical specifications for engineering properties, mix proportions, construction methods, and acceptance criteria. Systematic field testing and monitoring in pilot projects are essential to validate ISW grout performance. The accumulated experience should be formulated into detailed protocols and regulations to ensure quality assurance and safe, widespread implementation.

Advanced microstructure analysis can provide valuable insights into the composition–structure–property relationships of ISW grouting materials, guiding the optimization of their engineering properties. Some key ways in which advanced microstructure analysis aids in this optimization include:

(1) Scanning electron microscopy (SEM) coupled with energy-dispersive X-ray spectroscopy (EDS) can characterize the elemental distribution and chemical composition of different phases and reaction products in the grout's microstructure, revealing the role of specific ISW particles and hydration products in enhancing properties.

(2) Transmission electron microscopy (TEM) enables nanoscale examination of C-S-H gels and other nanoscale hydrates, facilitating an understanding of macro-scale strength and impermeability.

(3) Mercury intrusion porosimetry quantifies the pore size distribution, allowing for the study of pore parameters like total porosity, critical pore diameter, shape, and connectivity, which helps establish correlations between permeability and durability.

(4) XRD identifies crystalline reaction products, providing insights into strength development mechanisms by tracking formations such as calcium silicate hydrates and calcium aluminate hydrates at different curing ages.

(5) Fourier transform infrared spectroscopy detects chemical bonding information and can monitor the progress of pozzolanic reactions and hydration processes over time.

(6) Nuclear magnetic resonance spectroscopy analyzes silicate polymerization, revealing the cross-linking density of the C-S-H gel phase, which largely controls strength.

(7) Mass spectroscopy accurately identifies the ion types and concentrations leached from grout samples in durability testing, aiding in the evaluation of resistance mechanisms.

By correlating advanced microstructural observations with measured engineering properties, the optimal type and dosage of ISW can be determined to tailor grout behavior for specific applications. The micro–macro linkages provided by sophisticated analysis are invaluable for designing high-performance and durable ISW grouting materials.

Advanced techniques also offer opportunities to enhance ISW grouts. For example:

(1) Optimizing the particle size distribution through fine grinding and classification can improve particle packing density, rheology, and strength. Meanwhile, novel admixtures and nano-additives may regulate the setting time, reduce shrinkage, and enhance impermeability.

(2) Advanced sensors and monitoring technologies are utilized to monitor the rheological properties, uniformity, and curing process of the grouting process in real time for precise control and adjustment.

(3) Using 3D-printing technology, the geometry and microstructure of grouting materials can be precisely controlled to improve the mechanical properties of grouting and process efficiency.

(4) Microencapsulation technology encapsulates additives in tiny capsules, allowing for a precise controlled release effect that improves the performance and stability of the grouting material.

(5) Using computational simulation and numerical analysis methods, the grouting process can be accurately modeled and simulated to better understand material behavior and optimize the grouting design.

In the future, the integration of diverse industrial wastes into optimal grout design will be facilitated by advanced technical means, promoting sustainable infrastructure construction. More research should be directed towards the development of multifunctional grouts with self-sensing, crack-healing, and microbial properties for the creation of "smart tunnels". Resource utilization can be maximized through the recycling and reusing of excavated soil, demolition waste, and industrial waste. With persistent research efforts and interdisciplinary collaboration, the evolution of ISW grouting technology will benefit both the industry and the environment.

5. Conclusions

This comprehensive review of recent studies on utilizing industrial solid waste (ISW) in tunnel grouting materials demonstrates the promising potential of this sustainable approach. However, further research and development are needed to address the existing limitations and knowledge gaps. The main findings are as follows:

(1) The existing literature is well established on a variety of ISWs (including slag, fly ash, steel slag, red mud, silica fume, calcium carbide slag, etc.) as partial substitutes for cement in grout formulations. At the same time, depending on the ISW reactivity, ISW can be mixed up to 50%. Composite mineral admixtures can participate in the hydration reaction to produce strength-enhancing gel products, such as C-S-H and C-A-H. Therefore, although there is no uniform understanding of ISW as a grouting material, there is no doubt that the use of the right type of ISW in the right dosage can reduce the cost of grouting and provide an environmentally friendly recycling solution for the large quantities of ISW that are generated globally.

(2) The mechanical properties, fluidity, volume stability, impermeability, and durability of ISW grouts depend on the specific type and dosage of ISW. The development of strength relies on the pozzolanic reactivity of the ISW components. To achieve a balance between strength and workability loss, an optimal range of industrial solid waste (ISW) content is considered to be 20–50%. The particle size distribution and shape of ISW particles affect the rheology of the grout. It is crucial to optimize the mix design, considering the characteristics of ISW to meet the engineering requirements. Microstructure analysis reveals that ISW hydration and refinements in the micro-scale pore structure contribute to enhanced macro-scale properties.

(3) Before an extensive application of ISW grouts can take place, several technological barriers need to be addressed. The variability and uncertainty of ISW properties make it challenging to achieve consistent grout quality. There is a lack of standardized codes, specifications, and quality control procedures in this area. It is necessary to conduct rigorous assessments to evaluate the ecological impacts. The use of advanced computer simulation and 3D-printing techniques might enable the precise control of properties customized to local conditions.

In summary, the use of ISW as substitute raw materials in tunnel grouting offers significant potential for resource recycling and pollution mitigation. With ongoing research to address the current challenges, ISW grouting technology can be continually enhanced to deliver sustainable and high-performance solutions for underground construction globally. This review provides valuable insights to support further exploration in this emerging field.

Author Contributions: Conceptualization, B.J. and S.W.; methodology, M.W.; software, A.Z.; validation, S.H., S.W. and S.H.; formal analysis, S.H.; investigation, B.J. and M.W. and A.Z.; resources, M.W.; data curation, S.W. and A.Z.; writing—original draft preparation, B.J. and S.W.; writing—review and editing, B.J. and S.H.; visualization, M.W.; supervision, S.H.; project administration, A.Z.; funding acquisition, B.J. All authors have read and agreed to the published version of the manuscript.

Funding: This research was funded by the Science and Technology Research Project of Chongqing Education Commission, grant number KJQN202205802, KJQN202303436, KJQN202305802; and the Scientific Research Project of Chongqing Vocational Institute of Engineering, grant number 2022KJB09.

Data Availability Statement: Not applicable.

Conflicts of Interest: The authors declare no conflict of interest.

References

1. Guo, D.; Song, Z.; Xu, T.; Zhang, Y.; Ding, L. Coupling analysis of tunnel construction risk in complex geology and construction factors. *J. Constr. Eng. Manag.* **2022**, *148*, 04022097. [CrossRef]
2. Huang, J.; Liu, N.; Ma, Z.; Lu, L.; Dang, K. The construction stability of large section tunnel considering the deterioration of clay mechanical properties. *Front. Mater.* **2023**, *10*, 1135276. [CrossRef]
3. Zhang, M.; Yu, D.; Wang, T.; Xu, C. Coupling Analysis of Tunnel Construction Safety Risks Based on NK Model and SD Causality Diagram. *Buildings* **2023**, *13*, 1081. [CrossRef]
4. Qi, Q.; Nie, Y.; Wang, X.; Liu, S. Exploring the effects of size ratio and fine content on vibration compaction behaviors of gap-graded granular mixtures via calibrated DEM models. *Powder Technol.* **2023**, *415*, 118156. [CrossRef]
5. Li, Z.; Zhao, J.; Hu, K.; Li, Y.; Liu, L. Adaptability Evaluation of Rotary Jet Grouting Pile Composite Foundation for Shallow Buried Collapsible Loess Tunnel. *Appl. Sci.* **2023**, *13*, 1570. [CrossRef]
6. Fan, H.; Xu, Q.; Lai, J.; Liu, T.; Zhu, Z.; Zhu, Y.; Gao, X. Stability of the loess tunnel foundation reinforced by jet grouting piles and the influence of reinforcement parameters. *Transp. Geotech.* **2023**, *40*, 100965. [CrossRef]

7. He, Y.; Wang, H.; Zhou, J.; Su, H.; Luo, L.; Zhang, B. Water Inrush Mechanism and Treatment Measures in Huali Highway Banyanzi Tunnel—A Case Study. *Water* **2023**, *15*, 551. [CrossRef]

8. Liu, S.; Nie, Y.; Hu, W.; Ashiru, M.; Li, Z.; Zuo, J. The influence of mixing degree between coarse and fine particles on the strength of offshore and coast foundations. *Sustainability* **2022**, *14*, 9177. [CrossRef]

9. Zhang, W.; Wang, Y.; Song, M.; Shen, Z. Industrial structure upgrading, technological innovation and comprehensive utilisation of solid waste. *Technol. Anal. Strateg. Manag.* **2023**, *2023*, 2217719. [CrossRef]

10. Huang, C.; Zhang, J.-H.; Zhang, A.-S.; Li, J.; Wang, X.-Y. Permanent deformation and prediction model of construction and demolition waste under repeated loading. *J. Cent. South Univ.* **2022**, *29*, 1363–1375. [CrossRef]

11. Fares, G.; Alsaif, A.; Alhozaimy, A. Hybridization and cost-performance analysis of waste tire steel fibers into high-volume powdered scoria rocks-based ultra-high performance concrete. *J. Build. Eng.* **2023**, *72*, 106568. [CrossRef]

12. Siles-Castellano, A.B.; López-González, J.A.; Jurado, M.M.; Estrella-González, M.J.; Suárez-Estrella, F.; López, M.J. Compost quality and sanitation on industrial scale composting of municipal solid waste and sewage sludge. *Appl. Sci.* **2021**, *11*, 7525. [CrossRef]

13. Ragupathi, V.; Panigrahi, P.; Nagarajan, G.S. Supercapacitor active material from Recycling. *ECS J. Solid State Sci. Technol.* **2023**, *12*, 024001. [CrossRef]

14. Simoni, S.; Corrêa, J.A.; Ribeiro, C.J.C.; Abrunhosa, F.A.; Chira, O.P.A. Evaluation of the contamination of the soil and water of an open dump in the Amazon Region, Brazil. *Environ. Earth Sci.* **2021**, *80*, 113.

15. Janas, M.; Zawadzka, A. Assessment of the monitoring of an industrial waste landfill. *Ecol. Chem. Eng. S* **2018**, *25*, 659–669. [CrossRef]

16. Wang, N.; Chai, X.; Guo, Z.; Guo, C.; Liu, J.; Zhang, J. Hierarchy performance assessment of industrial solid waste utilization—Tracking resource recycling and utilization centers in China. *Environ. Sci. Pollut. Res.* **2023**, *30*, 83330–83340. [CrossRef]

17. Willson-Levy, R.; Peled, A.; Klein-BenDavid, O.; Bar-Nes, G. Development of One-part geopolymers based on industrial carbonate waste. *Constr. Build. Mater.* **2023**, *365*, 130009. [CrossRef]

18. Zhang, L.; Jia, X.; Yang, G.; Wang, D. High-volume upcycling of waste sediment and enhancement mechanisms in blended grouting material. *Adv. Mater. Sci. Eng.* **2022**, *2022*, 3225574. [CrossRef]

19. Liu, W.; Jiang, Y.; Zhao, Z.; Du, R.; Li, H. Material innovation and performance optimization of multi-solid waste-based composite grouting materials for semi-flexible pavements. *Case Stud. Constr. Mater.* **2022**, *17*, e01624. [CrossRef]

20. Wang, X.; Hu, J.; Zhang, X.; Wang, K.; Shen, D.; Liang, C. Experimental Study on Mineral Solid Waste Green Grouting Material Based on Electromagnetic Characteristic Detection. *Adv. Civ. Eng.* **2022**, *2022*, 6261429. [CrossRef]

21. Yao, T. *Study on the Effectiveness and Efficiency of Regulation on Resource Utilization of Industrial Solid Waste*; Shanxi University of Finance and Economics: Taiyuan, China, 2021.

22. Zhang, L.; Jia, X.; Wang, C.; Jiang, Q.; Sun, C. Characterization of Magnesium Potassium Phosphate Cement-Based Grouting Material Blended with High Volume Industrial Wastes. *Adv. Mater. Sci. Eng.* **2022**, *2022*, 7353985. [CrossRef]

23. Dun, Z.; Wang, M.; Zou, Y.; Ren, L. Laboratory Tests on Performance of Waste-Clay-Brick-Powder Cement Grouting Materials for Ground Improvement in Mine Goaf. *Adv. Eng. Mater.* **2022**, *24*, 2101575. [CrossRef]

24. Lin, C.; Dai, W.; Li, Z.; Sha, F. Performance and microstructure of alkali-activated red mud-based grouting materials under class F fly ash amendment. *Indian Geotech. J.* **2020**, *50*, 1048–1056. [CrossRef]

25. Song, W.; Zhu, Z.; Pu, S.; Wan, Y.; Huo, W.; Song, S.; Zhang, J.; Yao, K.; Hu, L. Synthesis and characterization of eco-friendly alkali-activated industrial solid waste-based two-component backfilling grouts for shield tunnelling. *J. Clean. Prod.* **2020**, *266*, 121974. [CrossRef]

26. Li, Y.; Debusschere, R.; Yuan, Q.; Li, J. Recycling of ground jet grouting waste as a supplementary cementitious material. *Resour. Conserv. Recycl.* **2023**, *194*, 106993. [CrossRef]

27. Shukla, N.; Harbola, M.; Sanjay, K.; Shekhar, R. Electrochemical fencing of Cr (VI) from industrial wastes to mitigate ground water contamination. *Trans. Indian Inst. Met.* **2017**, *70*, 511–518. [CrossRef]

28. Yan, J.; Karlsson, A.; Zou, Z.; Dai, D.; Edlund, U. Contamination of heavy metals and metalloids in biomass and waste fuels: Comparative characterisation and trend estimation. *Sci. Total Environ.* **2020**, *700*, 134382. [CrossRef]

29. Ceballos, E.; Cama, J.; Soler, J.M.; Frei, R. Release and mobility of hexavalent chromium in contaminated soil with chemical factory waste: Experiments, Cr isotope analysis and reactive transport modeling. *J. Hazard. Mater.* **2023**, *451*, 131193. [CrossRef]

30. Zhang, X.; Li, W.; Tang, Z.; Wang, X.; Sheng, D. Sustainable regenerated binding materials (RBM) utilizing industrial solid wastes for soil and aggregate stabilization. *J. Clean. Prod.* **2020**, *275*, 122991. [CrossRef]

31. Zhang, C.; Fu, J.; Yang, J.; Ou, X.; Ye, X.; Zhang, Y. Formulation and performance of grouting materials for underwater shield tunnel construction in karst ground. *Constr. Build. Mater.* **2018**, *187*, 327–338. [CrossRef]

32. Yu, Z.; Yang, L.; Zhou, S.; Gong, Q.; Zhu, H. Durability of cement-sodium silicate grouts with a high water to binder ratio in marine environments. *Constr. Build. Mater.* **2018**, *189*, 550–559. [CrossRef]

33. Zhang, G.; Liu, J.; Li, Y.; Liang, J. A pasty clay–cement grouting material for soft and loose ground under groundwater conditions. *Adv. Cem. Res.* **2017**, *29*, 54–62. [CrossRef]

34. Tian, K. *Study and Application on High Property Grouting Material Used in Synchronous Grouting of Shield Tunnelling*; Wuhan University of Technology: Wuhan, China, 2007.

35. Zhang, G. Research and Application of New Industrial Residue Two-Shot Grouting Materials. Master's Thesis, Wuhan University of Technology, Wuhan, China, 2007.
36. Ding, Y. Experimental Study on Physical Mechanical Behavior of Backfill Grouting Material in Shield Tunneling. Master's Thesis, Southeast University, Nanjing, China, 2017.
37. Xu, K. *The Research and Application on High-Performance Grouting Made by Shield Sediment*; Wuhan University of Technology: Wuhan, China, 2011.
38. Liang, X. *Study of Material Performance and Material Ratio Design in Synchronized Grouting Technology for Shield Tunneling on Watery Strata*; Chang'an University: Xi'an, China, 2009. (In Chinese)
39. Chen, Y. Investigation into the Characteristics and Strength Simulation Test of the Complex Two-Component in Synchronous Grouting of Shield Tunneling. Master's Thesis, Beijing Jiaotong University, Beijing, China, 2017.
40. Yuyou, Y.; Zengdi, C.; Xiangqian, L.; Haijun, D. Development and materials characteristics of fly ash-slag-based grout for use in sulfate-rich environments. *Clean Technol. Environ. Policy* **2016**, *18*, 949–956. [CrossRef]
41. Wang, S.; Wang, J.-F.; Yuan, C.-P.; Chen, L.-Y.; Xu, S.-T.; Guo, K.-B. Development of the nano-composite cement: Application in regulating grouting in complex ground conditions. *J. Mt. Sci.* **2018**, *15*, 1572–1584. [CrossRef]
42. Li, Z.; Zhang, J.; Li, S.; Gao, Y.; Liu, C.; Qi, Y. Effect of different gypsums on the workability and mechanical properties of red mud-slag based grouting materials. *J. Clean. Prod.* **2020**, *245*, 118759. [CrossRef]
43. Lin, C.; Dai, W.; Li, Z.; Wang, Y. Study on the inorganic synthesis from recycled cement and solid waste gypsum system: Application in grouting materials. *Constr. Build. Mater.* **2020**, *251*, 118930. [CrossRef]
44. Sha, F.; Jin, Q.; Liu, P. Development of effective microfine cement-based grouts (EMCG) for porous and fissured strata. *Constr. Build. Mater.* **2020**, *262*, 120775. [CrossRef]
45. Li, T.; Yue, Z.; Li, J.; Li, Q.; Li, Y.; Chen, G. Experimental study of improved cement silicate grouting material for broken surrounding rock. *J. Build. Eng.* **2023**, *74*, 106782. [CrossRef]
46. Zhang, C.; Fu, J.; Song, W. Mechanical model and strength development evolution of high content fly ash–cement grouting material. *Constr. Build. Mater.* **2023**, *398*, 132492. [CrossRef]
47. Shen, Q.; Jiang, R.; Cong, B.; Guo, B.; Shang, H.; Ji, X.; Li, L. Study on the effect of mineral admixtures on working and mechanical properties of the grouting material. *Front. Mater.* **2023**, *10*, 1197997. [CrossRef]
48. Sun, X.; Miao, H.; Li, H.; Cui, W.; Zhang, F.; Liu, X.; Song, H.; Li, S. Study on the effect of mixed aggregate gradation on the performance of cement-based grouting material. *Concrete* **2019**, *3*, 82–85.
49. Wang, M.; Wang, C.; Yu, J.; Li, Y.; Wen, P.; Fan, Q. Investigation of the grouting effect of blast furnace slag-based mortar on void road bases based on the grouting simulation test. *Constr. Build. Mater.* **2021**, *282*, 122567. [CrossRef]
50. Ren, C.; Wang, W.; Li, G. Preparation of high-performance cementitious materials from industrial solid waste. *Constr. Build. Mater.* **2017**, *152*, 39–47. [CrossRef]
51. Sha, F.; Li, H.; Pan, D.; Liu, H.; Zhang, X. Development of steel slag composite grouts for underground engineering. *J. Mater. Res. Technol.* **2020**, *9*, 2793–2809. [CrossRef]
52. Mirza, J.; Mirza, M.; Roy, V.; Saleh, K. Basic rheological and mechanical properties of high-volume fly ash grouts. *Constr. Build. Mater.* **2002**, *16*, 353–363. [CrossRef]
53. Das, D.; Rout, P.K. Synthesis of Inorganic Polymeric Materials from Industrial Solid Waste. *Silicon* **2023**, *15*, 1771–1791. [CrossRef]
54. Lin, R.; Yang, L.; Pan, G.; Sun, Z.; Li, J. Properties of composite cement-sodium silicate grout mixed with sulphoaluminate cement and slag powder in flowing water. *Constr. Build. Mater.* **2021**, *308*, 125040. [CrossRef]
55. Li, M.; Huang, G.; Cui, Y.; Wang, B.; Chang, B.; Yin, Q.; Zhang, S.; Wang, Q.; Feng, J.; Ge, M. Coagulation Mechanism and Compressive Strength Characteristics Analysis of High-Strength Alkali-Activated Slag Grouting Material. *Polymers* **2022**, *14*, 3980. [CrossRef]
56. Prentice, D.P.; Bernal, S.A.; Bankhead, M.; Hayes, M.; Provis, J.L. Phase evolution of slag-rich cementitious grouts for immobilisation of nuclear wastes. *Adv. Cem. Res.* **2018**, *30*, 345–360. [CrossRef]
57. Li, S.; Sha, F.; Liu, R.; Zhang, Q.; Li, Z. Investigation on fundamental properties of microfine cement and cement-slag grouts. *Constr. Build. Mater.* **2017**, *153*, 965–974. [CrossRef]
58. Ghouleh, Z.; Guthrie, R.I.; Shao, Y. Production of carbonate aggregates using steel slag and carbon dioxide for carbon-negative concrete. *J. CO2 Util.* **2017**, *18*, 125–138. [CrossRef]
59. Li, Z.; Lin, C.; Sha, F.; Zhang, Q.; Zhang, J.; Qi, Y.; Gao, Y. Durability of a new type of cement-based composite grouting material under effects of chemical corrosion. *Mater. Express* **2020**, *10*, 948–954. [CrossRef]
60. André, L.; Bacquié, C.; Comin, G.; Ploton, R.; Achard, D.; Frouin, L.; Cyr, M. Improvement of two-component grouts by the use of ground granulated blast furnace slag. *Tunn. Undergr. Space Technol.* **2022**, *122*, 104369. [CrossRef]
61. Gencel, O.; Karadag, O.; Oren, O.H.; Bilir, T. Steel slag and its applications in cement and concrete technology: A review. *Constr. Build. Mater.* **2021**, *283*, 122783. [CrossRef]
62. Sun, J.; Chen, Z. Effect of silicate modulus of water glass on the hydration of alkali-activated converter steel slag. *J. Therm. Anal. Calorim.* **2019**, *138*, 47–56. [CrossRef]
63. Yu, J.; Hua, S.; Qian, L.; Ren, X.; Zuo, J.; Zhang, Y. Development of steel slag-based solidification/stabilization materials for high moisture content soil. *J. Renew. Mater.* **2022**, *10*, 735.

64. Pei, S.-L.; Pan, S.-Y.; Li, Y.-M.; Chiang, P.-C. Environmental benefit assessment for the carbonation process of petroleum coke fly ash in a rotating packed bed. *Environ. Sci. Technol.* **2017**, *51*, 10674–10681. [CrossRef]

65. Trimurtiningrum, R.; Ekaputri, J.J. Geopolymer grout material. *Mater. Sci. Forum* **2016**, *841*, 40–47. [CrossRef]

66. Li, F.; Yang, Z.; Zhou, S.; Zhang, X.; Liu, D.; Su, Y.; Niu, Q. Fly-ash-based geopolymer modified by metakaolin for greener grouting material. *Proc. Inst. Civ. Eng. Eng. Sustain.* **2023**, *40*, 1–14. [CrossRef]

67. Huang, H.R.; Liao, Y.S.; Qunaynah, S.A.; Jiang, G.X.; Yuan, W.J. Effect of steel slag on shrinkage characteristics of calcium sulfoaluminate cement. *Mater. Sci. Forum* **2021**, *1036*, 263–276. [CrossRef]

68. Cui, P.; Wan, Y.; Shao, X.; Ling, X.; Zhao, L.; Gong, Y.; Zhu, C. Study on Shrinkage in Alkali-Activated Slag–Fly Ash Cementitious Materials. *Materials* **2023**, *16*, 3958. [CrossRef]

69. Deng, P.; Zheng, Z. Mechanical properties of one-part geopolymer masonry mortar using alkali-fused lead–zinc tailings. *Constr. Build. Mater.* **2023**, *369*, 130522. [CrossRef]

70. Zhang, J.; Guan, X.; Li, H.; Liu, X. Performance and hydration study of ultra-fine sulfoaluminate cement-based double liquid grouting material. *Constr. Build. Mater.* **2017**, *132*, 262–270. [CrossRef]

71. Guo, L.; Zhou, M.; Wang, X.; Li, C.; Jia, H. Preparation of coal gangue-slag-fly ash geopolymer grouting materials. *Constr. Build. Mater.* **2022**, *328*, 126997. [CrossRef]

72. Ruan, W.; Li, F.; Liao, J.; Gu, X.; Mo, J.; Shen, Y.; Zhu, Y.; Ma, X. Effects of water purifying material waste on properties and hydration mechanism of magnesium phosphate cement-based grouting materials. *Constr. Build. Mater.* **2022**, *349*, 128676. [CrossRef]

73. Duan, D.; Wu, H.; Wei, F.; Song, H.; Ma, Z.; Chen, Z.; Cheng, F. Preparation, characterization, and rheological analysis of eco-friendly geopolymer grouting cementitious materials based on industrial solid wastes. *J. Build. Eng.* **2023**, *78*, 107451. [CrossRef]

74. Zhang, H.; Dong, J.; Wei, C. Material properties of cement doped with carbonated steel slag through the slurry carbonation process: Effect and quantitative model. *Metall. Mater. Trans. B* **2022**, *53*, 1681–1690. [CrossRef]

75. Cao, H.; Gao, Q.; Zhang, X.; Guo, B. Research progress and development direction of filling cementing materials for filling mining in iron mines of China. *Gels* **2022**, *8*, 192. [CrossRef]

76. Xue, G.; Yilmaz, E.; Song, W.; Cao, S. Compressive strength characteristics of cemented tailings backfill with alkali-activated slag. *Appl. Sci.* **2018**, *8*, 1537. [CrossRef]

77. Sha, F.; Liu, P.; Ding, Y. Application investigation of high-phosphorus steel slag in cementitious material and ordinary concrete. *J. Mater. Res. Technol.* **2021**, *11*, 2074–2091. [CrossRef]

78. Deng, D.; Cao, G.; Liang, Y. Experimental Study on Lead-Smelting Slag as Paste Filling Cementing Material. *Geofluids* **2022**, *2022*, 6126881. [CrossRef]

79. Li, J.-F.; Liu, Y.-T.; Li, S.-J.; Song, Y. Experimental investigation of synchronous grouting material prepared with different mineral admixtures. *Materials* **2022**, *15*, 1260. [CrossRef] [PubMed]

80. Lahalle, H.; Benavent, V.; Trincal, V.; Wattez, T.; Bucher, R.; Cyr, M. Robustness to water and temperature, and activation energies of metakaolin-based geopolymer and alkali-activated slag binders. *Constr. Build. Mater.* **2021**, *300*, 124066. [CrossRef]

81. Bai, T.; Liang, Y.; Li, C.; Jiang, X.; Li, Y.; Chen, A.; Wang, H.; Xu, F.; Peng, C. Application and validation of fly ash based geopolymer mortar as grouting material in porous asphalt concrete. *Constr. Build. Mater.* **2022**, *332*, 127154. [CrossRef]

82. Wan, Y.; Hui, X.; He, X.; Xue, J.; Feng, D.; Chen, Z.; Li, J.; Liu, L.; Xue, Q. Utilization of flue gas desulfurization gypsum to produce green binder for dredged soil solidification: Strength, durability, and planting performance. *J. Clean. Prod.* **2022**, *367*, 133076. [CrossRef]

83. Sha, F.; Li, S.; Liu, R.; Li, Z.; Zhang, Q. Experimental study on performance of cement-based grouts admixed with fly ash, bentonite, superplasticizer and water glass. *Constr. Build. Mater.* **2018**, *161*, 282–291. [CrossRef]

84. Zhu, C.; Wan, Y.; Wang, L.; Ye, Y.; Yu, H.; Yang, J. Strength Characteristics and Microstructure Analysis of Alkali-Activated Slag–Fly Ash Cementitious Material. *Materials* **2022**, *15*, 6169. [CrossRef]

85. Liu, J.-H.; Zhou, Z.-B.; Wu, A.-X.; Wang, Y.-M. Preparation and hydration mechanism of low concentration Bayer red mud filling materials. *Chin. J. Eng.* **2020**, *42*, 1457–1464.

86. Zhang, W.; Zhu, X.; Xu, S.; Wang, Z.; Li, W. Experimental study on properties of a new type of grouting material for the reinforcement of fractured seam floor. *J. Mater. Res. Technol.* **2019**, *8*, 5271–5282. [CrossRef]

87. Carvalho, V.R.; Costa, L.C.B.; da Fonseca Elói, F.P.; da Silva Bezerra, A.C.; de Carvalho, J.M.F.; Peixoto, R.A.F. Performance of low-energy steel slag powders as supplementary cementitious materials. *Constr. Build. Mater.* **2023**, *392*, 131888. [CrossRef]

88. Li, C.; Hao, Y.F.; Zhao, F.Q. Preparation of fly ash-granulated blast furnace slag-carbide slag binder and application in total tailings paste backfill. *IOP Conf.* **2018**, *322*, 042003. [CrossRef]

89. Ehsani, A.; Nili, M.; Shaabani, K. Effect of nanosilica on the compressive strength development and water absorption properties of cement paste and concrete containing Fly Ash. *KSCE J. Civ. Eng.* **2017**, *21*, 1854–1865. [CrossRef]

90. Rehman, S.; Iqbal, S.; Ali, A. Combined influence of glass powder and granular steel slag on fresh and mechanical properties of self-compacting concrete. *Constr. Build. Mater.* **2018**, *178*, 153–160. [CrossRef]

91. Xue, G.; Fu, Q.; Xu, S.; Li, J. Macroscopic mechanical properties and microstructure characteristics of steel slag fine aggregate concrete. *J. Build. Eng.* **2022**, *56*, 104742. [CrossRef]

92. Gopinathan, S.; Anand, K.B. Properties of cement grout modified with ultra-fine slag. *Front. Struct. Civ. Eng.* **2018**, *12*, 58–66. [CrossRef]

93. Liu, J.; Guo, R. Applications of Steel Slag Powder and Steel Slag Aggregate in Ultra-High Performance Concrete. *Adv. Civ. Eng.* **2018**, *2018*, 1426037. [CrossRef]

94. Wei, X.; Ni, W.; Zhang, S.; Wang, X.; Li, J.; Du, H. Influence of the key factors on the performance of steel slag-desulphurisation gypsum-based hydration-carbonation materials. *J. Build. Eng.* **2022**, *45*, 103591. [CrossRef]

95. Li, Y.; Min, X.; Ke, Y.; Liu, D.; Tang, C. Preparation of red mud-based geopolymer materials from MSWI fly ash and red mud by mechanical activation. *Waste Manag.* **2019**, *83*, 202–208. [CrossRef]

96. Zhang, M.H. Microstructure, crack propagation, and mechanical properties of cement pastes containing high volumes of fly ashes. *Cem. Concr. Res.* **1995**, *25*, 1165–1178. [CrossRef]

97. Liu, Z.; Ni, W.; Li, Y.; Ba, H.; Li, N.; Ju, Y.; Zhao, B.; Jia, G.; Hu, W. The mechanism of hydration reaction of granulated blast furnace slag-steel slag-refining slag-desulfurization gypsum-based clinker-free cementitious materials. *J. Build. Eng.* **2021**, *44*, 103289. [CrossRef]

98. Liu, X.; Rong, Y.; Chen, X.; Chen, X.; Zhang, W. Recycling of Waste Stone Powder in High Fluidity Grouting Materials for Geotechnical Engineering Reinforcement. *Buildings* **2022**, *12*, 1887. [CrossRef]

99. Chen, Z.; Qiao, J.; Yang, X.; Sun, Y.; Sun, D. A review of grouting materials for pouring semi-flexible pavement: Materials, design and performance. *Constr. Build. Mater.* **2023**, *379*, 131235. [CrossRef]

100. Li, S.; Zhang, J.; Li, Z.; Liu, C.; Chen, J. Feasibility study on grouting material prepared from red mud and metallurgical wastewater based on synergistic theory. *J. Hazard. Mater.* **2021**, *407*, 124358. [CrossRef] [PubMed]

101. Moghadam, M.J.; Ajalloeian, R.; Hajiannia, A. Preparation and application of alkali-activated materials based on waste glass and coal gangue: A review. *Constr. Build. Mater.* **2019**, *221*, 84–98. [CrossRef]

102. Xu, J.; Kang, A.; Wu, Z.; Xiao, P.; Gong, Y. Effect of high-calcium basalt fiber on the workability, mechanical properties and microstructure of slag-fly ash geopolymer grouting material. *Constr. Build. Mater.* **2021**, *302*, 124089. [CrossRef]

103. ElKhatib, L.W.; Elkordi, A.; Khatib, J. Methods and surface materials repair for concrete structures—A review. *BAU J. Sci. Technol.* **2023**, *4*, 7. [CrossRef]

104. Xiang, J.; Liu, L.; Cui, X.; He, Y.; Zheng, G.; Shi, C. Effect of limestone on rheological, shrinkage and mechanical properties of alkali—Activated slag/fly ash grouting materials. *Constr. Build. Mater.* **2018**, *191*, 1285–1292. [CrossRef]

105. Wu, H.; Huang, B.; Shu, X.; Yin, J. Utilization of solid wastes/byproducts from paper mills in Controlled Low Strength Material (CLSM). *Constr. Build. Mater.* **2016**, *118*, 155–163. [CrossRef]

106. He, J.; Bu, X.; Bai, W.; Zheng, W.; Gao, Q.; Wang, Y. Preparation and properties of self-compacting alkali-activated slag repair mortar. *Constr. Build. Mater.* **2020**, *252*, 119034. [CrossRef]

107. Li, X.; Hao, J. New preparation of super-early-strength grouting materials by ternary complex system. *Adv. Cem. Res.* **2018**, *30*, 139–147. [CrossRef]

108. Borçato, A.G.; Thiesen, M.; Medeiros-Junior, R.A. Mechanical properties of metakaolin-based geopolymers modified with different contents of quarry dust waste. *Constr. Build. Mater.* **2023**, *400*, 132854. [CrossRef]

109. Li, M.; Gu, K.; Chen, B. Effects of flue gas desulfurization gypsum incorporation and curing temperatures on magnesium oxysulfate cement. *Constr. Build. Mater.* **2022**, *349*, 128718. [CrossRef]

110. Permeh, S.; Lau, K. Review of Electrochemical Testing to Assess Corrosion of Post-Tensioned Tendons with Segregated Grout. *Constr. Mater.* **2022**, *2*, 70–84. [CrossRef]

111. Bentz, D.P.; Garboczi, E.J. Modelling the leaching of calcium hydroxide from cement paste: Effects on pore space percolation and diffusivity. *Mater. Struct.* **1992**, *25*, 523–533. [CrossRef]

112. Sun, C.; Hui, R.; Qu, W.; Yick, S. Progress in corrosion resistant materials for supercritical water reactors. *Corros. Sci.* **2009**, *51*, 2508–2523. [CrossRef]

113. Zhao, Y.; Taheri, A.; Karakus, M.; Chen, Z.; Deng, A. Effects of water content, water type and temperature on the rheological behaviour of slag-cement and fly ash-cement paste backfill. *Int. J. Min. Sci. Technol.* **2020**, *30*, 271–278. [CrossRef]

114. Weldes, H.H.; Lange, K.R. Properties of soluble silicates. *Ind. Eng. Chem.* **1969**, *61*, 29–44. [CrossRef]

115. Gebauer, J.; Coughlin, R.W. Preparation, properties & corrosion resistance of composites of cement mortar and organic polymers. *Cem. Concr. Res.* **1971**, *1*, 187–210.

116. Wang, F.M.; Shi, M.S.; Li, H.J.; Zhong, Y.H. Experimental Study on the Anti-Permeability Properties of Polymer Grouting Materials. *Adv. Mater. Res.* **2011**, *284–286*, 1952–1955. [CrossRef]

117. Mangat, P.S.; Khatib, J.M. Influence of Fly Ash, Silica Fume, and Slag on Sulfate Resistance of Concrete. *ACI Mater. J.* **1993**, *92*, 542–552.

118. Opdyke, S.M.; Evans, J.C. Slag-cement-bentonite slurry walls. *J. Geotech. Geoenviron. Eng.* **2005**, *131*, 673–681. [CrossRef]

119. Zhang, S.; Wang, C.; Gao, G. Study of the Mechanical Properties of Alkali-Activated Solid Waste Cementitious Materials at the Interface. *ACS Omega* **2022**, *7*, 26531–26536. [CrossRef] [PubMed]

120. Zhao, F.-Q.; Liu, H.-J.; Hao, L.-X.; Li, Q. Water resistant block from desulfurization gypsum. *Constr. Build. Mater.* **2012**, *27*, 531–533. [CrossRef]

121. Yang, K.-H.; Song, J.-K.; Lee, J.-S. Properties of alkali-activated mortar and concrete using lightweight aggregates. *Mater. Struct.* **2010**, *43*, 403–416. [CrossRef]

122. Li, S.; Zhang, J.; Li, Z.; Gao, Y.; Qi, Y.; Li, H.; Zhang, Q. Investigation and practical application of a new cementitious anti-washout grouting material. *Constr. Build. Mater.* **2019**, *224*, 66–77. [CrossRef]
123. Porbaha, A.; Shibuya, S.; Kishida, T. State of the art in deep mixing technology. Part III: Geomaterial characterization. *Proc. Inst. Civ. Eng. Ground Improv.* **2000**, *4*, 91–110. [CrossRef]
124. Ruan, W.; Liao, J.; Mo, J.; Li, F.; Gu, X.; Ma, Y.; Zhu, Y.; Ma, X. Effects of red mud on properties of magnesium phosphate cement-based grouting material and its bonding mechanism with coal rock. *Ceram. Int.* **2023**, *49*, 2015–2025. [CrossRef]
125. Sha, F.; Fan, R.; Gu, S.; Xi, M. Strengthening effect of sulphoaluminate cementitious grouting material for water-bearing broken rocky stratum. *Constr. Build. Mater.* **2023**, *368*, 130390. [CrossRef]
126. Carde, C.; François, R.; Torrenti, J.-M. Leaching of both calcium hydroxide and CSH from cement paste: Modeling the mechanical behavior. *Cem. Concr. Res.* **1996**, *26*, 1257–1268. [CrossRef]
127. Yoshikawa, M.; Toda, T. In vivo estimation of periapical bone reconstruction by chondroitin sulfate in calcium phosphate cement. *J. Eur. Ceram. Soc.* **2004**, *24*, 521–531. [CrossRef]
128. Kourounis, S.; Tsivilis, S.; Tsakiridis, P.E.; Papadimitriou, G.D.; Tsibouki, Z. Properties and hydration of blended cements with steelmaking slag. *Cem. Concr. Res.* **2007**, *37*, 815–822. [CrossRef]
129. Karakurt, C.; Bayazıt, Y. Freeze-Thaw Resistance of Normal and High Strength Concretes Produced with Fly Ash and Silica Fume. *Adv. Mater. Sci. Eng.* **2015**, *2015*, 830984. [CrossRef]
130. Liu, Y.; Zhou, X.; Lv, C.; Yang, Y.; Liu, T. Use of Silica Fume and GGBS to Improve Frost Resistance of ECC with High-Volume Fly Ash. *Adv. Civ. Eng.* **2018**, *2018*, 7987589. [CrossRef]
131. Aiken, T.A.; Kwasny, J.; Sha, W.; Tong, K.T. Mechanical and durability properties of alkali-activated fly ash concrete with increasing slag content. *Constr. Build. Mater.* **2021**, *301*, 124330. [CrossRef]
132. Hou, D.; Li, D.; Hua, P.; Jiang, J.; Zhang, G. Statistical modelling of compressive strength controlled by porosity and pore size distribution for cementitious materials. *Cem. Concr. Compos.* **2019**, *96*, 11–20. [CrossRef]
133. Vegas, I.; Urreta, J.; Frías, M.; García, R. Freeze–thaw resistance of blended cements containing calcined paper sludge. *Constr. Build. Mater.* **2009**, *23*, 2862–2868. [CrossRef]
134. Slavik, R.; Bednarik, V.; Vondruska, M.; Nemec, A. Preparation of geopolymer from fluidized bed combustion bottom ash. *J. Mater. Process. Technol.* **2008**, *200*, 265–270. [CrossRef]
135. Wu, W.; Ma, S.; Wang, Y.; Wang, X.; Xu, L.; Ye, S. Evaluation of Sulfate Resistance of One-Part Alkali-Activated Materials Prepared by Mechanochemistry. *ACI Mater. J.* **2023**, *120*, 37–51.
136. Kanaan, D. *Developing of One-Part Alkali-Activated Self-Consolidating Concrete and Investigating Its Sulfate Attack Resistance*; Concordia University: Montreal, QC, Canada, 2021.
137. Fernández-Jiménez, A.; García-Lodeiro, I.; Palomo, A. Durability of alkali-activated fly ash cementitious materials. *J. Mater. Sci.* **2007**, *42*, 3055–3065. [CrossRef]
138. Luo, X.; Xu, J.; Bai, E.; Li, W. Systematic study on the basic characteristics of alkali-activated slag-fly ash cementitious material system. *Constr. Build. Mater.* **2012**, *29*, 482–486. [CrossRef]

Article

Experimental Study and Application of Controlled Low-Strength Materials in Trench Backfilling in Suqian City, China

Jingmin Xu [1,2,*], Qiwu Luo [1], Yong Tang [1], Zhibo Zeng [1] and Jun Liao [1]

1   China Construction Fifth Engineering Division Corp., Ltd., Changsha 410019, China
2   School of Transportation, Southeast University, Nanjing 211189, China
*   Correspondence: jingmin_xu@seu.edu.cn

Abstract: When backfilling narrow spaces, controlled low-strength materials (CLSM) can be used to achieve an effective backfilling effect. The pipeline engineering in Yahnghe Avenue of Suqian, China, provides a favorable on-site condition for the use of CLSM. However, no guidance exists for the determination of the material mixture ratio of CLSM for this geological condition. Laboratory tests were performed to investigate the basic physical parameters of excavated soil and the optimal mixture ratio of CLSM. Results indicate that the sand and silt account for 29.76% and 57.23% of the weight of excavated soil, respectively. As the water content increases (from 40% to 50%), the flowability of the CLSM approximately shows a linear increase (slumps values from 154.3 mm to 269.75 mm for 9% cement content), while its compressive strength shows a linear decreasing trend (from 875.3 KPa to 468.3 KPa after curing for 28 days); as the cement content increases (from 6% to 12%), the flowability approximately shows a linear decreasing trend (from 238.8 mm to 178.5 mm for 45% water content), while the compressive strength shows a linear increasing trend (from 391.6 KPa to 987.6 KPa after curing for 28 days). By establishing the relationship between compressive strength/flowability and the water–cement ratio, the optimal material ratio is determined to be 9% cement content and 40–43% water content. The engineering application results indicate that the use of CLSM can achieve efficient and high-quality backfilling effects for pipeline trenches. The findings of this research may provide a reference for the application of CLSM in fields with similar geological conditions.

Keywords: trench backfilling; controlled low-strength materials (CLSM); flowability; compressive strength

Citation: Xu, J.; Luo, Q.; Tang, Y.; Zeng, Z.; Liao, J. Experimental Study and Application of Controlled Low-Strength Materials in Trench Backfilling in Suqian City, China. *Materials* **2024**, *17*, 775. https://doi.org/10.3390/ma17040775

Academic Editor: Dimitrios Papoulis

Received: 10 December 2023
Revised: 2 February 2024
Accepted: 3 February 2024
Published: 6 February 2024

1. Introduction

In urban construction projects, the use of solid materials to backfill trenches is a common backfilling method. However, narrow spaces exist between excavation faces and structures (for instance, pipelines), and those spaces are difficult to backfill and compact. This can easily cause settlement and collapse during road operation, leading to the failure of the subgrade [1]. In addition, with traditional backfilling techniques it is also difficult to meet the requirements of rapid construction conditions and high backfilling quality [1].

To solve the above-mentioned engineering problems, a new type of fluidized cement–soil backfill technology known as the use of controlled low-strength materials (CLSM), is gradually being adopted in engineering [2]. According to the American Concrete Institute [3], CLSM is defined as a new backfilling material with high flowability, suitable compressive strength, and that can be self-compacted with less vibration under the action of self-weight. This technology mainly involves adding an appropriate amount (6–15%) of binding material (for instance, cement and fly ash), water, and/or curing agent to a specific soil mass, and they are mixed evenly to form a mixture with a certain fluidity. It is found that the CLSM adopts a higher proportion of cement (or fly ash) than other mixtures, such as the full-depth reclamation of pavements with cement [4], soil–cement [5] or cement-treated mixtures of RAP [6]. Then, the mixture is poured into the trench or other areas that need to be backfilled using a chute or pump. After solidification, a backfill body with a certain strength is formed. Therefore, the waste soil generated from pipeline excavation can be used to prepare CLSM on-site and then backfilled in the trench. This can simultaneously solve three engineering problems: soil treatment, procurement of filling material, and backfilling construction. The compressive strength and flowability of CLSM are important parameters [3]. Suitable flowability is crucial for self-compacting and self-leveling. The compressive strength in the early stage of curing should be sufficient for the resumption of traffic, and the final compressive strength should not be too high given the potential need for re-excavation in the future [7–9]. Therefore, it is of practical importance to study the key parameters governing the performance of CLSM and its application in engineering.

Due to numerous advantages, the performance and application of CLSM have been investigated extensively. For instance, Kaneshiro et al. [9] explored the use of controlled low-strength material in the backfilling of a pipeline trench. The authors selected the particle size distribution, water-to-solid ratio and binder-to-aggregate ratio as the performance control parameters and experimentally assessed the effects of these factors on the physical and mechanical properties of CLSMs. The obtained optimal mixture ratio was tested in a field test, demonstrating the feasibility of the proposed CLSM for trench backfilling. Finney et al. [10] reported four case studies in Northern California where native soils were used exclusively to produce CLSM for pipeline backfilling. The authors discussed the effects of the differences between native soils on the performance of CLSM and designed a specific mixture ratio for each project. Based on 115 literature articles, Ling et al. [2] summarized that the materials used for the production of CLSM varied from case to case, which in turn has a significant impact on the performance of CLSM and its application in the field. Similarly, Dalal et al. [11] gave a detailed discussion on the recent advances in the development of CLSM with different waste materials, and the effects of material properties on the flowability, strength, and hardening time of CLSM. Kaliyavaradhan et al. [12] presented a review on the strength and excavatability of strength material CLSM, and highlighted the need to consider the supply, transport of materials, and material properties before being used in engineering practice.

In summary, it can be seen that there have been many studies on the application of CLSM, indicating its potential advantages in construction projects. Although Blanco et al. [13] proposed a preliminary methodology for the design of optimized CLSM combining FEM simulation and experimental testing, the design of CLSM is primarily based on experimental approaches [14–19]. For instance, Fauzi et al. [14] adopted laboratory testing and an empirical response surface method to evaluate the effect of key parameters on the properties of CLSM mixes produced with unprocessed fly ash and recycled fine aggregate. The authors proposed mix design models to reveal the relationships between the mixing parameters and the CLSM properties. A review by Ibrahim et al. [15] highlighted the importance of cement content, curing time, and temperature on the strength of CLSM based on experimental results. Nevertheless, no guidance was provided to determine a suitable mixture ratio. Kuo et al. [16] experimentally studied the hardened properties and the durability of CLSM mixed with waste oyster shells, cement, sand, and fine aggregate, suggesting that waste oyster shells are effective in the replacement of sand. Türkel [18] adopted direct shear testing to assess the shear strength of CLSM materials (a mixture of high-volume fly ash, crushed limestone powder, and pozzolana cement), and concluded that CLSM mixtures have excellent shear strength properties compared to compacted soil. Importantly, Wang et al. [19] conducted a comprehensive review of the mechanical properties of green CLSM and have summarized the effects of the binding material types and contents, water content, and admixture on the compressive strength of CLSM. Results demonstrate the complex interplay between those parameters and the properties of CLSM and the highly non-linear influence of key parameters on the compressive strength of CLSM. These results suggest that numerous factors affect the strength of CLSM. In other words, there is still no effective instruction to determine the optimized control parameters in specific engineering applications; for instance, Suqian Yanghe Avenue, China.

In Suqian City, China, the excavated soil for the drainage pipeline trench of the YHDD-TJ1 section of Yanghe Avenue is high moisture content silt soil, which is not suitable for direct trench backfilling. To utilize the excavated soils and save the cost of backfilling material procurement, the excavated silt soil was used to produce CLSM for backfilling pipeline trenches. To determine the mixture ratio of CLSM and to ensure safe backfilling, an experimental study and field testing were conducted. First, the physical properties of the excavated soil were tested. This was followed by the experimental investigation of the effects of key control parameters (cement content, water content, and curing time) on the performance (flowability and compressive strength) of CLSM in Yanghe Avenue. Then, key control parameters were optimized using the experimental data. Finally, the obtained CLSM technique was tested in Yanghe Avenue of Suqian City, achieving an efficient backfilling effect. This paper provides a unique dataset of the properties of the CLSM materials for the high moisture content silt soil, which is spread throughout the East China region. The results may be useful for the further application of CLSM technology under similar soil conditions.

2. Materials and Methods

The YHDD-TJ1 section of the Yanghe Avenue Reconstruction Project in Suqian City, China has a total length of 6.881 km, starting from S325 Provincial Road and ending in Kangcheng Road. This section of the drainage project involved a large number of open-cut pipeline construction activities. After the pipeline was installed, the closed water test was conducted, and then the trench backfilling was carried out, as shown in Figure 1. The on-site investigation results show that the excavated soil from the Yanghe Avenue project pipeline trench is high moisture content silty soil. If the trench was directly backfilled using the excavated soil, the backfilling effect in narrow spaces would be poor. Therefore, CLSM technology was used in this project.

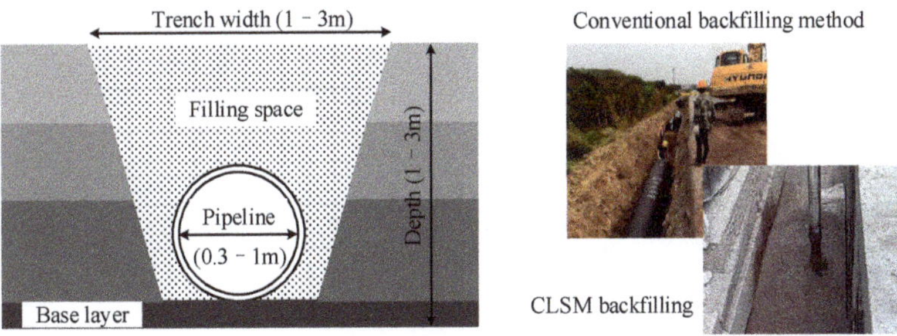

Figure 1. Illustration of the filling space and filling methods in the pipeline engineering.

The performance of CLSM has a significant impact on the successful application of this technology. The flowability of CLSM is important. On the one hand, greater flowability can ensure a better filling effect in narrow spaces. On the other hand, the higher the water content, the lower the strength of the CLSM after curing. If the strength is lower than the threshold, it may not meet the engineering requirement for the strength of the subgrade [3]. Therefore, it is necessary to have a good balance between the water content and cement content to meet both strength and flowability requirements. Based on extensive field-testing experiences, the flowability of CLSM within the range of 150–200 mm can meet the requirements of pumping and achieve self-compacting filling in narrow spaces. The strength of CLSM after solidification and the time to reach its initial setting strength need to be determined according to specific engineering requirements.

3. Experimental Design

3.1. Basic Physical Properties of the Excavated Soil

The excavated silt soil in the typical road sections was selected and tested using a densimeter and laser particle analyzer Microtrac S3500 (Microtrac MRB, York, PA, USA) to obtain the grain composition of the silt soil. The basic physical properties, including organic matter content and water content, of the excavated soil were measured through weighing. The particle size distribution is shown in Figure 2 and Table 1.

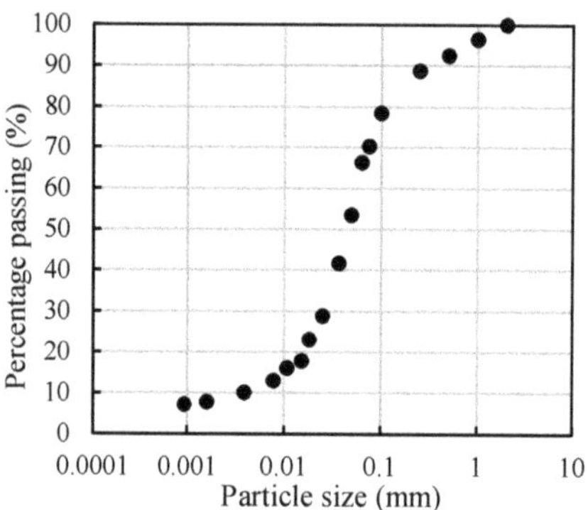

Figure 2. Particle size distribution curve.

Table 1. Particle size distribution of silty sand.

Particle (mm)	>0.25	0.25~0.075	0.075~0.005	<0.005
Percentage (%)	11.16	18.6	57.23	13.01

Table 1 shows that the sand content of the soil sample is 29.76%, and the silt content is 57.23%. It is calculated that the non-uniformity coefficient Cu of the grading index is 16.35 and the curvature coefficient Cc is 41.79. These results meet the condition of Cu $\geq$ 5, but do not satisfy the condition of Cc = 1–3. According to "Regulations for Highway Geotechnical Testing" [20], the particle size distribution of the soil sample in this study is between silty soil and sandy soil. Based on the testing results, it is found that the organic matter content Wu of the excavated soil is 0.28%, which is much smaller than 5% (having negligible effects on the production of CLSM). According to the classification of the soil, it can be classified as inorganic soil. Finally, the experimental results also show that the average water content of the excavated soil is about 10%.

3.2. Materials

To simplify the production procedure of CLSM, cement is used as the binding material. Therefore, water and cement were added to the excavated soil, and they were mixed evenly to prepare the CLSM. To investigate the effects of cement content, water content and curing time on the performance of the CLSM, the testing plan shown in Table 2 was designed. The cement content, which is defined as the ratio of the weight of cement to the weight of dry soil, is between 6% and 12%; water content, defined as the ratio of the weight of water to that of dry soil, is between 40% and 50%. The curing time of CLSM is set as 3 days, 7 days, and 28 days. Therefore, a total of 27 combinations were tested.

Table 2. Cement soil flow proportioning scheme.

Cement Content (%)	Water Content (%)	Curing Time (%)
6	40, 45, 50	3, 7, 28
9	40, 45, 50	3, 7, 28
12	40, 45, 50	3, 7, 28

3.3. Flowability Tests

To test the flowability of the CLSM, the trumpet shaped slump cone (see Figure 3a) with an upper opening of 100 mm, a lower opening of 200 mm, and a height of 300 mm, was used, along with 1 measuring cylinder, 1 tamping rod, and 2 steel rulers, each with a measuring range exceeding 300 mm. The mixture of the CLSM was prepared according to Table 2 and three parallel experiments on each group of ratios were conducted. The average value was used as the final result of flowability.

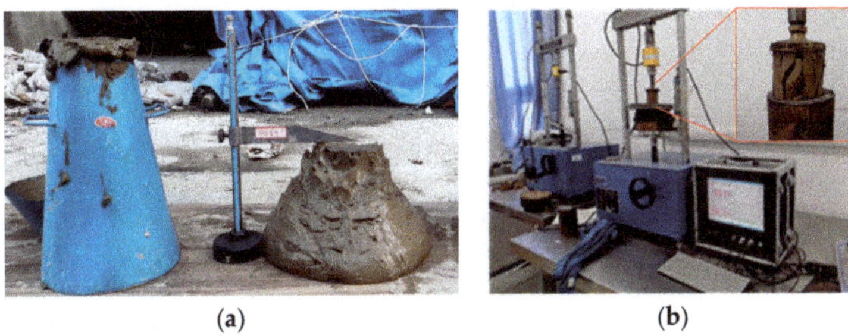

(**a**) (**b**)

Figure 3. (**a**) Slump cone and (**b**) CBR-2 testing equipment with damaged specimen.

3.4. Compressive Strength Testing

The CLSM was prepared according to Table 2, and specimens were made using man-made molds. The unconfined compressive strength of the CLSM specimens was tested using CBR-2 California bearing ratio equipment, as shown in Figure 3b. The unconfined compressive strength of the sample is defined as q_{u}, calculated as follows:

$$q_{\mathrm{u}} = F/A \tag{1}$$

where q_{u} is the unconfined compressive strength of the specimen (kPa); F is the maximum load at which the specimen fails (kN); and A is the cross-sectional area of the specimen (m^2).

In this paper, three parallel experiments on each group were performed and the average value was used. Therefore, a total of $27 \times 3 = 81$ samples were tested.

4. Experimental Results

The average unconfined compressive strength values and flowability values are presented in Table 3. To evaluate the repeatability and credibility of the results, relative standard deviation (RSD) values were also calculated. Table 3 shows that RSD values for all the data are below 10%, which is acceptable for this type of experiment.

Table 3. Experimental results for unconfined compressive strength, flowability, and RSD.

Water (%)	Cement (%)	3 Days q_{u} (KPa), RSD (%)	7 Days q_{u} (KPa), RSD (%)	28 Days q_{u} (KPa), RSD (%)	Flowability Value (mm), RSD (%)
	6	302, 1.00	347.3, 6.81	588.6, 4.02	173.5, 6.56
40	9	548, 5.67	637.6, 8.89	875.3, 4.02	154.25, 4.31
	12	871.3, 9.98	1043.6, 9.79	1230.3, 0.94	130.5, 4.74

Table 3. *Cont.*

Water (%)	Cement (%)	3 Days q_u (KPa), RSD (%)	7 Days q_u (KPa), RSD (%)	28 Days q_u (KPa), RSD (%)	Flowability Value (mm), RSD (%)
	6	261.3, 9.06	291.6, 4.95	391.6, 9.49	238.75, 1.05
45	9	390.3, 3.70	454.6, 6.94	560, 9.57	212, 4.74
	12	541.3, 4.37	624.3, 2.31	987.6, 2.69	178.5, 6.51
	6	171.3, 4.46	186.3, 10.39	275, 1.08	279.75, 1.02
50	9	191.7, 10.62	247.3, 4.67	468.3, 4.38	269.75, 1.30
	12	397, 6.54	448, 9.15	718.6, 10.33	240.25, 1.04

4.1. Flowability of CLSM

Figure 4 shows the variation of the flowability of the CLSM with respect to the water content. As the water content increases, the flowability of the CLSM approximately increases linearly. For instance, when the water content increases from 40% to 45%, the increase in flowability in all groups (6%, 9%, and 12% cement contents) of mixtures is about 38%; when the water content increases from 45% to 50%, the increments of flowability in three groups of mixtures are 18%, 27%, and 35%. This is because the dispersion effect of free water reduces the frictional resistance between solid particles and forms a lubricating water film that exists between solid particles. As the free water increases, the thickness of the water film also continuously increases, thus improving the flowability of the CLSM.

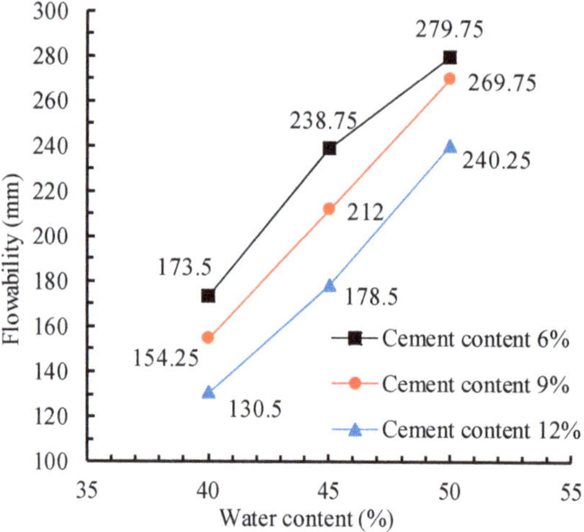

Figure 4. Flowability of the CLSM based on water content.

When the cement content increases from 6% to 9%, the flowability of the CLSM decreases by about 11%. Among them, when the water content is 50%, the flowability value remains nearly unchanged when the cement content increases from 6% to 9%, indicating that there is sufficient water in the CLSM to undergo a hydration reaction with the cement. When the ash–soil ratio increases from 9% to 12%, the flowability values continuously decrease from 269.75 mm to 212 mm and then to 154.25 mm, with a decrease of 11.1%, 15.8%, and 15.4%, respectively. Those results broadly agree well with previous studies [2], which highlights the critical factors (water content and material properties) affecting the flowability of CLSM.

4.2. Compressive Strength of CLSM

4.2.1. Effects of Water Content

Figure 5 shows the influence of water content on the unconfined compressive strength of CLSM. It is shown that when the water content increases from 40% to 50%, the compressive strength of the CLSM at different curing ages shows a linear decreasing trend, all by around 50%. When the cement content is 6%, the strength of CLSM decreases by 43.3% to 53.3%; when the water–cement ratio is 9%, the decrease in the compressive strength of the CLSM ranges from 46.5% to 65%; when the cement content is 12%, the decrease in strength ranges from 41.6% to 57%. Wang et al. [19] also reported the dominate role of the water–cement ratio in the compressive strength of CLSM, which broadly agrees with the results of this study. This aspect will be discussed in a later section.

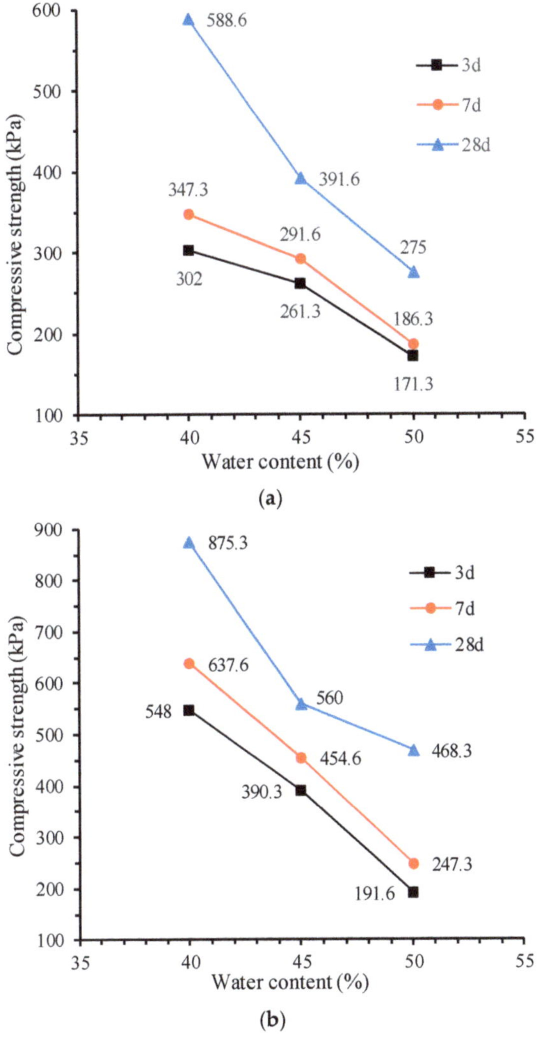

Figure 5. *Cont.*

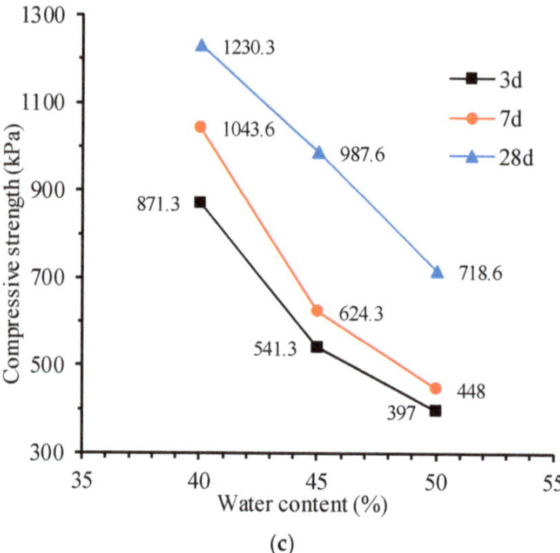

Figure 5. Influence of water content on the compressive strength of CLSM: (**a**) 6% cement content; (**b**) 9% cement content; and (**c**) 12% cement content.

During the experiment, it was found that when the water content was 50%, the water would separate from the mixture during the static process of the sample. At the same time, some samples showed volume shrinkage after curing, indicating that a high water content may bring difficulties in evenly mixing the CLSM. Special attention should be paid during on-site construction.

4.2.2. Effects of Cement Content

Figure 6 shows the effect of cement content on the compressive strength of CLSM. It is shown that as the cement content increases, the compressive strength of CLSM samples at the given curing age continues to increase. Moreover, when the cement content increases from 9% to 12%, the strength of CLSM specimens increases significantly. This is because when the cement content is small, the degree of solidification of cementitious soil is low, and at this point, the overall cementitious skeleton structure has not yet formed within the mixture; with the increase in cement content, the hydration products also continue to increase, and the cementitious skeleton structure is densified, which improves the compressive strength of cementitious soil (CLSM) samples. Again, the results are in good agreement with Wang et al. [19], who summarized that the strength of CLSM increases linearly with the cement–aggregates ratio.

4.2.3. Effects of Curing Time

Figure 7 shows the effect of curing time on the compressive strength of CLSM. As shown in Figure 7, the increase in compressive strength from 3 days to 7 days is limited, with a minimum increase of 8.76% and a maximum increase of 19.8%; the compressive strength increases significantly from 7 days to 28 days, with a minimum increase of 23.1% and a maximum increase of 69.5%. It is believed that the compressive strength of the CLSM sample increases approximately linearly with the logarithm of the curing time, as shown in Figure 7. As the curing time increases, the hydrolysis and hydration reaction of cement becomes more complete, increasing the number of hydration products and the strength of cementitious soil. In general, the compressive strength of cementitious soil will still increase after 28 days, and the trend of strength growth will gradually stabilize after exceeding 90 days [17–19].

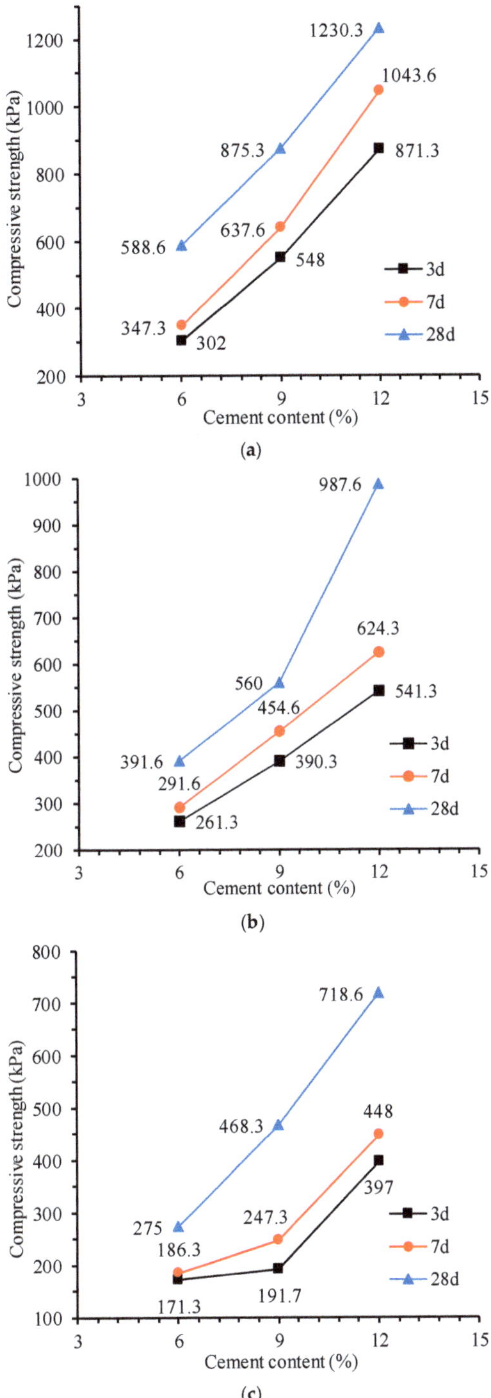

Figure 6. Influence of cement content on the compressive strength of CLSM: (**a**) 40% water content; (**b**) 45% water content; and (**c**) 50% water content.

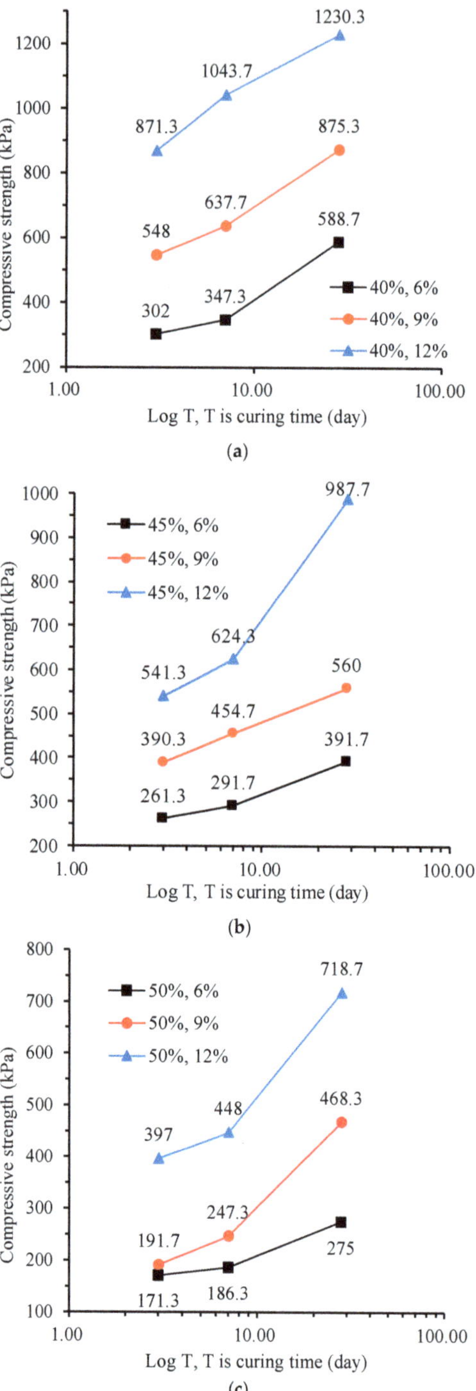

Figure 7. Influence of curing time on the compressive strength of CLSM: (**a**) 40% water content; (**b**) 45% water content; and (**c**) 50% water content.

4.3. Optimisation of the Ratio of Materials

The water–cement ratio refers to the ratio of the weight of water to that of cement. It is generally believed that the water–cement ratio is an important factor affecting the strength of cementitious soil [19]. Therefore, the compressive strength of CLSM is plotted against the water–cement ratio in Figure 8. Those results are the average values obtained from three parallel tests. Then, empirical equations can be established between the water–cement ratio and the unconfined compressive strength to determine the optimal material ratio of CLSM.

Based on the experimental results, the fitting equations for the relationship between the compressive strength at 3 days, 7 days, and 28 days and the water–cement ratio were obtained, as shown in Figure 8. The obtained equations are as follows:

(1)   For 6% cement content,

$$q_{u,3d} = -78.555m + 833.97 \tag{2}$$

$$q_{u,7d} = -96.795m + 1001 \tag{3}$$

$$q_{u,28d} = -188.64m + 1833.1; \tag{4}$$

(2)   For 9% cement content,

$$q_{u,3d} = -319.25m + 1972.5 \tag{5}$$

$$q_{u,7d} = -349.73m + 2194.8 \tag{6}$$

$$q_{u,28d} = -364.93m + 2458.7; \tag{7}$$

(3)   For 12% cement content,

$$q_{u,3d} = -564.68m + 2720.8 \tag{8}$$

$$q_{u,7d} = -709.13m + 3364.6 \tag{9}$$

$$q_{u,28d} = -609.13m + 3263.1 \tag{10}$$

where m is the water–cement ratio and $q_{u,3d}$, $q_{u,7d}$, and $q_{u,28d}$ are the compressive strength of the CLSM in the curing time of 3 days, 7 days, and 28 days, respectively. The linear relationship between the compressive strength and the water–cement ratio is broadly in good agreement with previous studies on CLSM summarized by Wang et al. [19]. However, it is worth noting that the above equations were obtained from the experimental data of the excavated soil in Suqian City with the water content between 40% and 50%. Therefore, care should be taken when using these equations in other cases.

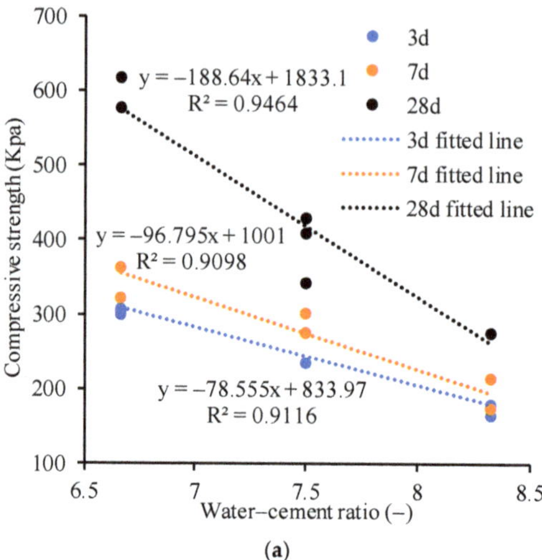

(**a**)

Figure 8. *Cont.*

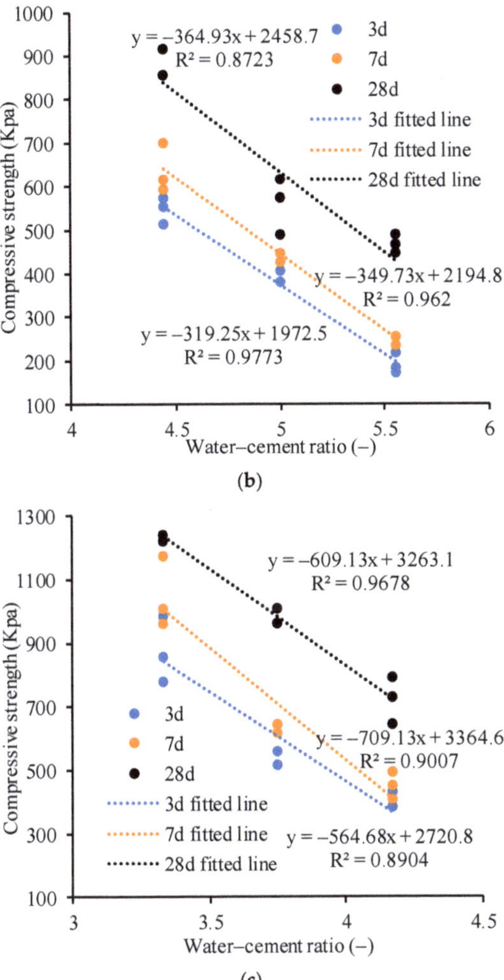

Figure 8. Influence of the water–cement ratio on the compressive strength of CLSM: (**a**) 6% cement content; (**b**) 9% cement content; and (**c**) 12% cement content.

Next, on the premise of meeting the compressive strength requirements, the flowability of CLSM also needs to satisfy the engineering requirements. By establishing the relationship between the water–cement ratio and flowability, as shown in Figure 9, the equations for 6%, 9%, and 12% cement content were obtained:

$$f_6 = 57.89m - 206.78 \tag{11}$$

$$f_9 = 103.49m - 307.75 \tag{12}$$

$$f_{12} = 130.65m - 306.87 \tag{13}$$

where f_6, f_9, and f_{12} are the flowability values (in mm) at 6%, 9%, and 12% cement content.

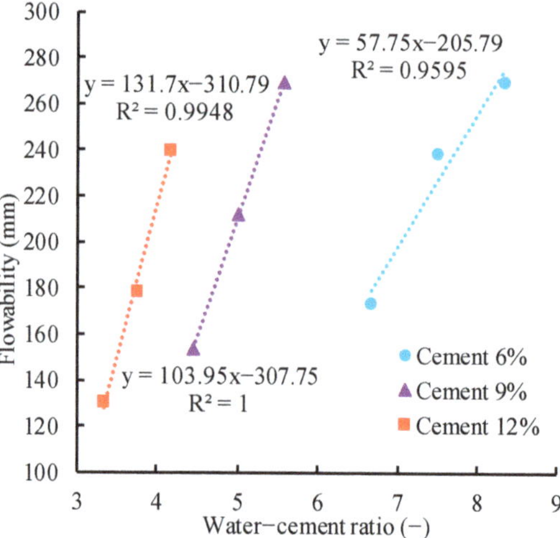

Figure 9. Relationship between water–cement ratio and flowability of CLSM.

According to the requirements of pipeline-trench backfilling engineering, combined with existing literature research and national standards [3,20], it is determined that the flowability of the CLSM should be within the range of 150 to 200 mm, and the unconfined compressive strength values at 3, 7, and 28 days should be no less than 200, 400, and 800 KPa.

Based on the above analysis and Equations, it is concluded that when the cement content is 9%, the water–cement ratio should be within the range of 4.40~4.8 (equivalent to the water content within the range of 39.6~43.2%). These parameters guarantee both the compressive strength and flowability of the CLSM to meet the engineering requirements. Furthermore, for the cement content of 12% and the water–cement ratio within the range of 3.5~3.88 (equivalent to the water content within the range of 42~46.6%), both the compressive strength and flowability of CLSM can also satisfy the engineering requirements. In order to reduce the amount of cement and to save engineering costs, a 9% cement content and a 40–43% water content were adopted in the Yanghe Avenue pipeline trench backfilling project.

5. Engineering Practice and Discussion
5.1. Preparation of and Backfilling with CLSM

The pipeline-trench backfilling project in the Yanghe Avenue (YHDD-TJ1 section) drainage engineering construction adopted the CLSM backfilling technology for trench backfilling. The CLSM was produced based on the material mixture ratio obtained from the experiments (9% cement content and 40–43% water content). Based on the strength and flowability characteristics of the CLSM, the equipment suitable for mixing and backfilling was selected, as shown in Figure 10.

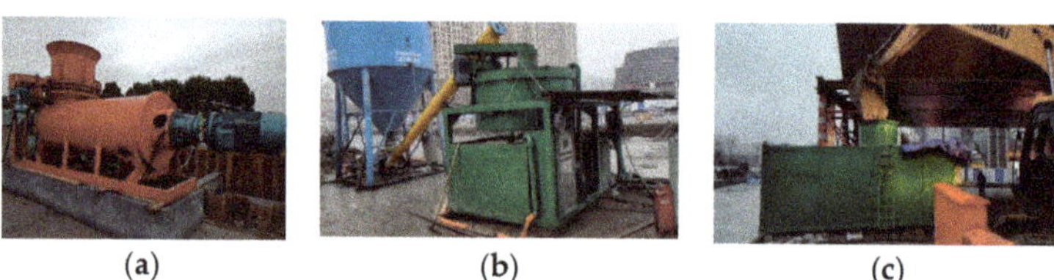

(a)   (b)   (c)

Figure 10. Equipment for the preparation and transportation of CLSM: (**a**) Soil metering device; (**b**) automated cement slurry preparation system; (**c**) intelligent CLSM mixing systems.

The construction equipment in the field mainly includes the following systems: (1) an automated cement slurry preparation system, which can achieve automatic and accurate weighing, automatic feeding according to the material proportion, and high-speed and efficient mixing and slurry production through settings; (2) the intelligent CLSM mixing system allows for a continuous feeding and mixing of materials. The soil filling speed can be adjusted through a frequency converter for various soil filling speeds; (3) the excavator is used to fill the soil, and stones with a diameter greater than 100 mm need to be removed; and (4) transportation equipment, which is mainly suitable for short distance pumping and pouring. Importantly, the automated cement-slurry preparation and intelligent CLSM mixing systems are crucial for the uniformity of the CLSM and backfilling quality. Those systems/facilities were specifically designed and manufactured for the preparation of CLSM and backfilling narrow spaces.

The main construction process can be summarized as follows: (1) preparation of raw materials, including determination of soil water content (alcohol combustion method, see Figure 11a), and sorting of debris such as stones; (2) flowability determination, which includes testing the flowability based on the optimal ratio combined with the natural water content of the soil material; (3) adjustment of the construction parameters of the mixing system, mainly by adjusting the frequency converter to control the filling speed of the cement slurry and soil materials; (4) cement soil mixing to form the CLSM; (5) the CLSM was transported to the site using a tubular chute, filling the trench under self-weight; (6) because the depth of the trench is around 1 m, trench backfilling was completed at one time point; and (7) the CLSM was levelled and immediately covered with plastic film after initial setting, as shown in Figure 11b.

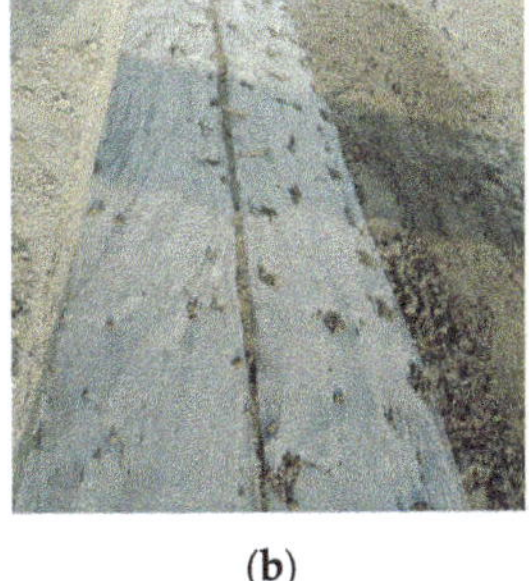

(a)   (b)

Figure 11. (**a**) Alcohol combustion method and (**b**) CLSM covered with plastic film.

5.2. Quality Detection

To test the construction quality and backfilling effect, samples were obtained by drilling and coring, and the strength of the samples was tested. The total length for sampling was 200 m, and five groups of samples were taken, with an interval of 50 m; three samples were taken from each group along the middle and boundary of the trench. The unconfined compressive strength of the samples after curing 3 days, 7 days, and 28 days was tested using CBR-2 testing equipment.

Experimental results show that the compressive strength values of the CLSM samples after 3 days, 7 days, and 28 days are all above 200, 400, and 800 KPa, respectively, and these results are in good agreement with the prediction of the proposed equations in Figure 8, indicating that the backfilling at different positions of the trench is relatively uniform and the strength of CLSM can fully meet the requirements of road engineering. The use of CLSM backfilling technology has achieved good backfilling effects for the trench backfilling in the YHDD-TJ1 section of the Yanghe Avenue Reconstruction Project in Suqian City, China.

5.3. Discussion

This study presents an application of CLSM in the trench backfilling project of Suqian, Yanghe Avenue, China. Cement was used as the only binding material, and the flowability and the comprehensive strength of CLSM were experimentally investigated. An optimal material mixture ratio was obtained, and the application result is satisfactory. The findings in this research may be useful for the trench backfilling community. However, in this study, limited samples were tested and only the compressive strength, flowability, and curing time were considered. The properties of CLSM were not exhaustively analyzed. Future work could consider the shear strength, porosity sorptivity, durability (dry–wet cycles, freeze–thaw cycles, and chemical corrosion), microstructure evolution, and chemical reaction mechanism.

6. Conclusions

This paper has presented an experimental and in situ testing study on the application of CLSM technology in the trench backfilling of the Yanghe Avenue of Suqian City, China. Cement was selected as the binding material. Laboratory tests and theoretical analysis were used to study the basic physical parameters of the excavated soil and the optimal material ratio of the CLSM. The backfilling quality of the trench in Yanghe Avenue of Suqian City, China, was also tested. Limitations are discussed and potential future work is suggested.

The flowability and unconfined compressive strength of CLSM are the most important properties. Experimental results show that they are strongly affected by water and cement content: as the water content increases, the flowability value of CLSM shows an approximately linear increase, while the unconfined compressive strength shows a linear decreasing trend; as the cement content increases, the flowability value of CLSM shows an approximately linear decrease, while the compressive strength linearly increases (for the same curing age). These results agree well with previous studies [21–23].

The relationships between the water–cement ratio of CLSM and the unconfined compressive strength/flowability were obtained. Based on the experimental data, the mixture ratio for producing CLSM in Suqian Yanghe Avenue was optimized. The results show that using 9% cement content and 40–43% water content can meet the requirement of both compressive strength and flowability of the CLSM. These parameters can also help save cement consumption and reduce construction costs.

The CLSM technique was used to backfill the pipeline trenches in the drainage project of Yanghe Avenue of Suqian, China. The performance of CLSM was evaluated by drilling and coring at a given curing time. The compressive strength of the specimens was tested using CBR-2 California bearing ratio equipment. The results show that the strength of CLSM samples can fully meet the requirements of road engineering, achieving good backfilling results. The obtained material mixture ratio of CLSM can also be used in other projects with similar soil conditions.

Author Contributions: Conceptualization, Q.L. and Y.T.; methodology, Z.Z.; investigation, J.L.; investigation, J.X. and Q.L.; writing—review and editing, J.X.; supervision, J.X.; funding acquisition, J.X.; revision—J.X. All authors have read and agreed to the published version of the manuscript.

Funding: This research has received funding from the National Natural Science Foundation of China (Grant No. 52108364).

Institutional Review Board Statement: Not applicable.

Informed Consent Statement: Not applicable.

Data Availability Statement: Data are contained within the article.

Conflicts of Interest: Authors Qiwu Luo, Yong Tang, Zhibo Zeng, and Jun Liao were employed by the company China Construction Fifth Engineering Division Corp., Ltd. All authors declare that this research was conducted in the absence of any commercial or financial relationships that could be construed as a potential conflict of interest.

References

1. Prum, S.; Jumnongpol, N.; Eamchotchawalit, C.; Kantiwattanakul, P.; Sooksatra, V.; Jarearnsiri, T.; Passananon, S. Guideline for Backfill Material Improvement for Water Supply Pipeline Construction on Bangkok Clay, Thailand. In Proceedings of the 4th World Congress on Civil, Structural, and Environmental Engineering (CSEE'19), Rome, Italy, 7–9 April 2019; pp. 7–9.
2. Ling, T.C.; Kaliyavaradhan, S.K.; Poon, C.S. Global perspective on application of controlled low-strength material (CLSM) for trench backfilling–An overview. *Constr. Build. Mater.* **2018**, *158*, 535–548. [CrossRef]
3. American Concrete Institute. *ACI-229R, Controlled Low Strength Materials (Reproved 2005)*; American Concrete Institute: Farmington Hills, MI, USA, 2005.
4. Fedrigo, W.; Nunez, W.P.; Visser, A.T. A review of full-depth reclamation of pavements with Portland cement: Brazil and abroad. *Constr. Build. Mater.* **2020**, *262*, 120540. [CrossRef]
5. Linares-Unamunzaga, A.; Pérez-Acebo, H.; Rojo, M.; Gonzalo-Orden, H. Flexural strength prediction models for soil–cement from unconfined compressive strength at seven days. *Materials* **2019**, *12*, 387. [CrossRef] [PubMed]
6. Fedrigo, W.; Núñez, W.P.; López, M.A.C.; Kleinert, T.R.; Ceratti, J.A.P. A study on the resilient modulus of cement-treated mixtures of RAP and aggregates using indirect tensile, triaxial and flexural tests. *Constr. Build. Mater.* **2018**, *171*, 161–169. [CrossRef]
7. Naik, T.R.; Singh, S.S. Permeability of flowable slurry materials containing foundry sand and fly ash. *J. Geotech. Geoenviron. Eng.* **1997**, *123*, 446–452. [CrossRef]
8. Dockter, B.A. Comparison of dry scrubber and class c fly ash in controlled lowstrength materials (CLSM) applications. In *The Design and Application of Controlled Low-Strength Materials (Flowable Fill), ASTM STP 1331*; Howard, A.K., Hitch, J.L., Eds.; American Society for Testing and Materials: West Conshohocken, PA, USA, 1998; pp. 13–26.
9. Kaneshiro, J.; Navin, S.; Wendel, L.; Snowden, H. Controlled low strength material for pipeline backfill—Specifications, case histories and lessons learned. In *Pipelines 2001: Advances in Pipelines Engineering and Construction*; American Society of Civil Engineers: Reston, VA, USA, 2001; pp. 1–13.
10. Finney, A.J.; Shorey, E.F.; Anderson, J. Use of native soil in place of aggregate in controlled low strength material (CLSM). In *Pipelines 2008: Pipeline Asset Management: Maximizing Performance of our Pipeline Infrastructure*; American Society of Civil Engineers: Reston, VA, USA, 2008; pp. 1–13.
11. Dalal, P.H.; Patil, M.; Iyer, K.K.; Dave, T.N. Sustainable controlled low strength material from waste materials for infrastructure applications: State-of-the-art. *J. Environ. Manag.* **2023**, *342*, 118284. [CrossRef] [PubMed]
12. Kaliyavaradhan, S.K.; Ling, T.C.; Guo, M.Z. Upcycling of wastes for sustainable controlled low-strength material: A review on strength and excavatability. *Environ. Sci. Pollut. Res.* **2022**, *29*, 16799–16816. [CrossRef] [PubMed]
13. Blanco, A.; Pujadas, P.; Cavalaro, S.H.P.; Aguado, A. Methodology for the design of controlled low-strength materials. *Appl. Backfill Narrow Trenches. Constr. Build. Mater.* **2014**, *72*, 23–30. [CrossRef]
14. Fauzi, M.A.; Arshad, M.F.; Nor, N.M.; Ghazali, E. Modeling and optimization of properties for unprocessed-fly ash (u-FA) controlled low-strength material as backfill materials. *Clean. Eng. Technol.* **2022**, *6*, 100395. [CrossRef]
15. Ibrahim, M.; Rahman, M.K.; Najamuddin, S.K.; Alhelal, Z.S.; Acero, C.E. A review on utilization of industrial by-products in the production of controlled low strength materials and factors influencing the properties. *Constr. Build. Mater.* **2022**, *325*, 126704. [CrossRef]
16. Kuo, W.T.; Wang, H.Y.; Shu, C.Y.; Su, D.S. Engineering properties of controlled low-strength materials containing waste oyster shells. *Constr. Build. Mater.* **2013**, *46*, 128–133. [CrossRef]
17. Liu, H.; Xiao, Y.; Liu, K.; Zhu, Y.; Zhang, P. Numerical simulation on backfilling of buried pipes using controlled low strength materials. *Appl. Sci.* **2022**, *12*, 6901. [CrossRef]
18. Türkel, S.E.L.Ç.U.K. Strength properties of fly ash based controlled low strength materials. *J. Hazard. Mater.* **2007**, *147*, 1015–1019. [CrossRef] [PubMed]
19. Wang, C.; Li, Y.; Wen, P.; Zeng, W.; Wang, X. A comprehensive review on mechanical properties of green controlled low strength materials. *Constr. Build. Mater.* **2023**, *363*, 129611. [CrossRef]
20. *JTG 3430—2020*; China Standard: Regulations for Highway Geotechnical Testing. National Standards of the People's Republic of China: Bejing, China, 2020.
21. Khadka, S.D.; Okuyucu, O.; Jayawickrama, P.W.; Senadheera, S. Controlled low strength materials (CLSM) activated with alkaline solution: Flowability, setting time and microstructural characteristics. *Case Stud. Constr. Mater.* **2023**, *18*, e01892. [CrossRef]
22. Wan, X.; Ding, J.; Jiao, N.; Zhang, S.; Wang, J.; Guo, C. Preparing controlled low strength materials (CLSM) using excavated waste soils with polycarboxylate superplasticizer. *Environ. Earth Sci.* **2023**, *82*, 214. [CrossRef]
23. Tran, T.Q.; Kim, Y.S.; Dang, L.C.; Do, T.M. A state-of-the-art review on the utilization of new green binders in the production of controlled low-strength materials. *Constr. Build. Mater.* **2023**, *393*, 132078. [CrossRef]

 *materials*

Article

Study on the Effect of Foam Stability on the Properties of Foamed Lightweight Soils

Hao Liu [1], Cong Shen [2], Jixin Li [1], Gaoke Zhang [2], Yongsheng Wang [1] and Huiwen Wan [2,*]

[1] China Construction Second Engineering Bureau Limited East China Branch, Shanghai 200135, China; liuhao-2@cscec.com (H.L.); lijixin@cscec.com (J.L.); wangyongsheng_@cscec.com (Y.W.)
[2] School of Materials Science and Engineering, Wuhan University of Technology, Wuhan 430070, China; shencong@whut.edu.cn (C.S.); gkzhang@whut.edu.cn (G.Z.)
* Correspondence: wanhw@whut.edu.cn; Tel.: +86-139-7100-7966

Abstract: The properties of prepared foamed lightweight soils (FLSs) using prefabricated foam requires high foam stability. This paper investigates the geometrical characteristics of different foam densities, different types of foaming agents in the air, and the presence of slurry. Then, it studies their effects on the pore structure and mechanical properties of FLS. Results show that with the increase in foam density the bleeding rate of foam in the air for 1 h increases and the foam with a foam density of 50 kg/m^3 is the most stable in the air. The stability of foam in slurry is not directly related to the property of foam in the air. The FLS prepared with the same foaming agent had the best performance with the FLS designed with a foam density of 50 kg/m^3, which had the smallest average pore size and the most minor pore size distribution, and had the highest compressive strength. Among the three different foaming agents, Type-S was the best, and the slurry had the lowest rate of increase in wet density after the defoaming test, indicating that the foam had the best stability in the cement slurry. The FLS prepared with the density of 50 kg/m^3 using the Type-S foaming agent and mixed with the slurry of cement, fly ash:slag:water = 105:105:140:227.5, was hardened to a mean pore size of 299 μm, and the 7 days, 28 days, and 56 days compressive strengths were 0.92 MPa, 2.04 MPa, and 2.48 MPa, respectively, which had the smallest average pore size and the highest compressive strength among the FLSs prepared using the three foaming agents.

Keywords: foamed lightweight soils; foam stability; pore structure

check for **updates**

Citation: Liu, H.; Shen, C.; Li, J.; Zhang, G.; Wang, Y.; Wan, H. Study on the Effect of Foam Stability on the Properties of Foamed Lightweight Soils. *Materials* **2023**, *16*, 6225. https://doi.org/10.3390/ma16186225

Academic Editor: René de Borst

Received: 25 August 2023
Revised: 12 September 2023
Accepted: 12 September 2023
Published: 15 September 2023

1. Introduction

Foamed lightweight soil (FLS) is a lightweight cementitious material containing a large number of closed tiny pores formed by physically preparing an aqueous solution of a foaming agent into a foam, then mixing it with a cement slurry in a certain proportion, pouring it, and finally curing it naturally [1–3]. By controlling the amount of prefabricated foam, the density of the FLS can be between 300 and 1800 kg/m^3 [4,5]. With its lightweight, good fluidity, ease of construction, high strength after hardening, and good integrity, FLS has been widely used in housing projects, railway and highway road foundations, military, civil engineering, airport runway end buffer paving, and other fields [6–8]. A large-scale test site for intelligent cars in Wuhan, China, has been successfully constructed using 860,000 m^3 of FLS instead of ordinary fill as fill material for the road base on soft ground such as lakes and fishing ponds [9]. However, the disadvantages of FLS are also obvious: low compressive strength and poor stability. Moreover, as the density of the FLS becomes lighter, its compressive strength decreases, and its stability worsens [10]. Therefore, it is essential to develop FLS with high stability and strength [11].

The low stability of the prefabricated foam is the main reason for the instability of the FLS slurry [12,13]. It is generally believed that the better the foam stability, the better the performance of the FLS can be obtained [14]. Many researchers have conducted many studies on improving foam stability [15,16]. Liquid foam is a non-equilibrium system consisting of internal air and an external liquid film, and its evolution is influenced by three main mechanisms: foam drainage, film rupture, and bubble coarsening [17,18]. All three mean to separate the surfactant solution and the gas, allowing the liquid foam to eventually break into different surfactant solutions and gases to reach equilibrium. Krämer prepared particle-stabilized foams using nanoparticles and applied them to FLS [12]. They found that the mechanical properties of FLS designed with particle-stabilized prefabricated foams were superior to those with reference foams. Panesar investigated the effect of composite and protein foaming agents on the properties of FLS and concluded that composite foaming agents were selected for protein foaming agents [19]. In addition, Jones also pointed out that stabilizers have an important influence on the stability, density, and strength of the bubbles in the FLS and the workability and water retention of the slurry [10].

There is little research on the presence of foam in fresh slurry and the conditions under which it exists in fresh slurry. The foam in the air differs considerably from that in the slurry. In fact, the foam in the slurry is not only subject to frictional forces from the particles of the cementitious material but also interacts with them to produce adsorption effects [20,21]. Xu modeled the force on the foam in the slurry from the rheological properties of the slurry and decomposed the foaming process into four external forces. Wan studied the rheological parameters of the slurry, the internal foam situation in the mixing, and hardening stages to demonstrate that low-cementitious ratio FLS is feasible. Therefore, a stable foam in the air does not necessarily lead to preparing FLS with excellent properties [22]. Only good compatibility between the foam and the cementitious material composition will give the foam in FLS longer-term stability [23,24].

However, many studies only discuss the relationship between foam properties and FLS sample properties or discuss the relationship between FLS slurry and FLS sample properties. There is less research to analyze and investigate the combination of the three. There is no simple positive correlation between the three, and it is necessary to find the connection between them through systematic experiments. Therefore, this paper investigates the design and influence of foam stability on FLS prepared with different foam properties. The foam stability is quantified into two parameters: foam density and foaming agent types. The stability of foam in the air is investigated through the performance of foam prepared with two different parameters. Then, the influence of two parameters on the stability of FLS slurry and samples are explored through the defoaming test, pore structure, and compressive strength tests of FLS. The results of the above studies are combined to establish the relationship between the stability of foam in air and the properties of FLS.

2. Materials and Methods

2.1. Raw Materials

The cementitious materials used in this study to prepare the FLS were Portland cement (PC), fly ash (FA), and granulated blast furnace slag (GBFS).

The cement is P·O 42.5 cement produced by China Conch Cement Co., Ltd. (Wuhu, China) with the physical properties shown in Table 1. The FA is Class F Grade II produced by Hubei Koneng Environmental Protection Co. (Wuhan, China). The GBFS use grade S95 produced by Hubei Jinshenglan Metallurgical Technology Co. (Xianning, China). The chemical composition of raw materials is shown in Table 2.

Table 1. Physical properties of P·O 42.5 cement.

Density/(kg/m³)	Specific Surface Area/(m²/kg)	Soundness of Cement/mm	Setting Time/min		Flexural Strength/MPa		Compressive Strength/MPa	
			Initial	Final	3 Days	28 Days	3 Days	28 Days
3100	340	2	170	235	5.6	8.7	28.1	50.4

Table 2. Chemical composition of raw materials (wt%).

Material	CaO	SiO$_2$	Al$_2$O$_3$	Fe$_2$O$_3$	MgO	SO$_3$	K$_2$O	Na$_2$O	TiO$_2$	LOI
PC	60.11	20.92	5.76	3.24	1.15	2.86	0.88	0.14	0.31	4.17
GBFS	39.92	31.23	14.12	0.78	7.34	2.23	0.61	0.72	0.76	−0.29
FA	0.44	57.64	21.49	6.52	1.77	0.37	3.42	0.12	0.93	6.85

Three foaming agents from different manufacturers were used in the test. The first is a ready-mixed compound reinforced foaming agent produced by Guangdong Shengrui Technology Co., Ltd. (Guangzhou, China) model JT-SRN2 (Type-S). The second is a compound foaming agent produced by Henan Huatai New Material Technology Co., Ltd. (Zhengzhou, China), model HTW-I (Type-H). The third is the mixed foaming agent produced by Wuhan Qianhesheng Building Material Development Co. (Wuhan, China) (Type-Q).

2.2. Specimen Preparation

This study is carried out on FLS for practical road-filling projects. The technical specifications of FLS must be 7 d compressive strength ≥ 0.5 MPa and 28 d compressive strength ≥ 1.0 MPa.

In this paper, the preparation of FLS specimens was achieved using the prefabricated foam method. The preparation of FLS is shown in Figure 1. The designed wet density of FLS in the test was 600 kg/m^3, and the water–binder ratio was 0.65 [25].

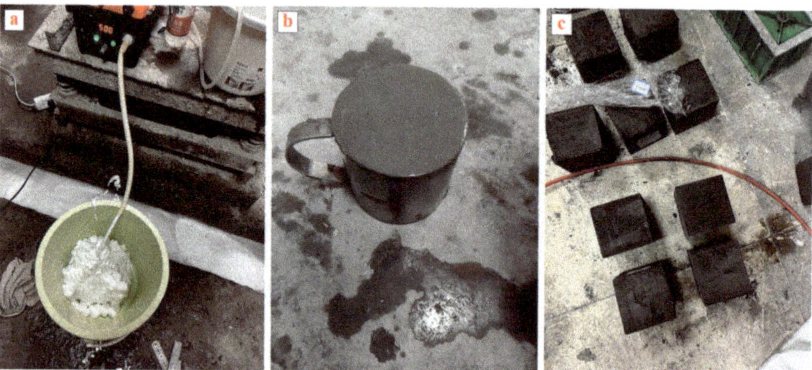

Figure 1. Preparation process of FLS: (**a**) Preparation of foam; (**b**) Preparation of FLS slurry; (**c**) Preparation of FLS sample.

The process of preparing the FLS can be divided into the following steps:

1. Preparation of the foam: First, a certain amount of foaming agent is weighed. According to the dilution multiple of the foaming agent, a required amount of water is weighed. The weighed water is poured into a bucket and then the foaming agent is poured into the bucket and mixed into an aqueous foaming solution using a glass rod. Next, the switch of the intelligent micro foaming machine is turned on and the parameters of the foaming machine are adjusted. A uniform and even foam flow comes out of the foam outlet tube. Afterwards, the density of the foam is calibrated in a 1 L container and tested at 50 ± 2 kg/m^3. Once the density of the foam meets the requirements, the foam is weighed in a drum using the mass method and set aside.
2. Preparation of the cementitious material slurry: The mixing water is weighed according to the designed ratio and divided equally into two parts. One is added directly to the bucket. The FA, GBFS, and PC are weighed and mixed according to the designed mix ratio, then added to the bucket containing one part of the water. During mixing, the other part of the water is poured into the bucket at a constant speed to make a slurry.

3. Preparation of the FLS: The prepared foam (in step 1) is added to the slurry (in step 2) at once, and then a mixer is used to mix the slurry. At the beginning of the mixing process, the mixer is placed on the upper surface so that the foam floating on the surface can dissolve in the slurry during the mixing processing and then move in a circular motion around the center of the barrel while oscillating up and down to ensure adequate mixing. After the slurry has been mixed evenly, the flow factor and wet density are measured, and the designed wet density of 600 kg/m^3 is reached before shaping and pouring.

4. Casting and curing: The prepared slurry (in step 3) is poured into the prepared $100 \times 100 \times 100$ mm^3 triplex mold using a beaker. The mold is lightly vibrated at half its height. The final pouring is finished about 1–2 cm above the mold to prevent collapse when casting. The surface of the poured mold is then covered with a layer of cling film. Due to the low early strength of the FLS, demolding after 24 h of maintenance will result in incomplete specimens, so the finished specimens are stored for about 48 h before demolding. When demolding, the extra pouring part is scraped off with a scraper. After demolding, the samples are placed in a plastic bag and sealed before being placed in the curing room.

2.3. Mix Design

The performance of the foams in the experiments was characterized using two parameters, namely, the type of foaming agent and the foam density. The specimens prepared for the study of foam density were numbered SF40, SF50, and SF60, and those prepared for the study of foaming agent types were numbered HD60, QD80, and SD100, respectively. The specific mixes are shown in Table 3.

Table 3. Mix ratios of FLS for different foam properties (kg/m^3).

No.	PC	FA	GBFS	Water	Foam Properties		
					Types	Density	Dilution Factor
SF40	105	105	140	227.5	S	40	100
SF50	105	105	140	227.5	S	50	100
SF60	105	105	140	227.5	S	60	100
HD60	105	105	140	227.5	H	50	60
QD80	105	105	140	227.5	Q	50	80
SD100	105	105	140	227.5	S	50	100

2.4. Test Methods

2.4.1. Foam Performance Tests

The physical properties of the foam are tested for the 1 h settling distance and bleeding rate as well as the shape and diameter of the foam.

The 1 h settling distance and bleeding rate of foams is carried out using a special measuring instrument. Figure 2 shows the measuring instrument.

The volume of the instrument is 5.015 L. After the foam has stood for 1 h, the settling distance of the buoy on the surface of the foam is first read through the scale on the wall of the vessel. Then, the foam liquid is passed through the glass tube in the lower part of the container, and its mass m_{1h} is weighed. The bleeding rate of the foam for 1 h is calculated according to the following Equation (1):

$$\varepsilon = \frac{m_{1h}}{\rho_1 V_1} \tag{1}$$

where ε is the 1 h bleeding rate of the foam, in %; m_{1h} is the mass of the foam flowing out after 1 h, in g; ρ_1 is the foam density, in g/mL; and V_1 is the volume of the container, which is 5.015 L.

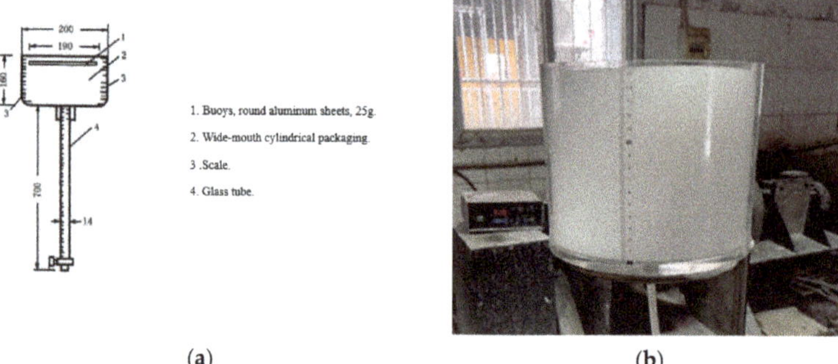

(**a**)   (**b**)

Figure 2. Shown is the 1 h settling distance and bleeding rate measuring instrument: (**a**) Instrument schematic; (**b**) Practical measurements.

The foam morphology and diameter were characterized using an optical microscope. A small amount of foam was prepared by aspirating it with a rubber-tipped dropper, then squeezing out a small amount onto a clean microscope slide, lightly covering the foam with a coverslip and placing it under the microscope. Adjust the magnification to 40× and observe the morphology and diameter of the foam.

2.4.2. Stability Tests of Foam in Slurry

The stability of the foam with the slurry can be evaluated using the rate of increase in wet density of the FLS slurry after the defoaming test. Firstly, its initial wet density ρ_0 is determined using a 1 L volumetric bucket, and then it is stirred by hand at a slow speed (stirring speed 60 r/min) for 1 min. After each stirring, the density ρ_x is measured. Repeating the stirring process 6 times, the maximum wet density ρ_6 is determined, and the rate of increase in wet density is calculated according to the following Equation (2):

$$\delta = \frac{\rho_6 - \rho_0}{\rho_0} \times 100\%$$

(2)

where δ is the rate of increase in wet density, in %; ρ_0 is the initial wet density of FLS, in g; and ρ_6 is the wet density of FLS after 6 mixes, in g.

2.4.3. Pore Structure Tests

A body microscope was used to observe the pore structure of the hardened FLS. The test samples were first cut into $20 \times 20 \times 20$ mm^3 sizes using cutting equipment, and the surface was smoothed with sandpaper to expose the original pore structure inside the samples. The sample was then placed in the middle of the viewing area of a body microscope, and the focus was adjusted to observe the pore structure of the sample.

2.4.4. Compressive Strength Tests

The compressive strength of FLS was tested according to standard GB/T 11969-2020 [26]. But, the samples were tested without water absorption or drying following standard CECS 249:2008 [27]. After the specimens reached the specified curing age, the samples' apparent naturally dried densities (quasi-dry densities) were measured first. Then, the machine (TYE-300D) conducted the compressive strength test, and the loading speed was 0.1 kN/s.

3. Results

3.1. Stability of Foam in the Air

According to the characteristics of raw materials, synthesis process, and solids content, the manufacturer of the foaming agent will recommend the optimal dilution times when using it. In this paper, Type-S is recommended 100 times, Type-Q is recommended 80 times, and Type-H is recommended 60 times.

The stability of foam in the air was controlled by varying the type of foaming agent and the foam density by measuring the 1 h settling distance and 1 h bleeding rate of the foam and characterizing the particle size and morphology of the prepared foam by observing the prepared foam using an optical microscope to characterize the foam stability in air.

3.1.1. Settling Distance and Bleeding Rate (1 h)

The results of the 1 h settling distance and 1 h bleeding rate of the three foaming agents used in the experiment under different foam density conditions are shown in Table 4.

Table 4. Performance of different foaming agents at different foam densities.

Item	Types of Foaming Agents								
	S (Dilution 100 Times)			H (Dilution 60 Times)			Q (Dilution 80 Times)		
Foam density (g/L)	40.5	49.2	60.3	42.2	50.2	60.0	41.4	51.2	60.2
Settlement (mm)	4.5	2	3	5	8	6	2	1	2
Bleeding rate (%)	70.19	72.66	76.72	68.69	76.84	82.17	58.00	59.82	63.19

The smaller the 1 h settling distance and bleeding rate, the more stable the foam is in the air and the lower the degree of foam drainage. From the test results, it can be found that as the foam density increases, there is a tendency for the 1 h settling distance and bleeding rate of each foaming agent to increase. This means that for the three foaming agents used in this test, the stability of foam in the air is all 40 kg/m^3 > 50 kg/m^3 > 60 kg/m^3. The 1 h settling distance and bleeding rate vary for foaming agents at the same foam density, with foam Type-Q > Type-H > Type-S stability.

3.1.2. Foam Density

The morphology of the foams prepared with Types-S with densities of 40 kg/m^3, 50 kg/m^3, and 60 kg/m^3 is shown in Figure 3 under an optical microscope.

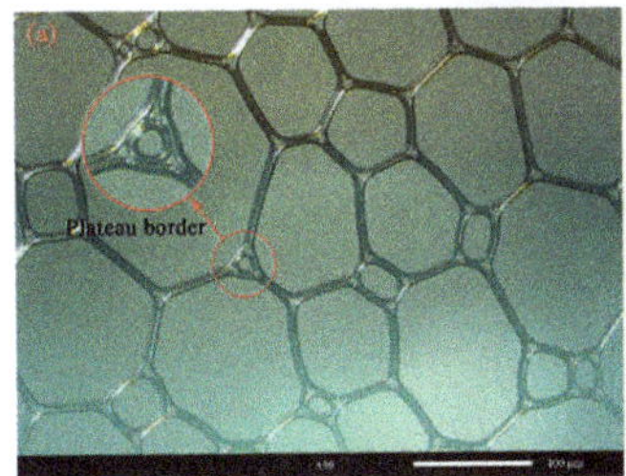

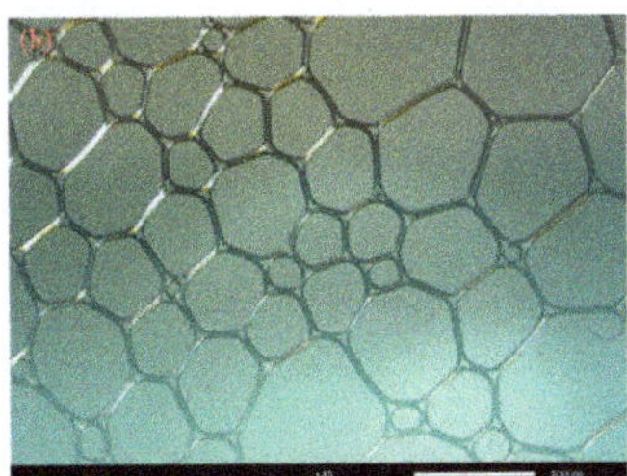
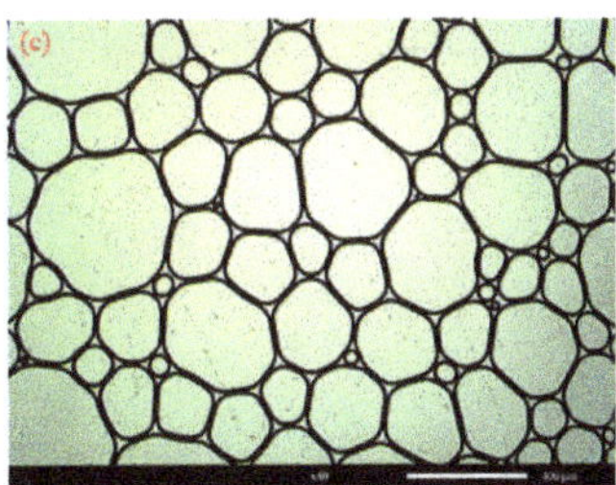

Figure 3. Foam morphology at different foam densities: (**a**) 50 kg/m^3; (**b**) 60 kg/m^3; (**c**) 40 kg/m^3.

It is apparent from Figure 3 that the shape of each group of foams is irregularly polygonal rather than spherical, as is commonly thought. This is because the foam density set in the experiments is relatively tiny. The prefabricated foams prepared using the foaming machine must be stacked to achieve the required foam density. When multiple foams are stacked together, the foams do not remain initially round and take on an irregular polygon shape. Moreover, it reaches its best stability when the number of sides reaches 6, which is determined by the draining effect of the foam film, the plateau boundary effect, as shown in the partial enlargement in Figure 3a.

The foam morphology in the figure was measured and counted using ImageJ (https://imagej.net/ij/), and the results of the particle size distribution of the foam are shown in Figure 4. The results of the average diameter of the foam and the average thickness of the foam film are shown in Table 5.

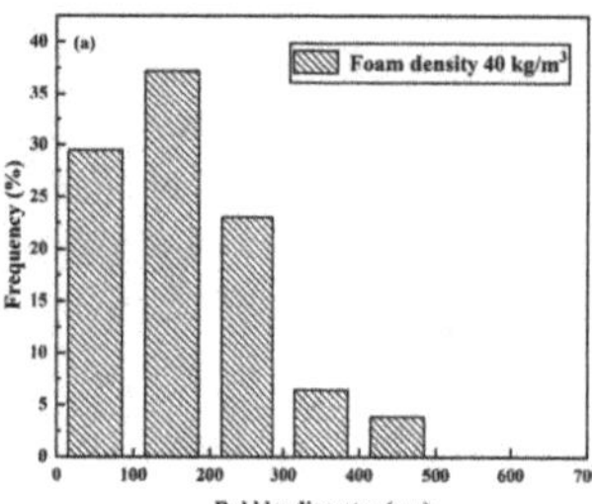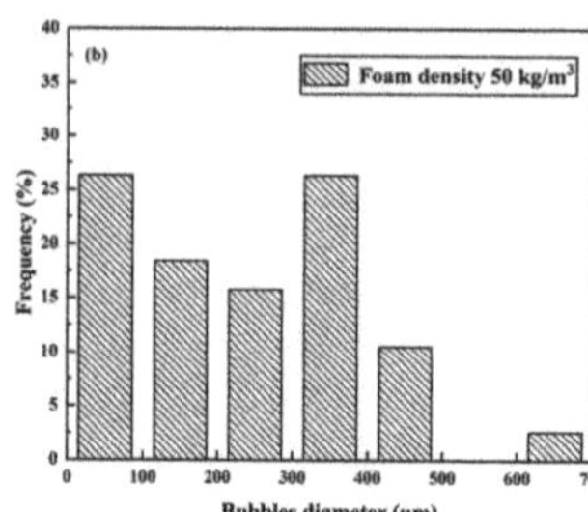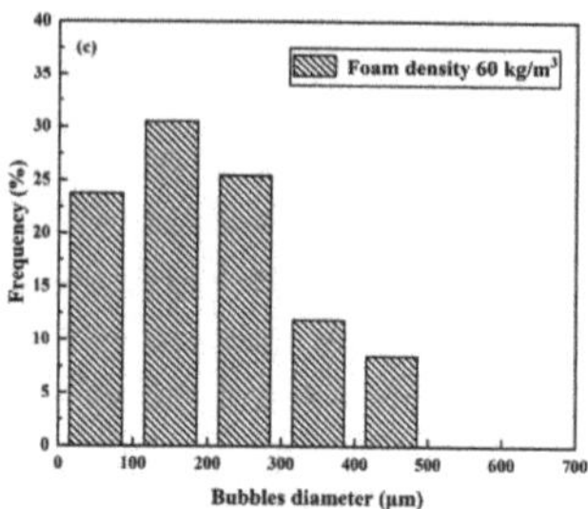

Figure 4. Foam particle size at different foam densities: (**a**) 40 kg/m³; (**b**) 50 kg/m³; (**c**) 60 kg/m³.

Table 5. Parameters of the foam at different foam densities.

No.	Average Diameter (μm)	Average Liquid Film Thickness (μm)
SF40	167.12	22.99
SF50	236.28	38.59
SF60	199.21	23.62

As can be seen in Figure 4 and Table 5, the average diameter of the foam with a foam density of 50 kg/m³ is 236.28 μm larger than that of the foam with a foam density of 40 kg/m³ at 167.12 μm and 60 kg/m³ at 199.21 μm, which is contrary to conventional knowledge. The foam with a foam density of 50 kg/m³ has a more extensive diameter distribution. This indicates that the diameter size of prefabricated foams prepared using foaming machines does not have a simple linear relationship with density. Instead, there is a maximum value, and when the foam density is greater or less than this maximum value the diameter of the prepared foam becomes smaller. In addition, the average film thickness of 38.59 μm for a foam with a density of 50 kg/m³ is also more significant than the film thickness of 22.99 μm for a foam with a density of 40 kg/m³ and 23.62 μm for a foam with a density of 60 kg/m³. The increase in foam density from 40 kg/m³ to 50 kg/m³ is mainly due to the rise in liquid film thickness, while the increase in foam density from 50 kg/m³ to 60 kg/m³ is mainly due to the decrease in individual foam diameter and thickness and the increase in overall liquid film mass to increase the density. Therefore, for foam densities between 40 kg/m³ and 60 kg/m³ the diameter of the prepared foam increases and then decreases and the average liquid film thickness of the foam increases and then decreases.

3.1.3. Foaming Agent Types

The morphology of the foams prepared using three foaming agents with a foam density of 50 kg/m³ under an optical microscope is shown in Figure 5. It can be seen that the morphology of the foams prepared by Type-Q and Type-H is similar to that of Type-S, which is also irregularly polygonal. Furthermore, although the foam densities of all three agents are 50 kg/m³, it is clear that their foam diameters and liquid film thicknesses are different.

Similarly, ImageJ was used to measure and count the foam morphology. The particle size distribution of the foam is shown in Figure 6, and the results of the foam's average diameter and the foam film's average thickness are shown in Table 6. Figure 6 and Table 6 show that the average pore size of the foam prepared in the QD80 group was the smallest at 121.75 μm, with a small range of foam diameter distribution. The average pore size of the foam prepared in the HD60 group was 147.69 μm, again more minor than that of the SD100 group. In addition, the QD80 and HD60 groups' average liquid film thicknesses were similar, around 22 μm. The small average diameter of the foam in the QD80 group and the small average liquid film thickness proves that the foam prepared in the QD80 group has the best stability in the air.

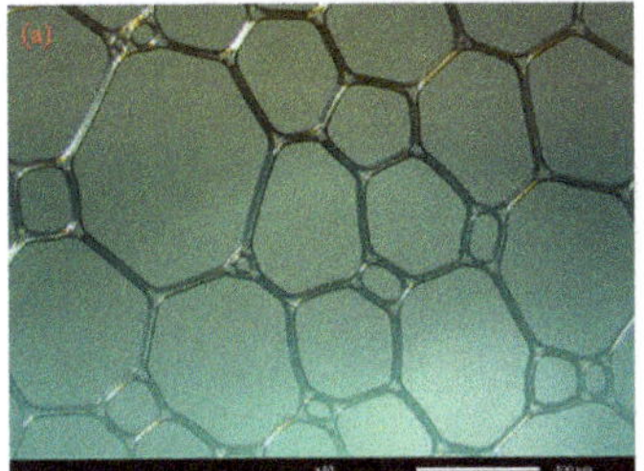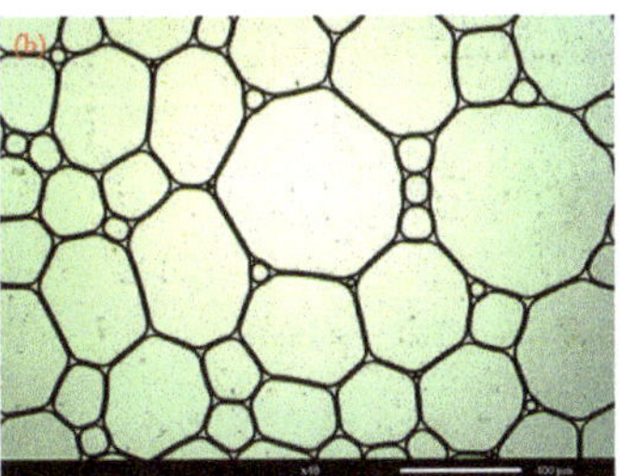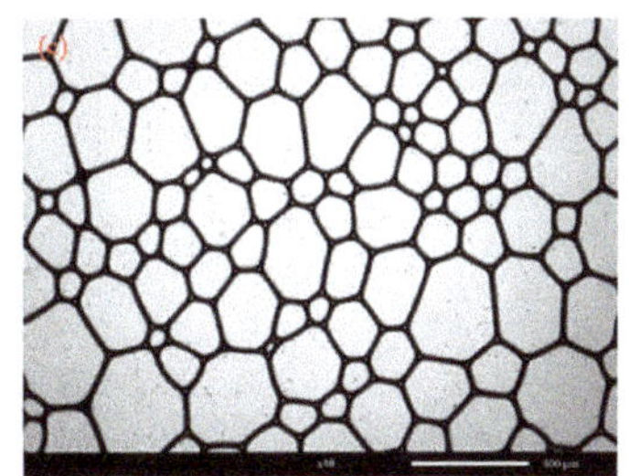

Figure 5. Foam morphology with different foaming agents: (**a**) SD100; (**b**) HD60; (**c**) QD80.

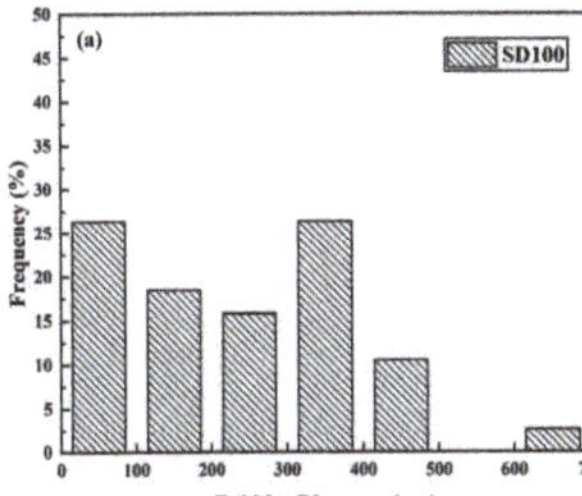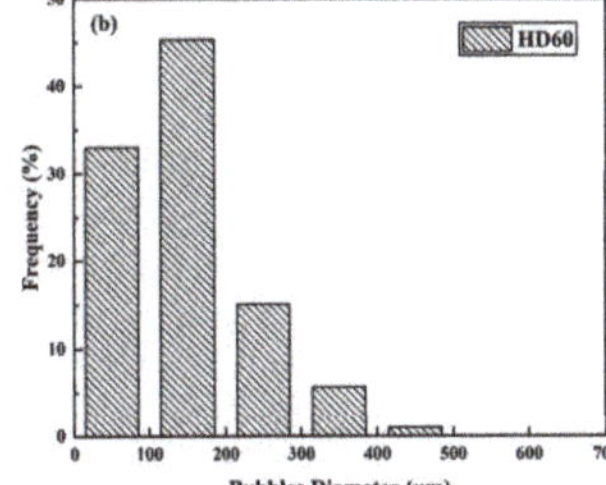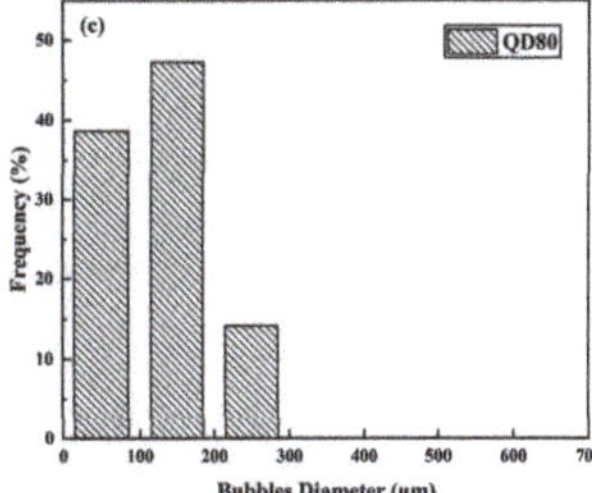

Figure 6. Foam particle sizes with different foaming agents: (**a**) SD100; (**b**) HD60; (**c**) QD80.

Table 6. Parameters of foams with different foaming agents.

No.	Average Diameter (μm)	Average Liquid Film Thickness (μm)
SD100	236.28	38.59
HD60	147.69	21.27
QD80	121.75	22.54

3.2. Stability of Foam in Slurry

The stability of the foam in the slurry system can be evaluated using the rate of increase of wet density after the defoaming test. Tables 7 and 8 shows the rate of increase in wet density after the defoaming test of FLS prepared with different foam density and types of foaming agents. The lower rate of increase in wet density indicates that the foam is less likely to burst in the slurry system, and the stability is better. The rate of increase in wet density in practical engineering applications is generally controlled at less than 10%.

Table 7. Rate of increase in wet density with different foam density.

Times	SF40		SF50		SF60	
	Measured Wet Density (kg/m^3)	Rate of Increase (%)	Measured Wet Density (kg/m^3)	Rate of Increase (%)	Measured Wet Density (kg/m^3)	Rate of Increase (%)
0	605	0	607	0	611	0
1	614	1.38	612	0.82	632	3.44
2	626	3.36	623	2.64	658	7.69
3	641	5.83	631	3.95	681	11.46
4	654	7.98	645	6.26	699	14.40
5	667	10.13	656	8.07	712	16.53
6	679	12.11	662	9.06	736	20.46

Table 8. Rate of increase in wet density with different types of foaming agents.

Times	SD100		QD80		HD60	
	Measured Wet Density (kg/m^3)	Rate of Increase (%)	Measured Wet Density (kg/m^3)	Rate of Increase (%)	Measured Wet Density (kg/m^3)	Rate of Increase (%)
0	607	0	609	0	625	0
1	612	0.82	612	0.38	637	1.81
2	623	2.64	619	1.53	651	4.05
3	631	3.95	635	4.16	668	6.77
4	645	6.26	668	9.57	683	9.16
5	656	8.07	692	13.50	698	11.56
6	662	9.06	727	19.25	713	13.96

As seen from Table 7, for the same foaming agent, SF50 has the lowest rate of increase in wet density, which indicates that the foam film formed by the foam is thicker, so the foam is not easy to break and has good stability after repeated stirring. Table 8 shows that under the condition of the same foam density, sample SD100 has the lowest rate of increase in wet density. This result differs from the stability of foam in the air in Table 5. This indicates that foams in the air are not necessarily stable in slurry. It is not possible to directly use the performance of the foam in the air to determine its performance in the slurry.

3.3. Pore Structure of FLS

3.3.1. Effect of Foam Density on Pore Structure

To investigate the effect of foam properties on the microstructure of FLS after hardening, samples of SF40, SF50, and SF60 at 28 d hydration age were analyzed using a body microscope. The pore structure diagrams with a magnification of 53.33 times are shown in Figure 7.

As seen in Figure 7, the hardened FLS specimens, provided the pore morphology is not polygonal, are instead mostly round. This is not the same state as the foam morphology in the air. This is because, in a design for wet density of 600 kg/m^3 FLS slurry, the volume share of the foam is theoretically between 60–70%. Suppose the foam is evenly dispersed around the cementitious material during the slurry mixing, forming the most compactly stacked structure. In that case, its volume share is approximately 78% more significant than the foam's theoretical volume share. Therefore, foams that have built up on each other in the air will regain the round shape of the foam itself when they enter the FLS slurry. In addition, it can be found that the SF50 group has a more homogeneous pore structure, and the pores are mainly regular in shape; the SF60 group has larger pores and more pores have undergone consolidation as well as irregularly shaped pores, while the SF40 group has formed more small-grained pores. There are also many large pores with larger grain sizes and there is more connectivity between the hardened pores.

ImageJ was used to count the pore structures in the graphs. Trainable Weka Segmentation v3.3.2 was first used to identify the holes in the images, then binarized, and finally, the pore size distribution was measured for the processed images. In general, the pore size distribution of FLS obeys a normal distribution, and the pore size distribution is fitted using the normal distribution function Gaussian with the following fitting equation:

$$f(D) = \frac{1}{\sqrt{2\sigma^2\pi}}e^{\left(-\frac{(D-\mu)}{2\sigma^2}\right)} \tag{3}$$

where $f(D)$ is the normal distribution probability function; D is the average diameter; μ is the mean deviation, and σ is the standard deviation. The standard deviation σ is related to the range of the pore size distribution, and the larger the value of its standard deviation σ, the more comprehensive the range of the pore size distribution. The fitted curves of the normal distribution of the pore size of FLS are shown in Figure 8. The statistical results are shown in Table 9.

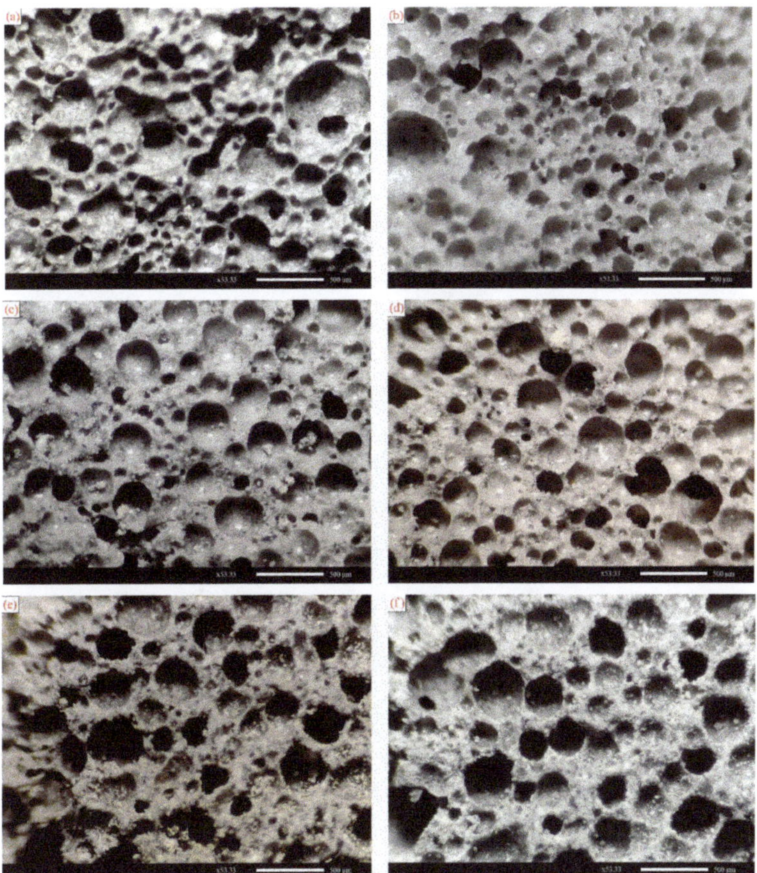

Figure 7. Pore structure diagram for different groups of FLS: (**a**,**b**) SF40 group; (**c**,**d**) SF50 group; (**e**,**f**) SF60 group.

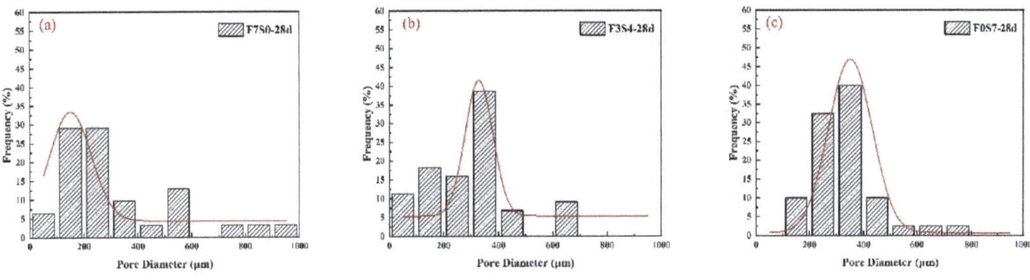

Figure 8. Pore size distribution of different groups of FLS: (**a**) SF40 group; (**b**) SF50 group; (**c**) SF60 group.

As shown in Figure 9, the pore size distribution of the SF40 group shows a more significant proportion of tiny pores (<200 μm), accounting for approximately 34.4%. However, as it has a similarly large proportion of large pores (>700 μm), accounting for about 9.6%, the average pore size of the SF40 group is more prominent, at 365.63 μm, and the pore size distribution is vast, with a significant standard deviation of 172. This indicates that the pore structure formed by the small pore size foams is also smaller after hardening, but some of the small foams will merge, resulting in more large foams and large pores. The pore size distribution of the SF60 group shows a higher proportion of medium pores (200–400 μm), accounting for 72.5% of the pores, with an average pore size of 324.52 μm and a standard deviation of 131. Therefore, the average pore size of the SF50 group is the smallest. This suggests that more of the SF60 group's foam was consolidated or destroyed during the forming and hardening. In summary, it can be seen that the SF50 group has the most minor pore structure and the best pore size distribution.

Table 9. Pore size parameters of the different groups of FLS.

No.	Average Diameter (μm)	Mean Deviation (μ)	Standard Deviation (σ)
SF40	365.64	302	172
SF50	299.37	328	101
SF60	324.52	350	131

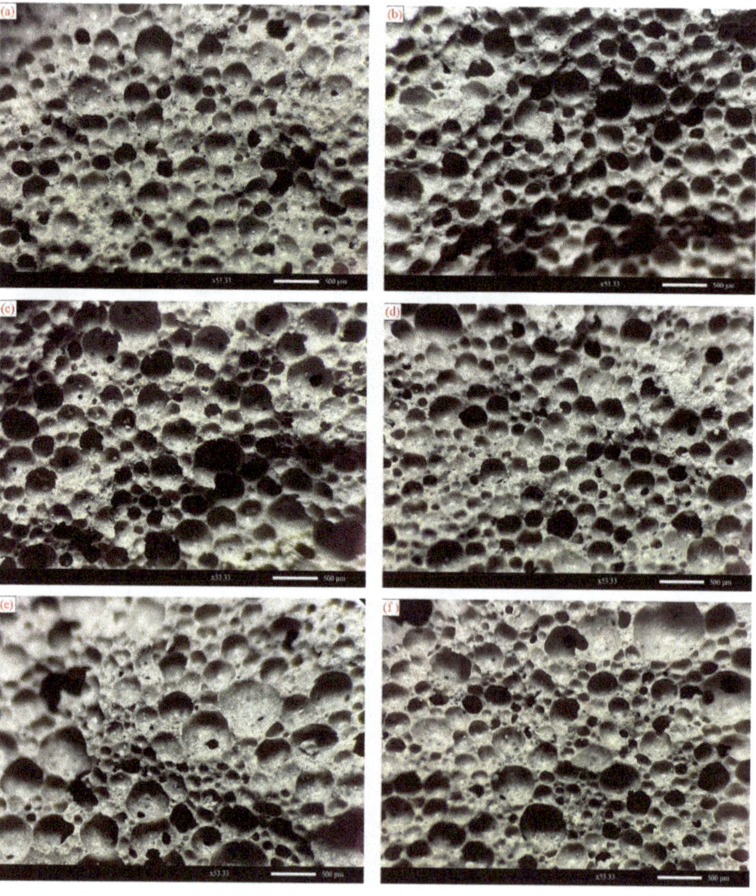

Figure 9. Pore structure diagrams of foam lightweight soils with different foam agents: (**a,b**) SD100 group; (**c,d**) HD60 group; (**e,f**) QD80 group.

3.3.2. Effect of Foaming Agent Types on the Pore Structure

As can be observed in Figure 9, the structure of the post-hardening holes in the SD100 and HD60 groups are primarily independent and do not have many interconnected holes. The QD80 group, on the other hand, has partially interconnected holes after hardening, and there are also cases where the larger holes contain smaller holes. However, overall, the pore structure formed after hardening in the SD100, HD60, and QD80 groups is circular. This indicates that foaming agent types does not excessively influence the shape of the pores after hardening.

ImageJ was used to count the pore structures in the figures, and finally, the pore size distribution was measured on the processed images; the results are shown in Figure 10. A standard distribution curve was fitted to the statistical aperture distribution and the results were obtained, as shown in Table 10.

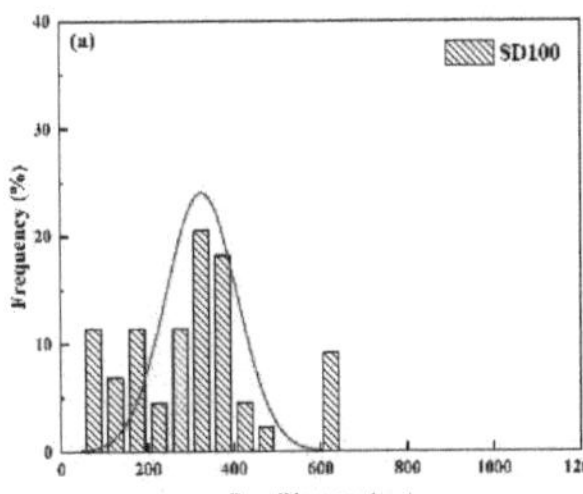
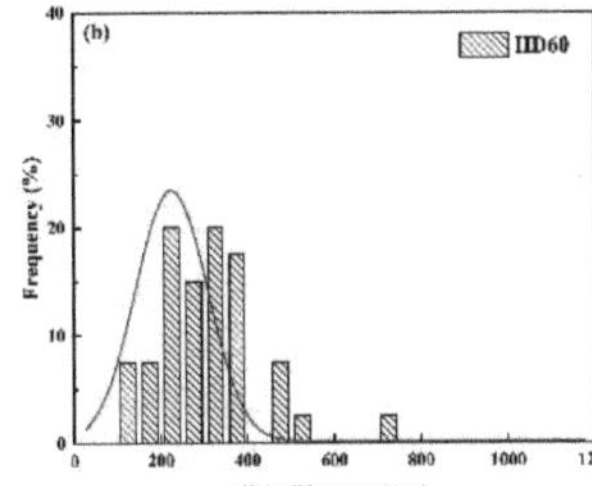
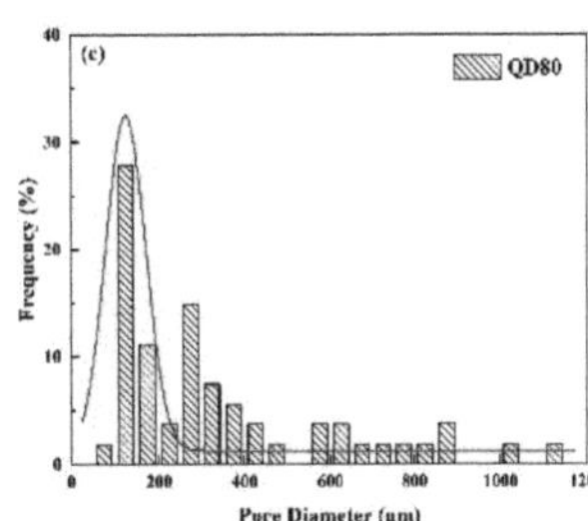

Figure 10. Pore size distribution of FLS with different foaming agents: (**a**) SD100 group; (**b**) HD60 group; (**c**) QD80 group.

Table 10. Pore size parameters of FLS under different foaming agents.

No.	Average Diameter (μm)	Mean Deviation (μ)	Standard Deviation (σ)
SD100	299.37	325	101
HD60	307.38	328	104
QD80	343.91	350	283

As can be seen from Figure 10 and Table 10, the pore size distribution of the QD80 group shows that the QD80 group has a high proportion of tiny pores (<200 μm), accounting for approximately 42.4%. The pore size distribution of the SD100 and HD60 groups is similar, with the majority of pores between 200 and 400 μm in diameter, with an average pore size of 299.37 μm and 307.37 μm, respectively. The average pore size was 299.37 μm and 307.38 μm, respectively, and their standard deviations were around 100. This indicates that the pores formed after the hardening of the foam in the SD100 and HD60 groups were similar in size and distribution. However, the foam diameter in the air for the QD80 and HD60 groups is about 140 μm, whereas the foam diameter in the air for the SD100 group is 236 μm. It can be observed that there is no direct linear correlation between the size of the pores formed by the foam after slurry hardening and the diameter of the foam in the air. This indicates that the size of the foam in the air cannot simply be used to determine the pore size of FLS after hardening.

3.4. Compressive Strength of FLS

3.4.1. Effect of Foam Density on the Compressive Strength

The 7 days, 28 days, and 56 days compressive strengths and compaction-to-density of FLS prepared with Type-S at foam densities of 40 kg/m^3, 50 kg/m^3, and 60 kg/m^3 are shown in Figure 11. As shown in Figure 11a, the compressive strength of FLS specimens tends to increase and then decrease with increasing foam density. Moreover, the compressive strengths at different ages are higher for the foam density of 50 kg/m^3 than for the foam density of 40 kg/m^3 than for the foam density of 60 kg/m^3. This indicates that the compressive strength is not necessarily higher for FLS with a wet density of around 600 kg/m^3, designed for a high theoretical wet density. The main influencing factors for compressive strength are the cementitious material's composition and the incorporated foam's properties. As can be seen from Figure 11b, similar to the effect of dilution multiplier on the compressive density ratio of foam lightweight soils, the increase between ages for specimens with different foam densities is about 150% from 7 days to 28 days and about 30% from 28 days to 56 days, which is not a significant difference. Again, it can be demonstrated that for the strength development of FLS, the cementitious material's composition plays the primary role, independent of the foam density of the foaming agent itself.

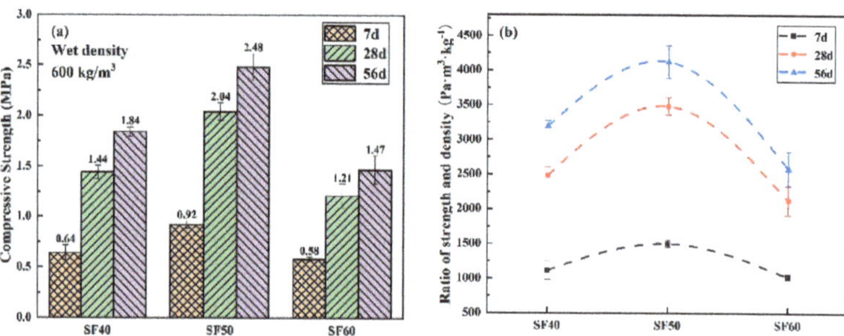

Figure 11. Strength of FLS at different foam densities: (**a**) compressive strength; (**b**) compression-to-density ratio.

3.4.2. Effect of Foaming Agent Types on the Compressive Strength

The 7 days, 28 days, and 56 days compressive strengths and compaction-to-density of FLS prepared with the three foam agents are shown in Figure 12. Figure 12a indicates that the Type-S group has the highest compressive strength at all ages, with compressive strengths of 0.92 MPa, 2.04 MPa, and 2.48 MPa at 7 d, 28 d, and 56 d, respectively. The result can be better explained by the size and distribution of the pore structure of FLS in the previous section of 3.3.2. And, as can be seen in Figure 12b, the rise in the Type-Q and Type-H groups from 7d to 28d is only 105% and 78%, which is lower than the 130% of the Type-S group at the same stage. This suggests that foaming agent types can affect strength development similarly. This may be related to some substances added to the foaming agent. As the substances contained within each foam agent were not identified in the study, it can only be speculated that the foam agent used in the Type-S may have been mixed with ingredients with a strength-enhancing effect.

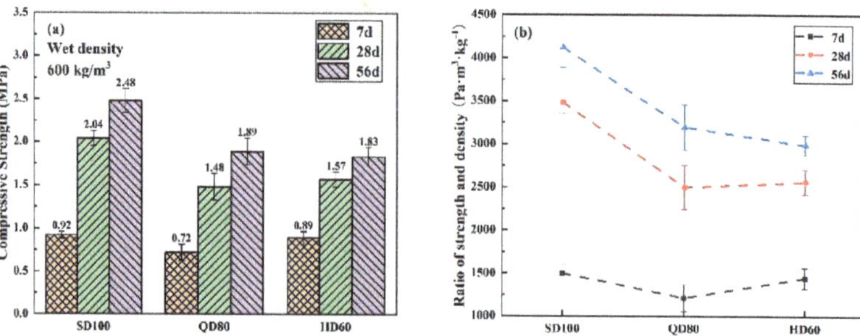

Figure 12. Strength of foam lightweight soils with different types of foaming agents: (**a**) compressive strength; (**b**) compression-to-density ratio.

3.5. Analysis of Foam Properties on the Role of FLS

3.5.1. Mechanism of Action of Foam Density

As can be summarized in the experiments and discussions in Section 3.1, the effect of foam density on foam properties is mainly achieved by influencing the prepared foam's average diameter and the foam film's thickness. For foam density, the foam's average diameter and the foam film's thickness tend to increase and then decrease at foam densities between 40 and 60 kg/m^3. According to a particular form of the Young–Laplace equation: Equation (4) shows that the pressure difference on the foam in the slurry is proportional to the surface tension of the liquid film and inversely proportional to the radius of the foam so that the larger the radius, the more stable the foam is, irrespective of other factors.

$$\Delta p = \frac{2\gamma}{R} \tag{4}$$

where Δp is the pressure difference to which the foam is subjected, γ is the surface tension of the foam film, and R is the foam's curvature radius, in this case, the foam radius. During the mixing of the foam with the slurry, particles of gelling material are adsorbed on the surface of the foam film, and these particles attached to the foam surface reduce the free energy of the specific surface, thus reducing the surface tension. Under the combined influence of these two factors, within a certain range (less than the critical foam diameter) foams with a larger average foam diameter and a thicker foam film are more stable in FLS slurry, as shown in Figure 13.

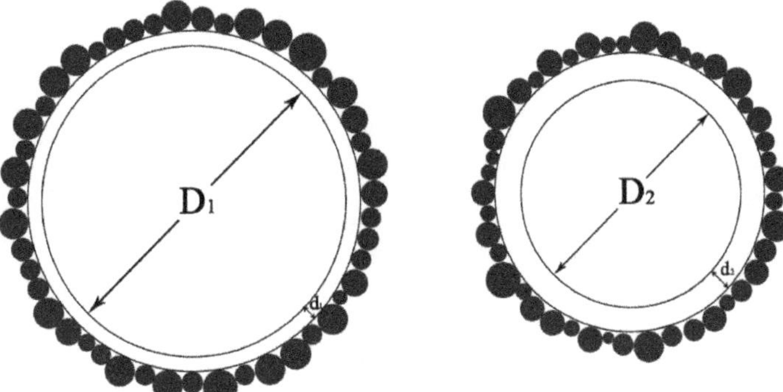

Figure 13. Presence of foam of different particle sizes in the slurry.

For FLS with a designed wet density of 600 kg/m^3, during the mixing and forming process, the foam first reverts from a stacked irregular polygon to a round shape after entering the slurry. Then, it adsorbs in contact with the cementitious material in the slurry, causing the cementitious material particles to adhere around the foam. During this process, the smaller radius foams or foams with a thinner film will merge and break up due to the foam's drainage action, leaving behind stable foams in the slurry. Finally, these foams form holes during the hardening process. Therefore, for the conclusion of foam density on FLS, it can be concluded that for FLS with a design wet density of 600 kg/m^3 a foam density of 50 kg/m^3 gives better performance.

3.5.2. Mechanism of Action of Foaming Agent Types

A situation like the one shown in Figure 14 occurs for a standard round foam in the air. According to the general Equation (5) of the Young–Laplace equation, in the absence of other conditions, the large radius of curvature will discharge to the small radius of curvature due to the pressure difference. b, c, and d are located at the junction of two foams where the radius of curvature tends to infinity, while point a is at the junction of three foams whose radius of curvature is related to the thickness of the liquid film. The liquid film at points b, c, and d will drain towards point a, eventually leading to a rupture of the liquid film of the foam and forming a combined large foam.

$$\Delta p = \gamma \left(\frac{1}{R_1} + \frac{1}{R_2} \right) \tag{5}$$

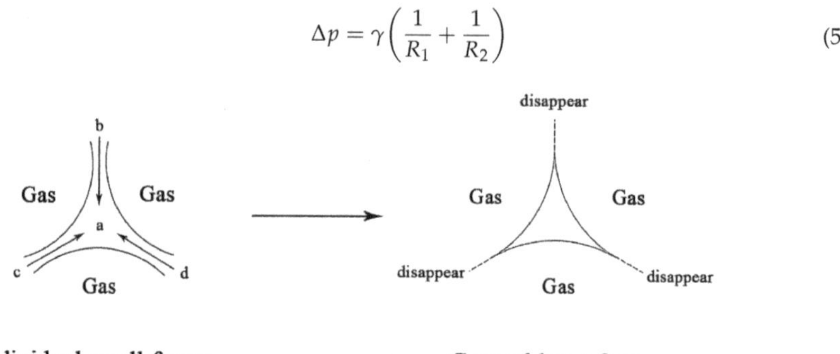

Figure 14. Consolidation within the foam.

The effect of foaming agent types on the properties of the foam can be summarized in two points: Firstly, it affects the diameter of the prepared foam. Secondly, it affects the surface tension of the foam. From Equation (5), it can be judged that the foam of Type-Q is more stable in air, on the one hand. On the other hand, it forms a smaller diameter foam because its surface tension is less than that of Type-H and Type-S. The previous experimental data shows that the internal conditions of the QD80 group of foamed lightweight clay slurry are pretty different from those of SD100 and HD60. It can be assumed that due to the small diameter and low surface tension of most of the QD80 foams, it is difficult to achieve uniform dispersion during mixing with the slurry, and some of the foams still agglomerate to form a "foam isolation zone". In this zone, the slurry is smaller and foamy, which makes it easier for the foam to merge and form more oversized foams. This also explains the tendency for the pore size distribution of the prepared FLS to be bipolar after hardening, despite the small diameter of the QD80 group foam in the air. However, although the HD60 and SD100 groups are relatively similar in terms of pore size distribution, there is still a difference in the compressive strength of FLS specimens, which may be related to the composition of the foaming agent itself or the charge carried by the liquid film, for which there is not yet a reasonable explanation.

Therefore, it can be concluded that the influence of foaming agent types on the FLS is a complex situation. Judging the foaming agent's goodness and the prepared FLS's quality is impossible simply from a few performances. However, one thing is clear: the foam prepared using the foaming agent is stable in the air but not necessarily in the FLS slurry.

In general, because manufacturers keep their technology secret, the internal composition of different foaming agents is not precise, and it is impossible to model the interaction between foam and slurry to establish the relationship between the three directly. However, using the same foaming agent, we can judge a positive correlation between the foam properties in the air, the properties in the FLS slurry, and the properties in FLS samples.

4. Conclusions

(1) The stability of foam in the air can be evaluated by using a 1h settling distance and bleeding rate. The stability of foam in the slurry can be evaluated by using the rate of increase in wet density after the defoaming test.

(2) The stability of foam in the air is related to the liquid film's size and thickness. However, the stability of foam in the slurry is additionally associated with compatibility. For the same type of foaming agent, the stability of foam of different foam density in the air is the same as that in the slurry, with 50 kg/m^3 as the best. For different types of foaming agents, the stability in the air is not the same as in the slurry for the same

foam density. Experimental measurements are necessary to determine the stability of foaming agent types.

(3) Foams with a 40–60 kg/m^3 density are irregularly polygonal in the air and return to their original round shape when entering the slurry. According to the Young–Laplace equation, foam with a larger diameter adsorbs more cementitious material particles on the surface during slurry incorporation and mixing, thus obtaining superior stability. Smaller-diameter foams will more readily undergo surface tension drainage and merge into more oversized-diameter foams.

(4) For the FLS with a design wet density of 600 kg/m^3, a foam density of 50 kg/m^3 gives better performance. After hardening, the samples prepared with foaming agent Types-S had the smallest average pore size of 299 μm and the highest compressive strength of 2.04 MPa at 28 d. The FLS prepared with the above three foaming agents met the technical specifications of the actual road-filling projects.

Author Contributions: Conceptualization, H.L., H.W. and G.Z.; methodology, H.L. and H.W.; validation, Y.W., H.W. and Y.W.; investigation, Y.W.; resources, J.L.; data curation, C.S.; writing—original draft preparation, C.S.; writing—review and editing, H.W.; visualization, G.Z.; supervision, H.W.; funding acquisition, Y.W., H.L. and J.L. All authors have read and agreed to the published version of the manuscript.

Funding: This research was funded by the China Construction Group Limited Project: "Integrated Construction Technology Research on Soft Ground Track Grade High Speed Road", grant number CSCEC-2019-Z-29.

Institutional Review Board Statement: Not applicable.

Informed Consent Statement: Not applicable.

Data Availability Statement: The data that support the findings of this study are available from the corresponding author upon reasonable request.

Conflicts of Interest: The authors declare no conflict of interest.

References

1. Chica, L.; Alzate, A. Cellular Concrete Review: New Trends for Application in Construction. *Constr. Build. Mater.* **2019**, *200*, 637–647. [CrossRef]
2. Amran, Y.H.M.; Farzadnia, N.; Abang Ali, A.A. Properties and Applications of Foamed Concrete; a Review. *Constr. Build. Mater.* **2015**, *101*, 990–1005. [CrossRef]
3. Narayanan, N.; Ramamurthy, K. Structure and Properties of Aerated Concrete: A Review. *Cem. Concr. Compos.* **2000**, *22*, 321–329. [CrossRef]
4. Chandni, T.J.; Anand, K.B. Utilization of Recycled Waste as Filler in Foam Concrete. *J. Build. Eng.* **2018**, *19*, 154–160. [CrossRef]
5. Singh, P.; Bhardwaj, S.; Bera, P.; Lone, T.; Karim, S.; Singh, S.K. Development of Forecasting Model for Prediction of Compressive Strength of Foamed Concrete Using Density with W/C Ratio and S/C Ratio by the Application of ANN. *IOP Conf. Ser. Earth Environ. Sci.* **2021**, *889*, 012039. [CrossRef]
6. Huang, J.; Su, Q.; Zhao, W.; Li, T.; Zhang, X. Experimental Study on Use of Lightweight Foam Concrete as Subgrade Bed Filler of Ballastless Track. *Constr. Build. Mater.* **2017**, *149*, 911–920. [CrossRef]
7. Wang, Y.; Wan, H.; Liu, H.; Zhang, G.; Xu, X.; Shen, C. Preparation and Properties of Low-Carbon Foamed Lightweight Soil with High Resistance to Sulphate Erosion Environments. *Materials* **2023**, *16*, 4604. [CrossRef] [PubMed]
8. Watabe, Y.; Noguchi, T. Site-Investigation and Geotechnical Design of D-Runway Construction in Tokyo Haneda Airport. *Soils Found.* **2011**, *51*, 1003–1018. [CrossRef]
9. Liu, H.; Liu, G.; Wang, H.; Wan, H.; Xu, X.; Shen, C.; Xuan, J.; He, Q. Preparation and Performance Study of Large Volume Foamed Lightweight Soil for an Intelligent Networked Vehicle Test Site. *Materials* **2022**, *15*, 5382. [CrossRef]
10. Jones, M.R.; Ozlutas, K.; Zheng, L. Stability and Instability of Foamed Concrete. *Mag. Concr. Res.* **2016**, *68*, 542–549. [CrossRef]
11. Feneuil, B.; Aimedieu, P.; Scheel, M.; Perrin, J.; Roussel, N.; Pitois, O. Stability Criterion for Fresh Cement Foams. *Cem. Concr. Res.* **2019**, *125*, 105865. [CrossRef]
12. Krämer, C.; Azubike, O.M.; Trettin, R.H.F. Reinforced and Hardened Three-Phase-Foams. *Cem. Concr. Compos.* **2016**, *73*, 174–184. [CrossRef]
13. Wu, K.; Han, H.; Li, H.; Dong, B.; Liu, T.; De Schutter, G. Experimental Study on Concurrent Factors Influencing the ITZ Effect on Mass Transport in Concrete. *Cem. Concr. Compos.* **2021**, *123*, 104215. [CrossRef]

14. Huang, Y.; Gong, L.; Shi, L.; Cao, W.; Pan, Y.; Cheng, X. Experimental Investigation on the Influencing Factors of Preparing Porous Fly Ash-Based Geopolymer for Insulation Material. *Energy Build.* **2018**, *168*, 9–18. [CrossRef]

15. Roussel, N.; Ovarlez, G.; Garrault, S.; Brumaud, C. The Origins of Thixotropy of Fresh Cement Pastes. *Cem. Concr. Res.* **2012**, *42*, 148–157. [CrossRef]

16. Dhasindrakrishna, K.; Pasupathy, K.; Ramakrishnan, S.; Sanjayan, J. Effect of Yield Stress Development on the Foam-Stability of Aerated Geopolymer Concrete. *Cem. Concr. Res.* **2020**, *138*, 106233. [CrossRef]

17. Koehler, S.A.; Hilgenfeldt, S.; Stone, H.A. Foam Drainage on the Microscale: I. Modeling Flow through Single Plateau Borders. *J. Colloid Interface Sci.* **2004**, *276*, 420–438. [CrossRef] [PubMed]

18. Koehler, S.A.; Hilgenfeldt, S.; Weeks, E.R.; Stone, H.A. Foam Drainage on the Microscale II. Imaging Flow through Single Plateau Borders. *J. Colloid Interface Sci.* **2004**, *276*, 439–449. [CrossRef]

19. Panesar, D.K. Cellular Concrete Properties and the Effect of Synthetic and Protein Foaming Agents. *Constr. Build. Mater.* **2013**, *44*, 575–584. [CrossRef]

20. Gu, G.; Xu, F.; Huang, X.; Ruan, S.; Peng, C.; Lin, J. Foamed Geopolymer: The Relationship between Rheological Properties of Geopolymer Paste and Pore-Formation Mechanism. *J. Clean. Prod.* **2020**, *277*, 123238. [CrossRef]

21. Shah, S.N.; Mo, K.H.; Yap, S.P.; Yang, J.; Ling, T.-C. Lightweight Foamed Concrete as a Promising Avenue for Incorporating Waste Materials: A Review. *Resour. Conserv. Recycl.* **2021**, *164*, 105103. [CrossRef]

22. Anderson, R.; Zhang, L.; Ding, Y.; Blanco, M.; Bi, X.; Wilkinson, D.P. A Critical Review of Two-Phase Flow in Gas Flow Channels of Proton Exchange Membrane Fuel Cells. *J. Power Sources* **2010**, *195*, 4531–4553. [CrossRef]

23. Bao, Z.; Niu, Z.; Jiao, K. Numerical Simulation for Metal Foam Two-Phase Flow Field of Proton Exchange Membrane Fuel Cell. *Int. J. Hydrogen Energy* **2019**, *44*, 6229–6244. [CrossRef]

24. Krämer, C.; Kowald, T.L.; Trettin, R.H.F. Pozzolanic Hardened Three-Phase-Foams. *Cem. Concr. Compos.* **2015**, *62*, 44–51. [CrossRef]

25. Shen, C.; Liu, H.; Wan, H.; Li, J.; Liu, P.; He, Q.; Xuan, J. Effect of Cementitious Material Composition on the Performance of Low-Carbon Foamed Lightweight Soil. *Buildings* **2023**, *13*, 759. [CrossRef]

26. *GB/T 11969-2020*; Test Methods of Autoclaved Aerated Concrete. State Administration for Market Regulation: Beijing, China, 2020.

27. *CECS 249:2008*; Technical Specification for Cast-In-Situ Foamed Lightweight Soil. Beijing China Planning Publishing House: Beijing, China, 2008.

 *materials*

Article

Study on Preparation and Performance of Foamed Lightweight Soil Grouting Material for Goaf Treatment

Zhizhong Zhao [1,2], Jie Chen [2,3,4], Yangpeng Zhang [2,4,5,*], Tinghui Jiang [2,4] and Wensheng Wang [6,*]

1   Guangxi Baining Expressway Co., Ltd., Nanning 533800, China; liuzyjlu2019@163.com
2   Guangxi Key Lab of Road Structure and Materials, Nanning 530007, China; cj_engineering@163.com (J.C.); jth1949@outlook.com (T.J.)
3   Guangxi Hetian Expressway Co., Ltd., Nanning 530022, China
4   Guangxi Transportation Science and Technology Group Co., Ltd., Nanning 530007, China
5   School of Traffic and Transportation Engineering, Changsha University of Science and Technology, Changsha 410114, China
6   College of Transportation, Jilin University, Changchun 130025, China
*   Correspondence: zyp_engineering@outlook.com (Y.Z.); wangws@jlu.edu.cn (W.W.)

Abstract: The harm goafs and other underground cavities cause to roads, which could lead to secondary geological hazards, has attracted increased attention. This study focuses on developing and evaluating the effectiveness of foamed lightweight soil grouting material for goaf treatment. The study examines the foam stability of different foaming agent dilution ratios by analyzing foam density, foaming ratio, settlement distance, and bleeding volume. The results show that there is no significant variation in foam settlement distance for different dilution ratios, and the difference in foaming ratio does not exceed 0.4 times. However, the bleeding volume is positively correlated with the dilution ratio of the foaming agent. At a dilution ratio of $60\times$, the bleeding volume is about 1.5 times greater than that at $40\times$, which reduces foam stability. Furthermore, an appropriate amount of sodium dodecyl benzene sulfonate improves both the foaming ability of the foaming agent and the stability of the foam. Additionally, this study investigates how the water–solid ratio affects the basic physical properties, water absorption, and stability of foamed lightweight soil. Foamed lightweight soil with target volumetric weights of 6.0 kN/m^3 and 7.0 kN/m^3 meet the flow value requirement of 170~190 mm when the water–solid ratio ranges are set at 1:1.6~1:1.9 and 1:1.9~1:2.0, respectively. With an increasing proportion of solids in the water–solid ratio, the unconfined compressive strength initially increases and then decreases after 7 and 28 days, reaching its maximum value when the water–solid ratio is between 1:1.7 and 1:1.8. The values of unconfined compressive strength at 28 days are approximately 1.5–2 times higher than those at 7 days. When the water ratio is excessively high, the water absorption rate of foamed lightweight soil increases, resulting in the formation of connected pores inside the material. Therefore, the water–solid ratio should not be set at 1:1.6. During the dry–wet cycle test, the unconfined compressive strength of foamed lightweight soil decreases, but the rate of strength loss is relatively low. The prepared foamed lightweight soil meets the durability requirements during dry–wet cycles. The outcomes of this study may aid the development of enhanced approaches for goaf treatment using foamed lightweight soil grout material.

Keywords: goaf treatment; grouting material; foamed lightweight soil; water–solid ratio

Citation: Zhao, Z.; Chen, J.; Zhang, Y.; Jiang, T.; Wang, W. Study on Preparation and Performance of Foamed Lightweight Soil Grouting Material for Goaf Treatment. *Materials* **2023**, *16*, 4325. https://doi.org/10.3390/ma16124325

Academic Editor: Carlos Leiva

Received: 5 May 2023
Revised: 4 June 2023
Accepted: 6 June 2023
Published: 12 June 2023

1. Introduction

The mining industry has been a crucial element of human civilization for millennia. Nevertheless, the issue of goaf treatment has become increasingly significant with the advancement of mining activities [1,2]. The goaf—the excavated space left behind after minerals are extracted—poses substantial risks to the environment, as well as to the safety of mining operations and transport, if not adequately treated [3–5]. Therefore, the development of an efficient goaf treatment method is imperative to secure the sustainable growth of the transportation infrastructure.

Researchers from both domestic and international institutions have taken an avid interest in studying the effects of underground cavities, such as goafs, on highways, due to their potential to cause significant damage [6,7]. With the swift growth of China's expressway construction, certain highways have been affected to differing degrees by secondary geological hazards stemming from coal mining subsidence in recent years. As a result, some scholars have initiated investigations into goaf impacts. Cao et al. discussed research on detecting goafs in mines, emphasizing the significance of precision exploration techniques and the adoption of established technology, such as 3D seismic exploration and transient electromagnetic method, for identifying presumed mining areas [8]. Their study also introduces the usage of 3D laser scanning technology to visualize concealed mined-out areas and analyze their formation mechanisms. Zhang et al. [9] examined the stress on the center of a goaf roof in a gypsum mine affected by faults, concluding that the mining stope's structural parameters are reasonable but faults can cause subsidence. To evaluate the stability of surface structures, accurate detection approaches for deserted goafs are indispensable. A combination of incremental, conventional, and seismic methodologies was employed to locate an abandoned goaf on China's Mu Shi expressway [10]. Han et al. analyzed the excavation stability and reinforcement measures of a cutting slope with a collapsed goaf roadway and mining face, finding that the collapsed mining face was the main factor affecting stability and proposing specific slope ratios for excavation as well as reinforcing methods [3]. In general, the technology for building highways in goaf regions remains in its early stages of development, with no established systematic theories or technological frameworks in place. As a result, the stability and strengthening of goaf roadways are crucial areas of inquiry.

After coal seams are extracted, abandoned goaf areas are created that disrupt the original state of equilibrium in the rock formations and generate secondary stresses [11,12]. As soon as the stress accumulates and exceeds the critical threshold of the surrounding rock, the goaf will become distorted and collapse, resulting in disasters such as roof fractures and collapse of overlying strata. This poses a risk to lives and properties while simultaneously leading to wastage of land resources. Moreover, uneven settlements of embankments in road construction may cause the roadbed and pavement structures to crack. Currently, reinforcement is the most effective method for treating goaf areas [13,14]. Grouting reinforcement is a widely employed treatment measure for both domestic and international highway goaf zones [15–17]. It involves injecting a configured slurry material into the rock and soil of the goaf area using appropriate equipment. The slurry has fluidity when it is freshly prepared. Once it reaches voids, cracks, or pores in the rock and soil, it solidifies and hardens gradually into a robust consolidation body that adheres to the original rock and soil, forming a whole. This technique improves the bearing capacity, impermeability, and deformation control of the rock and soil. Wang et al. proposed a modified theoretical formula for seepage grouting in goaf foundations that takes into account the fracture distribution characteristics and superposition effect of porous grouting. The formula was validated through laboratory tests and has great engineering significance, as it aids in designing optimal spacing between grouting holes to reduce residual deformation and activation deformation in goaf foundations [18]. When selecting grouting materials for reinforcement projects, it is important to consider their impact on the reinforcement effect and cost [19]. Researchers from around the world have conducted extensive research on grouting materials, mainly focusing on cement–clay slurry, single-cement slurry, cement–fly ash slurry, cement–water glass slurry, and other materials [20]. Using conventional grouting materials to fill goaf areas makes it difficult for the slurry to quickly accumulate, results in significant loss of slurry, and leads to low stone formation rates, thereby increasing the cost of treatment. New materials and technologies, including recycled rubber crumbs [21,22] and steel fiber [23], have been developed to enhance the performances of concrete. Additionally, temperature and humidity greatly influence the performance of grouting materials, particularly their dry-wetting resistance, which is highly relevant in actual engineering [24]. Lightweight fillers, compared to traditional grouting materials, offer several benefits such as being lightweight, possessing high strength and good seismic performance, being environmentally friendly and conserving resources [25–27]. Foamed lightweight soil grouting material is widely used in the field of civil engineering due to its excellent performance, including its high strength, low thermal conductivity, and good workability [28–30]. In recent years, researchers have applied this material to the goaf treatment field, providing advantages over traditional methods such as backfilling with solid materials or filling with water. It can be easily injected into the goaf through pipelines, filling up the space effectively, and thereby greatly reducing the risk of surface subsidence caused by goaf collapse [31,32].

This study aims to investigate the development and effectiveness of a foamed lightweight soil grouting material for the treatment of goaf. The prepared material will be used for the goaf treatment in the Guangxi Province, China. The study involves preparing the foamed lightweight soil grouting material for goaf treatment and discussing its performance. Firstly, the foam stability of different foaming agent dilution ratios is analyzed by testing the density, foaming ratio, settlement distance, and bleeding volume. Furthermore, the influence of the water–solid ratio on the basic physical properties of the material is investigated. Additionally, the water absorption and stability of the material are studied.

2. Materials and Methods

2.1. Raw Materials

This study investigates the use of cement, foaming agents, and water as the primary raw materials for foamed lightweight soil grouting material in goaf treatment. Cement acts as the main cementitious material in the preparation process, providing strength to the foamed lightweight soil through hydration and hardening reactions. Due to the large number of air bubbles in the foamed lightweight soil, the resulting material has relatively lower strength compared to ordinary soil. To ensure that the strength of the material meets project requirements, higher-quality cement is typically required compared to ordinary soil. Ordinary Portland Cement with a grade of 42.5 is selected for this experimental study due to its high early strength, fast setting time, and high density, which make it an effective cementitious material. The basic performance of the cement used in this study is listed in Table 1; it meets the requirements of "Common Portland Cement" (GB 175-2007) specifications in China.

Table 1. Basic properties of cement in this study.

Item	Density (kg/m^3)	Standard Consistence (%)	Initial Setting Time (min)	Final Setting Time (min)	Composition (%)				
					SiO$_2$	CaO	Al$_2$O$_3$	Fe$_2$O$_3$	LOI
Result	3117	28.7	390	535	21.14	60.43	6.11	2.54	1.02

In the preparation of foamed lightweight soil grouting material, a foaming agent is a necessary raw material that significantly affects the resulting material's various properties. For this study, a composite foaming agent obtained from Henan Huatai Cement Technology Co., Ltd (Zhengzhou, China) was selected, as it meets the requirements of Chinese standard "Technical Specification for Design and Construction of Cast-in situ Foamed Light-weight Soil Subgrade" (TJG F1001-2001). Section 3.1 provides details on the foaming agent's specific performance indices, such as foam density, settlement distance, foaming ratio, and bleeding volume at various dilution ratios.

Water is also a significant constituent of foamed lightweight soil grouting material, with its quality and consumption directly impacting the resulting material's properties. To avoid impacts on foam quality and material properties, blending water must be free from impurities such as oil or organic matter. Excessive impurities in blending water can negatively impact foam quality by reducing the effectiveness of foaming agents. Furthermore, harmful impurities in water can erode the cement paste when mixed, thereby influencing the properties of the resulting foamed lightweight soil grouting material. Additionally, water consumption affects the material's physical and mechanical properties and its durability. Therefore, to guarantee the quality of foamed lightweight soil grouting material, strict control of water quality and consumption during the preparation process is necessary. Moreover, the maximum particle size of soil aggregate should not exceed 5 mm, in accordance with the Chinese standard "Technical Specification for Cast-in situ Foamed Lightweight Soil" (CECS 249: 2008).

2.2. Preparation Method of Foamed Lightweight Soil Grouting Material

In this study, the prefabricated foam mixing method is used to prepare the foamed lightweight soil grouting material. The detailed preparation process of foamed lightweight soil grouting material is described as follows:

1. Step 1: According to the mix proportion scheme, the corresponding cement and water are weighted and then added to the mixer for the first mixing;
2. Step 2: According to the corresponding dilution ratio, the water and foaming agent are weighted for dilution and added into the foaming machine. The foaming agent diluent is introduced into air under pressure, thus generating a foam group;
3. Step 3: A certain amount of foam is weighted using an electronic balance, which is poured into the mixer. Then, it is mixed with the cement paste for 2 min;
4. Step 4: After the mixture is stirred evenly, the physical performance tests such as flow value, wet weight, etc. are conducted. In addition, the prepared foamed lightweight soil mixture is placed in the mold for manual vibration and molding. After the mold is removed, it is cured to the specified curing age according to the standard method, and then the subsequent performance test is carried out.

The main instruments used to prepare the foamed lightweight soil grouting material are a foaming machine and a mixer. The preparation process flow and procedure of foamed lightweight soil grouting material are shown in Figure 1.

The foaming machine and mixer are the primary instruments used to prepare foamed lightweight soil grouting material. The process flow and procedure for this preparation are illustrated in Figure 1. In accordance with the Chinese standard "Specification for Design of Highway Subgrades" (JTG D30-2015), the minimum and maximum construction wet densities of foamed lightweight soil grouting materials should be greater than 500 kg/m^3 and less than 1100 kg/m^3, respectively. To assess the effects of the water–solid ratio on the resulting material, water–solid ratios were varied between 1:1.6~1:2.0 for target volumetric weights of 6.0 kN/m^3 and 7.0 kN/m^3. The specific mix proportion scheme is detailed in Table 2.

Table 2. The mix proportion of foamed lightweight soil grouting material.

No.	Target Volumetric Weight (kN/m^3)	Water–Solid Ratio	Cement (kg/m^3)	Water (kg/m^3)	Foam (kg/m^3)
A0		1:1.6	356	223	21
A1		1:1.7	364	214	21
A2	6.0	1:1.8	372	207	21
A3		1:1.9	379	199	21
A4		1:2.0	386	193	22
B0		1:1.6	419	261	19
B1		1:1.7	429	252	19
B2	7.0	1:1.8	437	243	19
B3		1:1.9	446	234	20
B4		1:2.0	453	226	20

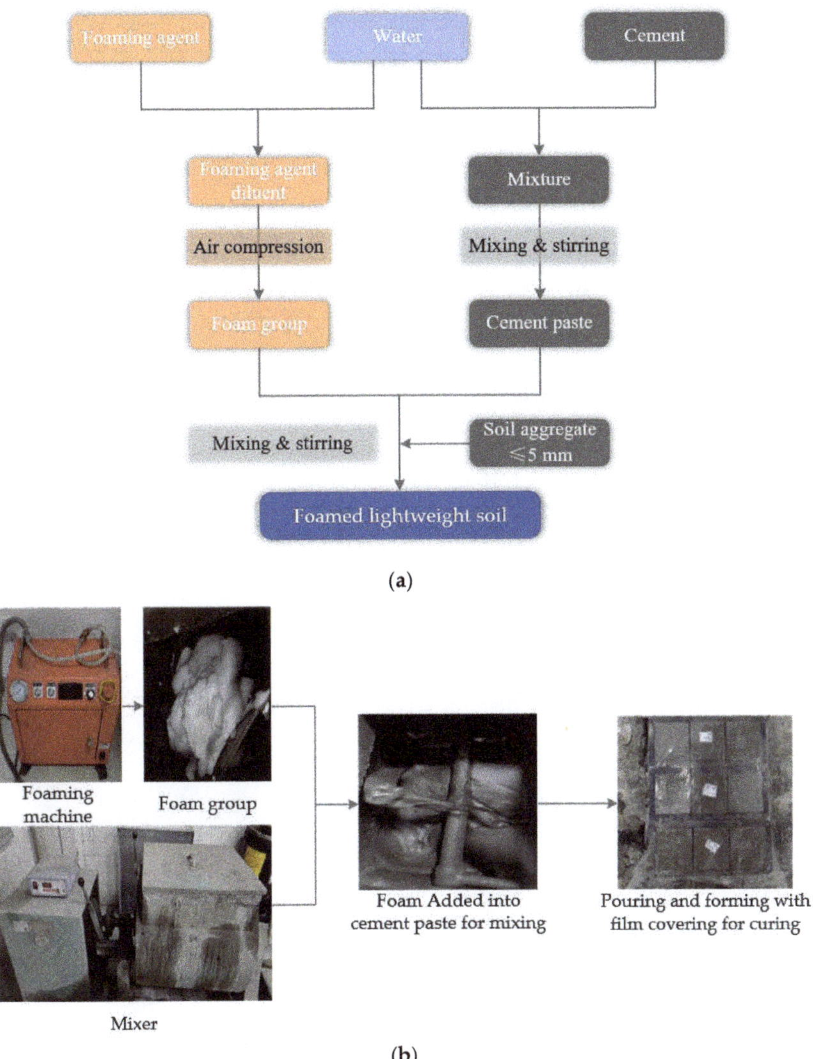

(**a**)

(**b**)

Figure 1. The preparation process of foamed lightweight soil grouting material: (**a**) Preparation process flow; (**b**) preparation process procedure.

2.3. Experimental Methods

2.3.1. Stability Test of Foam

Referring to the Chinese standard "Technical Specification for Foamed Concrete Application on Highway" (DB33/T 996-2015), the density and foaming ratio of foam could be tested. The test process was repeated three times for parallel experiments and the average value was taken. The foaming ratio (M) could be calculated as shown in Equation (1):

$$M = V/(m_2 - m_1)/\rho_0, \tag{1}$$

in which ρ_0 is the density of the foam solution (taken as 1.0 g/cm^3); m_1 is the mass of the empty dry stainless steel measuring cup (g); m_2 is the total mass of the stainless-steel measuring cup and foam (g); V is the volume of the stainless-steel measuring cup (cm^3)

The effectiveness of foamed lightweight soil depends on the quality of the foaming agent; therefore, this study subjected the foaming agent to a series of performance tests and optimized the preparation parameters. Settlement distance pertains to the foam settling distance after standing still for one hour, while bleeding volume refers to the amount of water that leaks out due to foam rupture during the same period. Both measurements were taken for 1 L of foam. Similarly, in accordance with the Chinese standard "Technical Specification for Foamed Concrete Application on Highway" (DB33/T 996-2015), settlement distances and bleeding volumes of foam can be examined using the following procedure: firstly, clean and dry the testing apparatus' glass container and glass tube, stainless steel measuring cup, and flat knife. Then, use the measuring cup to take a foam sample and fill the glass container with it. Level the sample slowly along the rim of the measuring cup using the flat knife. Cover the container with an aluminum float and wait for one hour. After that, record the settlement distance value (in mm) accurately to 0.5 mm from the scaled glass container. Next, open the small valve on the glass tube of the testing apparatus and transfer the liquid into a graduated cylinder, recording the bleeding volume (in mL) by reading the liquid volume with accuracy up to 1 mL. The device used to measure settlement distance and bleeding volume of foam is shown in Figure 2.

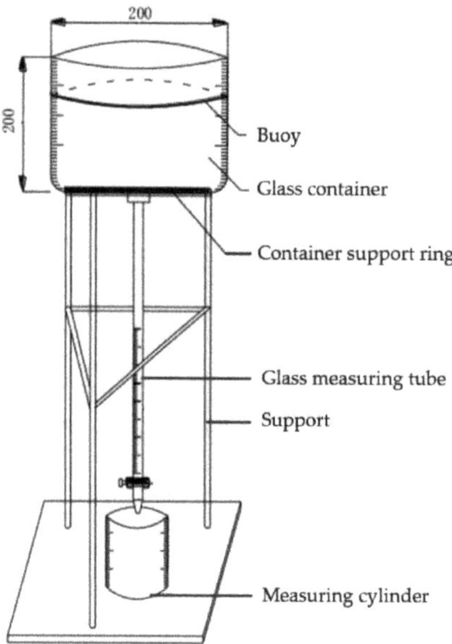

Figure 2. The testing device of settlement distance and bleeding volume of foam.

2.3.2. Performance Test of Foamed Lightweight Soil

In accordance with the Chinese standard "Technical Specification for Foamed Mixture Lightweight Soil Filling Engineering" (CJJ/T 177-2012), the flow value is typically assessed using the cylinder method, utilizing a cylinder with a diameter and height of 80 mm. The following procedure should be followed: clean and dry the hollow cylinder, stainless steel plate, measuring cup, and flat knife before placing the hollow cylinder horizontally on the stainless-steel plate. Use one measuring cup to scoop a sample and carefully transfer the sample into the hollow cylinder without overflowing it. Tap the outside of the hollow cylinder gently with your finger to distribute the sample evenly throughout the cylinder's entire length. Use a flat knife to slowly level the sample along the upper port plane of the hollow cylinder. Then, lift the hollow cylinder vertically with both hands and let the sample settle for one minute. Finally, measure the maximum horizontal diameter of the sample using a vernier caliper to obtain the actual flow value, illustrated in Figure 3.

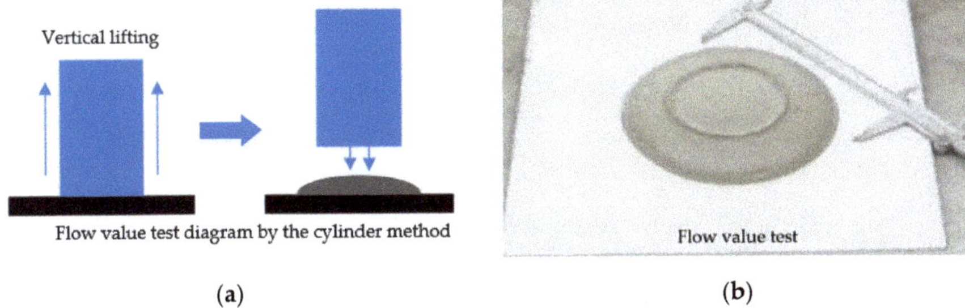

(**a**)　　　　　　　　　　　　　　　　　　(**b**)

Figure 3. The flow value test diagram: (**a**) Flow value test diagram by the cylinder method; (**b**) flow value test.

The unconfined compressive strength test of foamed lightweight soil grouting material assesses its capacity to withstand external pressure. In line with the Chinese standard "Foamed Concrete Block" (JC/T 1062-2007), a microcomputer-controlled electronic universal testing machine was used in this study to perform uniaxial compression tests on the samples. The unconfined compression strength test value of each sample represents the average of three test results.

Foamed lightweight soil, being a porous cement-based material, forms a large number of pores upon solidification, akin to autoclaved aerated concrete. Accordingly, the water absorption rate of foamed lightweight soil grouting material can be determined, drawing guidance from the "Test Methods of Autoclaved Aerated Concrete" (GB/T 11969-2020). Three specimens are cured for 28 days and transferred to a drying oven set at (60 ± 5) °C and (80 ± 5) °C for 24 h each. The specimens are then dried to a constant weight at (105 ± 5) °C, with their corresponding masses recorded as M_0. After the specimens have been cooled for 6 h at room temperature, they are immersed in a constant temperature water tank maintained at (20 ± 2) °C. Water is added to maintain a height of 1/3 of the specimen for 24 h, followed by adding water to reach 2/3 of the specimen. After an additional 24 h, water is added above 30 mm and kept for another 24 h. Finally, surface moisture is wiped off with a wet cloth, and the specimens are weighed immediately to obtain their mass, recorded as M. Using Equation (2), the mass water absorption rate (W) of the specimen is calculated.

$$W = (M - M_0)/M_0 \times 100\%, \tag{2}$$

The water stability of foamed lightweight soil grouting material was investigated using the dry–wet cycle test. After curing for 28 days, specimens were first dried in a (60 ± 5) °C oven for 48 h and then placed in a water tank with a constant temperature of (20 ± 2) °C, where they were fully immersed for 24 h, representing one dry–wet cycle. After repeating this process five times, we recorded the unconfined compressive strength as $q_{u,dw}$. The water stability coefficient (K_r), a dimensionless quantity, is defined as the ratio of the unconfined compressive strength value obtained from the dry–wet cycle test to that of the standard cured specimen at 28 days ($q_{u,28}$). K_r is computed using Equation (3).

$$K_r = q_{u,dw}/q_{u,28}, \tag{3}$$

3. Results and Discussions

3.1. Foam Performance Analysis of Foaming Agent Based on Stability

3.1.1. Effect of Dilution Ratio on Foam Stability

In this study, the foaming agent was diluted at different ratios (i.e., $40\times$, $50\times$, and $60\times$) before producing foam for the foamed lightweight soil grouting material. Following the "Technical Specification for Foamed Concrete Application on Highway" (DB33/T 996-2015), the foam density, foaming ratio, settlement distance, and bleeding volume were tested. Comparative results of foam stability at the three different dilution ratios are presented in Table 3 and Figure 4, including foam density, foaming ratio, settlement distance, and bleeding volume.

Table 3. The foam density and settlement distance at different dilution ratios.

Dilution Ratio	Foam Density (kg/m^3)	Settlement Distance (mm)
40×	30.7	0
50×	30.3	0
60×	30.3	0

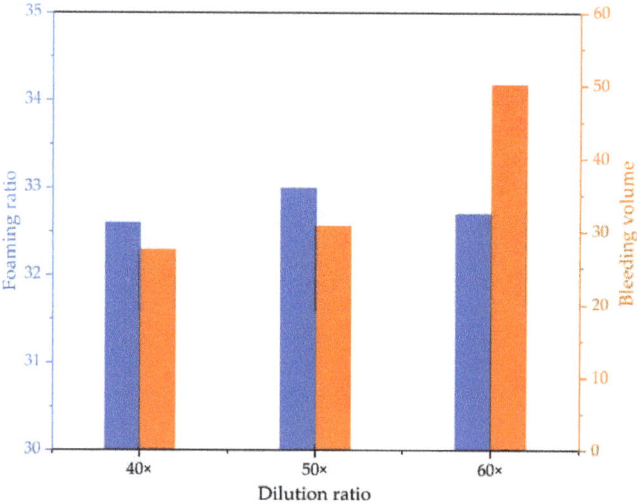

Figure 4. The foaming ratio and bleeding volume at different dilution ratios.

The experimental results presented in Table 3 demonstrate a relatively stable foam density for the foaming agent diluted at different ratios, with differences not exceeding 0.4 kg/m^3. Furthermore, Figure 4 shows that the prepared foam has a stable foaming ratio, with differences not exceeding 0.4 times. However, the one-hour settlement distance of the foam under the three different dilution ratios is negligible, indicating that foam stability cannot be assessed based on the settlement distance at this time. This is likely due to the relatively good foam stability resulting from the foaming agent. Despite some bubbles bursting, the foam mainly comprises neighboring small bubbles merging and connecting into large bubbles, resulting in insignificant sinking and settlement distance. Additionally, the bleeding volume result shown in Figure 4 indicates that as the dilution ratio of the foaming agent increases, the bleeding volume of the prepared foam significantly increases, leading to decreased foam stability.

3.1.2. Foam Stability Improvement

The analysis indicates that the foaming performance of the foaming agent is better when diluted 40 times. Sodium dodecyl benzene sulfonate is an anionic surfactant known for properties such as emulsification, foaming, and foam stabilization. To evaluate if sodium dodecyl benzene sulfonate is a viable foam stabilizer for the foaming agent, foam stability tests were performed by adding sodium dodecyl benzene sulfonate to the dilution solution. Sodium dodecyl benzene sulfonate powder was added to the foaming agent dilution solution to form solutions with concentrations of 0.1%, 0.2%, and 0.3%, followed by foam preparation. Comparative results of foam stability for the foaming agent at the three different dilution ratios with sodium dodecyl benzene sulfonate, including foam density, foaming ratio, settlement distance, and bleeding volume, are shown in Table 4 and Figure 5, respectively.

Table 4. The foam density and settlement distance at a 40× dilution ratio with different sodium dodecyl benzene sulfonate concentrations.

Concentration	Foam Density (kg/m^3)	Settlement Distance (mm)
0.1%	30.9	0
0.2%	30.2	0
0.3%	29.7	1

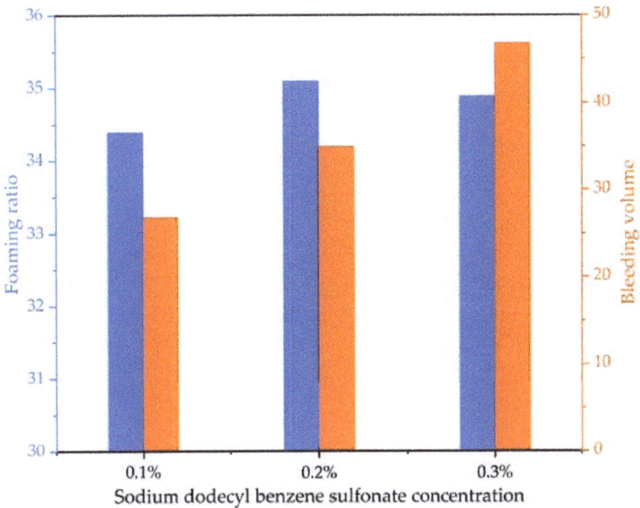

Figure 5. The foaming ratio and bleeding volume at a 40× dilution ratio with different sodium dodecyl benzene sulfonate concentrations.

A comparison of Figures 4 and 5 reveals that adding 0.2% sodium dodecyl benzene sulfonate to the foaming agent dilution solution increases the foaming ratio. The foaming ratio at a 50× dilution ratio rises to 35.1, indicating that sodium dodecyl benzene sulfonate enhances the foaming ability of the foaming agent. However, comparative analysis among Tables 3 and 4 as well as Figures 4 and 5 demonstrates that increasing sodium dodecyl benzene sulfonate concentration negatively affects the foam stability generated by the foaming agent. On one hand, the addition of sodium dodecyl benzene sulfonate increases settlement distance, with foam prepared from a foaming agent dilution solution at a dilution rate of 40× with sodium dodecyl benzene sulfonate concentration of 0.3% showing a 1 mm increase in settlement distance after 1 h (Table 4). On the other hand, the bleeding volume from foams prepared from a foaming agent dilution solution with different sodium dodecyl benzene sulfonate concentrations increases to varying degrees (Figure 5). Although sodium dodecyl benzene sulfonate does not significantly alter the properties of the foaming agent, it can function as a foam stabilizer for the foaming agent and improve the foaming ratio to a certain extent without significantly increasing the bleeding volume. Therefore, a sodium dodecyl benzene sulfonate concentration of 0.1% is more appropriate for enhancing the foaming ability of the foaming agent.

3.2. Performance Evaluation of Foamed Lightweight Soil Considering Water–Solid Ratios

This study investigates the effect of the water–solid ratio in foamed lightweight soil grouting material on its fundamental physical properties, such as wet density, flow value, and unconfined compressive strength. Moreover, it explores the water absorption and water stability of the foamed lightweight soil grouting material. Foamed lightweight soil grouting material is prepared using a dilution rate of 40× for the foaming agent and a concentration of 0.1% for sodium dodecyl benzene sulfonate. The resulting mixture proportions from Table 2 are used to prepare foamed lightweight soil specimens, which are then tested for their wet density, flow value, unconfined compressive strength, water absorption, and water stability.

3.2.1. Effect of Water–Solid Ratio on Flow Value

Flow value is used to measure the fluidity of foamed lightweight soil grouting material. The foamed lightweight soil grouting material has good fluidity and can achieve a self-compacting construction state without rolling and vibration. According to the Chinese standard "Technical Specification for Foamed Mixture Lightweight Soil Filling Engineering" (CJJ/T 177-2012), the flow value results are tested using the cylinder method, as shown in Table 5 and Figure 6.

Table 5. The flow value results of foamed lightweight soil at various water–solid ratios.

No.	Target Volumetric Weight (kN/m^3)	Water–Solid Ratio	Flow Value (mm)	Wet Density (kN/m^3)
A0		1:1.6	186	6.35
A1		1:1.7	186	6.64
A2	6.0	1:1.8	177	6.34
A3		1:1.9	170	6.58
A4		1:2.0	165	6.26
B0		1:1.6	239	6.58
B1		1:1.7	207	7.10
B2	7.0	1:1.8	224	7.30
B3		1:1.9	174	6.84
B4		1:2.0	188	6.22

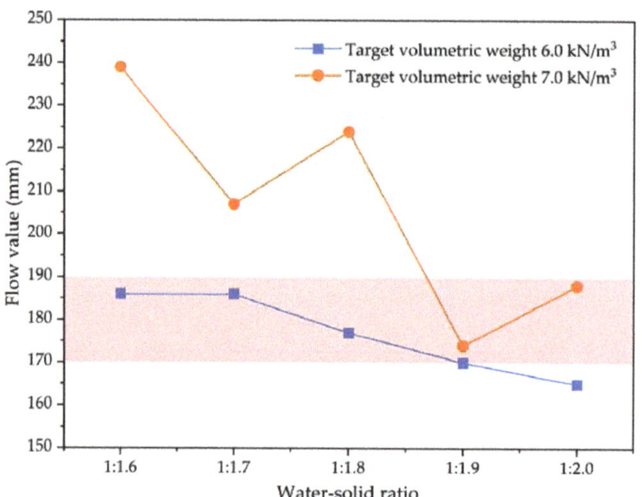

Figure 6. The flow value results of foamed lightweight soil at various water–solid ratios.

Based on the experimental results, at a target volumetric weight of 6.0 kN/m^3, an increase in solid proportion (ratio of cement) in the water–solid ratio parameter of foamed lightweight soil grouting material leads to a decrease in its flow value. This suggests that decreasing the water content has a significant impact on fluidity within this volumetric weight range, which aligns with existing literature [26]. However, at a target volumetric weight of 7.0 kN/m^3, the flow value of the foamed lightweight soil grouting material fluctuates with the water–solid ratio, unlike the pattern observed at 6.0 kN/m^3. Under higher volumetric weights, bubbles in the foamed lightweight soil break and gather, forming large pores that negatively affect fluidity, resulting in a downward trend of fluctuation. When the water–solid ratio is constant, increasing the amount of cement paste improves the fluidity of foamed lightweight soil grouting material with a target volumetric weight of 7.0 kN/m^3 compared to that with a target volumetric weight of 6.0 kN/m^3. According to the Chinese standard "Technical Specification for Design and Construction of Cast-in situ Foamed Lightweight Soil Subgrade" (TJG F1001-2012), the flow value of foamed lightweight soil grouting material in physical engineering should be between 170 and 190 mm. Thus, it can be concluded that a water–solid ratio of 1:1.6–1:1.9 meets the flow value requirement of foamed lightweight soil grouting material at a target volumetric weight of 6.0 kN/m^3. Similarly, a water–solid ratio of 1:1.9–1:2.0 meets the flow value requirement of foamed lightweight soil grouting material at a target volumetric weight of 7.0 kN/m^3.

3.2.2. Effect of Water–Solid Ratio on Unconfined Compression Strength

Unconfined compressive strength is a crucial indicator of the mechanical strength properties of foamed lightweight soil grouting material. It denotes the maximum vertical compressive stress that an unconfined foamed lightweight soil specimen can withstand. Due to the material's high porosity, it is vital to ensure that the compressive strength of foamed lightweight soil grouting material meets the necessary engineering construction standards at low densities. This concern is particularly relevant in practical applications. The unconfined compressive strength test results of foamed lightweight soil grouting material prepared for two distinct groups with target volumetric weights of 6.0 kN/m^3 and 7.0 kN/m^3 are illustrated in Figure 7.

The unconfined compressive strength test results in Figure 7 indicate that the unconfined compressive strength of the foamed lightweight soil specimens cured for 7 and 28 days increases initially and then decreases as the water–solid ratio changes from 1:1.6 to 1:2.0. The maximum unconfined compressive strength values for the specimens cured for 7 and 28 days occur when the water–solid ratio is between 1:1.7–1:1.8. The reduction in strength could be due to the formation of interconnected pores inside the foamed lightweight soil grouting material caused by the evaporation of water during the hardening process when the solid proportion in the water–solid ratio is small (i.e., the water proportion is larger). On the other hand, a large solid proportion of foamed lightweight soil grouting material could increase defects, leading to a decrease in the unconfined compressive strength due to increased friction between foam and cement paste and uneven distribution of foam in the paste. In addition, the unconfined compression strength of foamed lightweight soil materials at different target volumetric weights indicates that higher volumetric weight and cement content result in smaller strength than lower volumetric weight and cement content. This could be attributed to the generation of more internal stress in foamed lightweight soil materials with high cement content and volumetric weight, resulting in cracks and defects that reduce strength performance. Furthermore, high-volumetric weight cement materials could have larger porosity, affecting the compactness and strength of foamed lightweight soil materials. Conversely, low-cement content and low-volumetric weight foamed lightweight soil materials can reduce internal stresses and porosity effectively and improve material compactness, thereby enhancing their strength performance.

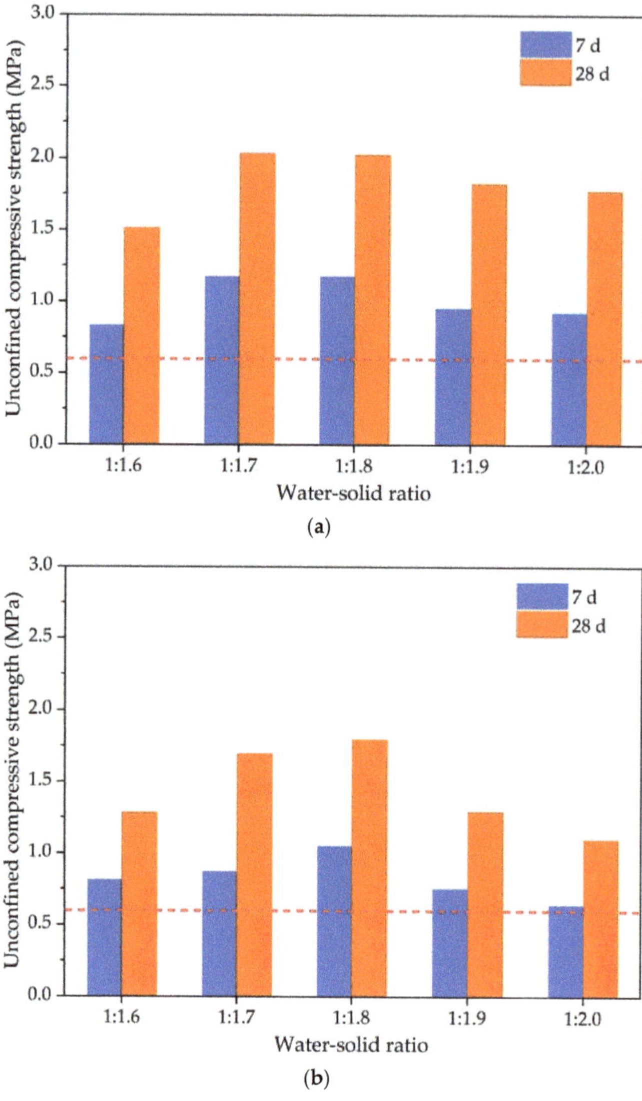

Figure 7. The unconfined compressive strength test results of the foamed lightweight soil grouting material: (**a**) Foamed lightweight soil with target volumetric weight of 6.0 kN/m^3; (**b**) foamed lightweight soil with target volumetric weight of 7.0 kN/m^3.

According to the Chinese standard "Technical Specification for Design and Construction of Cast-in situ Foamed Lightweight Soil Subgrade" (TJG F1001-2011), it is stipulated that for high-grade highways, when the top surface of the foamed lightweight soil subgrade is more than 0.8 m away from the bottom surface of the road, the 28-day unconfined compressive strength of the foamed lightweight soil should not be less than 0.6 MPa. Therefore, the abovementioned mix proportions of foamed lightweight soil grouting material with the target volumetric weight of 6.0 kN/m^3 and 7.0 kN/m^3 meet the strength requirements. The foamed lightweight soil specimens with the target volumetric weight of 6.0 kN/m^3 are prepared for the water absorption and water stability.

3.2.3. Effect of Water–Solid Ratio on Water Absorption

In this study, the water absorption is represented by water absorption rate, which refers to the ability of foamed lightweight soil grouting material to absorb water under the soaking state. Referring to the "Test Methods of Autoclaved Aerated Concrete" (GB/T 11969-2020), the water absorption rate results of foamed lightweight soil grouting material are tested. The measured mass water absorption of foamed lightweight soil grouting material is shown in Figure 8.

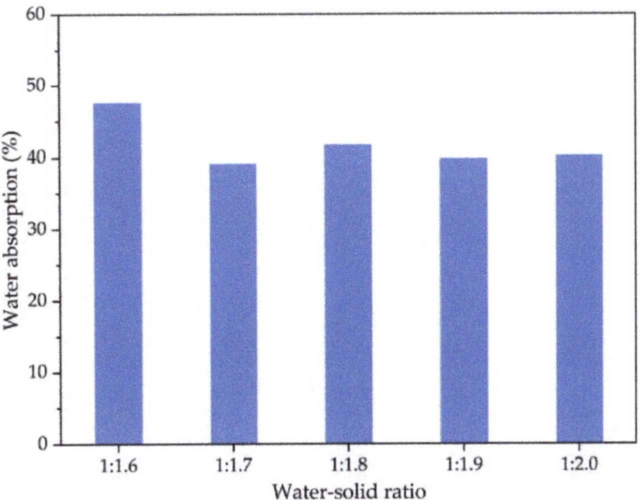

Figure 8. The mass water absorption results of foamed lightweight soil at various water–solid ratios.

Figure 8 shows that the water absorption rate of the foamed lightweight soil grouting material varies with changes in the water–solid ratio. At a water–solid ratio of 1:1.6, the water absorption rate is the highest at 47.6%. However, when the water–solid ratio is between 1:1.7 and 1:2.0, the water absorption rate is lower than that of foamed lightweight soil grouting material with a water–solid ratio of 1:1.6, and the water absorption rate of foamed lightweight soil grouting material with different water–solid ratios changes slightly. The internal pore connectivity of the foamed lightweight soil grouting material affects its water absorption rate. A greater number of connected pores results in a higher water absorption rate. Therefore, the similar water absorption rates of the foamed lightweight soil grouting material with water–solid ratios of 1:1.7–1:2.0 indicate similar internal pore conditions. The higher water absorption rate of the foamed lightweight soil grouting material with a water–solid ratio of 1:1.6 suggests that a high water proportion in the water–solid ratio can promote the formation of connected pores inside the foamed lightweight soil grouting material.

3.2.4. Effect of Water–Solid Ratio on Water Stability

Water stability describes a subgrade's resistance to the adverse effects of water after being invaded by it. The water stability coefficient typically characterizes a soil's water stability, with higher coefficients indicating better performance. In areas with frequent rainfall, roads are often exposed to water and erosion. If foamed lightweight soil grouting material is used as a filling or grouting material for subgrades, its water stability performance must meet high requirements. Therefore, this study uses the dry–wet cycle test to investigate the water stability of the foamed lightweight soil grouting material, and the calculated water stability coefficient results are presented in Figure 9.

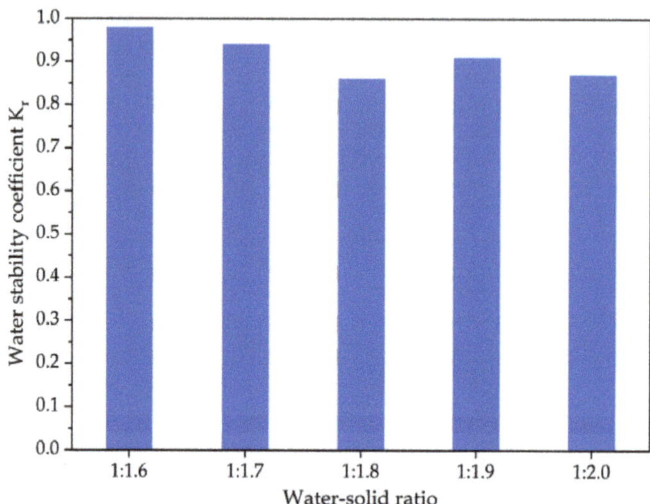

Figure 9. The water stability coefficient results of foamed lightweight soil at various water–solid ratios.

As shown in Figure 9, the dry–wet cycles results in a decrease in the unconfined compressive strength of foamed lightweight soil grouting material. This is mainly due to the fact that under dry–wet cycle conditions, the pore wall inside the foamed lightweight soil grouting material is constantly subjected to stress damage, resulting in a decrease in its unconfined compressive strength. From the water stability coefficient results of foamed lightweight soil grouting material at various water–solid ratios, it can also be seen that under the five mixing ratios, the water stability coefficients of foamed lightweight soil grouting material are all greater than 0.85, indicating that although there are many pores inside the foamed lightweight soil grouting material, the corresponding water stability is good. The water absorption results also showed that despite the changing water–solid ratio, the internal pore conditions are relatively stable, leading to good water stability. In future studies, more dry–wet cycles will be carried out to investigate the durability evaluation of foamed lightweight soil, which would provide more systematic research conclusions.

4. Conclusions

This study investigates the preparation and performance of foamed lightweight soil grouting material for goaf treatment. Initially, foam stability is analyzed by testing the density, foaming ratio, settlement distance, and bleeding volume of foam at different foaming agent dilution ratios. Furthermore, the impact of the water–solid ratio on the basic physical properties of foamed lightweight soil grouting material is examined. Additionally, the water absorption and water stability of the material are studied. Based on the study's findings, the following conclusions can be drawn:

(1) The dilution ratio of the foaming agent affects foam stability. The settlement distance of foam does not vary significantly for different dilution ratios, with no more than a 0.4 times difference in the foaming ratio. However, bleeding volume is positively correlated with the dilution ratio of the foaming agent. The bleeding volume of the foaming agent at a $60\times$ dilution ratio is approximately 1.5 times that at a $40\times$ dilution ratio, reducing the foam stability. Furthermore, an appropriate amount of sodium dodecyl benzene sulfonate is suitable for enhancing the foaming capacity of the foaming agent and stabilizing the foam.

(2) The water–solid ratio affects the flowability of foamed lightweight soil. The suitable water–solid ratios for producing a target volumetric weight of 6.0 kN/m^3 and 7.0 kN/m^3,

while satisfying the required flow value of 170~190 mm, range between 1:1.6~1:1.9 and 1:1.9~1:2.0, respectively.

(3) The unconfined compressive strength of foamed lightweight soil initially increases and then decreases after curing for 7 days and 28 days as the proportion of solids in the water–solid ratio increases. The maximum value is achieved at a water–solid ratio of 1:1.7~1:1.8. Additionally, the unconfined compressive strength values at 28 days are approximately 1.5~2 times higher than those at 7 days.

(4) Water absorption tests indicate that excessive water in the water–solid ratio leads to connected pores inside the foamed lightweight soil, resulting in a higher water absorption rate. Thus, a water–solid ratio of 1:1.6 should be avoided.

(5) Foamed lightweight soil experiences reduced unconfined compressive strength under dry–wet cycling conditions, but the rate of strength loss is minimal. Additionally, foamed lightweight soil prepared with various water–solid ratios satisfies the durability requirements under dry–wet cycling conditions.

In summary, foamed lightweight soil grouting material with a target density of 6.0 kN/m^3 demonstrates superior flowability, unconfined compressive strength, water absorption, and water stability when the water–solid ratio is set between 1:1.7~1:1.8. Further research involving micro analysis and modification mechanisms should be conducted to facilitate a more systematic evaluation of foamed lightweight soil in future studies.

Author Contributions: Conceptualization, Z.Z., Y.Z. and W.W.; Methodology, Z.Z., Y.Z., T.J. and W.W.; Validation, J.C. and T.J.; Formal Analysis, Z.Z., J.C. and T.J.; Investigation, Y.Z., J.C., T.J. and W.W.; Writing—Original Draft Preparation, Z.Z.; Writing—Review and Editing, Y.Z., J.C. and W.W.; Project Administration, Y.Z. and W.W.; Funding Acquisition, Y.Z. and W.W. All authors have read and agreed to the published version of the manuscript.

Funding: This research was funded by the key research and development Plan of the Guangxi Zhuang Autonomous Region: AB22080012; Nanning innovation and entrepreneurship leading talents "Yongjiang plan" funded project: 2019009; the key technology projects in the transportation industry of the Guangxi Zhuang Autonomous Region in 2022: item 13 of 2022.

Institutional Review Board Statement: Not applicable.

Informed Consent Statement: Not applicable.

Data Availability Statement: Not applicable.

Conflicts of Interest: The authors declare no conflict of interest.

References

1. Liu, Z.J.; Deng, D.Q.; Feng, J.F.; Wang, R.Z.; Fan, J.K.; Ma, Y.F. Study on Natural Settlement Index Characteristics of Iron-Bearing Tailings Applied to Goaf Filling Treatment. *Sustainability* **2022**, *14*, 739. [CrossRef]
2. Song, H.Q.; Xu, J.J.; Fang, J.; Cao, Z.G.; Yang, L.Z.; Li, T.X. Potential for mine water disposal in coal seam goaf: Investigation of storage coefficients in the Shendong mining area. *J. Clean. Prod.* **2020**, *244*, 118646. [CrossRef]
3. Han, C.P.; Zu, F.J.; Du, C.; Shi, L. Analysis of Excavation Stability and Reinforcement Treatment of the Cutting Slope under the Influence of Old Goaf. *Appl. Sci* **2022**, *12*, 8698. [CrossRef]
4. Wang, C.X.; Lu, Y.; Qin, C.R.; Li, Y.Y.; Sun, Q.C.; Wang, D.J. Ground Disturbance of Different Building Locations in Old Goaf Area: A Case Study in China. *Geotech. Geol. Eng.* **2019**, *37*, 4311–4325. [CrossRef]
5. Marian, D.P.; Onica, I.; Marian, R.R.; Floarea, D.A. Finite Element Analysis of the State of Stresses on the Structures of Buildings Influenced by Underground Mining of Hard Coal Seams in the Jiu Valley Basin (Romania). *Sustainability* **2020**, *12*, 1598. [CrossRef]
6. Guo, Q.B.; Li, Y.M.; Meng, X.R.; Guo, G.L.; Lv, X. Instability risk assessment of expressway construction site above an abandoned goaf: A case study in China. *Environ. Earth Sci.* **2019**, *78*, 588. [CrossRef]
7. Guo, Q.B.; Meng, X.R.; Li, Y.M.; Lv, X.; Liu, C. A prediction model for the surface residual subsidence in an abandoned goaf for sustainable development of resource-exhausted cities. *J. Clean. Prod.* **2021**, *279*, 123803. [CrossRef]
8. Cao, B.; Wang, J.; Du, H.; Tao, Y.B.; Liu, G.W. Research on comprehensive detection and visualize of hidden cavity goaf. *Sci. Rep* **2022**, *12*, 22309. [CrossRef]
9. Zhang, C.R.; Yang, X.C.; Ren, G.F.; Ke, B.; Song, Z.L. Instability of Gypsum Mining Goaf Under the Influence of Typical Faults. *IEEE Access* **2019**, *7*, 88635–88642. [CrossRef]

10. Zhang, S.K.; Jiang, P.; Lu, L.; Wang, S.; Wang, H.H. Evaluation of Compressive Geophysical Prospecting Method for the Identification of the Abandoned Goaf at the Tengzhou Section of China's Mu Shi Expressway. *Sustainability* **2022**, *14*, 3785. [CrossRef]
11. Zhou, X.D.; Yang, Y.L.; Zheng, K.Y.; Miao, G.D.; Wang, M.H.; Li, P.R. Study on the Spontaneous Combustion Characteristics and Prevention Technology of Coal Seam in Overlying Close Goaf. *Combust. Sci. Technol.* **2022**, *194*, 2233–2254. [CrossRef]
12. Li, X.L.; Wei, S.J.; Wang, J. Study on Mine Pressure Behavior Law of Mining Roadway Passing Concentrated Coal Pillar and Goaf. *Sustainability* **2022**, *14*, 3175. [CrossRef]
13. Yu, Y.; Bai, J.B.; Wang, X.Y.; Zhang, L.Y. Control of the Surrounding Rock of a Goaf-Side Entry Driving Heading Mining Face. *Sustainability* **2020**, *12*, 2623. [CrossRef]
14. Hebblewhite, B.K.; Lu, T. Geomechanical behaviour of laminated, weak coal mine roof strata and the implications for a ground reinforcement strategy. *Int. J. Rock. Mech. Min.* **2004**, *41*, 147–157. [CrossRef]
15. Li, L.S.; Qian, D.Y.; Yang, X.G.; Jiao, H.X. Pressure Relief and Bolt Grouting Reinforcement and Width Optimization of Narrow Coal Pillar for Goaf-Side Entry Driving in Deep Thick Coal Seam: A Case Study. *Minerals* **2022**, *12*, 1292. [CrossRef]
16. Zhang, Z.Y.; Shimada, H. Numerical Study on the Effectiveness of Grouting Reinforcement on the Large Heaving Floor of the Deep Retained Goaf-Side Gateroad: A Case Study in China. *Energies* **2018**, *11*, 1001. [CrossRef]
17. Dun, Z.L.; Wang, M.Q.; Zou, Y.F.; Ren, L.W. Laboratory Tests on Performance of Waste-Clay-Brick-Powder Cement Grouting Materials for Ground Improvement in Mine Goaf. *Adv. Eng. Mater.* **2022**, *24*, 2101575. [CrossRef]
18. Wang, H.; Li, Y.; Dun, Z.Y.; Cheng, J.H.; Dun, Z.L.; Wu, C.X. Seepage Grouting Mechanism for Foundations in Goaf Sites considering Diffusion Paths. *Geofluids* **2022**, *2022*, 8394811. [CrossRef]
19. Huang, S.J.; Zhao, G.M.; Meng, X.R.; Cheng, X.; Gu, Q.H.; Liu, G.; Zhu, S.K. Development of Cement-Based Grouting Material for Reinforcing Narrow Coal Pillars and Engineering Applications. *Processes* **2022**, *10*, 2292. [CrossRef]
20. Feng, X.; Xia, C.; Zhang, S.F.; Li, C.G.; Zhao, H.K.; Wu, J.F.; Chen, M.J.; Yan, J. Properties and Engineering Applications of a New Goaf Grouting Filling Material. *Geofluids* **2022**, *2022*, 6778076. [CrossRef]
21. Chalangaran, N.; Farzampour, A.; Paslar, N.; Fatemi, H. Experimental investigation of sound transmission loss in concrete containing recycled rubber crumbs. *Adv. Concr. Constr.* **2021**, *11*, 447–454. [CrossRef]
22. Chalangaran, N.; Farzampour, A.; Paslar, N. Nano Silica and Metakaolin Effects on the Behavior of Concrete Containing Rubber Crumbs. *CivilEng* **2020**, *1*, 264–274. [CrossRef]
23. Mansouri, I.; Shahheidari, F.S.; Hashemi, S.M.A.; Farzampour, A. Investigation of steel fiber effects on concrete abrasion resistance. *Adv. Concr. Constr.* **2020**, *9*, 367–374. [CrossRef]
24. Farzampour, A. Temperature and humidity effects on behavior of grouts. *Adv. Concr. Constr.* **2017**, *5*, 659–669. [CrossRef]
25. Shaheen, Y.B.; Eltaly, B.A.; Yousef, S.G.; Fayed, S. Structural Performance of Ferrocement Beams Incorporating Longitudinal Hole Filled with Lightweight Concrete. *Int. J. Concr. Struct. Mater.* **2023**, *17*, 21. [CrossRef]
26. Shi, X.N.; Huang, J.J.; Su, Q. Experimental and numerical analyses of lightweight foamed concrete as filler for widening embankment. *Constr. Build. Mater.* **2020**, *250*, 118897. [CrossRef]
27. Lim, S.K.; Tan, C.S.; Lim, O.Y.; Lee, Y.L. Fresh and hardened properties of lightweight foamed concrete with palm oil fuel ash as filler. *Constr. Build. Mater.* **2013**, *46*, 39–47. [CrossRef]
28. Allouzi, R.; Almasaeid, H.; Alkloub, A.; Ayadi, O.; Allouzi, R.; Alajarmeh, R. Lightweight foamed concrete for houses in Jordan. *Case Stud. Constr. Mat.* **2023**, *18*, e01924. [CrossRef]
29. Song, Y.; Lange, D. Influence of fine inclusions on the morphology and mechanical performance of lightweight foam concrete. *Cem. Concr. Comp.* **2021**, *124*, 104264. [CrossRef]
30. Kadela, M.; Kukielka, A.; Malek, M. Characteristics of Lightweight Concrete Based on a Synthetic Polymer Foaming Agent. *Materials* **2020**, *13*, 4979. [CrossRef]
31. Xu, C.Y.; Han, L.J.; Tian, M.L.; Wang, Y.J.; Jin, Y.H. Study on Foamed Concrete Used as Gas Isolation Material in the Coal Mine Goaf. *Energies* **2020**, *13*, 4377. [CrossRef]
32. Shrestha, R.; Redmond, L.; Thompson, J.J. Diagonal tensile strength and lap splice behavior of concrete masonry assemblies with lightweight grout. *Constr. Build. Mater.* **2022**, *344*, 128141. [CrossRef]

Article

Corrosion Behavior of Magnesium Potassium Phosphate Cement under Wet–Dry Cycle and Sulfate Attack

Linlin Chong [1,2], Jianming Yang [3,*], Jin Chang [1,*], Ailifeila Aierken [1,2], Hongxia Liu [1], Chaohuan Liang [1] and Dongyong Tan [1]

1 College of Civil Engineering, Changsha University, Changsha 410022, China
2 Innovation Center for Environmental Ecological and Green Building Materials of CCSU, Changsha University, Changsha 410022, China
3 School of Civil Engineering, San Jiang University, Nanjing 210012, China
* Correspondence: yjm_kk@163.com (J.Y.); changjin1906@126.com (J.C.)

Abstract: This paper investigated the influence of dry–wet cycles and sulfate attack on the performance of magnesium potassium phosphate cement (MKPC) as well as the effect of waterglass on MKPC. X-ray diffraction (XRD), TG-DTG, and scanning electron microscopy (SEM-EDS) were used to examine the phase composition and microstructure of MKPC. The results showed that the flexural and compressive strength of an MKPC paste increased initially and subsequently decreased in different erosion environments. The final strength of the M0 paste exposed to the SK-II environment was the highest, while that of the M0 paste exposed to the DW-II environment was the lowest. The final volume expansion value of MKPC specimens under four corrosion conditions decreased in the following order: DW-II, M0 > SK-II, M0 > DW-II, M1 > SK-I, M0 > DW-I, M0. Compared to the full-soaking environment, the dry–wet cycles accelerated sulfate erosion and the appearance of damages in the macro and micro structure of the MKPC paste. With the increase in the number of the dry and wet cycles, more intrinsic micro-cracks were observed, and the dissolution of hydration products was accelerated. Under the same number of dry–wet cycles, the strength test and volume stability test showed that the durability in a Na_2SO_4 solution of the MKPC paste prepared with 2% waterglass (M1) was superior to that of the original M0 cement. The micro analysis indicated that waterglass can improve the compactness of the microstructure of MPC and prevent the dissolution of struvite-K.

Keywords: magnesium potassium phosphate cement (MKPC); sulfate attack; dry–wet cycles; strength change; volume stability

Citation: Chong, L.; Yang, J.; Chang, J.; Aierken, A.; Liu, H.; Liang, C.; Tan, D. Corrosion Behavior of Magnesium Potassium Phosphate Cement under Wet–Dry Cycle and Sulfate Attack. *Materials* **2023**, *16*, 1101. https://doi.org/10.3390/ma16031101

Academic Editor: Jose M. Bastidas

Received: 9 December 2022
Revised: 15 January 2023
Accepted: 25 January 2023
Published: 27 January 2023

1. Introduction

As one of the most widely used construction materials, concrete is increasingly being used in difficult environments such as salt lakes, salty soil, collapsible soil, as well as in maritime engineering [1]. However, the concrete structures used under such conditions suffer severely from long-term corrosion caused by sulfate attack, which severely affects the normal performance of buildings [2–4]. Furthermore, differences in river and groundwater levels with the changes in season, as well as the splash zone, result in dry–wet cycles. Sulfates have a more severe effect on concrete structures when combined with wet–dry cycles. The combined effects of wet–dry cycles and sulfate pose a greater risk of damage to concrete structures. Magnesium phosphate cement (MPC) has been promoted for the rapid repair of concrete structures in various corrosive environments in order to extend their service life [5]. MPCs have excellent properties as compared to other cementitious materials, such as enhanced adaptability to environmental temperature, fast hardening, high early strength, low shrinkage, better adhesion, etc. [6–11].

Magnesium phosphate cements (MPC) are a class of inorganic cementitious materials containing phosphates as their binder phase, which are formed through neutralization reactions occurring in the cement [6]. In recent years, more attention has been paid to MPC-based repair materials. Research mainly focuses on its preparation technology, modification mechanisms, and repair in the normal environment. With the development of research on MPC, some studies have investigated the water stability [12–15], salt stability (through long-term soaking in a salt solution) [16–20], salt freeze–thaw durability of MPC-based materials [4,21,22] and the passivation of steel [23–26]. Some researchers determined that MPC, unlike common concrete, has excellent sulfate resistance. For example, Li et al. [14] found that the strength of MPC in a Na_2SO_4 solution was higher than that in water and in a NaCl solution. Yang et al. [18,19] revealed that the strength of MPC pre air curing for 28 d was 93% of the original strength after immersion in seawater and even slightly increased after immersion in a Na_2SO_4 solution for 360 days. MPCs used as coating materials for concrete also showed good sulfate resistance [27,28]. As a repair material used in marine concrete construction, MPC is also subjected to sulfate attack in a dry–wet cycling environment. However, only one study described the durability of MPC-based materials during dry–wet cycles in water and salt solutions. Li [29] examined the degradation of MKPC and the effects of fly ash and quartz sand under the same corrosion environment. An MKPC paste with quartz sand showed poor durability when it was treated with dry–wet cycles. By adding fly ash to the MKPC paste, the authors observed a drop in compressive strength and mass. Clearly, more work in this respect is needed to better predict the performance of MPC exposed to marine conditions.

This study aimed to provide further information about the corrosion behavior of an MKPC paste subjected to the effects of sulfate attack and dry–wet cycles by focusing on changes in compressive strength and flexural strength, volume deformation, and microstructure changes. Plain MKPC pastes and a paste with 2% waterglass subjected to dry–wet cycle and sulfate attacks were tested for about one year. A plain paste was also subjected to water immersion and water dry–wet tests for comparison. The phase composition as well as the microstructure changes were evaluated at the end of the test period by means of SEM, EDS TG-DTG, and XRD. This study provides useful information for improving the durability of MPC-based materials and their application in the reinforcement of concrete structures subject to both salt attack and dry–wet cycles.

2. Materials and Methods

2.1. Raw Materials

Dead burnt magnesium was obtained from the Huan ren Dong fang hong Hydropower Factory for Station Magnesite (Liaoning province, China). The specific surface area of magnesium oxide was 230 m^2/kg, its average particle size was 45.26 μm, and its chemical composition determined by X-ray fluorescence analysis is shown in Table 1.

Table 1. Chemical composition of the used MgO powder (wt.%).

Component	MgO	CaO	SiO$_2$	Al$_2$O$_3$	Fe$_2$O$_3$	Na$_2$O	K$_2$O	Others
Content/%	91.85	3.14	3.68	0.17	0.87	-	-	0.30

Industrial-grade potassium dihydrogen phosphates (KH_2PO_4, PDP) were purchased from Georgia Legislature Chemical Company (Lianyungang, Jiangsu province, China). They consisted of colorless or white crystals with a main particle size of 40/380–60/250 (mesh/μm). A composite retarder (CR) [13] was used as a setting retarder. Waterglass with Baume degree of 39.2–40.2 and modulus of 3.2–3.4 was from Yixing Credible Chemical Co., Ltd.

2.2. Specimen Preparation

The mass ratio of alkali component (MgO) to acid component (KH_2PO_4) was set at 3 according to previous results of superior mechanical performance [4,13,30]. Meanwhile, to obtain samples with both good mechanical strength and good workability, the mass ratio of water to solid (w/s) was 0.115. The solid contained MgO, KH_2PO_4, and the composite retarder (CR). The content of the composite retarder was 12.0% by weight of MgO, and the added waterglass (by weight of the solid) was 2 wt.%. The MKPC mix proportions are summarized in Table 2. During casting, MgO powder, PDP, and CR were firstly dry-mixed at a low speed for 1 min, and then water was added and stirred rapidly for another 3 min. Then, the fresh MKPC paste was quickly cast into specimens with dimensions of $40 \times 40 \times 160$ mm and $25 \times 25 \times 280$ mm to test the compressive strength and volume deformation. All the molds were covered with plastic sheets, and the samples were cured for 5 h at room temperature and then demolded. They were then cured indoors at a temperature of $20 \pm 5\ °C$ and a relative humidity of $60 \pm 5\%$ for 28 days.

Table 2. Mixture proportions, fluidity, and initial setting time of the MKPC pastes.

Samples	Solid Materials			Liquid Materials			Fluidity (mm)	Initial Setting Time (min)
	MgO	PDP	Retarder (CR/MgO)	Waterglass (WG/MgO)	Water (w/s Ratio)			
M0	3.0	1	12%	0	0.115		162	20.5
M1	3.0	0.12	12%	2%	0.115		165	22.0

2.3. Environmental Exposure Conditions

After natural curing for 28 days, a batch of MKPC specimens were cured for a longer time under natural conditions and used as reference samples. The remaining samples were divided into four groups and subjected to four different exposure conditions (listed in Table 3) up to 360 days. Part of the specimens were subjected to a full-soaking (i.e., SK) environment, while the other specimens were exposed to a dry–wetting cycle (i.e., DW) environment. In both environments, tap water without sodium sulfate and a 5% Na_2SO_4 solution were selected as the media, which were replaced with fresh ones every 30 days. In the DW environment, the specimens were first immersed in water or the Na_2SO_4 solution for 23 ± 0.5 h (until reaching a full water state) and then dried at 60 °C at a vacuum-drying temperature for 47 ± 0.5 h (to reach a constant mass state). After drying, the specimens were cooled at room temperature for 2 h. Each dry–wet cycle lasted for 3 days; 120 dry–wet cycles were carried out on the specimens in total, which equated to 360 days of immersion.

Table 3. Details of the exposure conditions.

Abbreviation	Fresh Solutions	External Environment	Exposure Period
SK-I	tap water	Full soaking, $20 \pm 5\ °C$	360 days
SK-II	5.0 wt.% Na_2SO_4 solution	Full soaking, $20 \pm 5\ °C$	360 days
DW-I	tap water	Dry–wet cycles, $20 \pm 5\ °C$	360 days (120 cycles)
DW-II	5.0 wt.% Na_2SO_4 solution	Dry–wet cycles, $20 \pm 5\ °C$	

2.4. Experimental Methods

2.4.1. Fluidity and Setting Time

Based on ASTM C1437-2007 [31], NLD-3 jump tables were used to measure the fluidity of the fresh MKPC paste. The MKPC paste initial setting time was measured and recorded following ASTM C191a-2001 [32], using a Vicar apparatus.

2.4.2. Mechanical Testing

According to ASTM C348-2008 [33] and ASTM C 349-2008 [34], three samples for each batch were used to evaluate the flexural strength by a WED-300 electronic universal testing machine at loading rates of 40 to 60 N/s, and six broken samples were then used for the compressive strength test at a loading rate from 2200 to 2600 N/s. The flexural strength and the compressive strength of the MKPC specimens over time were tested in a saturated-surface dry state. The residual strength ratios were calculated as Kf = Rfn/Rf0 and Kc = Rcn/Rc0, where Rf0 (Rc0) and Rfn (Rcn) are the flexural (compressive) strengths of the MKPC specimens (saturated-surface dry state) after 0 and t dry–wet cycles, respectively.

2.4.3. Volume Stability

MKPC paste bars with a size of 25 × 25 × 280 mm were measured every two months by using a Digimatic outside micrometer. The saturated-surface-dry length (L0) at the beginning of the corrosion test was measured, as well as the length of the specimens (Lt) after t dry–wet cycles. The effective length of the sample is shown here based on 250 mm. The deformation rate (εn) was calculated based on JC/T603–2004 [35] as follows:

$$\varepsilon_t = (L_t - L_0)/250 \times 100\% \tag{1}$$

2.4.4. Microanalysis

Following the strength tests, samples of the MKPC paste were collected from the broken specimens for microscopic analysis. To stop the hydration of the samples, they were submerged for at least 2 days in anhydrous alcohol. Then, they were dried to constant mass in a vacuum-drying chamber at 60 °C. Some parts were ground into powder and passed through a 45 μm sieve for TG and XRD analyses. The phase composition of the samples was determined using Rigaku Neo D/max RB X-ray diffractometers (Cu-Ka radiation in the 5° to 80° range with a step size of 0.02°, 40 kV, 100 mA). The powder samples were heated at 10 °C/min from 20 °C to 1100 °C using a NETZSCH STA 409 PC/PG thermal analyzer under the protection of N_2. MKPC paste samples were analyzed using a quantitative environmental scanning electron microscope (Quanta-200, American FEI Company) to determine their morphologies and by energy-dispersive X-ray spectroscopy (EDS).

3. Results and Discussion

3.1. Strength Development

Figure 1 depicts the development of flexural strength and compressive strength in the MKPC paste under different corrosive environments. As shown in Figure 1, the flexural strength and compressive strength of both M0 and M1 increased first to a peak and then decreased as the erosion grew in the four corrosion conditions, while they increased continuously under air curing. After 40 dry–wet cycles, both M0 and M1 showed a higher compressive strength than the air-cured samples, possibly due to better paste hydration resulting from corrosion-induced water adsorption [13,18,36]. The M0 paste subjected to the DW environment at first had a higher strength than the MKPC specimen subjected to the SK environment under the same hydration time. After 120 dry–wet cycles, the final strengths of the samples were in the order AC, M0 > SK-II, M0 > DW-I, M0 > SK-I, M0 > DW-II, M1 > DW-II, M0. The flexural strength and compressive strength of M0 exposed to the SK-II environment were 12 MPa and 63.5 MPa, the highest among those of all the specimens except for that of the air-cured samples. In contrast, the M0 paste exposed to the DW-II environment had the lowest strength. This indicated that wet–dry cycling appeared to be beneficial for improving the water resistance of MKPC, but detrimental for its sulfate resistance. During sulfate attack, the M1 paste reached the peak strength after 40 dry–wet cycles, while the M0 paste reached the peak strength after 20 dry–wet cycles, and the peak strength of the M1 paste was greater than that of the M0 paste. Under the DW-II environmental conditions, this could be explained by a more dense structure and a reduction in the dissolution of hydration products due to the addition of waterglass, resulting in an increase in the flexural and compressive strength and durability of MKPC.

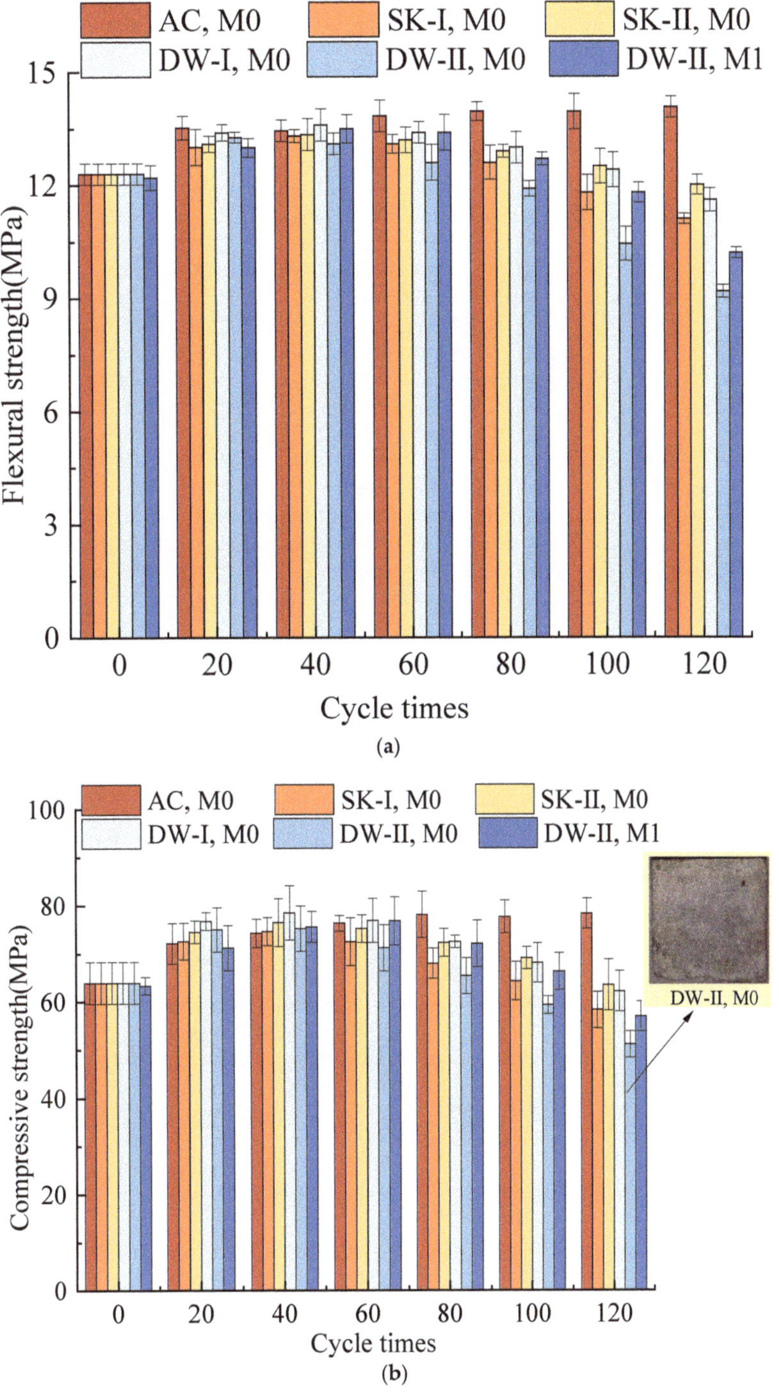

Figure 1. Flexural strength (**a**) and compressive strength (**b**) of MKPC mortars under different curing conditions.

The main raw material in MKPC (i.e., dead burned MgO powder) is obtained by calcining magnesite ($MgCO_3$) at above 1300 °C; so, the surfaces of the obtained particles are smooth and compact and have low water absorption [11,37]. Consequently, the water-to-cement ratio of the fresh MKPC paste necessary to reach the required consistency is low (Table 2), but it is not sufficient for a complete hydration of the MKPC paste. Therefore, there are still some unreacted phosphates in addition to the main hydration product $MgKPO_4 \cdot 6H_2O$ and many unreacted dead burned MgO particles in the hardened MKPC paste. Subsequent hydration in the MKPC paste is closely related to the environmental humidity [13,18,30]. For example, when an MKPC specimen is soaked in water or a salt solution, the infiltration of ambient water will lead to its further hydration. The newly generated MKP crystals will fill the micropores in the hardened MKPC paste, leading to an increase in the strength of the MKPC specimen [13,30]. This explains why the strengths of the MKPC specimens immersed in water and the 5% Na_2SO_4 solution increased first. Particularly, the strength of the MKPC specimen immersed in the 5% Na_2SO_4 solution was higher than that of the MKPC specimen immersed in water for the same period. This was due to the filling effect of some newly generated sulfate-containing crystals in addition to the MKP crystals [14,22].The process of vacuum-drying at 60 °C can accelerate the mutual penetration between unreacted acid and alkali components and promote the hydration process, so that more newly generated hydrates will fill the pores. This is the reason why the hardened MKPC paste subjected to dry–wet cycles with water and 5% Na_2SO_4 at first had higher strength than the MKPC specimen soaked in water and 5% Na_2SO_4 for the same hydration time. With the extension of corrosion, the hydration reaction in the hardened MKPC paste tends to stop, and the over-expansion and decomposition of the erosion products begin to play a leading role in the subsequent degradation of the MKPC specimens under sulfate and water attacks, as corrosion continues [18]. Compared to the M0 samples exposed to the SK-I environment, the M0 samples under dry–wet cycles with water were only intermittently in a water-rich environment, which decreased the hydrolysis and loss of the MKP in the specimens exposed to the DW-I environment and led to a higher strength. However, when the M0 samples were cured under the DW-II environment, the hardened MKPC paste absorbed the solution by capillary action. The salt crystallized as the water evaporated in the vacuum-heating stage. When the salt crystals in the capillaries absorbed water and converted to a state containing crystal water, a high crystallization pressure was produced. This could result in cracking and peeling of the surface of the hardened MKPC paste and corner damage, finally leading to a strength reduction in the MKPC specimens.

3.2. Strength Retention Coefficient

Figure 2 shows the strength retention coefficient of the MKPC paste under various corrosive conditions. Compared to the initial strength, the strength retention coefficient of the M0 paste under different corrosive environments increased first and reached a peak value at 60 dry–wet cycles. The strength residual rate of the specimens (M0) soaked in water and 5% Na_2SO_4 was 106.5% and 107.3% (flexural strength) and 113.3% and 117.5% (compressive strength). The strength residual rate of the specimens after 60 dry–wet cycles in water and 5% Na_2SO_4 was 108.9% and 102.4% (flexural strength) and 120.2% and 113.4% (compressive strength). After 120 dry–wet cycles, the strength residual rate of the specimens (M0) when soaked in water and 5% Na_2SO_4 was 90.2% and 97.56% (flexural strength) and 91.1% and 99.2% (compressive strength), the strength residual rate of the specimens after 120 dry–wet cycles in water and 5% Na_2SO_4 was 94.3% and 74.6% (flexural strength) and 97.2% and 79.9% (compressive strength). M0 soaked in the sulfate solution exhibited significantly better corrosion resistance than after soaking in water. It appears that M0 had a higher resistance to sulfate corrosion than to water corrosion, as previously reported [18]. However, the M0 paste under dry–wet cycles in the 5% Na_2SO_4 solution presented the lowest strength retention ratio in the whole corrosion test when the strength started to drop.

This indicates that dry–wet damage was dominant when the samples were exposed to the combined action of sulfate attack and wet–dry cycling.

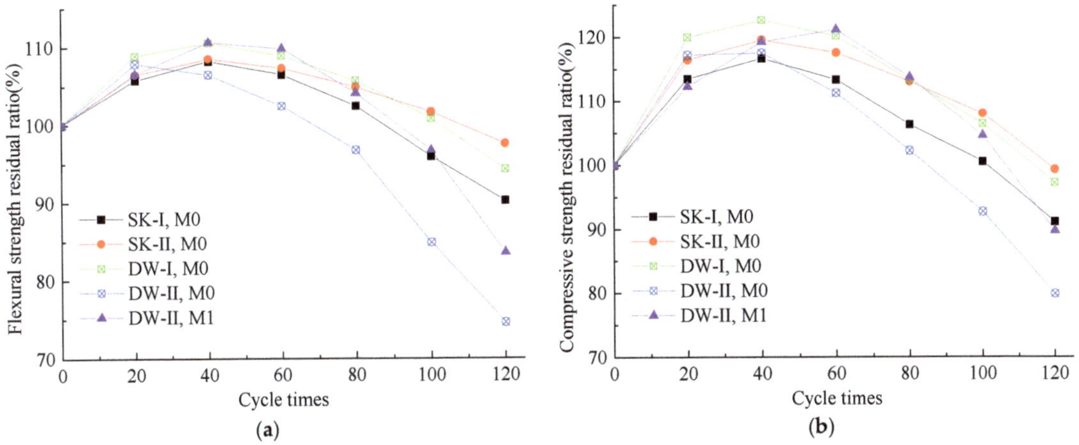

Figure 2. Flexural strength retention ratio (**a**) and compressive strength retention ratio (**b**) of MKPC mortars under different curing conditions.

According to the results of M0 and M1, the increase in the strength of the M1 samples took longer than that of M0, and the strength residual rates of the M1 samples were 83.6% (flexural strength) and 89.8% (compressive strength). The strength residual rates of the M1 samples were more than 5% higher than those of M0 under DW-II conditions, indicating that adding waterglass to the MKPC paste would improve its resistance to dry and wet sulfate cycles. When waterglass is added to MKPC, $H_3SiO_4^-$, $H_2SiO_4^{2-}$ and $HSiO_4^{3-}$ will react with Mg^{2+} in the MKPC paste to form magnesium silicate gels [12], which can fill the pores and make the structure dense. When the hardened MKPC paste was immersed in water, its dense initial structure prevented external water from infiltrating it and the dissolution of MKP and it could also prevent the infiltration of the sulfate solution which in turn reduced the formation of sulfate-containing crystals and the destruction of salt crystals when subjected to dry–wet cycles with the 5% Na_2SO_4 solution.

3.3. Volume Stability

Figure 3 shows the evolution of the deformation of the MKPC specimens exposed to the four corrosive environments. All specimens exhibited expansion deformation, and the value increased as the number of dry–wet cycles increased. This can be attributed to the increased degree of hydration in environments with excess liquid phase [36]. Furthermore, the unreacted MgO and H_2O formed $Mg(OH)_2$, resulting in volume expansion [10,38]. Typically, M0 exhibited an expansion deformation pattern consistent with its corrosion age, except in DW-II corrosion conditions. The specimens swelled rapidly within 60 dry–wet cycle, and the expansion rate was moderate. Compared to M0 cured under SK-II environments, the expansion of M0 under DW-II conditions was significantly lower when the number of dry–wet cycles was limited to 40. The expansion tended to be more pronounced when the number of dry–wet cycles exceeded 80. Initially, the MKPC specimens subjected to dry–wet cycling were significantly less deformed than the MKPC specimens soaked in water or the sulfate solution. This can be explained by the fact that the MKPC specimens soaked in the corrosion solution were always water-saturated, while those subjected to the dry–wet cycles were in a drying state, which means they would not undergo deformation due to crystal expansion as a result of water absorption. A further increase in the number of the dry–wet cycles would result in an expansion of the MKPC specimen. This can be attributed to the crystallization pressure of salt resulting from the water-saturated crystallization of the sulfate salt during drying, as well as the newly formed hydrated product (MKP) and the newly generated sulfate-containing phases. The volume expansion rate of M0 soaked in the 5% Na_2SO_4 solution was much higher than that of M0 soaked in water throughout the whole process. This can be attributed to the formation of new sulfate-containing crystals such as $MgK_2(SO_4)_2·6H_2O$ [19]. After 120 dry–wet cycles, the volume expansion rates were in the order DW-II, M0 (3.62‰) > SK-II, M0 (2.77‰) > DW-II, M1 (2.73‰) > SK-I, M0 (1.89‰) > DW-I, M0 (1.28‰).

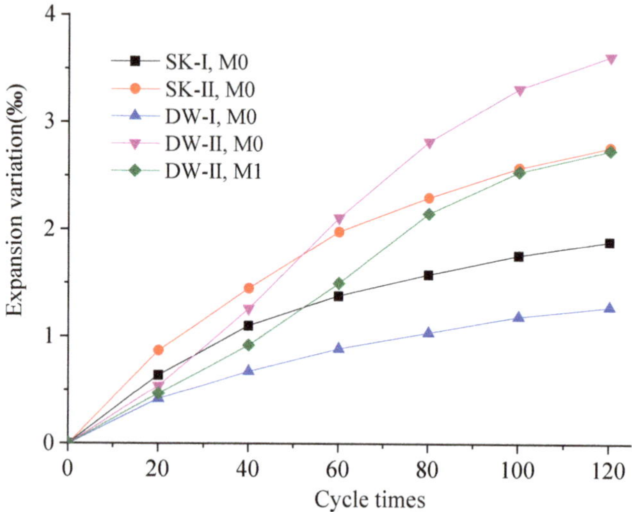

Figure 3. Deformation of the MKPC specimens under different corrosion conditions.

The volume expansion rate of M1 was less than that of M0 under the same conditions, indicating that the addition of waterglass could obviously improve the volume stability of the MKPC paste. By combining M1 with some water glass, magnesium silicate gels formed and could fill the pores, making the structure dense. As a result, external water could not infiltrate to the structure, thereby reducing the formation of sulfate-containing crystals, as well as the expansion and deformation caused by the crystallization pressure.

3.4. XRD Analysis

Figure 4 shows the XRD patterns of the MKPC samples exposed to the four corrosion environments. The patterns revealed that $MgKPO_46H_2O$ (MKP) was the primary hydration product, and a significant amount of unreacted MgO was still present. The main characteristic peak positions of the MKPC samples soaked in water and subjected to 120 dry–wet cycles in water were basically the same, indicating that no extra new phases formed. Compared to the M0 samples soaked in water, the MKP samples soaked in the sulfate solution exhibited a stronger diffraction peak intensity, suggesting excellent crystallinity of the hydration products. There were two new main characteristic peaks corresponding to $MgK_2(SO_4)_2·6H_2O$ and $Na_2SO_4·10H_2O$ in the MKPC sample subjected to dry–wet cycles in the 5% Na_2SO_4 solution. It appeared that the sulfate ion from the salt solution combined with the cation in the hardened MKPC paste to form a hydrate containing sulfate. $Na_2SO_4·10H_2O$ should be produced by water-saturated crystallization of sulfate in the pores during drying of the MKPC specimens. These new phases will increase the crystallization pressure of salt and cause volume expansion. Additionally, the main diffraction peak of MKP showed the weakest intensity, which could be due to its significant dissolution. These factors led to the worst structural degradation of the hardened MKPC specimens.

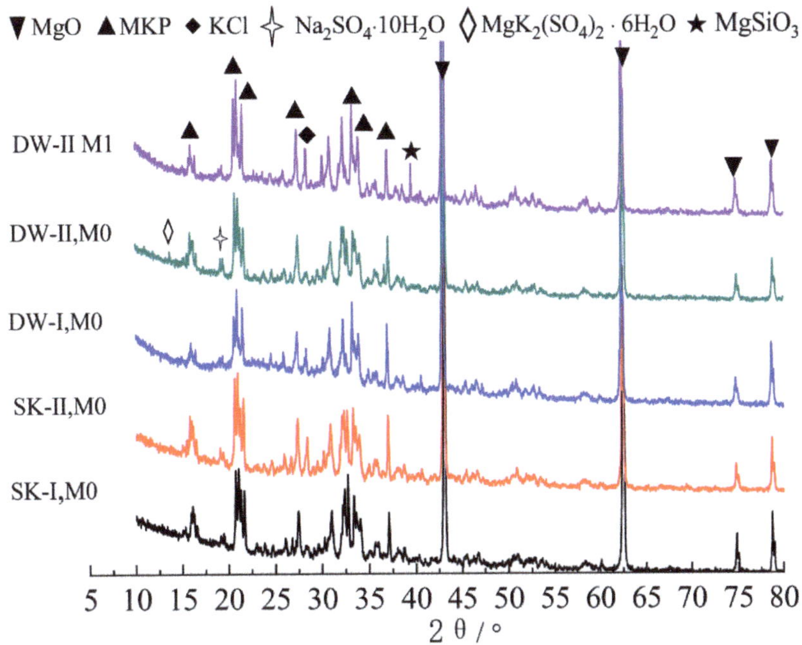

Figure 4. XRD curves of the MKPC samples under different corrosion conditions.

Compared to the M0 sample soaked in water, we observed more than one new main characteristic peak of $MgSiO_3$ at 39.8–40.0 $2\theta°$ in the M1 sample soaked in water, which should be due to the new phase produced by the reaction of silicate ions and Mg^{2+} in waterglass. Compared with the M0 sample subjected to 120 dry–wet cycles in 5% Na_2SO_4, the intensity of the main diffraction peak of MKP in M1 was obviously higher than that in M0. In addition, the main characteristic peaks of $Na_2SO_4 \cdot 10H_2O$ and $MgK_2(SO_4)_2 \cdot 6H_2O$ were missing in the M1 sample, which explains why the water-saturated crystallization of sulfate in the pores decreased during the drying process of the M1 specimen. The structural degradation of the M1 specimen caused by the volume expansion and the crystallization pressure of salt was relieved.

3.5. TG-DTG Analysis

Figure 5 exhibits the TG-DTG curves of MKPC after 360 days of immersion and 120 dry–wet cycles in water and the 5% Na_2SO_4 solution. At about 100 °C, the DTG curve illustrates a significant loss of mass due to the loss of crystalline water [39,40]. Moreover, an additional peak between 350–400 °C was observed when the MKPC paste was subjected to 120 dry–wet cycles in 5% Na_2SO_4, which could be due to the dehydration of the new sulfate-containing crystals; further study will be needed. Below 200 °C, the percentages of mass loss of the MPPC specimens were as follows: SK-II, M0 (11.56%) > DW-I, M0 (11.48%) > SK-I, M0 (11.43%) > DW-II, M1 (10.91%) > DW-II, M0 (9.92%). In the same corrosion conditions, the total mass loss below 200 °C in the M1 sample was obviously larger than that in M0, which confirmed that the hydrolysis and loss of MKP in M1 were more limited than that in M0, but the little mass loss at 300–400 °C in M1 was obviously less than that in M0.The lowest quantities of the hydration product MKP in the M0 samples cured under the DW-II environment indicated that hydrolysis and loss of MKP were enhanced due to the deterioration of the pore structure caused by the crystallization pressure of the salt, justifying the strength test results during this process.

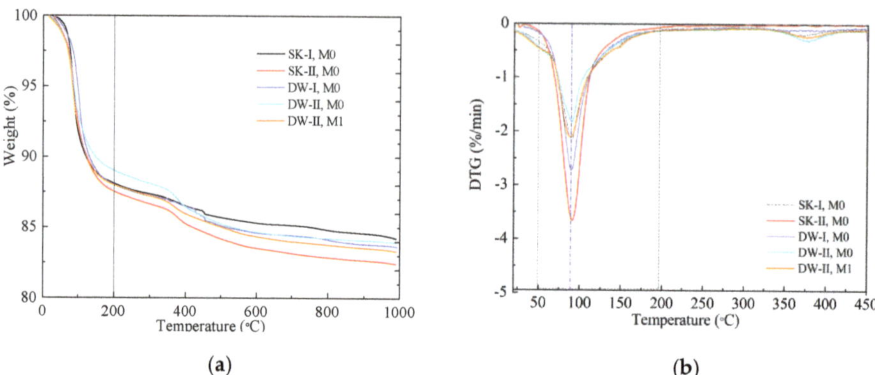

Figure 5. TG-DTG curves of the MKPC samples under different corrosion conditions. (**a**) TG, (**b**) DTG.

3.6. SEM-EDS Analysis

Figure 6a–f shows the microstructure of the MKPC samples after 12 months of exposure to the SK and DW environments in water and sulfate solution. The EDS results are shown in Table 4. When the MKPC paste was exposed to the SK environment in water (Figure 6a), good columnar crystals appeared tightly packed together in an orderly manner. These crystals (area A) were verified as MKP on the basis of a molar ratio of Mg/P/K of approximately one. Some Na and CI elements were also detected in area B, probably derived from the composite retarder. In the M0 sample exposed to the SK-II environment, distinct morphologies such as rod-like and needle-like crystals of hydrates were observed (Figure 6b,c); these hydrates may be the products of struvite-K dissolution and precipitation. As shown in Table 4, the components of Area C were O, Mg, P, K, Na, CI, and S elements, with a molar ratio of P to K of approximately 1:1. Based on the results of XRD, these elements should be derived from the hydration products and the remaining magnesium oxide. Among them, the presence of the S element confirmed that a sulfate-containing phase was mixed with the hydration products. On the fracture surface of the M0 sample subjected to the DW-I environment (Figure 6d), well-formed and randomly oriented rod-like crystals shown to be MKP by the EDS results were observed. The crystal surface was smooth without visible corrosion marks. The combination of the hydration products such as MKP and the remaining MgO made the microstructure dense, which could also explain the good compressive strength and deformation stability. After 120 dry–wet cycles in the sulphate solution, the microstructure showed in Figure 6e of the M0 samples was very loose, more large rod-like crystals were observed—some of them broken—and many amorphous phases were distributed among the crystals and adhered to their surfaces, indicating the dissolution of the hydration products in the MKPC past with the formation of pores [13]. Based on the high S content in area D, it was evident that SO_4^{2-} was involved in the hydration reaction, which resulted in new hydration products. The generation of $Mg(OH)_2$ occurred as MgO reacted with H_2O, resulting in a high internal stress that promoted the formation of intrinsic microcracks, further resulting in the decrease of strength and volume stability. These results are also in accordance to the XRD and TG-DTG results presented in Sections 3.3 and 3.4.

Moreover, the M1 samples showed well-compacted structures in comparison to the M0 samples under DW-II conditions, with many orderly MKP crystals forming prismatic shapes on a smooth, defect-free crystal surface (Figure 6f). Some amorphous phases were present in the crystals and adhered to their surfaces. The addition of waterglass improved the internal structures of MKPC and avoided the dissolution of MKP. Therefore, the M1 samples had higher resistance to sulfate corrosion-coupled wet and dry cycles. A magnesium silicate hydrate gel would also be formed by waterglass and magnesium ions [18], these results were confirmed by EDS in area E, and concurred with previously presented XRD results. Although the crystal size of the M1 samples was bigger and the samples were well crystallized, the amount of crystals was less than in the M0 samples cured under the SK environment, as seen from the TG-DTG curves; therefore, the strength and volume stability were lower.

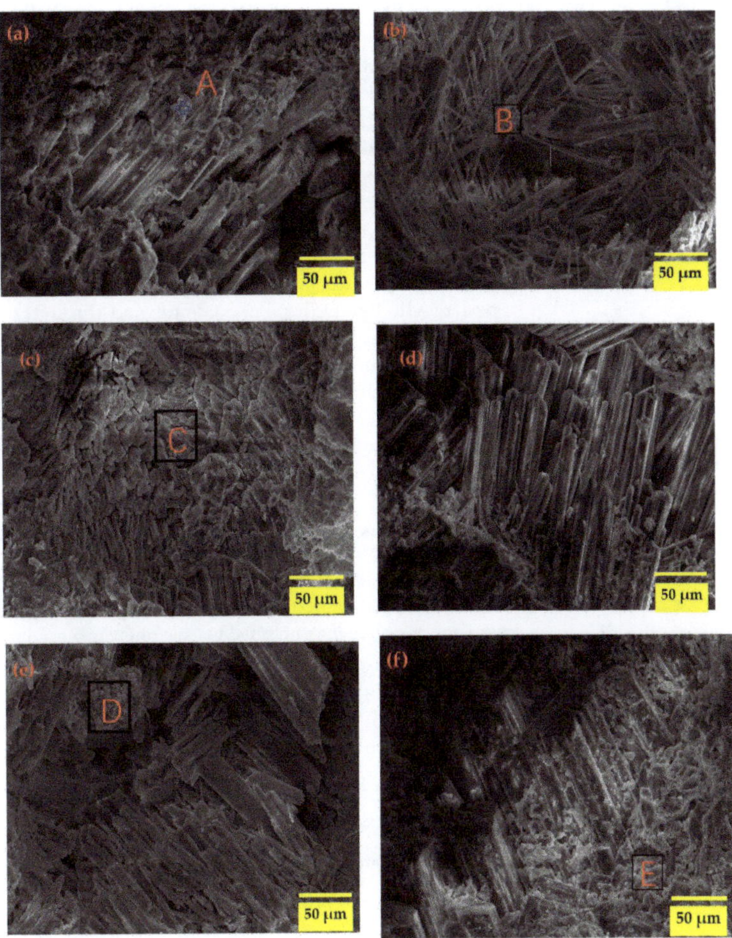

Figure 6. SEM photos of the MKPC samples under different corrosion conditions. (**a**) SK-I, M0, (**b**) SK-II, M0, (**c**) SK-II, M0, (**d**) DW-I, M0, (**e**) DW-II, M0, (**f**) DW-II, M1.

Table 4. EDS results of different areas and points in hardened MKPC pastes.

Element	Atomic Percentage				
	Point A	Area B	Area C	Area D	Area E
O	60.64	55.12	70.75	68.79	55.8
Na	0.72	2.12	3.25	5.43	7.36
Mg	13.39	16.62	16.43	6.71	12.12
Si	-	-	-	-	1.60
P	13.23	12.34	3.54	7.05	11.07
K	11.90	13.15	3.52	6.81	6.75
S	-	0.04	2.18	4.37	5.30
CI	0.12	0.61	0.23	0.87	-

4. Conclusions

In this study, the corrosion behavior of magnesium potassium phosphate cement under wet–dry cycle and sulfate attack was investigated. Conclusions can be drawn as follows:

1. After 120 dry–wet cycles, the final strengths of the MKPC paste were in the order AC, M0 > SK-II, M0 > DW-I, M0 > SK-I, M0 > DW-II, M1 > DW-II, M0. The dry–wet cycles in water caused less corrosion of the MKPC paste compared to soaking in water, and dry–wet damage played a dominant role when the samples were exposed to the combined action of 5% Na_2SO_4 sulfate and dry–wet cycles. The analysis by TG-DTG and SEM further showed that the dry–wet cycles led to the dissolution of struvite-K and the formation of more intrinsic micro-cracks.

2. All specimens showed volume expansion in the full soak and dry–wet cycle test in the Na_2SO_4 solution and water environments; the final volume expansion value of the MKPC specimens under four corrosion conditions was in the order DW-II, M0 > SK-II, M0 > DW-II, M1 > SK-I, M0 > DW-I, M0. Under the SK-II conditions, the M0 paste with the highest strength revealed a higher volume expansion. The new needle-like hydrate crystals formed through the dissolution–precipitation of struvite-K may account for the volume expansion.

3. Under the same number of dry–wet cycles, the strength test and volume stability test on the M1 paste confirmed that waterglass could effectively increase the durability under dry–wet cycling in the Na_2SO_4 solution. The micro analysis validated that waterglass can improve the compactness of the microstructure of MPC and prevent the dissolution of struvite-K.

Author Contributions: Conceptualization, L.C. and J.Y.; methodology, H.L., C.L. and D.T.; formal analysis, L.C. and J.C.; investigation, L.C.; resources, L.C.; data curation, H.L. and D.T.; writing—original draft preparation, L.C. and H.L.; writing—review and editing, A.A. and L.C.; supervision, L.C. and J.Y.; project administration, L.C. and J.C.; funding acquisition, L.C., A.A. and J.C. All authors have read and agreed to the published version of the manuscript.

Funding: This research was funded by the Changsha Municipal Natural Science Foundation, grant number kq2007089, the National Natural Science Foundation of China, grant number 42107166, the Hunan Provincial Natural Science Foundation, grant number 2021JJ40632, and the Education Department of Hunan Province, grant number Nos 21C0740 and 22B0825.

Institutional Review Board Statement: Not applicable.

Informed Consent Statement: Not applicable.

Data Availability Statement: Data are contained within the article.

Conflicts of Interest: The authors declare no conflict of interest.

References

1. Li, L.; Shi, J.; Kou, J. Experimental Study on Mechanical Properties of High-Ductility Concrete against Combined Sulfate Attack and Dry–Wet Cycles. *Materials* **2021**, *14*, 4035. [CrossRef] [PubMed]
2. Goncalves, J.; El-Bakkari, M.; Boluk, Y.; Bindiganavile, V. Cellulose nanofibres (CNF) for sulphate resistance in cement based systems. *Cem. Concr. Compos.* **2019**, *99*, 100–111. [CrossRef]
3. Gong, J.; Cao, J.; Wang, Y.-F. Effect of creep on the stress–strain relation of fly-ash slag concrete in marine environments. *Struct. Concr.* **2019**, *20*, 1076–1085. [CrossRef]
4. Ji, R.J.; Li, T.; Yang, J.M.; Xu, J. Sulfate Freeze–Thaw Resistance of Magnesium Potassium Phosphate Cement Mortar according to Hydration Age. *Materials* **2022**, *15*, 4192. [CrossRef] [PubMed]
5. Zhang, L.; Zhang, A.; Wang, Q.; Han, Y.; Li, K.; Gao, X.; Tang, Z. Corrosion resistance of wollastonite modified magnesium phosphate cement paste exposed to freeze-thaw cycles and acid-base corrosion. *Case Stud. Constr. Mater.* **2020**, *13*, e00421. [CrossRef]
6. Wagh, A. *Chemically Bonded Phosphate Ceramics: Twenty-First Century Materials with Diverse Applications*; Elsevier: Amsterdam, The Netherlands, 2016.
7. Andrade, A.; Schuiling, R.D. The chemistry of struvite crystallization. *Min. J.* **2001**, *23*, 37–46.

8.   Seehra, S.; Gupta, S.; Kumar, S. Rapid setting magnesium phosphate cement for quick repair of concrete pavements—Characterisation and durability aspects. *Cem. Concr. Res.* **1993**, *23*, 254–266. [CrossRef]
9.   Yang, Q.; Zhu, B.; Zhang, S.; Wu, X. Properties and applications of magnesia–phosphate cement mortar for rapid repair of concrete. *Cem. Concr. Res.* **2000**, *30*, 1807–1813. [CrossRef]
10. Jiang, H.Y.; Liang, B.; Zhang, L. Investigation of MPB with super early strength for repair of concrete. *J. Build. Eng.* **2001**, *4*, 196–198.
11. Chen, W.; Elkatatny, S.; Murtaza, M.; Mahmoud, A.A. Recent Advances in Magnesia Blended Cement Studies for Geotechnical Well Construction-A Review. *Front. Mater.* **2021**, *8*, 447. [CrossRef]
12. Shi, C.; Yang, J.; Yang, N.; Chang, Y. Effect of waterglass on water stability of potassium magnesium phosphate cement paste. *Cem. Concr. Compos.* **2014**, *53*, 83–87. [CrossRef]
13. Chong, L.; Yang, J.; Shi, C. Effect of curing regime on water resistance of magnesium–potassium phosphate cement. *Constr. Build. Mater.* **2017**, *151*, 43–51. [CrossRef]
14. Li, Y.; Shi, T.; Li, J. Effects of fly ash and quartz sand on water-resistance and salt-resistance of magnesium phosphate cement. *Constr. Build. Mater.* **2016**, *105*, 384–390. [CrossRef]
15. Fan, S.J.; Chen, B. Experimental research of water stability of magnesium alumina phosphate cements mortar. *Constr. Build. Mater.* **2015**, *94*, 164–171. [CrossRef]
16. Zhen, S.; Yang, J.; Zhang, Q.; Wang, Y.Q.; Wang, T.B. Research on Resistance of Magnesia-Phosphate Cement to Chloride Ion Erosion. *J. Build. Mater.* **2001**, *13*, 700–704. (In Chinese)
17. Ding, Z.; Li, Z.; Xing, F. Chemical durability investigation of magnesium phosphor silicate cement. *Key. Eng. Mater.* **2006**, *302*, 275–281. [CrossRef]
18. Yang, J.M.; Zeng, S.C. Experimental research on seawater erosion resistance of magnesium potassium phosphate cement pastes. *Constr. Build. Mater.* **2018**, *183*, 534–543. [CrossRef]
19. Yang, J.M.; Wang, L.M.; Jin, C.; Sheng, D. Effect of fly ash on the corrosion resistance of magnesium potassium phosphate cement paste in sulfate solution. *Constr. Build. Mater.* **2020**, *237*, 117639. [CrossRef]
20. Lahalle, H.; Patapy, C.; Glid, M.; Renaudin, G.; Cyr, M. Microstructural evolution/durability of magnesium phosphate cement paste over time in neutral and basic environments. *Cem. Concr. Res.* **2019**, *122*, 42–58. [CrossRef]
21. Yang, Q.B.; Zhang, S.Q.; Wu, X.L. Deicer-scaling resistance of phosphate cement-based binder for rapid repair of concrete. *Cem. Concr. Res.* **2002**, *32*, 165–168. [CrossRef]
22. Chong, L.L.; Yang, J.M.; Xu, Z.Z.; Xu, X.C.; Xu, X. Freezing and thawing resistance of MKPC paste under different corrosion solutions. *Constr. Build. Mater.* **2019**, *212*, 663–674. [CrossRef]
23. Wang, B.; Lu, K.; Han, C.; Wu, Q. Study on anti-corrosion performance of silica fume modified magnesium potassium phosphate cement-based coating on steel. *Case Stud. Constr. Mater.* **2022**, *17*, e01467. [CrossRef]
24. Wang, D.; Yue, Y.; Xie, Z.; Mi, T.; Yang, S.; McCague, C.; Qian, J.; Bai, Y. Chloride-induced depassivation and corrosion of mild steel in magnesium potassium phosphate cement. *Corros. Sci.* **2022**, *206*, 110482. [CrossRef]
25. Wang, D.; Yue, Y.; Mi, T.; Yang, S.; McCague, C.; Qian, J.; Bai, Y. Effect of magnesia-to-phosphate ratio on the passivation of mild steel in magnesium potassium phosphate cement. *Corros. Sci.* **2020**, *174*, 108848. [CrossRef]
26. Wang, D.; Liu, Z.; Yang, C. Passivation behaviour of mild steel embedded in magnesium potassium phosphate cement-calcium sulphoaluminate cement blended paste. *Constr. Build. Mater.* **2022**, *347*, 128537. [CrossRef]
27. Jun, L.; Yongsheng, J.; Linglei, Z.; Benlin, L. Resistance to sulfate attack of magnesium phosphate cement-coated concrete. *Constr. Build. Mater.* **2019**, *195*, 156–164. [CrossRef]
28. Li, J.; Ji, Y.; Huang, G.; Zhang, L. Microstructure evolution of a magnesium phosphate protective layer on concrete structures in a sulfate environment. *Coatings* **2018**, *8*, 140. [CrossRef]
29. Li, Y.; Shi, T.; Li, Y.; Bai, W.; Lin, H. Damage of magnesium potassium phosphate cement under dry and wet cycles and sulfate attack. *Constr. Build. Mater.* **2019**, *210*, 111–117. [CrossRef]
30. Li, T.; Chen, G.; Yang, J.; Chong, L.; Hu, X. Influence of curing conditions on mechanical properties and microstructure of magnesium potassium phosphate cement. *Case Stud. Constr. Mater.* **2022**, *17*, e01264. [CrossRef]
31. *ASTM C1437 2007*; Standard Test Method for Flow Hydraulic Cement Mortars. ASTM International: West Conshohocken, PA, USA, 2007.
32. *ASTM C191-2001a*; Standard Test Method for Time of Setting of Hydraulic Cement by Vicat Needle, Using the Vicar Apparatus. ASTM International: West Conshohocken, PA, USA, 2001.
33. *ASTM C348 2008*; Standard Test Method for Flexural Strength of Hydraulic Cement Mortars. ASTM International: West Conshohocken, PA, USA, 2008.
34. *ASTM C349 2008*; Standard Test Method for Compressive Strength of Hydraulic Cement Mortars, Using Portions of Prisms Broken in Flexure. ASTM International: West Conshohocken, PA, USA, 2008.
35. *Chinese Standard JC/T603–2004*; Standard Test Method for Drying Shrinkage of Mortar. ASTM International: West Conshohocken, PA, USA, 2004. Available online: https://www.chinesestandard.net/PDF/English.aspx/JCT603-2004 (accessed on 12 January 2023).

36. Shi, J.; Zhao, J.; Chen, H.; Hou, P.; Kawashima, S.; Qin, J.; Zhou, X.; Qian, J.; Cheng, X. Sulfuric acid-resistance performances of magnesium phosphate cements: Macro-properties, mineralogy and microstructure evolutions. *Cem. Concr. Res.* **2022**, *157*, 106830. [CrossRef]

37. Soudée, E.; Péra, J. Influence of magnesia surface on the setting time of magnesia–phosphate cement. *Cem. Concr. Res.* **2002**, *32*, 153–157. [CrossRef]

38. on of Superpave gyratory compaction and its influence on mechanical properties of asphalt mixtures. *J. Mater. Civ. Eng.* **2022**, *35*, 04022453. [CrossRef]

39. Zhang, S.; Shi, H.-S.; Huang, S.-W.; Zhang, P. Dehydration characteristics of struvite-K pertaining to magnesium potassium phosphate cement system in non-isothermal condition. *J. Therm. Anal. Calorim.* **2013**, *111*, 35–40. [CrossRef]

40. Li, Y.; Shi, T.; Chen, B.; Li, Y. Performance of magnesium phosphate cement at elevated temperatures. *Constr. Build. Mater.* **2015**, *91*, 126–132. [CrossRef]

Article

Evaluating the Effects of Polyphosphoric Acid (PPA) on the Anti-Ultraviolet Aging Properties of SBR-Modified Asphalt

Yanling Xu [1], Kaimin Niu [1,2], Hongzhou Zhu [1,3,*], Ruipu Chen [1] and Li Ou [1]

1   School of Civil Engineering, Chongqing Jiaotong University, Chongqing 400074, China
2   Research Institute of Highway Ministry of Transport, Beijing 100088, China
3   National and Local Joint Engineering Laboratory of Transportation and Civil Engineering Materials, Chongqing Jiaotong University, Chongqing 400074, China
*   Correspondence: zhuhongzhouchina@cqjtu.edu.cn

Abstract: The ultraviolet (UV) aging of asphalt is an important factor affecting the long-term performance of asphalt pavement, especially in high altitude cold regions. The current studies have reported that styrene butadiene rubber-modified asphalt (SBRMA) has a good cracking resistance at low temperatures. In addition, polyphosphoric acid (PPA) is an effective modifier that can enhance the anti-UV aging properties of asphalt. However, the understanding of the improvement mechanism of PPA on the anti-aging of SBRMA remains unclear. Therefore, this study aimed to evaluate the effect of PPA on the UV aging resistance of SBRMA. The rheological properties of PEN90 asphalt(90#A), SBRMA, and PPA/SBR modified (PPA/SBR-MA) before and after UV aging were evaluated by dynamic shear rheometer (DSR) and bending beam rheometer (BBR) tests. The molecular weight and chemical structure of 90#A, SBRMA, and PPA/SBR-MA were determined by Fourier transform infrared spectroscopy (FTIR) and gel permeation chromatography (GPC), and the interaction and modification mechanism of the modifiers were analyzed. The rheological analysis shows that the high and low temperature performances of SBRMA are improved by adding PPA, and PPA also significantly reduces the sensitivity of SBRMA to UV aging. The microscopic test results show that PPA has a complex chemical reaction with SBRMA, which results in changes in its molecular structure. This condition enhances SBRMA with a more stable dispersion system, inhibits the degradation of the polymer macromolecules of the SBR modifier, and slows down the aging process of base asphalt. In general, PPA can significantly improve the anti-UV aging performance of SBRMA. The Pearson correlations between the aging indexes of the macro and micro properties are also significant. In summary, PPA/SBRMA material is more suitable for high altitude cold regions than SBRMA, which provides a reference for selecting and designing asphalt pavement materials in high altitude cold regions.

Keywords: polyphosphoric acid; styrene butadiene rubber-modified asphalt; ultraviolet aging; rheological properties; modification mechanism

Citation: Xu, Y.; Niu, K.; Zhu, H.; Chen, R.; Ou, L. Evaluating the Effects of Polyphosphoric Acid (PPA) on the Anti-Ultraviolet Aging Properties of SBR-Modified Asphalt. *Materials* **2023**, *16*, 2784. https://doi.org/10.3390/ma16072784

Academic Editors: Giovanni Polacco and Gilda Ferrotti

Received: 20 February 2023
Revised: 24 March 2023
Accepted: 28 March 2023
Published: 30 March 2023

1. Introduction

Asphalt pavement is a major structure of highways globally because of its significant advantages, such as driving comfort, low noise, easy repair, and renewable nature [1,2]. After the asphalt pavement is opened to traffic, the surface asphalt is directly exposed to ultraviolet (UV) light, oxygen, water, and other factors in the external atmospheric environment. These factors lead to its continuous aging; that is, the chemical composition and microstructure of the asphalt are changed and will continue to deteriorate. Meanwhile, the repeated action under dynamic stress induced by traffic loading aggravates the aging of the asphalt concrete pavement, which is caused by climate factors including UV light and thermal oxidation [3–5]; this, in turn, leads to early diseases such as loose cracks and pits. Then, the durability of the asphalt concrete is seriously affected [6–9]. In the western region of China, the crack rate of asphalt concrete pavements is much higher than that in the inland region due to the low temperature and large temperature difference in the high-altitude area. The main reason for this is that the high-intensity UV radiation accelerates the photoaging process of asphalt materials in road applications. Studies have reported that UV radiation has a greater effect on asphalt cracking resistance at low temperatures than that of thermal oxidation and water aging [10]. Therefore, using asphalt material with good low temperature and UV aging properties is key to improving the durability of asphalt pavements in high altitude cold regions.

Since its discovery and use, the SBR-modified asphalt (SBRMA) has been well-known for its good low temperature performance. The adhesion between the asphalt and aggregate is enhanced because SBR can greatly improve the low temperature ductility and cohesion. Thus, asphalt pavements paved with this material has better crack resistance at low temperatures and the production method of SBR is also simple and low cost; consequently, SBRMA is widely used in cold areas [11]. In the high-cold and -altitude areas of western China, styrene butadiene styrene-modified asphalt, SBRMA, and rubber asphalt are commonly used, among which SBRMA is the most widely used type [12]. However, problems have gradually emerged with the increment in repealed loads and frequency, in which it lacks permanent deformation resistance and has seriously degrading anti-aging properties. Meanwhile, its poor storage stability seriously affects the use and development of SBRMA [13,14].

Many scholars have added polyphosphoric acid (PPA) modifier into SBRMA and improved the pavement performance of PPA/SBR-MA under synergy to overcome the defects of SBRMA. PPA can chemically modify asphalt. Specifically, a stable chemical bond is formed in the blending process of asphalt and the modifier, which enhances the compatibility and other properties. The current research indicates that PPA/SBR-MA has a better high temperature performance and fatigue resistance than base asphalt and SBRMA. The viscosity of asphalt is increased, the penetration degree is decreased, and the softening point is improved after the PPA modifier is added into SBRMA.

From the perspective of rheology, the high temperature rutting resistance of SBRMA is enhanced, the elastic modulus is significantly improved, and the phase angle is decreased with the addition of PPA [15]. The thermal storage stability and thermal oxygen aging resistance are also improved [16,17]. However, other studies have found that PPA gives SBRMA more light components and it becomes more sensitive to thermal and oxygen aging [18]. From the micro perspective, PPA is found to significantly improve the adhesion ability, high temperature elasticity, and rutting resistance of SBRMA through reverse flocculation; the incorporation of PPA also reduces the size of the polymer dispersed phase, which forms a more ideal microstructure and significantly enhances the compatibility of SBR and asphalt [19]. In regard to the mechanism of PPA improving the thermal oxygen aging resistance of asphalt, it has been suggested that PPA causes a gelation effect on asphalt and can disperse agglomerates of asphaltene micelles, which significantly reduces the impact of aging on the asphalt properties [20,21]. In addition, Han et al. [22] argued that a chemical reaction between PPA and the reactive groups in SBS-modified asphalt resulted in a tighter cross-linked network structure of the SBS-modified asphalt. Additionally, Ge et al. [23] suggested that PPA can ease the decomposition process and improve the stability of asphalt. Wei et al. [24] suggested that PPA can improve the aging resistance of asphalt primarily because it is not easily oxidized. However, studies on the effect of PPA on the low temperature cracking resistance of SBRMA have drawn different conclusions. Some scholars believe that PPA can reduce the low temperature cracking resistance of SBRMA [25], or that the low temperature cracking resistance first increases and then decreases with the rise in the PPA content [19]. Hao et al. [26] studied the cracking property of a PPA/SBR asphalt mixture at a low temperature through a thermal stress constrained sample test and semicircular bending test. They found that the low temperature cracking resistance of SBRMA can be effectively improved by the addition of PPA. In addition, PPA can also enhance the anti-UV aging properties of SBRMA [27].

In general, the addition of PPA can compensate for the deficiency of the permanent deformation resistance at high temperatures and the poor storage stability. However, the research about the anti-UV aging performance of asphalt by the addition of PPA was ignored. This study aimed to investigate the effect of PPA on the anti-UV aging behavior of SBRMA at the macro- and micro-scales, which determine whether PPA/SBRMA material is more suitable for high altitude cold regions than SBRMA. The anti-UV aging properties of PPA/SBR-MA, SBRMA, and 90#A were compared, and the excellent anti-UV aging properties of PPA/SBR –MA were analyzed from its chemical composition and molecular structure. Second, the effects of UV aging on the rheological properties were studied. Several anti-aging indexes were applied to compare the UV aging resistance of 90#A, SBRMA, and PPA/SBR-MA. The relationship between the changes in the chemical structure, molecular weight structure, and rheological parameters of the three asphalts in the UV aging process was further discussed.

2. Materials and Methods

2.1. Materials

In this study, the 90#A was obtained from the Karamay refinery of PetroChina. The basic technical indexes of 90A are shown in Table 1. The polymer modifier SBR was also purchased from a local company it is a white powder, and the specific technical specifications are shown in Table 2. The polyphosphoric acid (PPA) was purchased from a local supplier; its physical state is a viscous, colorless, and transparent liquid, and the basic technical indexes are shown in Table 3.

Table 1. Basic technical indexes of 90#A.

Parameter	Results
Penetration (25 °C, 0.1 mm)	88.1
Softening point (°C)	45.4
Ductility (15 °C, cm)	>150
C (%)	87.02
H (%)	9.31
O (%)	3.12
N (%)	0.51
S (%)	0

Table 2. Basic technical indexes of SBR.

Parameter	Results
Granularity (16-mesh, %)	98.6
Bound styrene content (%)	23.0
Mooney viscosity (ML)	55.5
Tensile strength (MPa)	25.0
Breaking elongation (%)	378
300% Constant tensile stress (MPa)	15.0

Table 3. Basic technical indexes of PPA.

Parameter	Results
Phosphoric acid (H_3PO_4, %)	115.2
Phosphorus pentoxide (P_2O_5, %)	83.5
Chloride (Cl, %)	<0.0003
Fe (%)	0.0014
As (%)	0.009
Heavy metal (Pb, %)	<0.002

2.2. Preparation of Asphalt

The manufacturer recommends that the reasonable dosage of PPA polyphosphate is 0.5–2.0 wt%, but the range is too large and has little significance for practical application guidance. Many domestic and international studies have shown that the appropriate dosage of PPA with SBS- and SBR-modified asphalt is 0.75–1.25% dosages by weight of the matrix asphalt [19,26,28]. In this study, an orthogonal test was used to select the optimal combination dosage and preparation process of PPA and SBR, and all the additive dosages are based on the matrix asphalt. The rate is generally 4000–4500 r/min to prevent the raw material reaching a temperature too high during shearing and to make the modified material mix shear homogeneous. The mixed shear process and the optimal mixing ratio were determined by orthogonal test after the PPA was added into the SBR asphalt binder. The optimum mixture ratio was 4 wt% SBR + 1 wt% PPA, and the mixed shear time for the PPA/SBR binder was 40 min. The temperature was 170 °C in the PPA/SBR-MA asphalt preparation process.

(1)   Preparation of SBRMA

First, the 90#A was heated at 150 °C to a liquid state. Then, at a constant temperature of 160 °C, 4.0 wt% SBR modifier was continuously added to 90#A, and the blend was evenly stirred at a low speed for 15 min to make it fully swollen. Next, a high-speed shearing procedure (4000–4500 r/min) was performed for the blend at 165 °C for 40 min and stirred with a mixer for 30 min (speed of 600–700 r/min), followed by a curing procedure at 165 °C for 1 h to incubate constantly. Finally, the SBRMA was prepared.

(2)   Preparation of PPA/SBR-MA

First, the 90#A was heated at 150 °C to liquid state. At a constant temperature of 160 °C, 4.0 wt% SBR modifier was continuously added to 90#A, and the blend was evenly stirred at a low speed for 15 min to make it fully swollen. Next, a high-speed shearing procedure (4000–4500 r/min) was performed for the blend at 165 °C for 30 min. Then, the temperature was kept at 170 °C. Thereafter, 1% PPA modifier was added to the asphalt binder for high-speed shear treatment, and the shearing rate was 4000–4500 r/min for 40 min. Stirring was performed at 160 °C for 30 min (600–700 r/min). Finally, the PPA/SBR-MA was prepared.

2.3. Testing Methods

Figure 1 shows the experimental framework of this study. The macro properties of the asphalt binder were characterized by a basic properties test and a rheological test. A GPC test and FTIR were performed to explain the mechanism and to characterize the microscopic aging effect.

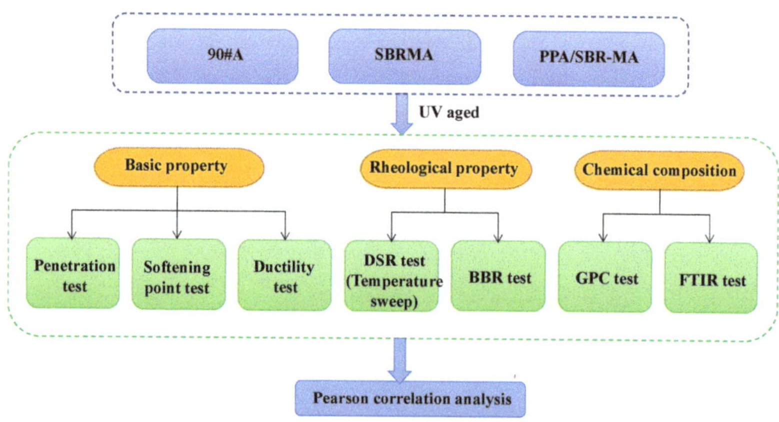

Figure 1. The designed process of experimental program.

2.3.1. Ultraviolet (UV) Aging Procedure

First, the density of each sample was measured for 90#A, SBRMA, and PPA/SBR-MA. Then, various unaged samples were uniformly spread on a steel plate (diameter of 140 mm). The thickness of each sample was controlled at 3 mm according to the relational calculation between the mass and volume. Some test samples are shown in Figures 2 and 3.

Figure 2. Asphalt before UV aging.

Figure 3. Asphalt after UV aging.

The laboratory modeling UV aging experiment equipment of the asphalt was an opaque chamber. Two high-pressure lamb lights, with a light intensity of 500 W/m^2, were installed on the top of the box. In the summer of the alpine region, the road temperature is often 30–40 °C higher than the air temperature and can reach 50–60 °C due to the unusually strong solar UV with the strong heat absorption ability of the asphalt mixture. Therefore, the UV aging chamber was kept ventilated and the constant temperature was 60 °C during the experiment. The radiation intensity of the sample surface in the simulated UV aging chamber was 75 W/m^2 and measured by the light intensity instrument. The outdoor radiation intensity in this study was calculated based on the amount of solar radiation received in the Tibet region. Some studies have shown that the asphalt performance changes most obviously around 4 months [29]. Thus, an outdoor time of 4 months was selected to design the indoor UV radiation time. The indoor simulation time was 19.125 days based on the energy-equivalence of indoor and outdoor radiation. Thus, the indoor simulation time was 20 days.

2.3.2. Standard Physical Properties Test

The basic properties of the three asphalts, including the penetration (25 °C, 5 s, 100 g), softening point, and ductility (5 mm/min, 5 °C), were conducted, respectively, in reference to the Chinese standard JTG-E20-2011 [30]. The basic properties aging indexes RP, SPI, and RD of the three asphalts were calculate by Equations (1)–(3) [31,32].

$$RP = \frac{P_{aged}}{P_{virgin}} \times 100\% \tag{1}$$

$$SPI = \frac{(SP_{aged} - SP_{virgin})}{SP_{virgin}} \times 100\% \tag{2}$$

$$RD = \frac{D_{aged}}{D_{virgin}} \times 100\% \tag{3}$$

2.3.3. Temperature Sweep Test

The rheological properties of the three types of asphalt samples were characterized by the high temperature sweep test following the AASHTO-T315 method [33]. The type of dynamic shear rheometer used is DHR-2, manufactured by the America TA Instruments Company. The strain level is limited by 1.25%. The test temperatures were selected from 52 °C to 76 °C with the interval of 6 °C. The complex modulus G* and phase angle δ are recorded as viscoelastic parameters. The rutting factor aging index (RAI) at 64 °C was selected to evaluate the anti-UV aging performance of the different asphalts [34]. The RAI was calculated according to Equation (4). Notably, the anti-aging effect is better when the value is closer to 1.

$$RAI = \frac{(G^*/\sin \delta)_{aged}}{(G^*/\sin \delta)_{virgin}} \tag{4}$$

2.3.4. Bending Beam Rheometer Test (BBR)

The BBR test was used to characterize the low temperature creep behavior of the asphalt binder, which was performed at -12 °C and -18 °C in accordance with AASHTO-T313 [35] in this study. In the test, the BBR was manufactured by the Superpave Company. The deformation was generated by a load of 980 ± 50 mN applied to the asphalt trabecular span. The average values of the flexural creep stiffness modulus (S) and the relaxation parameter (m) at 60 s were taken as the low temperature rheological parameters. They were used to analyze the low temperature rheological behaviors of the asphalt before and after UV aging. According to the studies by Li [36] and Xu [37], the ratio λ(λ = m/S) can be used to determine the low temperature performance when it is different between m and S. In general, the low temperature performance increases with the high λ value. Therefore, the ratio λ was determined in this study. The low temperature aging index (I_λ) was used to quantify the influence of UV aging on the low temperature performance of the asphalt binder. The low temperature aging index I_λ of the three asphalts was calculated according to Equation (5). Notably, the anti-UV aging effect is better when the value is closer to 1.

$$I_\lambda = \frac{\lambda_{\text{aged}}}{\lambda_{\text{virgin}}} \tag{5}$$

2.3.5. Fourier Transform Infrared Spectroscopy (FTIR)

The FTIR test is an analytical tool used to characterize the different functional groups of chemical components. An IRprestge21 Fourier transform infrared spectrometer produced by Shimadzu Corporation of Japan was used to conduct infrared spectrum tests on 90#A, SBRMA, and PPA/SBR-MA before and after aging. The wave number ranges between 4000 and 400 cm^{-1}, and the scan time was 32 times.

The absorption peaks of C=O and S=O are approximately 1700 and 1030 cm^{-1}, respectively, while the absorption peaks of the modified asphalt are about 966 cm^{-1} to identify the peak of SBR [38,39]. Some studies suggest that sulfoxides are also susceptible to additional oxidation to sulfones; then, the peak value of the sulfoxide group around 1030 cm^{-1} is not a reliable indicator of oxidative aging [38]. Meanwhile, the content of S element in Karamay based asphalt is 0. Thus, the sulfoxide group index was ignored in the quantitative analysis of the current study. After baseline correction, the ICO and BI indexes were computed as Equations (6) and (7) [40–42], and the aging index of the chemical structure (CR) was calculated according Equation (8) [4].

$$\text{ICO} = \frac{\text{area\ around\ 1700\ cm}^{-1}}{\text{area\ around\ 1460\ cm}^{-1} + \text{area\ around\ 1375\ cm}^{-1}} \tag{6}$$

$$\text{BI} = \frac{\text{area\ around\ 966\ cm}^{-1}}{\text{area\ around\ 1460\ cm}^{-1} + \text{area\ around\ 1375\ cm}^{-1}} \tag{7}$$

$$\text{CR} = \frac{\text{Index\ of\ asphalt\ after\ UV\ aged} - \text{Index\ of\ asphalt\ before\ UV\ aged}}{\text{Index\ of\ asphalt\ before\ UV\ aged}} \tag{8}$$

2.3.6. Gel Permeation Chromatography Test (GPC)

Agilent 1260 GPC50 was adopted to determine the molecular weight distribution of the samples. In this study, tetrahydrofuran was used as the mobile phase solvent, and the solution passed through the columns at a flow rate of 1 mL/min. The columns were kept at 35 °C throughout the test, and the sample size was 100 μL each time. Polystyrene was made as the standard sample. Based on the theory of the GPC test, the large molecular size (LMS), medium molecular size (MMS), and small molecular size (SMS) were eluted successively with an increasing elution time. The chromatogram of each bitumen sample was divided into 13 sections according to the equal elution time between the beginning and end periods, with the first 5 sections representing LMS, sections 6 to 10 representing MMS, and the last 4 sections representing SMS [43–45]. The GPC curve is shown in Figure 4, and the three divided areas are expressed as percentages. The common average molecular weight includes the number average molecular weight (M_n), weight average molecular weight (M_w), and dispersion coefficient d. These parameters were calculated according to Equations (9)–(11) [46].

$$M_n = \frac{\sum_{i=1}^{n} N_i \times M_i}{\sum_{i=1}^{n} N_i} \tag{9}$$

$$M_W = \frac{\sum_{i=1}^{n} W_i \times M_i}{\sum_{i=1}^{n} W_i} \tag{10}$$

$$d = \frac{M_n}{M_W} \tag{11}$$

where M_i is the molecular mass; N_i is the number of molecules of M_i; W_i is the weight fraction of each type of molecule.

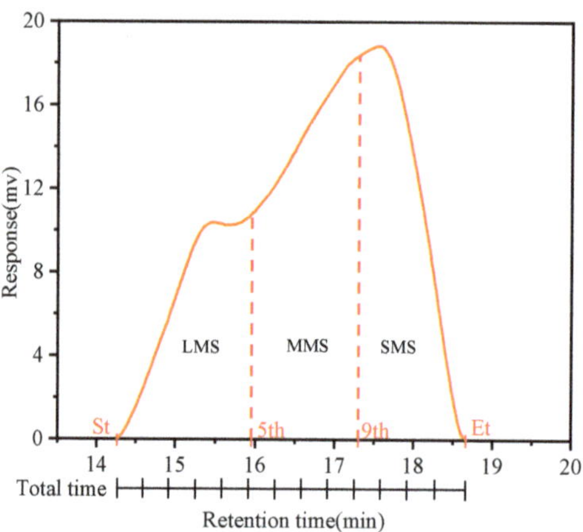

Figure 4. Methodology of analyzing molecular size.

To further quantify the effect of UV aging on the molecular weight, the change rate of the molecular weight was used as the evaluation index, which was calculated according to Equations (12) and (13).

$$I_{M_n} = \frac{(M_n)_{\text{aged}} - (M_n)_{\text{virgin}}}{(M_n)_{\text{virgin}}} \times 100\% \tag{12}$$

$$I_{M_W} = \frac{(M_W)_{\text{aged}} - (M_W)_{\text{virgin}}}{(M_W)_{\text{virgin}}} \times 100\% \tag{13}$$

3. Results and Discussions

3.1. Standard Physical Properties

As shown in Figure 5a, the penetration of the three asphalts is decreased after UV aging compared to before UV aging. The penetration of PPA/SBR is obviously smaller than that of 90#A and SBRMA, which illustrates that PPA/SBR-MA has the highest hardness among the three asphalts. The sequence of the RP is PPA/SBR-MA, SBRMA,90#A, which shows that PPA/SBR-MA has the best UV aging resistance. The analysis results indicate that PPA can increase the hardness of SBRMA, but significantly reduces the sensitivity of the penetration to UV aging.

Figure 5b shows that the softening point of 90#A is the smallest, followed by that of SBRMA, and that of PPA/SBR-MA is the largest. Therefore, the addition of SBR can slightly increase the softening point of 90#A, while PPA can significantly increase the softening point of SBRMA and enhance its high temperature performance. After UV aging, the penetration of the three asphalts is decreased. The softening point of asphalt is increased because the oxidation of asphalt can increase the asphaltene. The sequence of the SPI is 90#A, SBRMA, PPA/SBR-MA, which shows that PPA/SBR-MA has the best UV-aging resistance among the three asphalts. PPA/SBR-MA has the best high temperature stability and the least influence of UV aging among the three asphalts. The analysis results show that PPA can improve the high temperature performance of SBRMA and reduce the effect of UV aging on it.

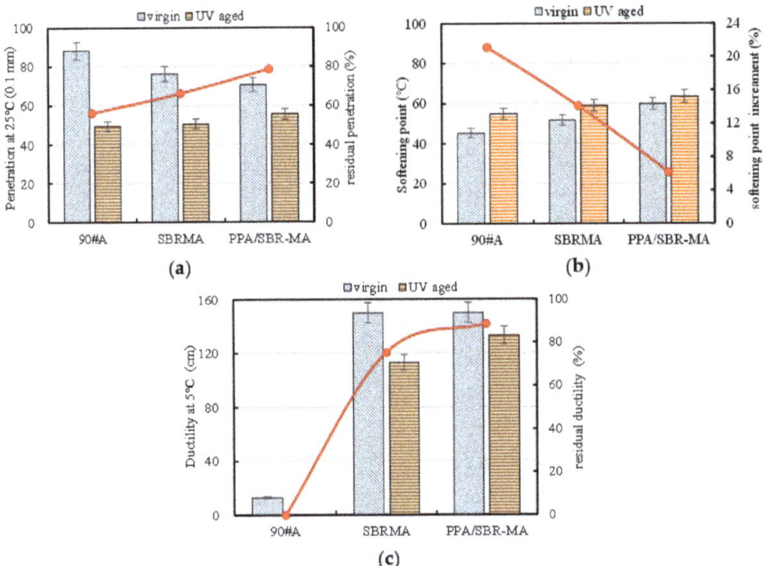

Figure 5. Basic property test results. (**a**) Penetration; (**b**) Softening point; (**c**) Ductility.

The ductility results of the three asphalts at 5 °C are shown in Figure 5c. Figure 5c indicates that the low temperature performance increases significantly with the addition of SBR, which is also the most important advantage of SBRMA. The ductility of SBRMA and PPA/SBR-MA are both greater than 150 cm (value equals 150 cm in calculation) before UV aging, which proves that the addition of PPA may enhance the low temperature performance or have no influence on the low temperature properties of SBRMA. The specific results will be further determined from the subsequent low temperature rheological test (BBR). The ductility of the three asphalts all decreased after UV aging. The RD was used to characterize the sensitivity of the low temperature flexibility on the UV aging. The sequence of the RD is PPA/SBR-MA > SBRMA > 90#A. The results show that PPA/SBRMA has a better low temperature flexibility than 90#A and SBRMA after UV aging, which indicates that PPA can significantly reduce the sensitivity of the low temperature flexibility of SBRMA to UV aging.

3.2. High Temperature Rheological Properties Analysis

3.2.1. Phase Angle

The phase angle mainly reflects the viscoelasticity of the asphalt binder. The elasticity and high temperature performance of asphalt are higher when the phase angle is smaller. As seen from Figure 6a, the phase angle increases with the rise in the test temperature. SBR and PPA contribute to the reduction in the phase angle. For example, at 64 °C, the phase angles δ of 90#A, SBRMA, and PPA/SBR-MA are 86.78°, 83.56°, and 76.59°, respectively. The phase angle δ of SBRMA is reduced by 3.7% compared with that of 90#A. The phase angle δ of PPA/SBR-MA is decreased by 11.7% compared with that of 90#A and 8.3% compared with that of SBRMA. Therefore, the results suggest that PPA enhances the high temperature performance of SBRMA.

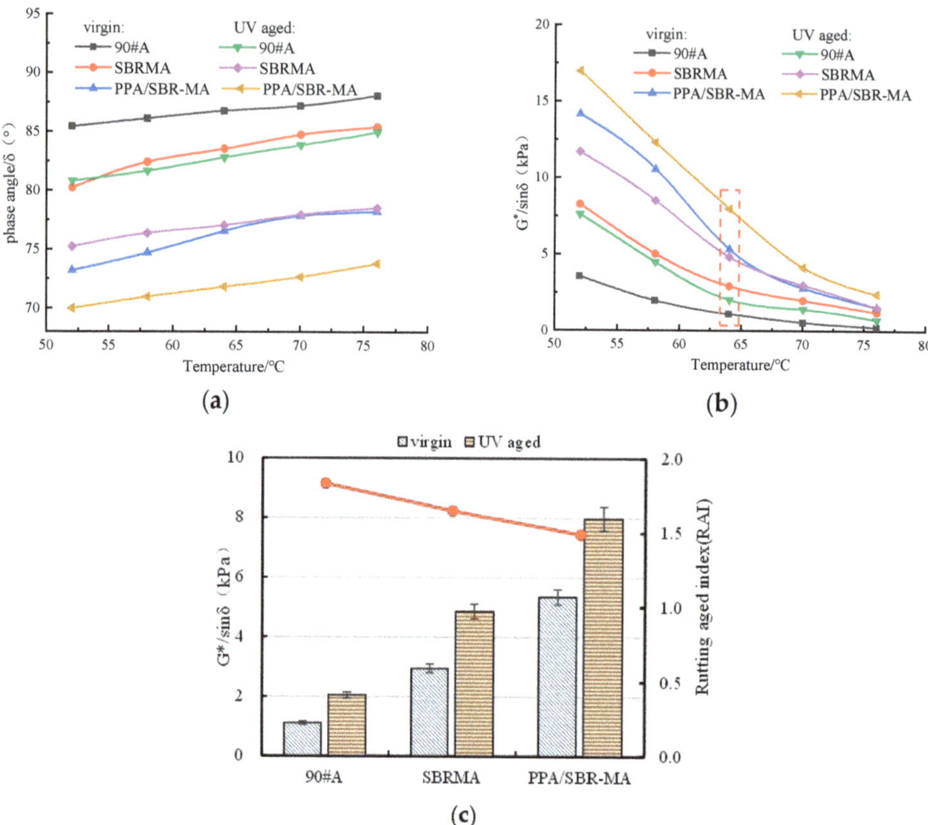

Figure 6. High temperature rheological properties results. (**a**) Phase angle of asphalt (δ); (**b**) Rutting factor (G*/sin δ); (**c**) Rutting factor and aged index at 64 °C.

The phase angle curves of the three asphalts all move downward after UV aging. The phase angles become smaller after UV aging, which indicates that the asphalts are hardened. The decreasing order of the values of the phase angle variable is SBRMA > PPA/SBR-MA > 90#A. For example, at 64 °C, the phase angles of 90#A, SBRMA, and PPA/SBR-MA before UV aging are 82.85°, 77.06°, and 71.84°, respectively. Meanwhile, the phase angles of the three asphalts are reduced by 4.5%, 7.8%, and 6.2%, respectively, after UV aging. The results indicate that SBRMA has a poorer anti-UV aging performance. The addition of PPA also reduces the sensitivity of the phase angle of SBRMA to UV aging.

3.2.2. Rutting Factor

Figure 6b shows that the rutting factor decreases with the increasing temperature. When the test temperature is higher than 64 °C, the decreasing trend of the rutting factor of 90#A and SBRMA slows down. However, the decreasing trend of the rutting factor of PPA/SBR-MA slows down when the temperature is higher than 70 °C. The rutting factor of PPA/SBR-MA is obviously higher than those of SBRMA and 90#A. For example, at 64 °C, the rutting factor of SBRMA is increased by 163.4% compared with that of 90#A, and the rutting factor of PPA/SBR-MA is increased by 377.7% and 81.4% compared with those of 90#A and SBRMA, respectively. This result shows that PPA can obviously increase the high temperature performance of SBRMA and remedy the shortage of permanent deformation resistance. The rutting factor curves of the three asphalts move upward, and the rutting factor increases significantly after UV aging. The rutting factor shows a decreasing trend when the test temperature reaches 70 °C. Figure 6c shows the RAI of 90#A, SBRMA, and PPA/SBR-MA asphalts at 64 °C. The decreasing order of the RAI is 90#A > SBRMA> PPA/SBRMA. This result indicates that PPA/SBR-MA has the best high temperature stability under UV aging among the three asphalts.

In summary, the analysis results show that PPA can significantly improve the high temperature performance of SBRMA and reduce the effect of UV aging on it.

3.3. Low Temperature Rheological Properties Analysis

Two test temperatures of −12 °C and −18 °C were used to test the 90#A, SBRMA, and PPA/SBRMA before and after UV aging. As shown in Figure 7a, the creep stiffness S(60) of 90#A decreases with the addition of SBR and PPA before UV aging, and S(60) of PPA/SBRMA is the minimum. The S(60) of the three asphalts are 61.9, 50.0, and 49.0 MPa, respectively, at −12 °C. Meanwhile, the values are 248.0, 194.5, and 172.0 Mpa at −18 °C. Therefore, the temperature stress of the asphalt increases significantly with the drop in temperature. The low temperature flexibility of SBRMA is better than 90#A, and the creep stiffness reduction rates of −12 °C and −18 °C are 19.2% and 21.6%, respectively. The flexibility of PPA/SBR-MA at low temperature is better than that of SBRMA, and the reduction rates of the creep stiffness at −12 °C and −18 °C are 2% and 11.6%, respectively. This result implies that the flexibility of PPA/SBR-MA is more obvious when the temperature is lower.

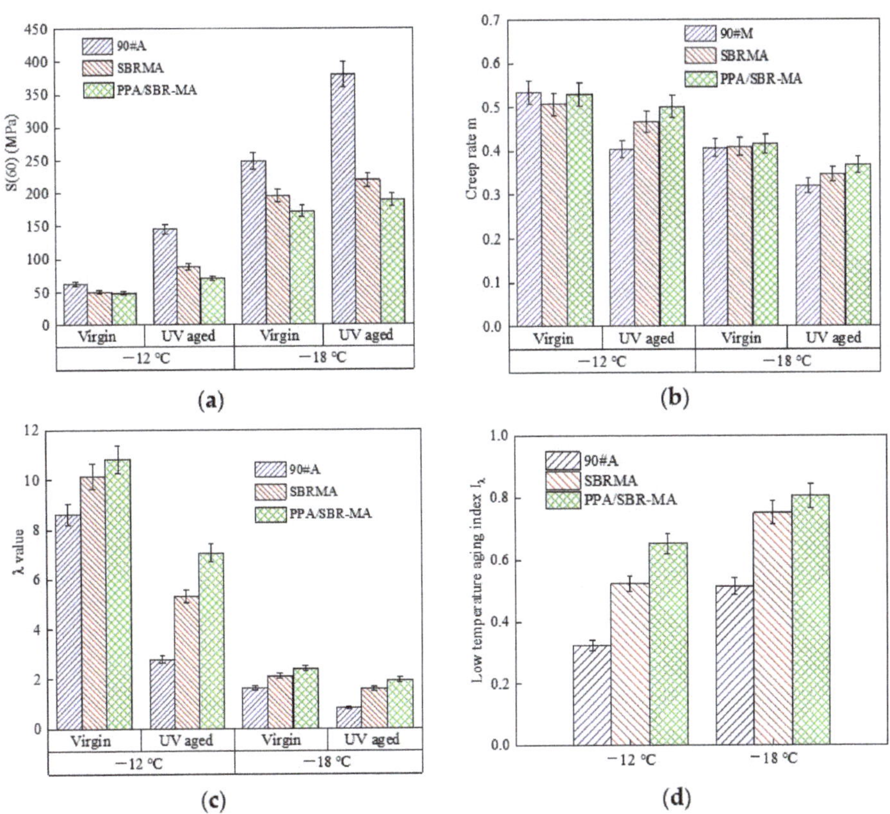

Figure 7. Low temperature rheological properties results. (**a**) Stiffness Modulus S(60); (**b**) Creep rate m; (**c**) The ratio λ; (**d**) Low temperature aging index.

As shown in Figure 7b, the 90#A has a better in situ stress relaxation ability at -12 °C, and the m values of SBRMA and PPA/SBR-MA are equivalent to, or even slightly lower, than that of 90#A. At -18 °C, the advantage of the stress relaxation capacity of the modified asphalt gradually becomes obvious. With the addition of SBR and PPA, the m values of the modified asphalt increase successively to 0.406, 0.409, and 0.416. However, the increase is small. The results reflect that PPA can significantly improve the low temperature performance of SBRMA and reduce the sensitivity to UV aging. In addition, the lower the temperature, the more pronounced the effect.

Figure 7c shows that the λ value increases gradually with the addition of the SBR modifier and PPA, which indicates that the SBR modifier and PPA can enhance the low temperature performance of asphalt. The λ values of the three asphalts all decrease after UV aging. As displayed in Figure 7d, the aging index values of the low temperature properties at -12 °C and -18 °C are as follows: 90#A > SBRMA > PPA/SBR-MA. The above-mentioned analysis shows that PPA/SBR-MA has the best flexibility and stress relaxation ability before and after aging. The sensitivity of each parameter to UV aging also decreases with the decline in temperature, which is consistent with the reaction law of S(60) and m. The results show that the addition of PPA enhances the low temperature cracking resistance of SBRMA and reduces the sensitivity of UV aging. The effect of UV aging on the low temperature cracking resistance of asphalt can be better reflected by using the low temperature aging index I_λ.

3.4. Chemical Structure Analysis

In this study, the molecular structure changes in the three asphalts were analyzed by FTIR test before and after UV aging, and the anti-UV aging mechanism of PPA/SBR-MA was explained.

Figure 8 shows the FTIR spectra of the PPA. Figure 9a shows that only marginal differences are found between 90#A and SBRMA. A new absorption peak of SBRMA and PPA/SBR-MA appears at 966 cm^{-1} because of the SBR characteristic peak vibration caused by the butadiene block vibration. Therefore, only physical interactions occur with the addition of SBR particles. This result agrees with the finding of a previous study [47]. A new absorption peak of PPA/SBR-MA appears at 1150 cm^{-1} compared with SBR-MA, which is mainly due to the stretching vibration of P–O and P=O [48,49]. However, the peak strength is low because the dosage of PPA is only 1%. As seen from Figures 8 and 9a, the PPA infrared spectrum also shows the P=O stretching vibration frequency at 1223 cm^{-1}, and the peak value disappears after PPA is added. Therefore, PPA chemically modifies the base asphalt [50,51], which is also an important reason why PPA can improve the compatibility of SBRMA. The characteristic peaks of PPA/SBR-MA and SBRMA either do not produce absorption peaks or the absorption peaks disappear. However, the intensity of the characteristic peaks changes. For example, the ICO and BI indexes are different after the addition of PPA, which indicates that the functional group of the asphalt binder does not change, but some chemical bonds change with the occurrence of chemical reactions [47]. Thus, whether chemical reactions occur between SBR and PPA needs further study.

The influence of PPA on the anti-aging properties of the asphalt binders was determined by calculating the carbonyl group (ICO), SBR index (BI), and chemical structure aging index (RI) to quantitatively analyze the changes in the chemical bonds of the asphalt binders after UV aging. The analysis results of these indexes are shown in Figure 10. The carbonyl index (ICO) values of the three asphalts increase after UV aging, which is mainly caused by the oxygen absorption of the unsaturated carbon chain. The decreasing order of the ICO is 90#A > SBRMA > PPA/SBR-MA. Figure 10b shows that the BI index values of SBRMA and PPA/SBR-MA decrease after UV aging. After UV aging, the BI index value of SBRMA decreases from 0.031 to 0.011, while that of PPA/SBR-MA decreases from 0.045 to 0.028. The chemical structural aging index (CR) of PPA/SBR-MA is much smaller than those of 90#A and SBRMA, which indicates that the oxygen absorption of the saturated carbon chain and the degradation of the SBR modifier occur in the UV aging process. The addition of PPA effectively slows down the development and enhances the anti-UV aging property for SBRMA, which may be attributed to the complex chemical reaction between PPA and SBRMA. Thus, the results lead to the changes in the molecular structure, which effectively block polymer degradation and free radical diffusion. The reason for this is that the progression of UV aging is accompanied by the deep diffusion of free radicals and any delay in free radical diffusion can help delay the aging process [52].

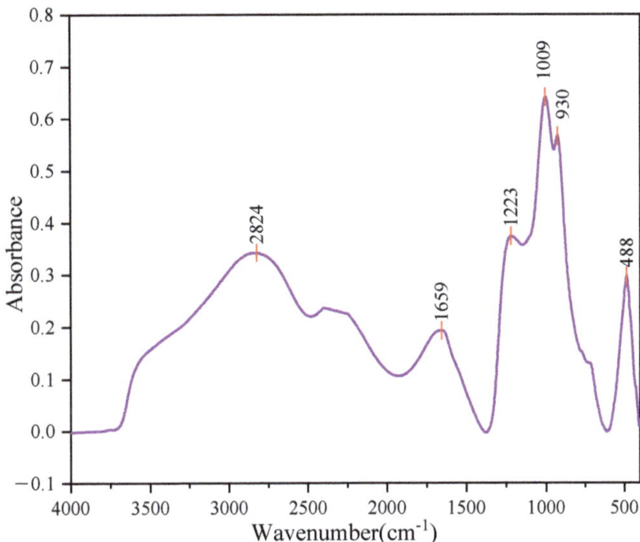

Figure 8. FTIR spectrum of. polyphosphoric acid (PPA).

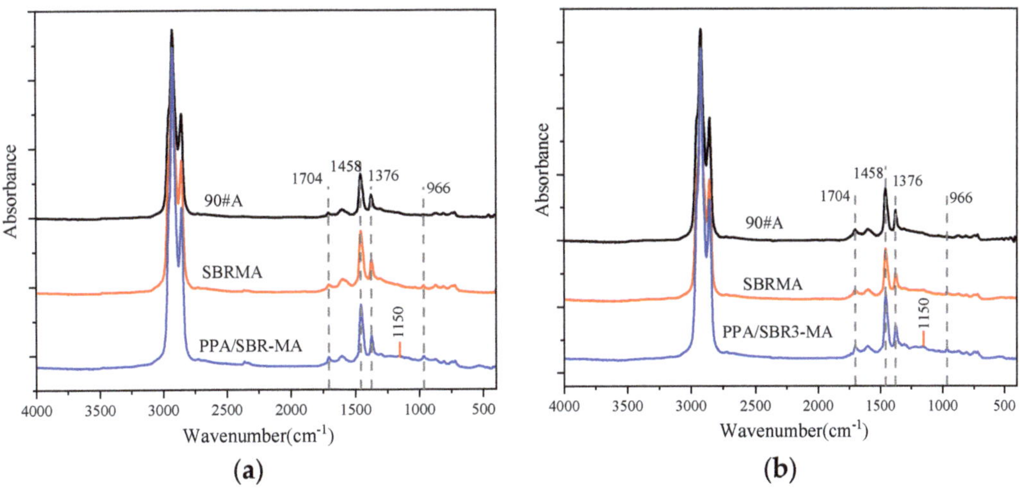

Figure 9. FTIR spectra of asphalt. (**a**) virgin; (**b**) UV aged.

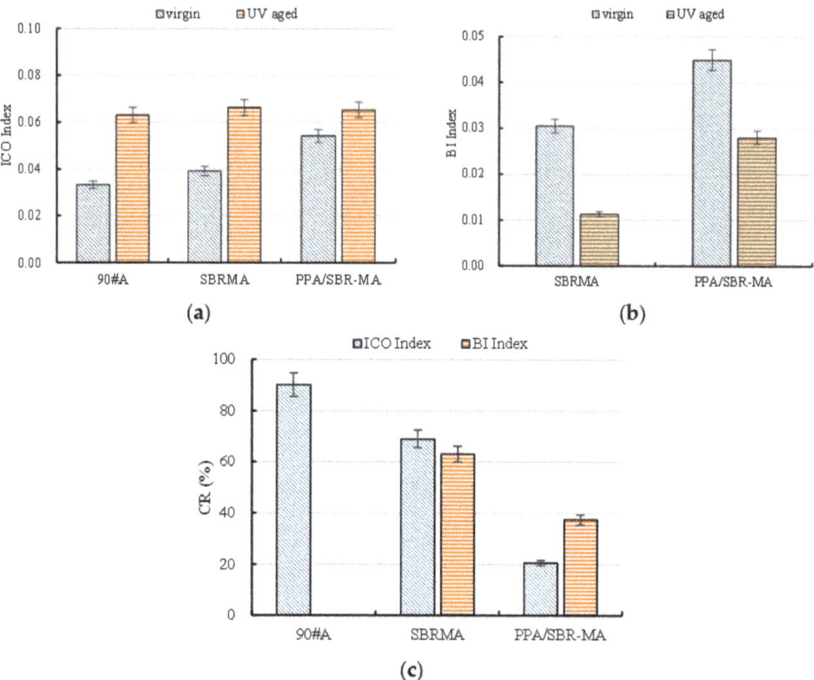

Figure 10. Chemical structure index of the three asphalts. (**a**) Carbonyl index; (**b**) SBR index; (**c**) Chemical structure aging indexes.

3.5. Molecular Size Evolution Analyses

Figure 11 shows that the addition of SBR significantly increases the macromolecules of 90#MA. Meanwhile, the addition of PPA results in a decrease in the LMS and an increase in the SMS, which implies that the addition of PPA transforms the heavyweight polymers to lightweight polymers [53]. The molecular weights of 90#A, SBRMA, and PPA/SBR-MA after UV aging decrease, which qualitatively reflects the changing trend of the molecular weights of asphalt before and after aging. The molecular weight distribution in the LMS range shows an obvious shoulder peak, and the percentage composition increases after UV aging. Therefore, UV aging would lead to intermolecular polymerization and promote an increase in the content of substances with large molecular weight. The molecular weight distribution of the three asphalts changes significantly after UV aging. The LMS values of the three asphalt cements are increased, but their MMS and SMS are decreased.

Figure 12a,b show that the LMS increment of 90#A is the largest (31.79%), the LMS increment of PPA/SBR-MA is the smallest (14.13%), and the increment of LMS of SBRMA is in the middle (18.53%). Similarly, the increment of the MMS shows the same trend. The values are −9.36%, −7.27%, and −1.83% for 90#A, SBRMA, and PPA/SBR-MA, respectively. However, PPA/SBR-MA has the largest reduction in SMS (−6.08%), followed by 90#A (−5.33%), and SBRMA has the smallest (−0.52%). The reason is that 90#A may be primarily based on the transformation of the small and medium molecules to large molecules and the evaporation of small molecules in the UV aging process. Compared with 90#A, SBRMA improves the degradation of the SBR modifier polymer and increases the content of the medium molecule and small fraction. As a result, the change in the SMS is minimal. Furthermore, the addition of PPA inhibits the decomposition of the SBR polymer into small and medium molecules, and slows down the UV aging process of the base asphalt. Therefore, the conversion of medium molecules into large molecules is reduced, and the aging process is concentrated on the conversion of small molecules into medium molecules. This results in the smallest change in the LMS and MMS and the largest decrease in the SMS.

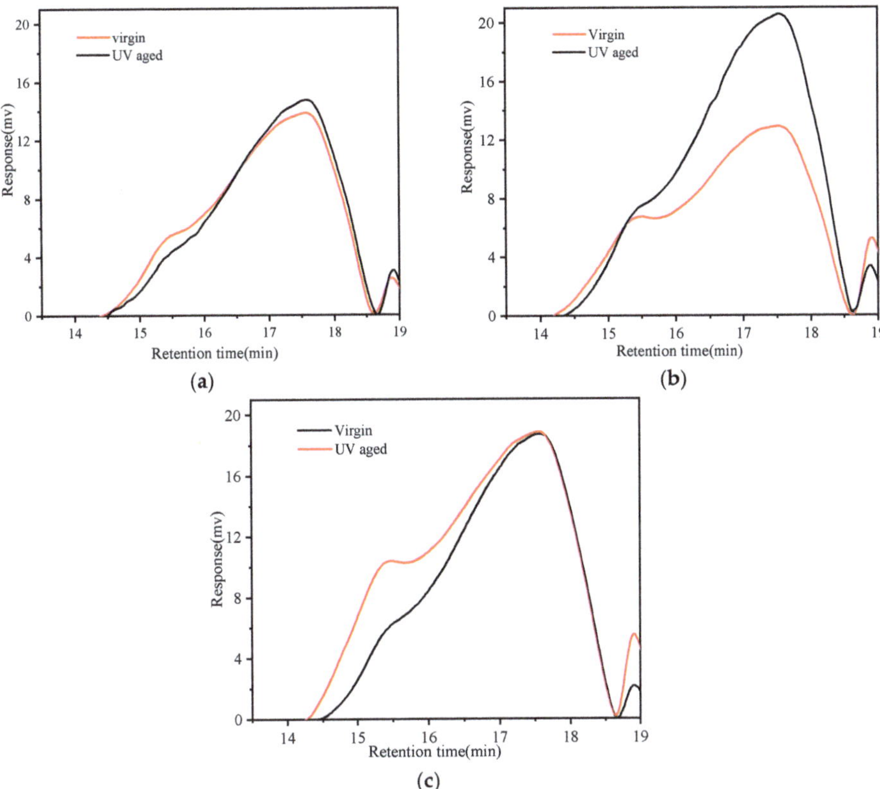

Figure 11. GPC curves of the three asphalts. (**a**) 90#A; (**b**) SBRMA; (**c**) PPA/SBR-MA.

As shown in Table 4, the size rule of the parameters M_n, M_w, and d before UV aging is SBRMA >90#A > PPA/SBR-MA. The addition of the SBR modifier leads to a significant increase in heavy weight molecules, and the addition of PPA reduces the amount of SBRMA. The reason for this is that the addition of PPA induces a loss of hydrogen bonding and the decomposition of asphaltene, which result in a greater dispersion of smaller asphaltene domains [54]. The parameters M_n, M_w, and d of 90#A show an increasing trend after UV aging. Figure 12 also shows that the oxidation of components with medium and low molecular weights and the evaporation of components with low molecular weight mainly occur in the UV aging of base asphalt, which is consistent with the conclusions of previous studies [52,55]. The M_n value of SBRMA shows a decreasing trend after UV aging, but the M_w value shows the opposite trend. The main reason is that the polymerization of small and medium molecules and the degradation of large molecules of polymer-modified asphalt occur in the aging process, including the degradation of the polymer modifier. As a result, the polymerization of small and medium molecules increases the recombination fraction. The degradation of macromolecules and modifiers leads to the migration of the recombination to light components [56,57]. The change trend of M_n and M_w for PPA/SBR-MA after UV aging is consistent with that of SBRMA. However, the change rate of PPA/SBR-MA is significantly less than that of SBRMA. The addition of PPA significantly slows down the UV aging process of SBRMA and inhibits the degradation of the SBR modifier macromolecules. It greatly reduces the conversion of small to large molecules.

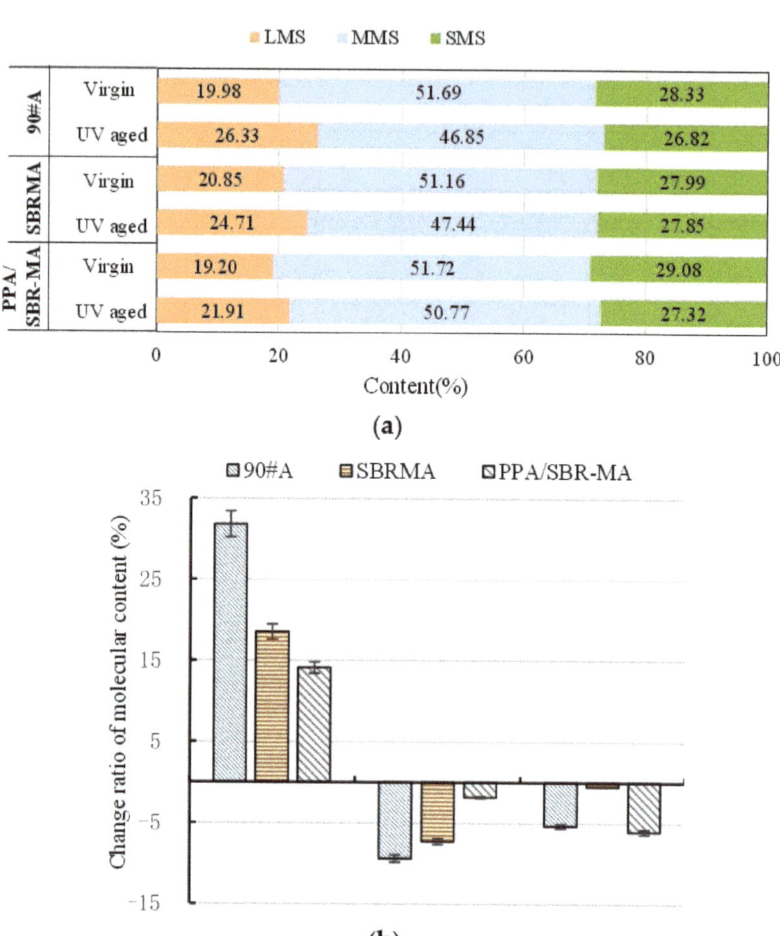

Figure 12. Molecular distribution analysis results. (**a**) Molecular distribution analysis results; (**b**) Change rate in molecular size.

Table 4. Results of molecular weight analysis.

Sample	Index				
	M_n	M_w	d	I_{Mn} (%)	I_{Mw} (%)
90#-virgin	909	3634	3.998		
90#-UV aged	1124	5215	4.640	23.65	43.51
SBR-virgin	998	3996	4.004		
SBR-UV aged	950	4966	5.227	−4.81	24.27
PPA/SBR-virgin	905	3460	3.823		
PPA/SBR-UV aged	889	3905	4.393	−1.77	12.86

Comparing the molecular distribution and weight changes before and after aging proves that PPA gives the SBRMA a more stable dispersion system, inhibits the decomposition of the SBR modifier polymer, delays the aging process of base asphalt, improves the stability of SBRMA asphalt, and thus results in a better anti-UV aging performance.

3.6. Relationship Analysis of Microcosmic Property and Macroscopical Property

The Pearson correlation coefficient "r" of quantitative statistical analysis is often used in variable correlation analysis. The correlation between the macro and micro aging indexes were analyzed by the method in this study. As displayed in Table 5, with the exception of the micro aging index I_{SMS}, the other parameters have good correlation with the macro aging index, which indicates that I_{SMS} is unsuitable for evaluating the aging effect and is only applicable to the mechanism explanation of the aging process. In particular, the correlation between I_{LMS}, I_{Mw}, and the macro performance index parameters is above 0.9, which implies that the distribution of large and average molecular weights has a good correlation with the macro performance. This deduction is consistent with previous research conclusions [43,47,58]. Thus, I_{LMS} and I_{Mw} can be used as the main evaluation indexes of the micro aging effect, and CR, CIO and I_{Mn} follow as the secondary indexes. The correlation analysis results of the macro and micro aging indexes suggest that the micro test can not only explain the micro action mechanism, but can also predict the macro performance.

Table 5. Pearson correlation results.

Macro-Aging Index	Micro-Aging Index					
	I_{LMS}	I_{MMS}	I_{SMS}	I_{Mn}	I_{Mw}	CR_{ICO}
RP	−0.9412	0.9823	−0.1848	−0.7754	−0.9783	−0.9870
SPI	−0.9505	0.9765	−0.1568	−0.7931	−0.9838	−0.9821
DR	−0.9949	0.8052	0.2573	−0.9720	−0.9719	−0.8212
RAI	0.9694	−0.9597	0.0894	0.8326	0.9937	0.9670
I_λ (−12 °C)	−0.9875	0.9307	7.05×10^{-5}	−0.8788	−0.9997	−0.9403
I_λ (−18 °C)	−0.9979	0.8261	0.2222	−0.9629	−0.9798	−0.8413

4. Conclusions

The macro and micro properties of 90#A, SBRMA with 4% SBR modifier, and PPA/SBR-MA with 4 wt% SBR + 1 wt% PPA additives before and after UV aging were studied in this work. The key findings are as follows:

(1) The penetration of asphalt decreases with the addition of PPA, which proves that PPA/SBR-MA becomes harder and more resistant to deformation. The increase in the softening point indicates a higher temperature stability of asphalt. The incorporation of PPA significantly reduces the influence of UV aging on the consistency, high temperature stability, and low temperature flexibility of SBRMA according to the aging indexes of the penetration, softening point, and ductility.

(2) The addition of PPA significantly improves the high temperature deformation resistance of SBRMA before UV aging according to the analysis of the phase angle δ and rutting factor $G^*/\sin \delta$. The high temperature performance of the three asphalts is improved after UV aging. However, the rutting factor aging index (RAI) of 90#A, SBRMA, and PPA/SBR-MA are 1.83, 1.65, and 1.49, respectively, indicating that PPA can reduce the sensitivity of the high temperature rheological parameters of SBRMA to UV light.

(3) Based on the variations of S(60) and m before and after UV aging, it is found that UV aging will crack the low temperature properties of asphalt according to the change rule of these values. However, PPA/SBR-MA has the best flexibility and stress relaxation ability, which implies that PPA enhances the low temperature cracking resistance of SBRMA. The low temperature aging indexes I_λ of 90#A, SBRMA, and PPA/SBR-MA are 0.32, 0.52, and 0.65 at −12 °C temperature, and 0.51, 0.75, and 0.81 at −18 °C, respectively, which indicates that PPA can significantly reduce the UV aging sensitivity of SBRMA.

(4) The FTIR and GPC tests show that the addition of PPA gives the SBRMA a more stable dispersion system. UV aging mainly leads to the oxygen absorption of the saturated carbon chain and the degradation of the SBR modifier, along with the evaporation of small molecules. PPA can significantly enhance the anti-UV aging property of SBRMA

because it mainly inhibits the degradation of the SBR modifier and delays the aging process of base asphalt.

(5) The microscopic aging indexes I_{LMS} and I_{Mw} after UV aging are significantly correlated with the macroscopic aging indexes RP, SPI, DR, RAI, and I_λ by the analysis of the Pearson correlation coefficient. They can be used as the main evaluation indexes of the microscopic aging effect. The ICO and I_{Mn} have a good correlation with the macro performance aging index. However, I_{SMS} is unsuitable for evaluating the aging effect and is only applicable to the mechanism explanation of the aging process. Microscopic experiments can not only explain the microscopic mechanism of action, but can also predict the macroscopic properties.

Overall, based on the results of the experimental data in this study, PPA can improve the high and low temperature performance and UV aging resistance of SBRMA, and PPA/SBR-MA is more suitable for high altitude cold regions than SBRMA.

Author Contributions: Conceptualization, methodology, laboratory testing, data curation, writing original draft, Y.X.; Writing—review and editing, Project administration, K.N.; Writing—review and editing, Supervision, H.Z.; Writing—review and editing, R.C.; Investigation, Visualization, L.O. All authors have read and agreed to the published version of the manuscript.

Funding: This research is supported by the Construction Program of Chongqing Graduate Student Union Training Base (JDLHPYJD2020014), School-enterprise Cooperation Project (2012070187). The authors appreciate their financial support gratefully.

Institutional Review Board Statement: Not applicable.

Informed Consent Statement: Not applicable.

Data Availability Statement: Not applicable.

Conflicts of Interest: The authors declare no conflict of interest.

References

1. Yu, H.; Bai, X.; Qian, G.; Wei, H.; Gong, X.; Jin, J.; Li, Z. Impact of Ultraviolet Radiation on the Aging Properties of SBS-Modified Asphalt Binders. *Polymers* **2019**, *11*, 1111. [CrossRef]
2. Chen, Z.; Zhang, H.; Duan, H. Investigation of ultraviolet radiation aging gradient in asphalt binder. *Constr. Build. Mater.* **2020**, *246*, 118501. [CrossRef]
3. Zhang, H.; Yu, J.; Feng, Z.; Xue, L.; Wu, S.P. Effect of aging on the morphology of bitumen by atomic force microscopy. *J. Microsc.* **2012**, *246*, 11–19. [CrossRef] [PubMed]
4. Cheraghian, G.; Wistuba, M.P. Effect of Fumed silica nanoparticles on ultraviolet aging resistance of bitumen. *Nanomaterials* **2021**, *11*, 454. [CrossRef]
5. Jing, R.; Varveri, A.; Liu, X.; Scarpas, A.; Erkens, S. Ageing effect on chemo-mechanics of bitumen. *Road Mater. Pavement Des.* **2021**, *22*, 1044–1059. [CrossRef]
6. Eberhardsteiner, L.; Füssl, J.; Hofko, B.; Handle, F.; Hospodka, M.; Blab, R.; Grothe, H. Towards a microstructural model of bitumen ageing behaviour. *Int. J. Pavement Eng.* **2015**, *16*, 939–949. [CrossRef]
7. Ferrotti, G.; Baaj, H.; Besamusca, J.; Bocci, M.; Cannone-Falchetto, A.; Grenfell, J.; Hofko, B.; Porot, L.; Poulikakos, L.; You, Z.J.M.; et al. Comparison between bitumen aged in laboratory and recovered from HMA and WMA lab mixtures. *Mater. Struct.* **2018**, *51*, 150. [CrossRef]
8. Zhang, C.; Yu, J.; Xu, S.; Xue, L.; Cao, Z.J.C. Influence of UV aging on the rheological properties of bitumen modified with surface organic layered double hydroxides. *Constr. Build. Mater.* **2016**, *123*, 574–580. [CrossRef]
9. Wei, H.; Zhang, H.; Li, J.; Zheng, J.; Ren, J. Effect of loading rate on failure characteristics of asphalt mixtures using acoustic emission technique. *Constr. Build. Mater.* **2023**, *364*, 129835. [CrossRef]
10. Zhang, S.Y.; Cui, Y.A.; Wei, W.W. Low-temperature characteristics and microstructure of asphalt under complex aging conditions. *Constr. Build. Mater.* **2021**, *303*, 124408. [CrossRef]
11. Yildirim, Y. Polymer modified asphalt binders. *Constr. Build. Mater.* **2007**, *21*, 66–72. [CrossRef]
12. Yu, J.-Y.; Zhang, H.-L.; Sun, P.; Zhao, S.-F. Laboratory performances of nano-particles/polymer modified asphalt mixtures developed for the region with hot summer and cold winter and field evaluation. *Road Mater. Pavement Des.* **2020**, *21*, 1529–1544. [CrossRef]
13. Li, Q.S.; Zhang, H.L.; Chen, Z.H. Improvement of short-term aging resistance of styrene-butadiene rubber modified asphalt by Sasobit and epoxidized soybean oil. *Constr. Build. Mater.* **2021**, *271*, 121870. [CrossRef]

14. Liao, M.J.; Liu, Z.H.; Gao, Y.L.; Liu, L.; Xiang, S.C. Study on UV aging resistance of nano-TiO_2/montmorillonite/styrene-butadiene rubber composite modified asphalt based on rheological and microscopic properties. *Constr. Build. Mater.* **2021**, *301*, 124108. [CrossRef]
15. Babagoli, R.; Jalali, F.; Khabooshani, M. Performance properties of WMA modified binders and asphalt mixtures containing PPA/SBR polymer blends. *J. Thermoplast. Compos. Mater.* **2021**, *36*, 274–306. [CrossRef]
16. Zhang, F.; Hu, C.B. The research for SBS and SBR compound modified asphalts with polyphosphoric acid and sulfur. *Constr. Build. Mater.* **2013**, *43*, 461–468. [CrossRef]
17. Zhang, F.; Yu, J.Y. The research for high-performance SBR compound modified asphalt. *Constr. Build. Mater.* **2010**, *24*, 410–418. [CrossRef]
18. Domingos, M.D.I.; Faxina, A.L. High-temperature properties and modeling of asphalt binders modified with sbr copolymer and ppa in the multiple stress creep and recovery (MSCR) test. *Appl. Rheol.* **2016**, *26*, 53830.
19. Liang, P.; Liang, M.; Fan, W.Y.; Zhang, Y.Z.; Qian, C.D.; Ren, S.S. Improving thermo-rheological behavior and compatibility of SBR modified asphalt by addition of polyphosphoric acid (PPA). *Constr. Build. Mater.* **2017**, *139*, 183–192. [CrossRef]
20. Gao, L.N.; Cai, N.N.; Fu, X.H.; He, R.; Zhang, H.G.; Zhou, J.Y.; Kuang, D.L.; Liu, S.W. Influence of PPA on the Short-Term Antiaging Performance of Asphalt. *Adv. Civ. Eng.* **2021**, *2021*, 6628778. [CrossRef]
21. Zhang, F.; Hu, C.B. Influence of aging on thermal behavior and characterization of SBR compound-modified asphalt. *J. Therm. Anal. Calorim.* **2014**, *115*, 1211–1218. [CrossRef]
22. Han, D.Q.; Hu, G.S.; Zhang, J.T. Study on Anti-Aging Performance Enhancement of Polymer Modified Asphalt with High Linear SBS Content. *Polymers* **2023**, *15*, 256. [CrossRef]
23. Ge, D.D.; Yan, K.Z.; You, L.Y.; Wang, Z.X. Modification mechanism of asphalt modified with Sasobit and Polyphosphoric acid (PPA). *Constr. Build. Mater.* **2017**, *143*, 419–428. [CrossRef]
24. Wei, J.G.; Shi, S.; Zhou, Y.M.; Chen, Z.Y.; Yu, F.; Peng, Z.Y.; Duan, X.R. Research on Performance of SBS-PPA and SBR-PPA Compound Modified Asphalts. *Materials* **2022**, *15*, 2112. [CrossRef]
25. Liu, S.J.; Zhou, S.B.; Peng, A.H. Evaluation of polyphosphoric acid on the performance of polymer modified asphalt binders. *J. Appl. Polym. Sci.* **2020**, *137*, 48984. [CrossRef]
26. Hao, P.W.; Zhai, R.X.; Zhang, Z.X.; Cao, X.J. Investigation on performance of polyphosphoric acid (PPA)/SBR compound-modified asphalt mixture at high and low temperatures. *Road Mater. Pavement Des.* **2019**, *20*, 1376–1390. [CrossRef]
27. Chen, P.F.; Zhang, Z.M.; Li, Y.M. Anti-ultraviolet aging properties of polyphosphoric acid/styrene-butadiene rubber composite modified asphalts. *China Synth. Rubber Ind.* **2020**, *43*, 60–65.
28. Jafari, M.; Babazadeh, A. Evaluation of polyphosphoric acid-modified binders using multiple stress creep and recovery and linear amplitude sweep tests. *Road Mater. Pavement Des.* **2016**, *17*, 859–876. [CrossRef]
29. Wang, J.N. *Study on Aging Mechanism and Pheologic Behavior of Asphalt under Simulate Ultraviolet*; Harbin Institute of Technology: Harbin, China, 2008.
30. *JTG E20-2011*; Standard Test Methods of Bitumen and Bituminous Mixture for Highway Engineering. China Communications Press: Beijing, China, 2011.
31. Geng, J.; Meng, H.; Xia, C.; Chen, M.; Lu, T.; Zhou, H. Effect of dry-wet cycle aging on physical properties and chemical composition of SBS-modified asphalt binder. *Mater. Struct.* **2021**, *54*, 120. [CrossRef]
32. Geng, J.; Chen, M.; Xia, C.; Liao, X.; Chen, Z.; Chen, H. Aging characteristics of crumb rubber modified asphalt binder and mixture with regenerating agent. *Constr. Build. Mater.* **2021**, *299*, 124299. [CrossRef]
33. *ASTM-D7175*; Standard Test Method for Determining the Rheological Properties of Asphalt Binder. Using a Dynamic Shear Rheometer. ASTM: West Conshohocken, PA, USA, 2015.
34. Camargo, I.G.D.; Ben Dhia, T.; Loulizi, A.; Hofko, B.; Mirwald, J. Anti-aging additives: Proposed evaluation process based on literature review. *Road Mater. Pavement Des.* **2021**, *22*, S134–S153. [CrossRef]
35. *ASTM-D6648*; Standard Test Method for Determining the Flexural Creep Stiffness of Asphalt Binder Using the Bending Beam Rheometer (BBR). ASTM: West Conshohocken, PA, USA, 2001.
36. Li, J.; Xiao, F.; Amirkhanian, S.N. Storage, fatigue and low temperature characteristics of plasma treated rubberized binders. *Constr. Build. Mater.* **2019**, *209*, 454–462. [CrossRef]
37. Xu, N.; Wang, H.N.; Chen, Y.; Miljkovic, M.; Feng, P.N.; Ding, H.Y. Thermal storage stability and rheological properties of multi-component styrene-butadiene-styrene composite modified bitumen. *Constr. Build. Mater.* **2022**, *322*, 126494. [CrossRef]
38. Hung, A.M.; Fini, E.H. Absorption spectroscopy to determine the extent and mechanisms of aging in bitumen and asphaltenes. *Fuel* **2019**, *242*, 408–415. [CrossRef]
39. Hou, X.D.; Lv, S.T.; Chen, Z.; Xiao, F.P. Applications of Fourier transform infrared spectroscopy technologies on asphalt materials. *Measurement* **2018**, *121*, 304–316. [CrossRef]
40. Poulikakos, L.; Wang, D.; Porot, L.; Hofko, B. Impact of asphalt aging temperature on chemo-mechanics. *RSC Adv.* **2019**, *9*, 11602–11613.
41. Li, C.; Xie, X.F.; Wang, L.; Guo, Y.Y.; Zhang, L.; Xue, Z.H. Evaluation of the effect of thermal oxygen aging on base and SBS-modified bitumen at micro and macroscales. *Constr. Build. Mater.* **2022**, *324*, 126623. [CrossRef]
42. Hosseinnezhad, S.; Zadshir, M.; Yu, X.; Yin, H.; Sharma, B.K.; Fini, E. Differential effects of ultraviolet radiation and oxidative aging on bio-modified binders. *Fuel* **2019**, *251*, 45–56. [CrossRef]

43. Kim, K.W.; Burati Jr, J.L.; Park, J.S. Methodology for defining LMS portion in asphalt chromatogram. *J. Mater. Civ. Eng.* **1995**, *7*, 31–40. [CrossRef]
44. Wang, Z.; Xu, X.; Wang, X.; Jinyang, H.; Guo, H.; Yang, B. Performance of modified asphalt of rubber powder through tetraethyl orthosilicate (TEOS). *Constr. Build. Mater.* **2021**, *267*, 121032. [CrossRef]
45. Liu, J.; Qi, L.; Wang, X.; Li, M.; Wang, Z. Influence of aging induced by mutation in temperature on property and microstructure development of asphalt binders. *Constr. Build. Mater.* **2022**, *319*, 126083. [CrossRef]
46. Li, Z.Z.; Liu, H.J.; Chen, W.X.; Li, Y.; Zhao, Z.P.; Yin, Y.P.; Li, M.Y. Influence of residual SB di-block in SBS on the thermo-oxidative aging behaviors of SBS and SBS modified asphalt. *Mater. Struct.* **2022**, *55*, 23. [CrossRef]
47. Cheng, P.; Li, Y.; Zhang, Z.J. Effect of phenolic resin on the rheological, chemical, and aging properties of SBR-modified asphalt. *Int. J. Pavement Res. Technol.* **2021**, *14*, 421–427. [CrossRef]
48. Liu, H.; Zhang, M.; Wang, Y.; Chen, Z.; Hao, L. Rheological properties and modification mechanism of polyphosphoric acid-modified asphalt. *Road Mater. Pavement Des.* **2020**, *21*, 1078–1095. [CrossRef]
49. Wang, F.; Huang, T.; Xin, G.; Mu, M.; Shen, Q.J. Study on Conventional and Rheological Properties of Corn Stalk Bioasphalt/PPA Composite Modified Asphalt. *Adv. Civ. Eng.* **2021**, *2021*, 7928189. [CrossRef]
50. Qian, C.; Fan, W.; Ren, F.; Lv, X.; Xing, B. Influence of polyphosphoric acid (PPA) on properties of crumb rubber (CR) modified asphalt. *Constr. Build. Mater.* **2019**, *227*, 117094. [CrossRef]
51. Masson, J.; Gagné, M.; Robertson, G.; Collins, P. Reactions of polyphosphoric acid and bitumen model compounds with oxygenated functional groups: Where is the phosphorylation? *Energy Fuels* **2008**, *22*, 4151–4157. [CrossRef]
52. Hung, A.; Fini, E.H. Surface morphology and chemical mapping of UV-aged thin films of bitumen. *Sustain. Chem. Eng.* **2020**, *8*, 11764–11771. [CrossRef]
53. Lin, P.; Huang, W.D.; Tang, N.P.; Xiao, F.P.; Li, Y. Understanding the low temperature properties of Terminal Blend hybrid asphalt through chemical and thermal analysis methods. *Constr. Build. Mater.* **2018**, *169*, 543–552. [CrossRef]
54. Orange, G.; Dupuis, D.; Martin, J.; Farcas, F.; Such, C.; Marcant, B. Chemical modification of bitumen through polyphosphoric acid: Properties-micro-structure relationship. In Proceedings of the 3rd Eurasphalt and Eurobitume Congress, Vienna, Austria, 12–14 May 2004.
55. Zaidullin, I.; Petrova, L.; Yakubov, M.; Borisov, D.N. Variation of the composition of asphaltenes in the course of bitumen aging in the presence of antioxidants. *Russ. J. Appl. Chem.* **2013**, *86*, 1070–1075. [CrossRef]
56. Li, Y.; Feng, J.; Wu, S.; Chen, A.; Kuang, D.; Gao, Y.; Zhang, J.; Li, L.; Wan, L.; Liu, Q. Review of ultraviolet ageing mechanisms and anti-ageing methods for asphalt binders. *J. Road Eng.* **2022**, *2*, 137–155. [CrossRef]
57. Zhou, Y.; Chen, J.; Zhang, K.; Guan, Q.; Guo, H.; Xu, P.; Wang, J. Study on aging performance of modified asphalt binders based on characteristic peaks and molecular weights. *Constr. Build. Mater.* **2019**, *225*, 1077–1085. [CrossRef]
58. Baek, S.H.; Kim, H.H.; Doh, Y.S.; Kim, K.W. Estimation of high-temperature properties of rubberized asphalt using chromatograph. *KSCE J. Civ. Eng.* **2009**, *13*, 161–167. [CrossRef]

materials

Article

DIC-Enhanced Identification of Bodner–Partom Model Parameters for Bitumen Binder

Marek Klimczak *[iD], **Marcin Tekieli, Piotr Zieliński** [iD] and **Mateusz Strzępek**

Faculty of Civil Engineering, Cracow University of Technology, Warszawska 24 Street, 31-155 Cracow, Poland
* Correspondence: marek.klimczak@pk.edu.pl

Abstract: Bitumen binder is a component of asphalt mixtures that are commonly used as the materials constituting the upper layers of a pavement's structure. Its main role is to cover all the remaining constituents (aggregate, filler and other possible additives) and create a stable matrix, in which they are embedded due to the adhesion forces. The long-term performance of bitumen binder is crucial to the holistic behavior of the layer made of the asphalt mixture. In this study, we use the respective methodology to identify the parameters of the well-established Bodner–Partom material model. For the purposes of its parameters identification, we carry out a number of the uniaxial tensile tests with different strain rates. The whole process is enhanced with a digital image correlation (DIC) to capture the material response in a reliable way and to provide deeper insight into the experiment results. The obtained model parameters were used to compute numerically the material response using the Bodner–Partom model. Good agreement between the experimental and numerical results was observed. The maximum error for the elongation rates equal to 6 mm/min and 50 mm/min is of order of 10%. The novel aspects of this paper are as follows: the application of the Bodner–Partom model to the bitumen binder analysis and the DIC-enhancement of the laboratory experiment.

Keywords: digital image correlation; bitumen binder; numerical modeling

Citation: Klimczak, M.; Tekieli, M.; Zieliński, P.; Strzępek, M. DIC-Enhanced Identification of Bodner–Partom Model Parameters for Bitumen Binder. *Materials* **2023**, *16*, 1856. https://doi.org/10.3390/ma16051856

Academic Editors: Wensheng Wang, Qinglin Guo and Jue Li

Received: 30 December 2022
Revised: 21 February 2023
Accepted: 22 February 2023
Published: 24 February 2023

1. Introduction

Bitumen binder is one of the main constituents of asphalt mixture that is a common type of material used in civil engineering. In addition to bitumen binder, the asphalt mixture is made of aggregate, filler and other possible additives. In this study, we focus solely on the performance of bitumen binder. This is due to the fact that its behavior determines in the most significant manner the overall long-term response of the asphalt mixture. In the context of the numerical modeling, the approximation of the aggregate behavior as the elastic seems to be sufficient [1–4]. On contrary, the response of the bitumen binder can be regarded as elastic only in a very limited load range. Regardless of the aging, temperature, moisture and other phenomena influencing the performance of bitumen binder, its mechanical response exhibits both elastic and inelastic character in a typical range of applications.

Many researchers have contributed to the study of bitumen binder mechanical performance. We refer to excellent review books [5,6], presenting the current state-of-the-art in the asphalt modeling in a detailed manner. There is also a considerable number of the scientific papers, which also provide a comprehensive description of the latest achievements in this active research field [7–9]. Despite the fact that bitumen binder is a well-known material, emerging new technologies of its production influence its performance [10,11]. Consequently, the material models used for the description of its mechanical response also need to be validated and updated when necessary.

This can be illustrated with the example of neat bitumen binder and its modified counterpart. In Figure 1, one can observe the exemplary elongation–force plots obtained in a ductilometer test. On the left (Figure 1a), the results obtained for a neat bitumen binder are presented. After the maximum force was measured, the softening phenomenon was clearly visible. On the right (Figure 1b), the exemplary results obtained for a styrene-butadiene–styrene (SBS)-modified binder are shown. The plot character is different in this case. Due to the presence of the SBS modifier, after the first peak and the corresponding softening phase, a hardening phenomenon can be also observed. Both of the tests were finally performed up to rupture. It should be noted that the mechanical response to the applied load (elongation with a constant strain rate) is clearly different for these two binder types.

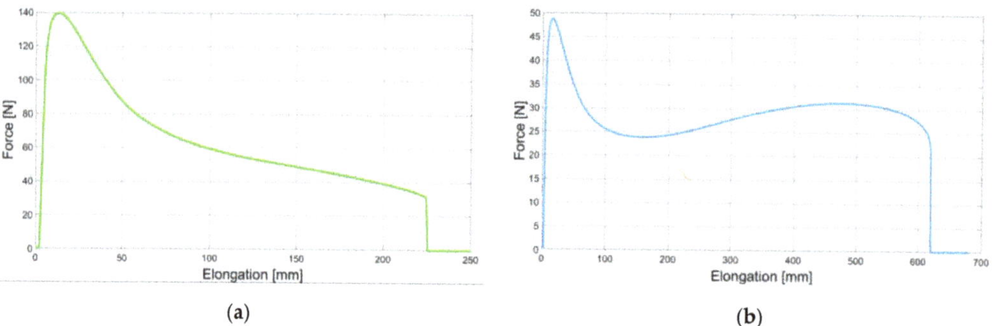

Figure 1. Exemplary ductilometer test results: (**a**) for neat bitumen binder; (**b**) for SBS-modified bitumen binder (own research results).

In the context of the numerical modeling, different constitutive equations should be adopted in order to adequately simulate the behavior of these materials. This example shows the complexity of the tackled problem in its general overview. As remarked in another study [5], there is no one general model capable of capturing all the phenomena occurring in the bitumen binder. One has to restrict the range of analysis developing even a very complex material model.

In this study, we focus on a small displacement range of the bitumen binder performance. This is justified by the mechanistic approach [12,13] to asphalt pavement structure design, in which the observed displacement quantities under a single-tire load are relatively small. Thus, the numerical analysis in the small displacement range seems to be sufficient in this case.

The main constitutive models used for the numerical modeling of bitumen binder at the continuum level in the small displacement range are far beyond the elasticity. Elastic computations are used only as the initial tests or as the part of the large multiscale framework [14]. When the research focus is on the possibly wide range of the bitumen binder performance, the assumption on its elastic response is naturally insufficient.

As reported in a previous study [15], the nonlinear mechanical asphalt behavior is influenced by the temperature, microvoid evolution as well as the rate-dependent viscoplastic hardening. In this paper, we do not investigate all of the consequences of the aforementioned phenomena. This is due to the fact that only relatively short tests (up to 1 min) at low strain levels were modeled.

In such a restricted setting, two main groups of models can be roughly distinguished:

- Viscoelastic models, which account for the recoverable asphalt deformations;
- Viscoplastic models, which account for the unrecoverable asphalt deformations.

A viscous character of the asphalt response can be easily demonstrated. For instance, in a ductilometer tests, the increase of the strain rate undisputedly results in the increase of the force value measured.

The assumption on the recovery mode of the deformation is a difference between viscoelastic and viscoplastic models. Schapery's nonlinear viscoelasticity model [16], originating from polymer analysis, is one of the well-established models also used for the purpose of bitumen binder modeling. Linear viscoelastic Burgers or generalized Maxwell models are also frequently used for typical binder types [17]. Their advantage is a clear mechanical interpretation of the basic versions and a relatively simple numerical implementation, consequently. In the case of their versions consisting of more than one basic element, the physical interpretation of model parameters becomes infeasible. Thus, these parameters are identified using the curve-fitting.

Pure viscoplastic behavior of bitumen binder is typically modeled using Perzyna's theory [18]. It is a classical strain rate-dependent plasticity model proposed in the 1960's, which is based on the viscoplastic flow rule.

In many latest papers (see e.g., [15,19]), the extensions of Schapery's [16] and Perzyna's [18] theories are developed. This is mainly due to the new microstructural observations. For instance, in another study [15], the viscoplastic Perzyna-type model accounted for the observation that only the undamaged subdomain carries the load. In this paper, viscoelasticity and viscoplasticity were additionally coupled with a viscodamage phenomenon.

Apart from these two major groups (and their combinations) of constitutive models used in the context of the bitumen binder modeling, one can also distinguish a group of so-called unified material models [20,21], in which the overall inelastic response is modeled together, introducing the internal variable concept. These models originate from metallurgy [20,21] and were tested on a variety of metal alloys. However, they are able to simulate the typical phenomena observed in the rheology, i.e., creep and relaxation. As such, they were also applied to other kinds of materials, including technical fabrics [22] and rubber-toughened plastics [23]. Two representatives of this group of constitutive models are the Chaboche [21] and Bodner–Partom models [20]. The applicability of the latter to the numerical modeling of bitumen binder is studied in this paper. A more elaborate discussion on the reasons for this model selection is provided in the forthcoming chapter. Basically, there are three main reasons:

- It is feasible for nonlinear asphalt behavior modeling;
- Its numerical implementation is very effective due to the explicit time-integration scheme; it is particularly profitable in the case of the transient analysis—no Newton-Raphson iterations are necessary in the time-stepping algorithm.
- Its material parameters can be identified in a physically based manner, whereas typically, only a curve-fitting is used.

The aforementioned advantages of the model make it potentially robust and trustworthy in the case of asphalt numerical modeling. Moreover, the Bodner–Partom model exhibits its superiority to other popular approaches (e.g., the Burgers or Maxwell models) due to its physically based parameters and the correspondingly reliable identification procedure. In the case of the typically used models (e.g., Burgers or Maxwell), only their basic versions have a clear mechanical interpretation. Their generalized versions are the series of the number of such basic elements. The identification of such generalized model parameters is based on the curve-fitting procedure. The number of basic elements and their parameters are not determined on the basis of physical observations but are assessed in the least squares sense. Consequently, one can obtain an excellent fit of such models, but the reproducibility of the results is very limited. These parameters can be used to model only a specific specimen, for which they were identified.

The purpose of our study is to present the procedure of Bodner–Partom model parameter identification for bitumen binder. To our best knowledge, this has never been conducted before for this material solely. A similar study devoted to the whole asphalt mixture response modeled with the Bodner–Partom model can be found [24]. We point out the main difference between our study and this paper in Section 2.3.

Additionally, we present the proposed DIC enhancement of the whole methodology. It provides a more detailed insight into the analyzed material. In addition to the correction of some experiment results, the DIC can be also used to obtain the information outreaching the possibilities of the traditional test machines.

This paper is organized as follows. Section 2.1 describes samples preparation. Section 2.2 provides a brief description of the Bodner–Partom model. Section 2.3 presents the methodology for model parameter identification. In Section 3, results for the selected bitumen binder are presented and the results of the experiments and finite element calculations are compared. Discussion on the results is provided in Section 4 and concise conclusions are presented in Section 5.

2. Materials and Methods

2.1. Samples Preparation

In order to identify Bodner–Partom model parameters, the results of several uniaxial tensile tests are necessary. For the purposes of this study, such tests were performed using a ductilometer VIATECO located in the laboratory of the Chair of Highway, Railway and Traffic Engineering of Civil Engineering Faculty (Cracow University of Technology). The specimens were made of neat bitumen binder 35/50 (see Table 1 for the parameters specified by the manufacturer) and prepared for the test according to PN-EN 13398:2017-12 [25]. Subsequently, two types of experiment were performed. The first type was a shortened (continued only within the small strains range) version of the standard test performed according to [25]. Namely, the specimens were kept with a full immersion in a water at a constant temperature of 10 °C. The presence of water disabled the DIC application in this case. This was due to serious image quality deterioration because of water motion, reflexes and other undesired aspects. Thus, the second experiment type was the uniaxial tensile test without the presence of water and performed at a room temperature of 19.5 °C.

Table 1. Neat bitumen binder 35/50 parameters.

Penetration at 25 °C (0.1 mm)	Softening Point R&B (°C)	Fraas Breaking Point (°C)	Ductility (cm)	Specific Gravity (kg/m^3)
45	55.8	−16	>100	1023

In Figure 2, one can observe a specimen with two additional markers stuck on and a virtual mesh (in green) generated in the image for the DIC purposes.

Figure 2. A specimen prepared for the DIC-enhanced test.

Due to the difference in temperature, the results of these two experiment types cannot be fully quantitatively compared to each other. For the sole purpose of material parameter identification, a standard experiment type is sufficient. We demonstrate a larger potential of the DIC-enhanced test, however.

The specimens used for this test need special treatment. In order to increase the image contrast, the specimens were sprayed with a black matt paint to create a very thin (but almost full) film. After drying out, a white paint was sprayed on to create a pattern for DIC purposes.

Both experiment types were performed for 4 elongation rates: 6 mm/min, 50 mm/min, 120 mm/min and 240 mm/min. Three specimens were used for every test in order to reduce the measurement error possibility. The results obtained for every triple were averaged for further processing.

2.2. Digital Image Correlation

Optical measurements were carried out using a digital single-lens camera (DSLR) Nikon D5300 equipped with a high-quality lens Sigma 17–50 mm f/2.8 EX DC OS HSM with negligible radial distortion in order to ensure high-quality photos, which are the basis for further processing. Photos of the sample's surface were taken at constant intervals using an intervalometer. Each photo was 6000 pixels horizontally and 4000 pixels vertically. The camera was mounted on a tripod equipped with a micrometer head and set perpendicularly with the lens axis to the surface of the tested sample. The photos were taken with a focal length of about 24 mm with an ISO-100 value and the shutter speed was set in the range from 1/50 to 1/25 s. To ensure the ability to use such a high shutter speed and to avoid discoloration related to the color temperature strong LED light source with a temperature range of 4000–5700 K was used.

In order to ensure the possibility of determining displacements of structures in the form of subsets of a digital image of the sample surface, the digital image correlation method (DIC) was used. This method was implemented in the proprietary CivEng Vision 1.0 software developed at CUT (Cracow, Poland) by one of the paper's author [26,27] for optical measurements and was used to process all of the photos taken. In general, the idea of using cross-correlation to measure shifts in datasets is to compare fragments of the digital image (photo) with its surroundings to determine the displacement of the examined fragment in the Cartesian coordinate system. The displacement in pixels can then be easily converted to real values in the metric system by analyzing the parameters of the lens, its distance from the sample surface or by rescaling the known size of the object included in the calibration image. By measuring the displacement of two independent subsets of the image in this way, the values of the engineering strains can be determined on a given basis. By covering the surface of the sample in the image with a grid of markers (subsets), it is also possible to determine the map (field) of displacements and, in the next step, the field of deformations. For markers tracking in this work, zero mean-normalized cross-correlation (ZNCC) was used. To ensure adequate measurement resolution at the level of traditional sensors (strain gauges), a subpixel measurement was used, obtained by interpolating pixel values to intermediate values assigned to subpixels.

2.3. Bodner–Partom Model

The Bodner–Partom model [20] was introduced in the 1970's for purposes of metal alloy modeling. Its initial version was based on thermodynamic and microstructure behavior observations. The model has been actively developed since its introduction and one can find a variety of its modifications. In this paper, we briefly recall its basic version (c.f. [22,23]).

As was remarked, the model originates from metal alloy analysis. However, there is no assumption on the internal structure of the analyzed material. There are no parameters corresponding to the internal structure character. It was successfully used for the materials that exhibit in the range of small strains the same phenomena as metal alloys (e.g., polymers) but have a different internal structure. Due to the general character of the Bodner–Partom model, there were no additional assumptions necessary in the analysis of the bitumen binder. Regardless of the internal microstructure, the macroscale phenomena were modeled after the identification of the specific model parameters for the bitumen binder. Having restricted the analysis to the small strains range, we were able to model its behavior in a reliable way compared to the laboratory experiments. Identified parameters are obviously different than those present in the literature for the metal alloys, but they allowed for the reproducibility of the experiments using finite element analysis without any numerical instabilities.

The restriction to the small strain range of analysis is justified using the mechanistic approach to the design of pavement structures. Therein, small strains are observed.

In the literature, one can find the attempt of the Bodner–Partom model's application to the modeling of cement-emulsified asphalt mixture [24]. There are two main differences between the approach developed in this study and that paper.

Firstly, we solely analyze the behavior of the bitumen binder. This is motivated by the multiscale modeling of the asphalt mixture, where all the constituents are analyzed using dedicated material models. The overall response, calculated at the macroscale, is the effective one. The results of our study can be reproduced for a variety of asphalt mixtures (different gradation scales, aggregate morphologies, etc.), since the bitumen binder parameters for the Bodner–Partom model were identified. Another approach was used in [24], where the effective model parameters were obtained for the asphalt mixture. Any modification to the asphalt mixture would require a new parameter identification to be valid for potential numerical modeling purposes.

Secondly, the identification procedure used in that study [24] is based on nonlinear curve fitting. In our study, we follow the methodology presented in other studies (e.g., [22,23]). It is a physically based approach, with many successful applications in the literature [20,22,23].

The Bodner–Partom model belongs to the group of the so-called unified models, i.e., the overall inelastic deformation (rate-dependent and rate-independent part) is modeled together, without decomposition into respective components. Additionally, there is no assumption on the existence of the yield surface in the model. Contrarily, it is assumed that the elastic deformations are accompanied with the inelastic ones regardless of the observed stress level. The lack of yield condition simplifies the numerical modeling process. The inelastic reconstruction of the material during the modeled process is characterized by the internal variable approach. One can observe an increasing number of material parameters when additional phenomena are accounted for in the model. Herein, we present a basic form of the Bodner–Partom constitutive equations that is applicable to the short-term process at the constant temperature.

The total strain rate $\dot{\varepsilon}$ is decomposed in an additive manner into the elastic $\dot{\varepsilon}_e$ and the inelastic $\dot{\varepsilon}_{vp}$ strain rate terms:

$$\dot{\varepsilon} = \dot{\varepsilon}_e + \dot{\varepsilon}_{vp}. \tag{1}$$

In the case of the uniaxial test, the inelastic strain rate term is expressed in the Bodner–Partom model as:

$$\dot{\varepsilon}_{vp} = \frac{2}{\sqrt{3}} D_0 sgn(\sigma) exp\left[-\frac{1}{2}\left(\frac{R+D}{\sigma}\right)^{2n}\frac{n+1}{n}\right], \tag{2}$$

where D_0 is the limiting strain rate (1/s), σ is the stress (Pa), R and D denote the isotropic and kinematic hardening functions, respectively. Dimensionless constant n is the strain rate sensitivity parameter. Both isotropic and kinematic hardening functions are defined by the initial value problems (see [22,23] for more details). For the sake of brevity, we present their integrated forms herein, i.e.,

$$\begin{aligned}
R &= R_1\left[1 - exp\left(-m_1 W^I\right)\right] + R_0 exp\left(-m_1 W^I\right), \\
X &= \sqrt{\tfrac{3}{2}} D_1 sgn(\sigma)\left[1 - exp\left(-m_2 W^I\right)\right], \\
D &= \sqrt{\tfrac{2}{3}} X sgn(\sigma), \\
W^I &= \sigma \varepsilon_{vp}.
\end{aligned} \tag{3}$$

In (3), X can be understood as the auxiliary variable, R_1 stands for the limiting value of the isotropic hardening (Pa), R_0 stands for the initial value of the isotropic hardening (Pa), D_1 is the limiting value of the kinematic hardening, m_1 and m_2 are the coefficients for isotropic and kinematic hardening (1/Pa), respectively. W^I is the inelastic work.

For the sake of the model parameters identification, a functional f_1, which links the stress and the inelastic strain rate, is defined:

$$f_1(\dot{\varepsilon}_{vp}) = \frac{\sigma}{R + D} \tag{4}$$

In the case of the Bodner–Partom model, it has the following form:

$$f_1 = \left[\frac{2n}{n+1} ln\left(\frac{2D_0}{\sqrt{3}\dot{\varepsilon}_{vp}}\right)\right]^{-\frac{1}{2n}}. \tag{5}$$

In the presented version of the Bodner–Partom model, one has to specify 7 parameters for the uniaxial tests: D_0, D_1, R_0, R_1, m_1, n, n. In addition, the Young's modulus E (Pa) needs to be specified in order to model the elastic deformation.

2.4. Identification of Bodner–Partom Model Parameters

As mentioned in Section 2.1, identification of model parameters is based on the series of the experimental results and their further processing. For the sake of clarity, only the selected or averaged results are typically presented in the manuscript during the methodology explanation.

In Figure 3, one can observe a typical stress–strain relationship for various elongation rates. Herein, the results of the standard experiment are presented.

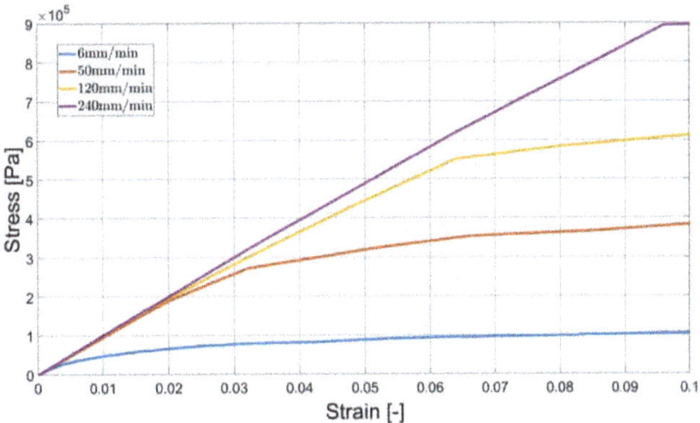

Figure 3. Stress–strain curves for different elongation rates (standard experiment type).

Taking into account the initial (linear) relationship, we compute the Young's modulus ($E = 10,000,000$ Pa). Then, the inelastic strains are computed subtracting from the total strain ε its elastic part:

$$\varepsilon_{vp} = \varepsilon - \frac{\sigma}{E} \tag{6}$$

The results are shown in Figure 4.

An initial yield stress (being also the technical elasticity limit) σ_{y02} is specified on the basis of Figure 4. Namely, the stress occurring at the moment of the inelastic strain equal to 0.2% is found. Using these values, which are specified for every elongation rate, one defines the relationship between the initial yield stress and the strain rate (see Figure 5).

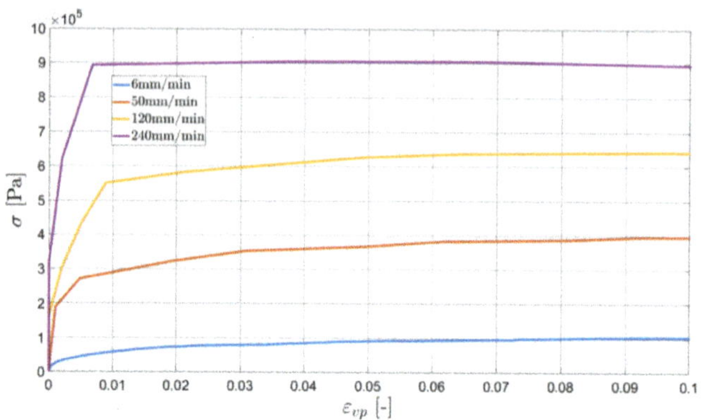

Figure 4. The relationship between the stress and the inelastic strain for different elongation rates (standard experiment type).

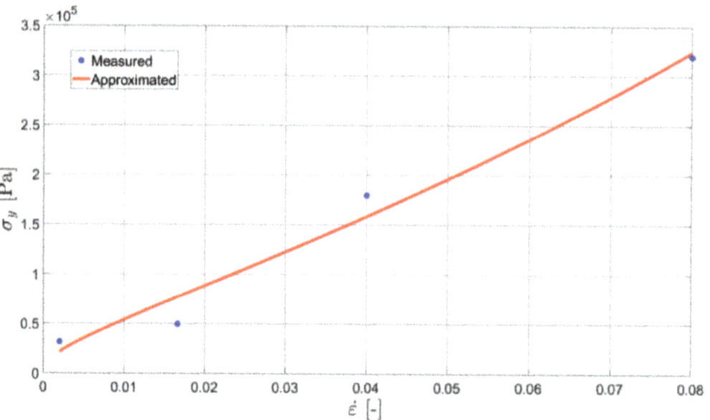

Figure 5. The relationship between the initial yield stress and the strain rate (standard experiment type).

Assuming that for the value of $\varepsilon_{vp} = 0.2\%$, which was specified for the σ_{y02} assessment, the isotropic hardening function R is approximately equal to its initial value R_0 and the kinematic hardening is negligible, one can define a closed-form expression for the technical elasticity limit rearranging Equations (4) and (5) for $\dot{\varepsilon}_{vp} = \dot{\varepsilon}$ at the initial level of inelastic strains ($\varepsilon_{vp} = 0.2\%$):

$$\sigma_{y02} = \left[\frac{2n}{n+1} ln\left(\frac{2D_0}{\sqrt{3}\dot{\varepsilon}_{02}}\right)\right]^{-\frac{1}{2n}} R_0 \tag{7}$$

In Equation (7), $\dot{\varepsilon}_{02}$ denotes the value of the total strain rate at the moment when $\varepsilon_{vp} = 0.2\%$. Since the strain rates are very low, the value of the limiting strain rate D_0 is assumed to be equal to 1 s^{-1}. Under this assumption, one can find the values of the parameters n and R_0. As suggested in previous studies [22,23], the Marquardt–Levensberg algorithm [28] was used to fit their values in the least-squares sense. The approximation of σ_y for the specific data set is shown in Figure 4.

In order to identify m_1 and m_2 parameters, one has to define a relationship between the yield stress $\sigma(\varepsilon_{vp})$ and the hardening work rate function γ, which is defined as:

$$\gamma = \frac{d\sigma}{dW^I} = \frac{d\sigma}{d\varepsilon_{vp}}\frac{1}{\sigma} \tag{8}$$

Rearranging Equations (3)$_{\text{III}}$ and (4), one obtains

$$\gamma = f_1[m_1(R_1 - R) + m_2(D_1 - D)] \tag{9}$$

Such a relationship, $\gamma\text{-}\sigma(\varepsilon_{vp})$, needs to be created for every elongation rate and the obtained values of m_1 and m_2 parameters have to be averaged. The results plotted for the elongation rate equal to 6 mm/min in the standard experiment are presented in Figure 6.

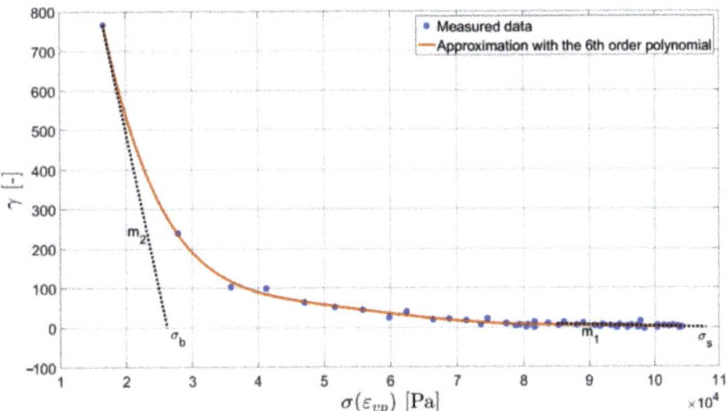

Figure 6. The relationship between the hardening work rate function and the yield stress.

Parameters m_2 and m_1 are the slopes of the tangents to the hardening work rate function corresponding to the inelastic strain initiation (m_2) and to its saturation (m_1). Yield stress values at the intersection points of these tangents with the horizontal axis are σ_b and σ_s (see Figure 5). They are used for the assessment of the remaining two parameters of the Bodner–Partom model.

Finally, parameters D_1 and R_1 are identified using the following expressions:

$$D_1 = \frac{\sigma_b m_2}{f_1(\dot{\varepsilon}_{vp,02})(m_2 - m_1)} - \frac{\sigma_s m_1}{f_1(\dot{\varepsilon}_{vp,2})(m_2 - m_1)} - R_0 \tag{10}$$

$$R_1 = \frac{\sigma_s m_2}{f_1(\dot{\varepsilon}_{vp,2})(m_2 - m_1)} - \frac{\sigma_b m_1}{f_1(\dot{\varepsilon}_{vp,02})(m_2 - m_1)} + R_0 \tag{11}$$

Equations (10) and (11) were obtained for $\gamma = 0$ inserted to (9) and computations for the initial plastic strains ($\varepsilon_{vp} = 0.2\%$) and the plastic strains for which the hardening is saturated ($\varepsilon_{vp} = 2\%$), respectively. The value of $\dot{\varepsilon}_{vp,2}$ was computed numerically. $\dot{\varepsilon}_{vp,02}$ was assumed to be equal to $\dot{\varepsilon}_{02}$, as stated before.

2.5. DIC Enhancement of the Identification Procedure

The specimen shape is shown in Figure 2. Slightly different shapes are also used in ductilometer tests. All of them allow for the approximated uniaxial tensile tests. In fact, the elongation measurements refer to the points in the central axis of the specimen. In terms of the DIC technique, it is equivalent to a single pair of markers used as shown in Figure 7.

In reality, the strain state of the specimen is far from being homogeneous. In Figure 8, the maps of three components of the strain tensor (ε_{xx}, ε_{yy}, ε_{xy}) are presented. In the uniaxial tensile test, one is particularly interested in the ε_{xx} component.

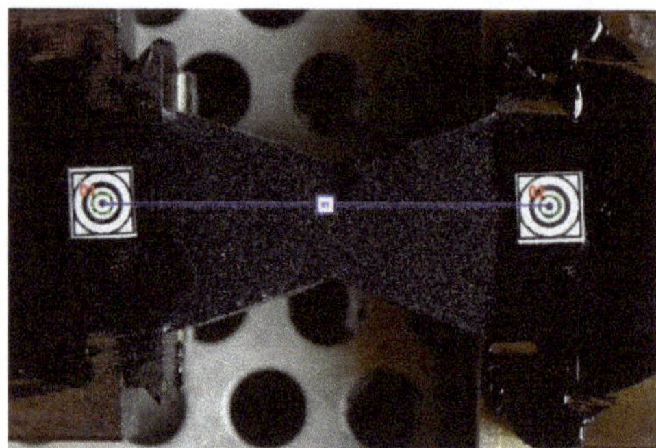

Figure 7. A single pair of markers used for the elongation measurement with the DIC technique.

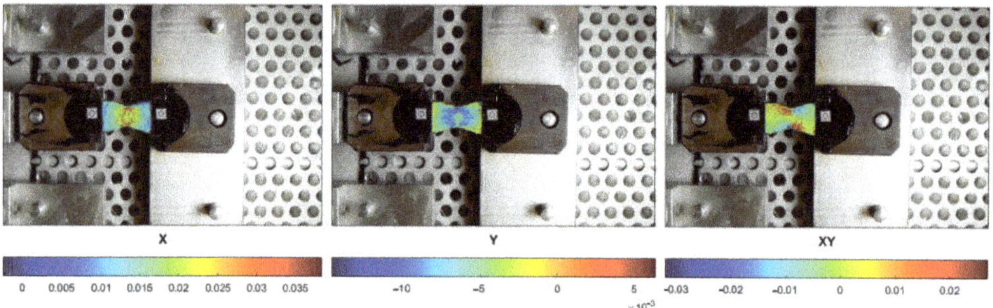

Figure 8. The maps of the strain tensor components: ε_{xx}, ε_{yy}, ε_{xy}, consecutively.

It can be easily noticed that even along the central axis and in its closest neighborhood, the ε_{xx} component is not equal. In order to improve the accuracy of the measurements, we used the elongations of the specimen averaged along its cross-section as shown in Figure 9.

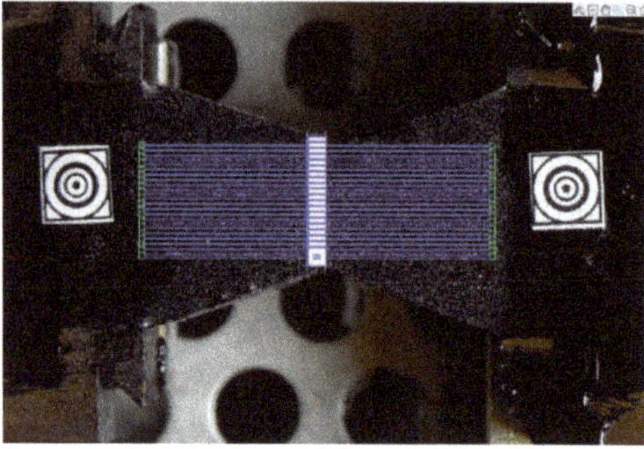

Figure 9. Twenty virtual tensometers placed along the specimen cross-section.

Technically, a number of virtual tensometers (we used 20 in this study) was applied. Obtained elongation values were averaged and used for the strain computations (see Figure 3).

The DIC technique was also used for the Poisson ratio assessment, which could not be carried out using only the ductilometer test results. Namely, a pair of perpendicular virtual tensometers was placed on the specimen, as shown in Figure 10.

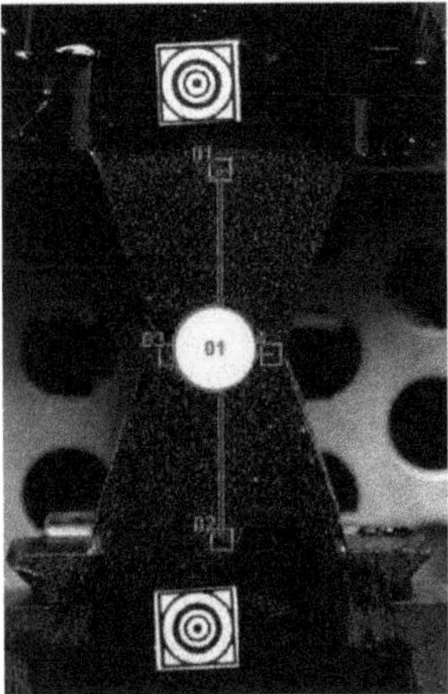

Figure 10. A pair of the perpendicular virtual tensometers used for the Poisson ratio assessment.

In this study, the value of approximately 0.35 was assessed, which can be also found in the literature [14]. We computed it averaging the values obtained for every three specimens and for every elongation rate.

Summing up, the enhancement of the standard methodology due to the DIC application consists of obtaining more reliable input data and its broader scope.

Specifically, we enhanced the methodology of Bodner–Partom model parameter identification with the DIC for several reasons:

- We were capable of computing the strain state not only between the two outermost points along the central axis of the specimen (as is the case of the standard ductilometer test results) but also within its whole surface and computing the averaged response along the specimen cross-section, consequently. In such a manner, we could eliminate a typical hidden assumption on the homogeneity of the specimen response along its cross-section;
- The application of the DIC technique allows for the recording of the laboratory test results with a user-defined frequency, which can overcome the limitations of the testing machine used.
- The parameters of the B–P model identified using the standard methodology would be enough for the further numerical tests only in 1D. Application of the DIC also enables the identification of the Poisson ratio, which is crucial to numerical modeling in 2D and 3D space.

3. Results

3.1. Results of the Identification Procedure

Below, we present the obtained values of Bodner–Partom parameters for the standard experiment type (Table 2) and its modified version due to DIC application (Table 3). Let us remind that the specimen in the first test was immersed in water at a temperature of 10 °C, whereas the second test was performed without the presence of water in a ductilometer, at a temperature equal to 19.5 °C.

Table 2. Bodner–Partom model parameters of bitumen binder 35/50 (at a temperature of 10 °C).

E (Pa)	v (-)	n (-)	D_0 (1/s)	D_1 (Pa)	R_0 (Pa)	R_1 (Pa)	m_1 (1/Pa)	m_2 (1/Pa)
1×10^7	0.35	0.1611	1	2.2541×10^4	1.2784×10^5	6.1822×10^5	3.2×10^{-4}	0.0803

Table 3. Bodner–Partom model parameters of bitumen binder 35/50 (at the temperature of 19.5 °C).

E (Pa)	ν (-)	n (-)	D_0 (1/s)	D_1 (Pa)	R_0 (Pa)	R_1 (Pa)	m_1 (1/Pa)	m_2 (1/Pa)
1.05×10^6	0.35	0.1623	1	2.2702×10^3	1.3423×10^4	6.1706×10^4	0.0031	0.8330

One can observe an obvious dependence of the obtained results on the temperature at which the test was performed. Even the elastic response, which is characterized using Young's modulus, is significantly different. A drop of about one order of magnitude (89.5%) is observed for this parameter at the room temperature. A similar relationship can be noticed for the parameters D_1 (a drop of 89.9%), R_0 (a drop of 89.5%), R_1 (a drop of 90.0%) that characterize the initial (R_0) and limiting (D_1 and R_1) values of the hardening functions. Higher values of parameters m_1 (increase of 89.7%) and m_2 (increase of 90.4%) denote earlier saturation of these functions at the higher temperature (19.5 °C). Only the parameter n does not exhibit a strong sensitivity to the temperature at the analyzed range (increase of 0.7%). As is discussed in the next section, the Bodner–Partom model is highly sensitive to the parameter n changes. Thus, one should not assume that this value is constant at every temperature, even in such a narrow temperature range (10÷19.5 °C).

Performing more standard experiments (with the presence of water), one could investigate the relationship between the model parameters and the temperature. We performed the ductilometer test at the temperature of 10 °C because this is a standard value used [25].

3.2. Numerical Results

In order to validate the Bodner–Partom model with application to bitumen binder, we present the comparison of two of the experiments and their numerical counterparts. Namely, we identified Bodner–Partom model parameters and used them for the self-developed finite element procedure. As an exemplary result of such a comparison, we present the DIC-enhanced ductilometer tests performed with the elongation rates equal to 6 mm/min and 50 mm/min (see Figures 11 and 12, respectively).

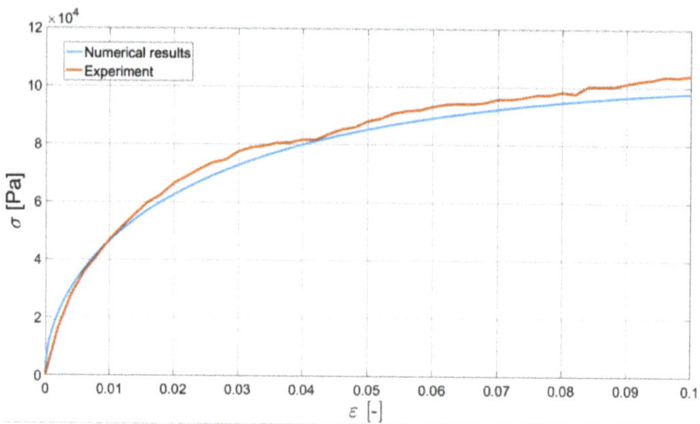

Figure 11. The comparison of the real experiment and its numerical simulation. Elongation rate is equal to 6 mm/min and the DIC-enhanced results are presented for the experiment.

For these specific tests, the differences between the experimental and numerical results are of the maximum order of 10%. A similar characteristic can be also observed for the other elongation rates. Some discrepancies between the measured and modeled material response can be explained by a constant ductilometer measurement frequency. For the higher elongation rates, at the intervals with the larger gradients of the $\sigma - \varepsilon$ curve observed, the measurement frequency may be too small to capture the nonlinear material response to the applied monotonic strain.

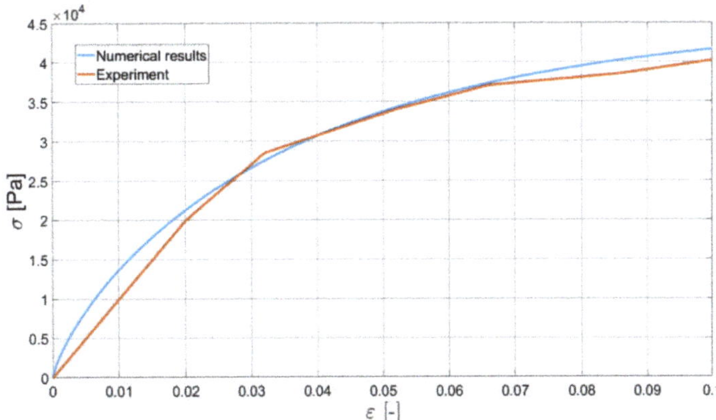

Figure 12. The comparison of the real experiment and its numerical simulation. Elongation rate is equal to 50 mm/min and the DIC-enhanced results are presented for the experiment.

The obtained numerical results, in some sense, smoothen the bitumen binder response. No oscillations typical of real experiments were observed. The initial results confirm the applicability of the Bodner–Partom model to the numerical analysis of bitumen binder.

The sensitivity analysis of the model revealed that the final results are particularly sensitive to the value of parameter n. Thus, special attention needs to be paid to its identification. The improvement can be obtained in a twofold manner. First of all, a number of experiments with various elongation rates can be substantially increased. Secondly, algorithms other than Marquardt–Levensberg [28] can be applied to increase the reliability of the function (see Figure 5) approximation.

Additionally, the manufacturer of the analyzed neat bitumen 35/50 provides some functional properties in [29]. For instance, the results of the dynamic shear rheometer tests (DSR) performed on the specimens subjected in advance to the rolling thin film oven test (RTFOT) are presented in [29].

Complex stiffness modulus $|G*|$ describes the ability of the bitumen binder to resist the applied load. It can be further decomposed into the elastic and viscous part that describe the mode of material response, respectively. The phase angle δ defines a dominant component. When it is equal to $0°$, the material behaves elastically. Approaching $90°$, it denotes that the material behaves like a Newtonian fluid. Both mechanical parameters are functions of temperature and load rate. In ([29], p. 86), the results for two frequencies are presented. It can be observed that, in particular, the complex stiffness modulus increases with the increase in load rate.

Presented (see Section 3.2 in [29]) results define the range of the applicability of the analyzed bitumen binder type to some extent. As suggested by the manufacturer [29], it can be used in the case of the base and binder courses made of asphalt concrete. Its application to wearing courses is not recommended.

4. Discussion

In this study, the initial results of the Bodner–Partom model's application to the numerical modeling of bitumen binder are presented. The whole process is enhanced with a DIC technique, which provides a more reliable input data for the model parameters identification. The obtained results confirm that the adopted constitutive model can be applied to the modeling of neat bitumen binder in the analyzed temperature range. As demonstrated in Figures 10 and 11, the Bodner–Partom model can properly describe the behavior of such a complex material as binder bitumen with good agreement with the experimental results. The difference between the modeled and measured response of bitumen binder to the applied elongation rates (see Figures 10 and 11) is of a maximum order of 10%. This confirms the applicability of the proposed model. The numerical efficiency of the time-stepping algorithm should additionally be underlined. The explicit Euler integration scheme is free of Newton–Raphson iterations, which makes the finite element code faster.

Looking at the identified Bodner–Partom parameters, which were assessed at different temperatures due to DIC limitations (water in ductilometer), one can observe a strong relationship between their values and the temperature. It makes identified parameter application limited to a specific temperature. In order to overcome this problem, so-called shift factors are used [5] to transfer the solution obtained at a given temperature to another. Alternatively, one can use the extensions of the Bodner–Partom model that accounts for thermal effects.

Carrying out a greater number of standard ductilometer experiments (with the presence of water) at various temperatures, one could study the relationship between the model parameters and the temperature.

As mentioned at the end of Section 3.2, the model is very sensitive to the value of parameter n. This part of the identification procedure also needs to be processed in order to increase the accuracy of the whole procedure.

We address this study to the further multiscale analyses of the selected asphalt mixtures. With the Bodner–Partom parameters identified, it is possible to account for this model for the binder phase of the asphalt mixture and apply other (e.g., linear elastic) models for the aggregate and possible additives. Finally, the effective macroscale response of the whole asphalt mixture can be studied. A number of possible multiscale approaches [5,14] can be applied.

5. Conclusions

Concluding the findings of this study:

- The constitutive Bodner–Partom model is feasible of capturing the response of the bitumen binder properly in the range of small strains (this restriction is justified with a future application of the obtained results to the mechanistic pavement design method);
- The physically based identification procedure of model parameters [22,23] can be also successfully applied to bitumen binder;
- In the specific case of neat bitumen binder 35/50, the differences between the experimental and numerical results are at the acceptable level (maximum error is of order of 10%);
- The DIC technique demonstrates its usefulness in the enhancement of the identification procedure due to the increase of the input data quality;
- Further research efforts should be made in order to increase the accuracy of the crucial model parameter's identification and the application of the extended versions of the Bodner–Partom model that account for the thermal effects;
- The obtained results can be applied to the multiscale analyses of different asphalt mixtures.

Author Contributions: Conceptualization, M.K. and M.T.; methodology, M.K., M.T., P.Z. and M.S.; validation, M.K., M.T., P.Z. and M.S.; investigation, M.K., M.T., P.Z. and M.S.; writing—original draft preparation, M.K. and M.T.; writing—review and editing, M.K., M.T., P.Z. and M.S.; funding acquisition, M.K. All authors have read and agreed to the published version of the manuscript.

Funding: The APC was funded by the Dean of the Civil Engineering Faculty (Cracow University of Technology).

Institutional Review Board Statement: Not applicable.

Informed Consent Statement: Not applicable.

Data Availability Statement: Not applicable.

Conflicts of Interest: The authors declare no conflict of interest. The APC funder had no role in the design of the study; in the collection, analyses or interpretation of data; in the writing of the manuscript or in the decision to publish the results.

References

1. Aigner, E.; Lackner, R.; Pichler, C. Multiscale prediction of viscoelastic properties of asphalt concrete. *J. Mater. Civ. Eng.* **2009**, *21*, 771–780. [CrossRef]
2. Kim, Y.-R.; Souza, F.V.; Teixeira, J.E. A two-way coupled multiscale model for predicting damage-associated performance of asphaltic roadways. *Comput. Mech.* **2013**, *51*, 187–201. [CrossRef]
3. Mo, L.; Huurman, M.; Wu, S.; Molenaar, A. 2D and 3D meso-scale finite element models for ravelling analysis of porous asphalt concrete. *Finite Elem. Anal. Des.* **2008**, *44*, 186–196. [CrossRef]
4. Mitra, K.; Das, A.; Basu, S. Mechanical behavior of asphalt mix: An experimental and numerical study. *Constr. Build. Mater.* **2012**, *27*, 545–552. [CrossRef]
5. Kim, Y.-R. *Modeling of Asphalt Concrete*, 1st ed.; McGraw Hill: New York, NY, USA, 2009.
6. Islam, M.-R.; Tarefder, R.-A. *Pavement Design: Materials, Analysis, and Highways*; McGraw Hill: New York, NY, USA, 2020.
7. Huang, W.; Wang, H.; Yin, Y.; Zhang, X.; Yuan, J. Microstructural Modeling of Rheological Mechanical Response for Asphalt Mixture Using an Image-Based Finite Element Approach. *Materials* **2019**, *12*, 2041. [CrossRef]
8. Neumann, J.; Simon, J.-W.; Mollenhauer, K.; Reese, S. A framework for 3D synthetic mesoscale models of hot mix asphalt for the finite element method. *Build. Mater.* **2017**, *148*, 857–873. [CrossRef]
9. Qu, X.; Wang, D.; Wang, L.; Huang, Y.; Hou, Y.; Oeser, M. The State-of-the-Art Review on Molecular Dynamics Simulation of Asphalt Binder. *Adv. Civ. Eng.* **2018**, *2018*, 4546191. [CrossRef]
10. Huang, Y.; Liu, Z.H.; Liu, J.Y.; Yan, D.H.; Yu, H.N. Laboratory study on mechanical properties of composite pavement under partial compression-shear load. *Constr. Build. Mater.* **2022**, *356*, 129240. [CrossRef]
11. Liu, G.Q.; Yang, T.; Li, J.; Jia, Y.S.; Zhao, Y.L.; Zhang, J.P. Effects of aging on rheological properties of asphalt materials and asphalt-filler interaction ability. *Constr. Build. Mater.* **2018**, *168*, 501–511. [CrossRef]
12. Asphalt Institute. *Thickness Design-Asphalt Pavements for Highways and Streets*; Asphalt Institute: Lexington, KY, USA, 1999.
13. *Shell Pavement Design Manual (SPDM 3.0)*; Shell: London, UK, 1998.
14. Wimmer, J.; Stier, B.; Simon, J.W.; Reese, S. Computational homogenisation from a 3D finite element model of asphalt concrete—linear elastic computations. *Finite Elem. Anal. Des.* **2016**, *110*, 43–57. [CrossRef]
15. You, T.; Al-Rub, R.K.; Darabi, M.; Masad, E.; Little, D. A thermo-viscoelastic–viscoplastic–viscodamage constitutive model for asphaltic materials. *Int. J. Solids Struct.* **2011**, *1*, 191–207. [CrossRef]
16. Schapery, R.A. On the characterization of nonlinear viscoelastic materials. *Polym. Eng. Sci.* **1969**, *9*, 295–310. [CrossRef]
17. Woldekidan, M.F. Response Modelling of Bitumen, Bituminous Mastic and Mortar. Ph.D. Thesis, Delft University of Technology, Delft, The Netherlands, 2011.
18. Perzyna, P. Thermodynamic theory of viscoplastcity. *Adv. Appl. Mech.* **1971**, *11*, 313–354.
19. Huang, C.W.; Abu-Al Rub, R.K.; Masad, E.A.; Little, D. Three-Dimensional Simulations of Asphalt Pavement Permanent Deformation Using a Nonlinear Viscoelastic and Viscoplastic Model. *J. Mater. Civ. Eng.* **2011**, *23*, 56–68. [CrossRef]
20. Bodner, S.R.; Partom, Y. Constitutive equations for elastic–viscoplastic strain-hardening materials. *J. Appl. Mech.* **1985**, *42*, 385–389. [CrossRef]
21. Chaboche, J.L. Constitutive equations for cyclic plasticity and cyclic viscoplasticity. *Int. J. Plast.* **1989**, *5*, 247–302. [CrossRef]
22. Kłosowski, P.; Zagubień, A.; Woznica, K. Investigation on rheological properties of technical fabric "Panama". *Arch. Appl. Mechnics* **2004**, *73*, 661–681. [CrossRef]
23. Ambroziak, A.; Kłosowski, P. Determining the viscoplastic parameters of rubber-toughened plastics. *Task Q.* **2008**, *12*, 35–43.
24. Zhang, C.; Cao, X.; Jiao, S.; Xu, X.; Zhi, T.; Fu, Y. Viscoelastoplastic Compaction Properties of Cement-emulsified Asphalt Mixture Based on Bodner-Partom Model. *China J. High. Transp.* **2019**, *32*, 41–48.
25. Polski Komitet Normalizacyjny. *PN-EN 13398:2017-12. Asfalty i Lepiszcza Asfaltowe—Oznaczanie Nawrotu Sprężystego Asfaltów Modyfikowanych*; Polski Komitet Normalizacyjny: Warsaw, Poland, 2018. (In Polish)
26. Tekieli, M.; De Santis, S.; De Felice, G.; Kwiecień, A.; Roscini, F. Application of Digital Image Correlation to composite reinforcement testing. *Compos. Struct.* **2017**, *160*, 670–688. [CrossRef]
27. Kwiecień, A.; Krajewski, P.; Hojdys, Ł.; Tekieli, M.; Słoński, M. Flexible Adhesive in Composite-to-Brick Strengthening—Experimental and Numerical Study. *Polymers* **2018**, *10*, 356. [CrossRef] [PubMed]
28. Marquardt, D.W. An Algorithm for Least Squares Estimation of Nonlinear Parameters. *J. Soc. Ind. Appl. Math.* **1963**, *11*, 431–441. [CrossRef]
29. Błażejowski, K.; Wójcik-Wiśniewska, M.; Baranowska, W.; Ostrowski, P. *Bitumen Handbook 2021*; ORLEN Asfalt sp. z o. o.: Płock, Poland, 2021.

Article

Sensitivity and Reliability Analysis of Ultrasonic Pulse Parameters in Evaluating the Laboratory Properties of Asphalt Mixtures

Xiaoshu Tan [1], Chunli Wu [1], Liding Li [1,*], He Li [1], Chunyu Liang [1], Yongchao Zhao [2], Hanjun Li [2], Jing Zhao [2] and Fuen Wang [2]

1   College of Transportation, Jilin University, Changchun 130025, China; tanxs20@mails.jlu.edu.cn (X.T.); clwu@jlu.edu.cn (C.W.); lihe326532558@163.com (H.L.); liangcy@jlu.edu.cn (C.L.)
2   Jilin China Railway Expressway Co., Ltd., Changchun 130052, China; 1593266955m@sina.cn (Y.Z.); lihanjun@crecg.com (H.L.); 18813098377@163.com (J.Z.); wangfuen@crecg.com (F.W.)
*   Correspondence: lild17@mails.jlu.edu.cn; Tel.: +86-0431-8509-5446

Abstract: The ultrasonic test is a promising non-destructive testing technique for evaluating the properties of asphalt mixtures. To investigate the applicability and reliability of ultrasonic testing technology (UTT) in evaluating the performance of asphalt mixtures, ultrasonic tests, indirect tensile tests, compression tests, and dynamic modulus tests were carried out at various temperatures. Subsequently, the distribution characteristics of ultrasonic traveling parameters for asphalt mixtures were analyzed. The variation of ultrasonic pulse velocity and amplitude in dry and wet states with temperature was studied. Then, the correlation between the ultrasonic parameters and both the volume parameters and the mechanical performance parameters of asphalt mixtures was revealed, and the functional relationship between ultrasonic pulse velocity and compressive strength was established. Finally, the reliability of predicting high-frequency dynamic modulus by ultrasonic velocity was verified. The laboratory tests and analysis results indicate that both ultrasonic pulse velocity and amplitude in dry and wet conditions show a decreasing trend with an increase in temperature. Ultrasonic parameters are greatly influenced by asphalt content and mineral aggregate content of 9.5~13.2 mm and 13.2~16 mm. The dynamic modulus at a high-frequency load can be predicted by using ultrasonic velocity, and predicting the results for OGFC and SMA mixtures deduced by using the UPV at a high-frequency load have higher reliability.

Keywords: asphalt mixture; performance evaluation; ultrasonic testing technology; ultrasonic pulse velocity; grey correlation analysis; dynamic modulus

Citation: Tan, X.; Wu, C.; Li, L.; Li, H.; Liang, C.; Zhao, Y.; Li, H.; Zhao, J.; Wang, F. Sensitivity and Reliability Analysis of Ultrasonic Pulse Parameters in Evaluating the Laboratory Properties of Asphalt Mixtures. *Materials* **2023**, *16*, 6852. https://doi.org/10.3390/ma16216852

Academic Editor: Jorge Carvalho Pais

Received: 19 September 2023
Revised: 17 October 2023
Accepted: 23 October 2023
Published: 25 October 2023

1. Introduction

Ultrasonic testing technology (UTT) has been widely adopted in the performance evaluation of pavement materials due to its nondestructive characteristics. The ultrasonic test is important not only for quickly obtaining material performance parameters but also for reducing parallel testing errors. A large number of studies on stiffness modulus, dynamic modulus, damage, crack, segregation, fatigue, and healing time of asphalt mixtures adopted the ultrasonic pulse parameters (UPP), which has gradually attracted the attention of pavement engineering researchers [1,2].

Some studies pointed out that there was a strong correlation between the UPP in asphalt mixtures and the indirect stiffness modulus of asphalt mixtures [3–5]. In addition, McGovern demonstrated that ultrasonic pulse velocity (UPV) could be adopted to calculate the dynamic modulus of asphalt mixtures under a high-frequency load [6]. Moreover, the dynamic modulus calculated by UPV has a functional relationship with the dynamic modulus obtained by both tension–compression tests [7] and the 2S2P1D (two springs, two parabolic elements, and one dashpot) model [8]. UTT is favorable for determining key rheological parameters (parabolic element parameters and glassy modulus) [9] and improving the prediction accuracy of the dynamic modulus master curve [10].

Additionally, UTT has distinct advantages in assessing the damage and healing performance of asphalt mixtures [11]. Cheng et al. [12] and You et al. [13] used the UTT to evaluate the damage properties of asphalt mixtures subjected to multiple freeze–thaw cycles and multiple coupling factors and concluded that the UTT is a viable method for investigating the potential damage of asphalt mixtures. Pan et al. [14] and Franesqui et al. [15] evaluated the feasibility of the UTT in studying the crack depth of asphalt mixtures and asphalt pavements and reported that the UTT is a promising technology for damage detection. However, it should be noted that the calculated results of crack depth need to be calibrated in the practical application according to the gradation type of asphalt mixtures [15,16]. Not only can using the UTT before and after damage or healing for the same sample save money and time, but it can also provide more comparable and reliable test results.

On the other side, the UTT can also be implemented to investigate the fatigue performance of asphalt mixtures [17]. According to the travel velocity of longitudinal and transverse waves in an asphalt mixture, more performance parameters of asphalt mixtures, such as the Lamé constants, Poisson's ratio, and modulus, can be deduced. These parameters can then be used to create a function for predicting the fatigue life of asphalt mixtures [18,19], which can benefit the rapid prediction of fatigue life for asphalt mixtures.

As previously stated, numerous studies have demonstrated that UTT serves an incredibly essential function in quickly evaluating asphalt mixture performance. However, the center of all these researches is the reliability and sensitivity of ultrasonic parameters to elaborate the performance of asphalt mixtures. It is well known that the UPV is greatly affected by temperature, material types, material properties, and damage state. Asphalt is a multi-component, temperature-sensitive material with several components. Its internal void, asphalt content, and characteristics all have an influence on ultrasonic travel speed in the asphalt mixture [20]. A key topic that road engineering researchers should explore is how to accurately evaluate the performance of asphalt mixes based on the UPP. However, the influence of a sample's temperature, dry and saturated states, and gradation on test results is rarely considered in UTTs current research on the evaluation of asphalt mixture performance. The sensitivity of various ultrasonic parameters in evaluating the performance of asphalt mixtures is also worthy of attention.

Therefore, in this research, four grades of asphalt mixtures were fabricated using a Superpave gyratory compactor (SGC), and the UPP of these asphalt mixtures at the dryness, saturation, and different freeze–thaw cycles were measured. Furthermore, the distribution characteristics of UPP on different graded asphalt mixtures at room temperature were analyzed using the normal distribution. Then, according to the results of the Marshall tests, splitting tests, and compression tests, the correlation between the three ultrasonic parameters and both the volume index of the asphalt mixtures and the mechanical performance indexes was analyzed by using the grey correlation algorithm. The functional relationship between the compressive strength at different temperatures and the UPV was established. Finally, the reliability of predicting high-frequency dynamic modulus by ultrasonic velocity was evaluated. It was expected to be a reference for the evaluation of the performance of asphalt mixtures by ultrasonic parameters. The main aims of the research are as follows:

- To establish the distribution function of the UPP for different asphalt mixtures and evaluate the reliability and sensitivity of three ultrasonic parameters in analyzing the performance of asphalt mixtures.
- To study the variation of three ultrasonic parameters in asphalt mixtures at dry and saturated states at different temperatures so as to provide a basis for formulating a standardized method for the UTT of asphalt mixtures.
- To analyze the correlation between three ultrasonic parameters and the asphalt aggregate ratio, aggregate content, volume indexes, and mechanical performance indexes at different temperatures; to study the internal mechanism affecting the ultrasonic travel

speed in asphalt mixtures; and then to establish the functional relationship between the UPV and compressive strength of asphalt mixtures with temperature.

- To deduce the dynamic modulus of four asphalt mixtures at high-frequency load by using UPV and comparing them with the fitting results of the Sigmoid model, and then to analyze the feasibility of using ultrasonic to calculate the high-frequency dynamic modulus of asphalt mixtures.

2. Materials and Methods

2.1. Materials and Sample Preparation

To study the ultrasonic transmission characteristics of different aggregate gradations of asphalt mixtures in various environments, four grades of asphalt mixtures were prepared by SGC. The gradations of four types of asphalt mixtures, named AC (Suspended dense asphalt mixture), SUP (Superpave graded asphalt mixture), SMA (Skeleton dense asphalt mixture), and OGFC (Skeleton void asphalt mixture), are exhibited in Table 1. Corresponding optimum asphalt aggregate ratios are 4.8%, 4.9%, 5.1%, and 5.5%, respectively. Asphalt mixtures are mainly composed of asphalt, coarse aggregate, fine aggregate, mineral powder, etc. In the fabricating process, high-performance asphalt was adopted as the binder, and its performance parameters are exhibited in Table 2. Basalt aggregate was applied as aggregate to build the skeleton of asphalt mixtures, and limestone mineral powder with a size less than 0.075 mm was adopted as filler.

Table 1. Gradation of four grades of asphalt mixtures.

Gradation Types	Sieve Size									
	0.075 mm	0.15 mm	0.3 mm	0.6 mm	1.18 mm	2.36 mm	4.75 mm	9.5 mm	13.2 mm	16 mm
AC (%)	6	10	13.5	19	26.5	37	53	76.5	95	100
SUP (%)	3	5	12	15	22	35	50	70	91	100
SMA (%)	10	12	13	16	19	20.5	27	62.5	95	100
OGFC (%)	4	5.5	7.5	9.5	12	16	21	70	95	100

Table 2. Properties of asphalt binder used by the research.

Properties	Standard Range	Test Value
25 °C Penetration (0.1 mm)	60~80	67.9
Ductility (cm) at 5 °C	$\geq$20	26.8
Softening point TR and B (°C)	$\geq$65	76.4
Elastic recovery (%) at 25 °C	$\geq$85	90.3
After aging test		
Aging mass loss (%)	$\leq\pm$0.8	0.26
25 °C Penetration ratio (%)	$\geq$60	81.5
Residual ductility (cm) at 5 °C	$\geq$10	14.4

2.2. Testing and Analytical Methods

2.2.1. Strategy of the Research

The strategy of this research is shown in Figure 1.

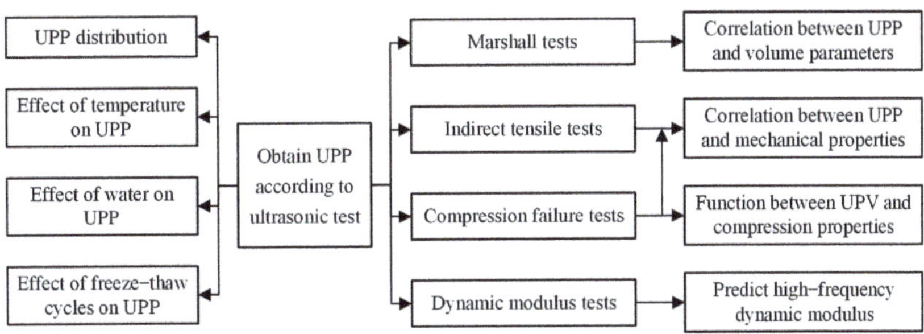

Figure 1. The strategy of this research.

2.2.2. Ultrasonic Tests of Asphalt Mixtures

To eliminate the influence of the uneven surface of mixture specimens on the UPP, the ultrasonic test specimens are obtained by core drilling sample and cutting. The corresponding sample acquisition and testing processes are shown in Figure 2. The ultrasonic testing instrument is a ZBL-U520/510 non-metallic ultrasonic monitor produced by Beijing Zhibolian Technology Co., Ltd. (Beijing, China). The UPP in four types of asphalt mixtures was measured at the dryness from −20 °C to 60 °C (at 10 °C increments) and at saturated water from 10 °C to 60 °C (at 10 °C increments). In addition, to study the influence of freeze–thaw cycle and damage from UPP, the UPP of four grades of asphalt mixtures with various freeze–thaw cycles was tested at 60 °C water saturation. The freeze–thaw cycle process was −18 °C for 16 h and 60 °C for 8 h.

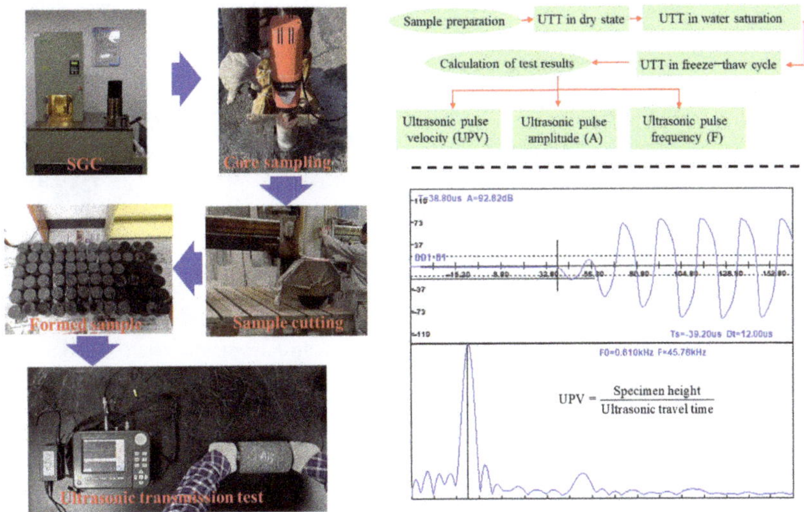

Figure 2. Sample preparation and ultrasonic transmission tests.

2.2.3. Splitting and Compression Failure Tests of Asphalt Mixtures

Splitting and compression tests were implemented to determine the mechanical performance of asphalt mixtures at low, room, and high temperatures. Both tests adopt a displacement loading mode. The loading rate of the splitting test at low temperature (−10 °C) is 1 mm/min, and that of the splitting tests at 20 °C and compression test (from 10 °C to 50 °C at 10 °C increments) is 2 mm/min. The two tests are shown in Figure 3a,b, respectively, and the corresponding mechanical properties are calculated by Equations (1)–(6) [21].

$$\sigma_I = 0.006287 \times P_I / h \tag{1}$$

$$\varepsilon_I = Y_I \times (0.0307 + 0.0936 \times \mu)/(17.94 - 0.314 \times \mu) \tag{2}$$

$$S_I = P_I \times (3.588 - 0.0628 \times \mu)/(h \times Y_I) \tag{3}$$

$$R_C = 4P_C/(\pi d^2) \tag{4}$$

$$\varepsilon_C = Y_C/h \tag{5}$$

$$S_C = R_C/\varepsilon_C \tag{6}$$

where σ_I and R_C are the indirect tensile strength and compressive strength, respectively; P_I and P_C are the indirect tensile failure load and compression failure load, respectively; ε_I and ε_C are the indirect tensile failure strain and compressive failure strain, respectively; Y_I and Y_C are the vertical deformation of indirect tensile failure and compression failure specimens, respectively; h and d are the specimen's height and diameter, respectively; μ is Poisson's ratio, 0.25 at $-10\ ^\circ\text{C}$ and 0.35 at $20\ ^\circ\text{C}$.

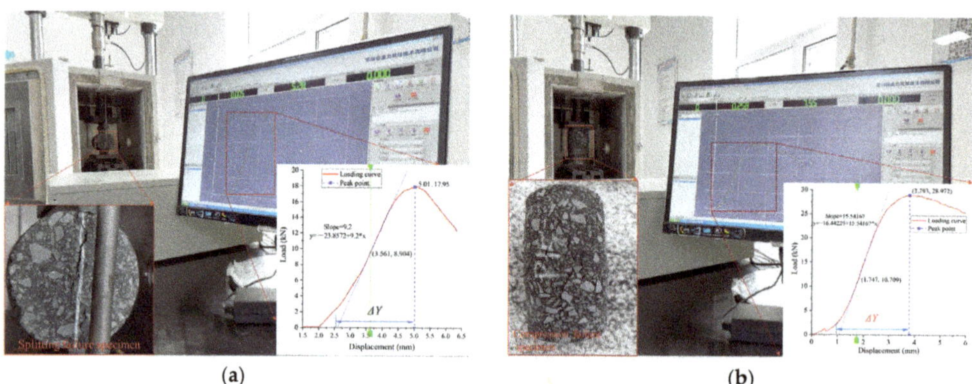

(**a**) (**b**)

Figure 3. Properties tests of asphalt mixtures: (**a**) splitting test; (**b**) compression failure test.

2.2.4. Dynamic Modulus Tests of Asphalt Mixtures

Relevant research pointed out that the essence of the ultrasonic travel process in asphalt mixtures is the vibration and movement of the high-frequency sound wave, which can reflect the high-frequency dynamic modulus G^* of mixtures. The corresponding calculation method is depicted in Equation (7) [22]. According to the viscoelastic theory, G^* of mixtures at different frequencies can be constructed into a master curve. To study the applicability of UPV in evaluating the high-frequency G^* of asphalt mixtures, the dynamic modulus tests at 10 °C, 20 °C, and 30 °C were carried out [23], and the test's process is shown in Figure 4, and G^* can be deduced by Equation (8).

$$E = v^2 \rho \frac{(1+\mu)(1-2\mu)}{(1-\mu)} \tag{7}$$

$$G^* = \frac{4P_i h}{\pi d^2 \Delta Y_i} \tag{8}$$

where E is the modulus of mixture deduced by the UPV; v is the UPV; ρ is the geometric density of mixtures; P_i is the mean value of load in the last 5 cycles; ΔY_i is the mean value of recoverable deformation in the last 5 cycles.

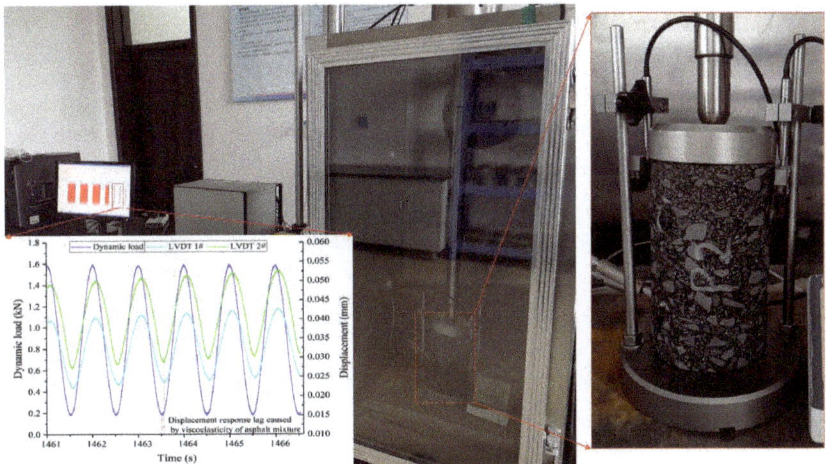

Figure 4. Dynamic modulus testing process of four graded asphalt mixtures.

2.2.5. Grey Correlation Analysis of Asphalt Mixtures Properties

The grey correlation analysis method is a method of evaluating the degree of correlation between factors based on the similarity of their development trends, which can effectively compensate for the shortcomings of some linear correlations. The traveling characteristics of ultrasonic pulse in asphalt mixtures are closely related to the volume indexes and mechanical property indexes of asphalt mixtures. Therefore, the grey correlation theory is adopted to evaluate the interrelation between the volume index and the UPP of asphalt mixtures in various environments, and its calculation method is shown in Equations (9) and (10) [24,25].

$$C_i = \frac{1}{n} \sum_{k=1}^{n} \xi(y_0(k), y_i(k)) \tag{9}$$

$$\xi(y_0(k), y_i(k)) = \frac{\min\limits_{i}\min\limits_{k}|y_0(k) - y_i(k)| + 0.5 \times \max\limits_{i}\max\limits_{k}\left|y_0(k) - y_i(k)\right|}{\left|y_0(k) - y_i(k)\right| + 0.5 \times \max\limits_{i}\max\limits_{k}\left|y_0(k) - y_i(k)\right|} \tag{10}$$

where C_i represents the grey correlation grade; $\xi(y_0(k), y_i(k))$ is the correlation coefficient between the reference and comparative sequences; $y_0(k)$ is the ultrasonic parameter sequence for reference; $y_i(k)$ are the volume parameters or mechanical property parameters sequence for comparison.

Before the grey correlation analysis, each parameter sequence needs to be normalized. In order not to change the variation characteristics of the original data, the normalization method is given as Equation (11).

$$Y_i' = y_i(k) / \frac{1}{n} \sum_{k=1}^{n} y_i(k) \quad k = 1, 2, \cdots, n \tag{11}$$

where Y_i' is the normalized sequence; $y_i(k)$ is the original data, n is the length of the sequence.

3. Research Results and Discussion

3.1. Traveling Characteristics of Ultrasonic Pulse in Mixtures

3.1.1. Distribution of UPP and Influence of Gradation

In the initial study, the UPP, including velocity, amplitude, and frequency of four grades of asphalt mixes at room temperature were tested by the UTT. Because the ultrasonic test is a non-destructive testing technology, a large number of testing data can be collected in the initial test, providing a foundation for the early analysis of the UPP in asphalt mixtures. To analyze the distribution characteristics and discreteness of the three ultrasonic indexes, histograms of the frequency distribution of the three ultrasonic indexes are plotted, as shown in Figure 5. Subsequently, the normal distribution function is adopted to analyze the distribution characteristics of ultrasonic parameters, and the P–P diagram is employed to evaluate the reliability of the normal distribution.

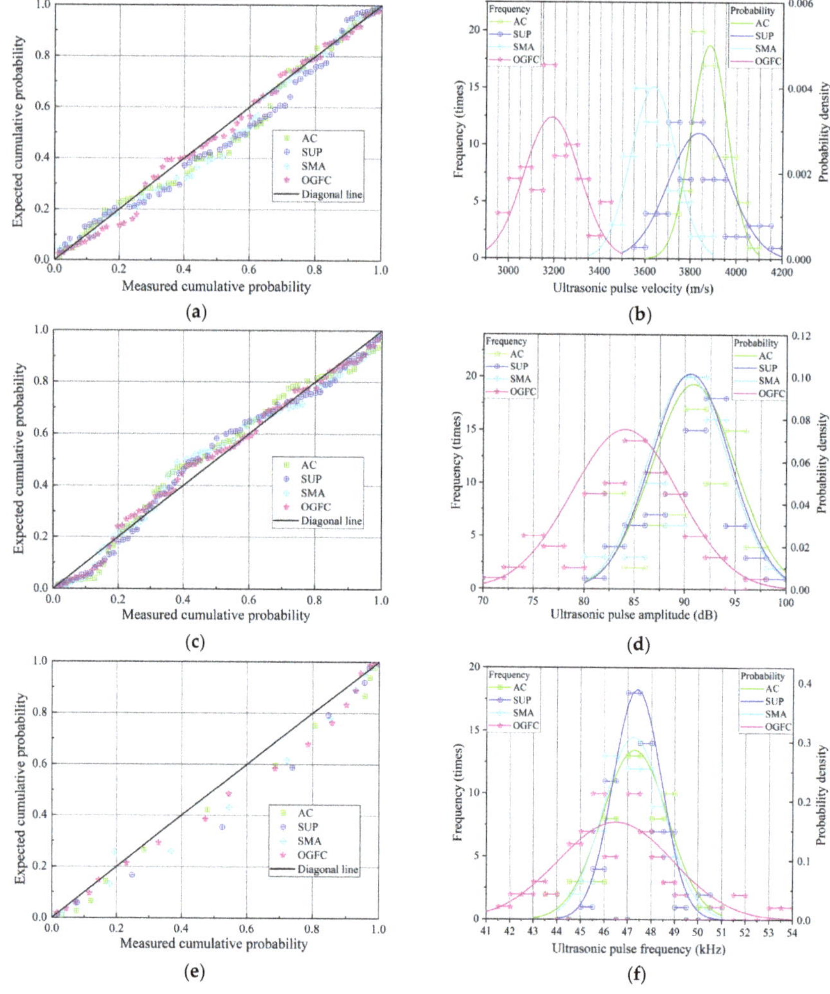

Figure 5. Probability distribution of ultrasonic pulse parameters of asphalt mixtures: (**a**) P–P diagram of pulse velocity; (**b**) frequency and probability distribution of pulse velocity; (**c**) P–P diagram of pulse amplitude (**d**) frequency and probability distribution of pulse amplitude; (**e**) P–P diagram of pulse frequency; (**f**) frequency and probability distribution of pulse frequency.

The actual cumulative probability is essentially consistent with the theoretical cumulative probability, as depicted in Figure 5a,c,e, implying that the ultrasonic parameters of the four grades of asphalt mixes tested in the study can better meet the normal distribution. Table 3 describes the normal distribution analysis results of ultrasonic parameters for the four grades of asphalt mixes. As seen in Table 3 and Figure 5b,c,f, there are certain differences in ultrasonic parameters due to differences in gradation type, void ratio (VV), or asphalt content. Furthermore, it was discovered that UPV is sensitive to the gradation change of mixtures, which means that UPV can be better employed to investigate the performance of asphalt mixtures with various aggregate gradations.

Table 3. Normal distribution results of ultrasonic parameters.

Gradation Types	Ultrasonic Pulse Velocity (m/s)		Ultrasonic Pulse Amplitude (dB)		Main Frequency of Ultrasonic Pulse (kHz)	
	μ	σ	μ	σ	μ	σ
AC	3878.33	78.477	90.80	4.121	47.27	1.394
SUP	3833.64	134.737	90.49	3.919	47.38	1.026
SMA	3633.69	97.921	90.23	3.954	47.22	1.295
OGFC	3187.71	119.728	84.04	5.295	46.48	2.425

Note: μ and σ is the characteristic parameter of the normal distribution.

Moreover, the UPV of suspension-dense mixtures (AC and SUP) is significantly faster than that of skeleton-dense mixtures (SMA) and skeleton-gap mixtures (OGFC), as shown in Table 3. The amplitude and frequency of the suspension-dense mixture and skeleton-dense mixture are much faster than those of the skeleton-gap mixture. However, when compared to UPV and amplitude, frequency has low sensitivity, implying that frequency is not suitable for characterizing the asphalt mixtures' performance.

3.1.2. Effect of Temperature on UPP in Four Grades of Mixtures

The movement of ultrasonic pulses in asphalt mixtures relies on the particle vibration of the medium, as shown in Figure 6. Testing temperature changes affect an asphalt mixture's performance, which in turn affects medium particle vibration, which in turn affects the movement of ultrasonic waves in the asphalt mixture. Figure 7 shows the variation of the UPV and amplitude with specimen temperature in four types of asphalt mixtures. The UPV continues to decline with rising temperatures and exhibits a pattern of initially slowing down and then accelerating its decline. According to the findings, asphalt continuously softens as the temperature rises, which prevents sound waves from moving through asphalt mixtures. The performance variety of the asphalt mixture with temperature is reduced at low temperatures, making the UPV less sensitive to variations in the specimen's temperature.

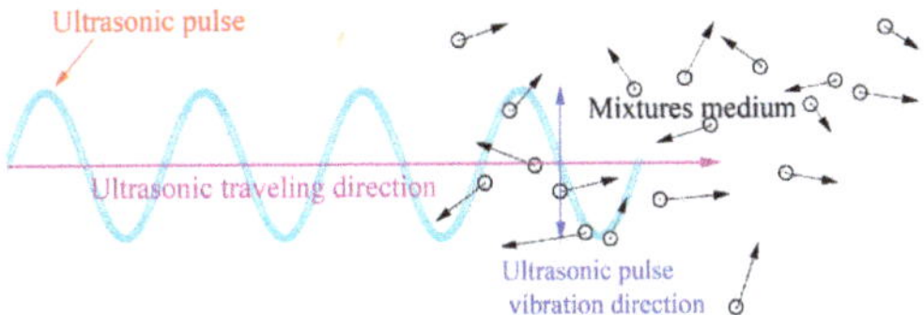

Figure 6. Traveling mechanism of ultrasonic pulse in mixture.

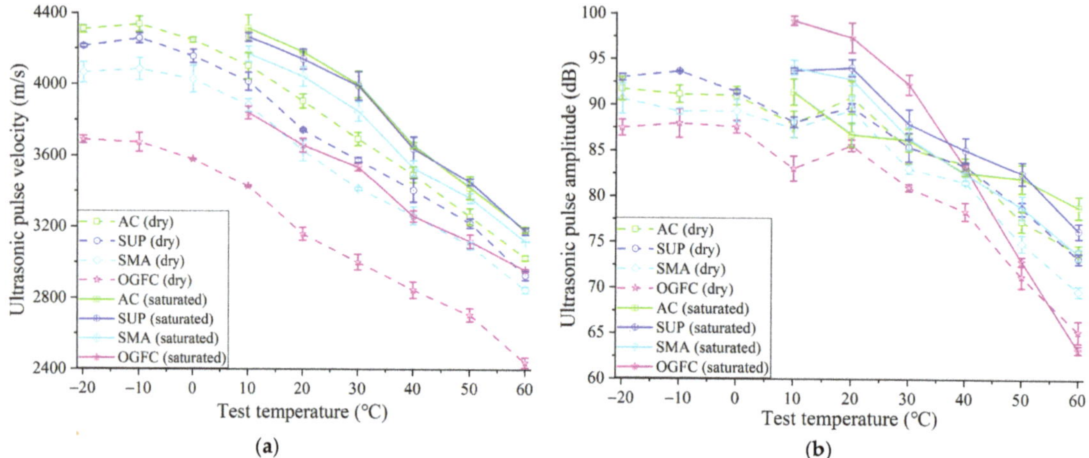

Figure 7. Effects of temperature and dry–wet state on ultrasonic pulse characteristics in asphalt mixtures: (**a**) variation of pulse velocity with temperature; (**b**) variation of pulse amplitude with temperature.

The attenuation of ultrasonic waves in mixtures can be directly identified by the ultrasonic pulse amplitude. The diffraction and reflection of the ultrasonic pulse are increased in the asphalt mixture due to the voids and viscosity of the asphalt medium, which causes the sound energy of the ultrasonic wave to attenuate during movement. The amplitude is less impacted by the temperature change when the temperature is low. The rapid attenuation of sound energy in the asphalt mixture is brought on by the viscous resistance of the asphalt mixture increasing as a result of a temperature rise. Additionally, it can be seen that the OGFC mixture with the highest void ratio has the greatest ability to dissipate ultrasonic pulse energy and the smallest ultrasonic pulse amplitude for SMA.

Table 4 shows the variation rate of UPV and amplitude with temperature. When the temperature rises from −20 °C to 60 °C at dryness, the UPV of AC, SUP, SMA, and OGFC mixtures decreases by 29.69%, 30.47%, 29.89%, and 33.87%, respectively. The corresponding amplitude decreases by 19.42%, 21.19%, 23.12%, and 25.42%, respectively. It can be found that the UPV and amplitude of OGFC mixtures are most affected by temperature. And it is well known that after the temperature rises, the sound in the air will spread faster. Additionally, the SMA has the highest asphalt content. This means that the influence of temperature on the UPV and amplitude of the asphalt mixture is not only due to the rise of air temperature in the mixture's voids but also the increase of asphalt viscous resistance. The subsequent researches focuses on the internal factors affecting the transmission of ultrasonic pulse in asphalt mixtures.

Table 4. The variation rate of UPV and amplitude with temperature.

Gradation Types	Variation Rate of UPV (%)			Variation Rate of Amplitude (%)		
	From −20 °C to 60 °C (Dry)	From 10 °C to 60 °C (Dry)	From 10 °C to 60 °C (Water)	From −20 °C to 60 °C (Dry)	From 10 °C to 60 °C (Dry)	From 10 °C to 60 °C (Water)
AC	29.69	26.16	26.14	19.42	15.89	13.73
SUP	30.47	27.02	25.47	21.19	16.83	18.60
SMA	29.89	26.59	25.15	23.12	20.27	21.60
OGFC	33.87	28.81	22.95	25.42	21.45	36.23

3.1.3. Effect of Water on UPP in Four Grades of Mixtures

Because there are some voids in the asphalt mixture, the internal open voids are occupied by water after immersion, which will affect the transmission of ultrasonic pulse in the asphalt mixture. To study the ultrasonic movement characteristics in the asphalt mixture at water saturation, the mixture specimens were first immersed in water to allow them to absorb sufficient water. Then, the UPP of the mixture specimens at water saturation was tested.

Figure 7a shows the UPV of asphalt mixtures at water saturation. The UPV exhibits a decreasing variation at water saturation with an increase in test temperature. In contrast with the UPV results in the dry, it can be found that the UPV of the four types of asphalt mixtures clearly increases after water saturation. The voids in the asphalt mixture are replaced by water, and the UPV in the water is faster than that in the voids, so the ultrasonic transmission for the entire specimen after water saturation is accelerated.

Figure 7b shows the comparison of ultrasonic pulse amplitude in dry and saturated environments. The ultrasonic pulse amplitude after water saturation is higher than in dry conditions. But, this trend fluctuates due to the complicated internal structure of the asphalt mixture, and the variation at various temperatures is noticeably different. Table 4 shows that the UPV of the AC, SUP, SMA, and OGFC mixtures decreases by 26.14%, 25.47%, 25.15%, and 22.95%, respectively, as the temperature increases from 10 °C to 60 °C at water saturation. And the corresponding amplitude decreases by 13.73%, 18.60%, 21.60%, and 36.23%, respectively. As the temperature rises, the decreased value of UPV of asphalt mixture at water saturation is lower than that of dry samples (as shown in Table 4). This might be explained by the fact that void water temperature has a greater impact on UPV than air temperature. When the temperature rises, the disordered movement of the mixture of medium particles intensifies (as shown in Figure 6), which counteracts the orderly directional movement of the ultrasonic waves, resulting in the reduction of the transmitted sound energy, and then the amplitude decreases with the increase in temperature.

The experimental temperature and the dry or wet state of specimens should be strictly controlled when using ultrasonic testing technology to study the performance of asphalt mixtures. Reasonable control variables, particularly in the freeze–thaw cycle testing process, should be considered to obtain reliable test results.

3.1.4. Effect of Freeze–Thaw Cycles on UPP in Four Grades of Mixtures

From the above test results, the test temperature and the dry or wet state of the specimen have a significant impact on the transmission characterization of the ultrasonic pulse. To avoid this influence during the freeze–thaw cycle test, the testing conditions were controlled at 60 °C water saturation.

Figure 8 shows the variation of UPV and amplitude for four grades of asphalt mixtures with the number of freeze–thaw cycles. The UPV and amplitude generally show a decreasing trend with an increase in freeze–thaw cycles, and the decreasing range of the OGFC mixture with the largest voids is the most obvious. The UPV and amplitude were both reduced by 18.12% and 30.75% after 14 freeze–thaw cycles, respectively.

Due to the effect of frost-heaving force, the voids in the asphalt mixture increase continuously as the freeze–thaw cycles increase, resulting in water entering the specimen. The transmission speed of ultrasonic waves in water is slower than in aggregate, which can reduce the UPV of the asphalt mixture specimen and significantly increase ultrasonic energy loss, as shown in Figure 8b.

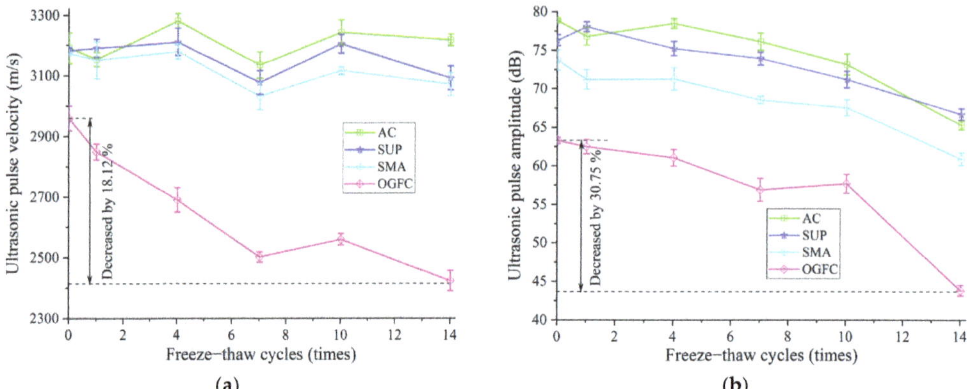

Figure 8. Effect of freeze–thaw cycle on UPP in asphalt mixture: (**a**) variation of pulse velocity with freeze–thaw cycle; (**b**) variation of pulse amplitude with freeze–thaw cycle.

3.2. Correlation between UPP and Both Volume Indexes and Performance Indexes

From the above analysis, it can be inferred that the ultrasonic transmission characteristics of mixtures are closely related to the properties of mixtures. To quantitatively analyze the correlation grade, the grey correlation theory is applied to calculate the correlation between the UPP and VV, voids in the mineral aggregate (VMA), voids filled with asphalt (VFA), asphalt content, indirect tensile properties at $-10\ ^\circ\text{C}$ and $20\ ^\circ\text{C}$, resistance to compression properties at $20\ ^\circ\text{C}$ and $50\ ^\circ\text{C}$, and aggregate content for each grade. This can lay the foundation for establishing the functional relationship between ultrasonic characteristics and mixture properties.

3.2.1. Correlation between UPP and Volume Parameters

Due to the discreteness of the fabricating process for mixture specimens, there are still some differences in the volume indexes for the same grading of mixture specimens. Therefore, the grey correlation grade between the UPP and the volume indexes (VV, VMA, and VFA), bulk density (γ_b), apparent density (γ_a), apparent dry density (γ_s), and geometric density (γ_v) were calculated, as shown in Figure 9.

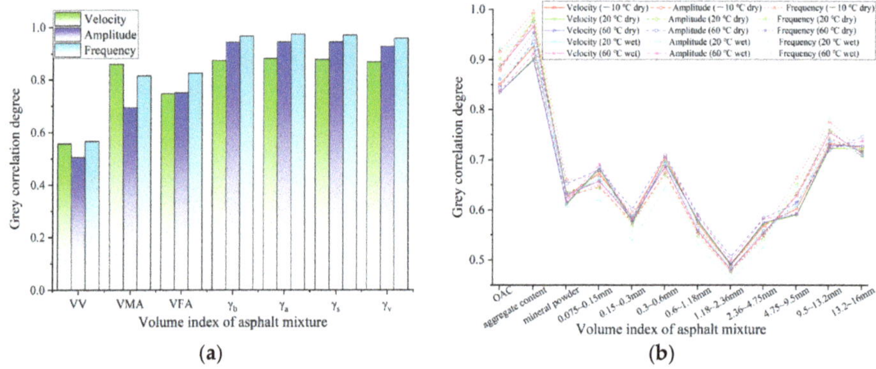

Figure 9. Correlation between UPP and volume parameters of asphalt mixtures: (**a**) correlation between UPP and volume indexes; (**b**) correlation between UPP and aggregate content.

The analysis results show that the VMA, VFA, γ_b, γ_a, γ_s, and γ_v of asphalt mixtures have a closer correlation with the UPP than VV. Among them, the correlation degree of the four types of density is the highest, which can also be reflected in Equation (7). There is a functional relationship between the UPV and the density of asphalt mixtures.

Moreover, from the correlation results of VMA and VFA, it can be inferred that there seems to be a close relationship between the UPP and mineral aggregate content and asphalt content. To further analyze and investigate the influence of asphalt and aggregate content on the UPP, the grey correlation between gradation, asphalt aggregate ratio, and the UPP at $-10\ ^\circ$C, 20 $^\circ$C, 60 $^\circ$C dry environment, and 20 $^\circ$C and 60 $^\circ$C water saturation is calculated, as shown in Figure 9b.

Figure 9b shows that the UPP is closely related to the asphalt aggregate ratio and aggregate content of mixtures, which means that the transmission of ultrasonic pulse in asphalt mixtures is mainly affected by the asphalt content and the mineral materials content. Further, the correlation between the aggregate content at each grade and the ultrasonic parameters is analyzed. It can be observed that the UPP has a higher correlation with the mineral aggregate content of 9.5~13.2 mm and 13.2~16 mm. The increase of aggregate content with the large particles in the asphalt mixture will be beneficial to the transmission of the ultrasonic pulse.

3.2.2. Correlation between UPP and Properties Parameters of Asphalt Mixture

According to Equation (7), the transmission of an ultrasonic pulse in an asphalt mixture is related to the mechanical properties parameters of the asphalt mixture. To analyze the relationship between the UPP and mechanical properties of asphalt mixtures, the correlation between the UPP and properties parameters of asphalt mixtures at low, normal, and high temperatures is calculated by the grey correlation algorithm. The analysis results are shown in Figure 10.

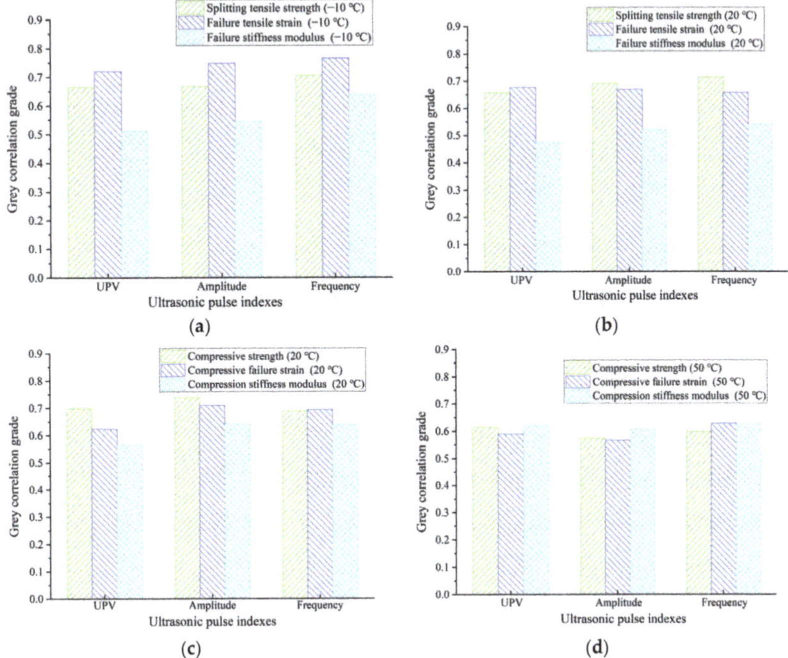

Figure 10. Correlation between UPP and properties parameters of asphalt mixtures: (**a**) correlation with low temperature (−10 $^\circ$C) performance; (**b**) correlation with splitting performance at room temperature (20 $^\circ$C); (**c**) correlation with compression performance at room temperature (20 $^\circ$C); (**d**) correlation with compression performance at high temperature (50 $^\circ$C).

As can be seen from Figure 10a, the three pulse parameters have a good correlation with splitting parameters at low temperatures ($-10\ °C$), and it can be found that the UPP has a better correlation with low-temperature splitting strength and strain compared with modulus. Similar to the low-temperature properties, the UPP also has a good correlation with the compressive strength parameters and the splitting resistance parameters at room temperature. Moreover, the strength and strain have a higher correlation, as shown in Figure 10b,c. This means that constructing the functional relationship between strength and UPP has good reliability.

It can be seen from Figure 10d that the correlation between UPP and compression parameters at high temperature decreases, which is primarily caused by the asphalt softening at higher temperatures and the rise in factors affecting the transmission of ultrasonic in asphalt mixtures. At room temperature and high temperature, it can be found that the compressive strength has a better correlation with UPP than the compressive strain and modulus. Therefore, the functional relationship between compressive strength and UPV will be established in the subsequent analysis.

3.2.3. Functional Relationship between UPV and Compression Properties

According to the above analysis, the UPP has an excellent correlation with the strength of asphalt mixtures at different temperatures. Moreover, the strength of four kinds of asphalt mixtures changes with the variation of temperature, and the existing relationship (Equation (7)) characterizes the functional relationship between the mixtures' modulus and the ultrasonic UPV. It is difficult to reflect the relationship between the strength (or modulus) varying with temperature and UPV.

Therefore, in this paper, the compressive strength from $10\ °C$ to $50\ °C$ (at $10\ °C$ increments) was tested by uniaxial compression tests, and a scatter diagram between the UPV and compressive strength at five temperatures was constructed, as shown in Figure 11. Figure 11 shows that with the increase of UPV, the compressive strength of asphalt mixtures shows an approximate parabolic growth from slow to fast. Thus, the quadratic polynomial (Equation (12)) is adopted to build the functional relationship between compressive strength and UPV at dry. The fitting results are shown in Figure 11.

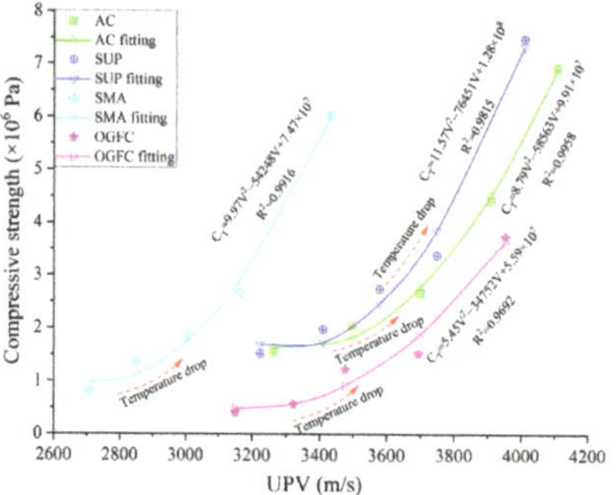

Figure 11. Functional relationship between compressive strength and UPV.

From Figure 11 and the correlation coefficient R^2, it can be found that the quadratic polynomial can better characterize the quantitative relationship between the UPV (v) and compressive strength (CT) for four grades of asphalt mixtures at various temperatures. However, it can also be clearly found that the fitting parameters of asphalt mixtures with different gradations are quite different. This is mainly because there are great differences in asphalt content and skeleton type for different graded asphalt mixtures, and the influence of temperature change on their compressive strength and UPV is obviously different.

$$E(T) = Av(T)^2 + Bv(T) + C \tag{12}$$

where A, B, and C are the fitting parameters.

3.3. Application of UPV to Predict High-Frequency Dynamic Modulus of Asphalt Mixtures

The transmission of ultrasonic pulse in asphalt mixtures is essentially the vibration transmission of mass points with high-frequency vibration in asphalt mixtures. Relevant studies pointed out that the dynamic modulus at high-frequency loads of asphalt mixtures could be deduced by Equation (7) based on the ultrasonic test results [4]. Therefore, the dynamic modulus tests of four kinds of mixtures at 10 °C, 20 °C, and 30 °C are carried out, and the dynamic master curves of four kinds of mixtures are constructed. The dynamic modulus of four kinds of asphalt mixtures at high-frequency loads is calculated by the Sigmoid model (as shown in Equation (13)) and compared with the dynamic modulus deduced by UPV [26].

$$\log(G^*) = (\log(G_{\max}) - \log(G_{\min})) / \left[\left(1 + \lambda e^{\beta + \gamma \log \omega}\right)^{\frac{1}{\lambda}} \right] + \log(G_{\min}), \qquad (13)$$

where $\log(G^*)$ is the logarithm of the dynamic modulus; $\log(G_{\min})$ and $\log(G_{\max})$ are logarithms of static and glassy modulus to be fitted, respectively. λ, β, and γ are the parameters to be fitted to demonstrate the morphological characteristics of the Sigmoid model curve; $\log(\omega)$ is the logarithm of the angular frequency.

Figure 12 shows the change of dynamic modulus of four kinds of mixtures with test temperature and frequency. As can be seen from Figure 11, the dynamic modulus of the mixtures increases with the rise of loading frequency. As the test temperature increases, the dynamic modulus of mixtures decreases.

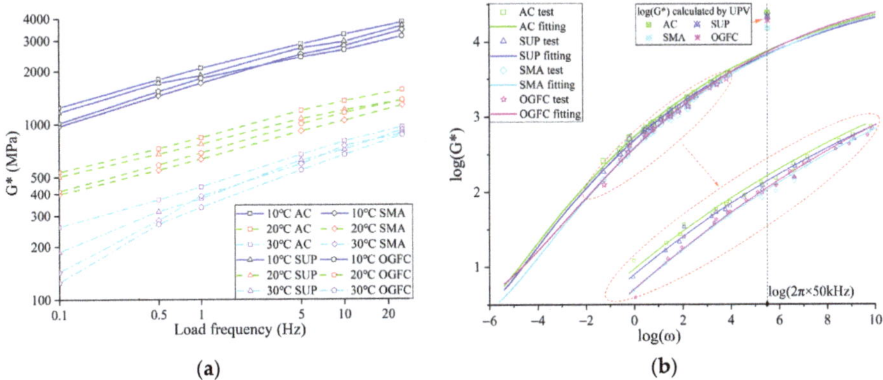

Figure 12. Dynamic modulus test results and main curves fitting comparison: (**a**) dynamic modulus; (**b**) fitting and prediction of dynamic modulus.

To further calculate the dynamic modulus of four kinds of mixtures at high-frequency load and compare the dynamic modulus calculated by the Sigmoid model with that calculated by UPV, the main curves of the dynamic modulus of four kinds of mixtures at 20 °C are established. Subsequently, the main curves are fitted by the Sigmoid model (Equation (13)) [27]. The fitting results are shown in Table 5 and Figure 12. Figure 12 illustrates that the Sigmoid model can better characterize the variation of dynamic modulus of asphalt mixtures with loading frequency.

Table 5. Fitting results of the Sigmoid model parameters of four asphalt mixtures.

Mixs	Fitting Parameters						Dynamic Modulus Comparison		
	$Log\ (G_{min})$	$Log\ (G_{max})$	β	γ	λ	R^2	Log (Calculated Value of Model)	Log (Calculated Value of UPV)	Prediction Error (%)
AC	0.69	4.938	−0.607	−0.138	−0.846	0.9966	3.877	4.384	11.58
SUP	0.675	4.912	−0.599	−0.136	−0.861	0.9956	3.831	4.336	11.63
SMA	0.45	4.935	−0.548	−0.147	−0.712	0.9967	3.824	4.178	8.47
OGFC	0.433	4.884	−0.471	−0.169	−0.428	0.9942	3.863	4.284	9.84

To compare and analyze the relationship between the dynamic modulus calculated by UPV and that obtained by the Sigmoid model, the dynamic modulus of asphalt mixtures at 50 kHz is calculated according to the fitting results, and the error between the dynamic modulus obtained by the two algorithms is calculated, as shown in Table 5. The calculation results in Table 5 indicate that the dynamic modulus at higher frequency can be deduced by using the UPV, but the predictive accuracy of the two algorithms is different for different graded asphalt mixtures, in which the predicting errors of SUP and AC are larger. The dynamic modulus of the OGFC and SMA mixture deduced by using the UPV at a high-frequency load has higher reliability.

4. Conclusions

The following conclusions can be distilled from the aforementioned testing and analysis results:

- Ultrasonic traveling parameters in asphalt mixtures present a good normal distribution, in which the UPV is the most sensitive to the gradation change of asphalt mixtures and has obvious advantages in evaluating the performance of asphalt mixtures. The sensitivity of frequency is poor.
- With the increase in temperature, both the UPV and amplitude in dry and wet mixtures samples demonstrate a downward trend. Among them, OGFC mixtures with skeleton-gap structures have the largest decline. The maximum decreases in UPV and amplitude are 33.87% and 36.23%, respectively. The disordered movement of the mixture of medium particles intensifies after the temperature rises, which counteracts the orderly directional movement of the ultrasonic waves, resulting in the reduction of the transmitted sound energy. In addition, compared with other graded mixtures, freeze–thaw cycles also have the greatest impact on OGFC. After 14 freeze–thaw cycles, the UPV and amplitude decreased by 18.12% and 30.75%, respectively.
- Mixture density, asphalt content, and the content of 9.5~13.2 mm and 13.2~16 mm mineral aggregate have a greater impact on ultrasonic parameters. Increasing the content of coarse aggregates can facilitate the transmission of ultrasonic waves in asphalt mixtures.
- The ultrasonic parameters have a good correlation with the mechanical properties of the mixture. According to the quadratic function, the UPV can be applied to predict the compression strength at different temperatures, but the corresponding prediction functions of different mixtures are different. The existing equation can be employed to calculate the high-frequency dynamic modulus by UPV. The dynamic modulus of the OGFC and SMA mixture deduced by using the UPV at a high-frequency load has higher reliability.

Author Contributions: Conceptualization, X.T., C.W. and L.L.; methodology, X.T., L.L. and Y.Z.; validation, H.L. (He Li) and J.Z.; formal analysis, X.T.; investigation, X.T.; supervision, C.W.; writing—original draft preparation, X.T.; writing—review and editing, L.L., F.W. and H.L. (Hanjun Li); project administration, C.W.; funding acquisition, C.W. and C.L. All authors have read and agreed to the published version of the manuscript.

Funding: This study was supported by the Technology Development Program of the Jilin Province (No. 20230203137SF) and Key Technology Projects for Transportation in Jilin Province (Science and Technology) (No. 2023ZDGC-1-3).

Institutional Review Board Statement: Not applicable.

Informed Consent Statement: Not applicable.

Data Availability Statement: Some or all data, models, or codes that support the findings of this study are available from the corresponding author upon reasonable request.

Acknowledgments: We gratefully acknowledge the financial support of the above funds and the researchers of all reports cited in our paper.

Conflicts of Interest: The authors declare no conflict of interest.

Abbreviations

UTT	Ultrasonic testing technology
UPP	Ultrasonic pulse parameters
UPV	Ultrasonic pulse velocity
2S2P1D	Two springs, two parabolic elements, and one dashpot
SGC	Superpave gyratory compactor
AC	Suspended dense asphalt mixture
SUP	Superpave graded asphalt mixture
SMA	Skeleton dense asphalt mixture
OGFC	Skeleton void asphalt mixture
VV	Void ratio
VMA	Mineral aggregate
VFA	Voids filled with asphalt
γ_b	Bulk density
γ_a	Apparent density
γ_s	Apparent dry density
γ_v	Geometric density

References

1. Najim, K.B.; Hall, M.R. Mechanical and dynamic properties of self-compacting crumb rubber modified concrete. *Constr. Build. Mater.* **2012**, *27*, 521–530. [CrossRef]
2. Zhang, W.; Akber, M.A.; Hou, S.; Bian, J.; Zhang, D.; Le, Q. Detection of Dynamic Modulus and Crack Properties of Asphalt Pavement Using a Non-Destructive Ultrasonic Wave Method. *Appl. Sci.* **2019**, *9*, 2946. [CrossRef]
3. Arabani, M.; Kheiry, P.T.; Ferdowsi, B. Use of Ultrasonic Pulse Velocity (Upv) for Assessment of Hma Mixtures Behavior. *Iran. J. Sci. Technol. Trans. Civ. Eng.* **2012**, *36*, 111–114.
4. Majhi, D.; Karmakar, S.; Roy, T.K. Reliability of Ultrasonic Pulse Velocity Method for Determining Dynamic Modulus of Asphalt Mixtures. *Mater. Today Proc.* **2017**, *4*, 9709–9712. [CrossRef]
5. In, C.-W.; Kim, J.-Y.; Kurtis, K.E.; Jacobs, L.J. Characterization of ultrasonic Rayleigh surface waves in asphaltic concrete. *NDT E Int.* **2009**, *42*, 610–617. [CrossRef]
6. McGovern, M.E.; Behnia, B.; Buttlar, W.G.; Reis, H. Concrete Testing: Characterisation of oxidative ageing in asphalt concrete—Part 2: Estimation of complex moduli. *Insight Non-Destr. Test. Cond. Monit.* **2013**, *55*, 605–609. [CrossRef]
7. Di Benedetto, H.; Sauzéat, C.; Sohm, J. Stiffness of Bituminous Mixtures Using Ultrasonic Wave Propagation. *Road Mater. Pavement* **2011**, *10*, 789–814. [CrossRef]
8. Mounier, D.; Di Benedetto, H.; Sauzéat, C. Determination of bituminous mixtures linear properties using ultrasonic wave propagation. *Constr. Build. Mater.* **2012**, *36*, 638–647. [CrossRef]
9. Larcher, N.; Takarli, M.; Angellier, N.; Petit, C.; Sebbah, H. Towards a viscoelastic mechanical characterization of asphalt materials by ultrasonic measurements. *Mater. Struct.* **2014**, *48*, 1377–1388. [CrossRef]
10. Tavassoti-Kheiry, P.; Boz, I.; Chen, X.; Solaimanian, M. Application of Ultrasonic Pulse Velocity Testing of Asphalt Concrete Mixtures to Improve the Prediction Accuracy of Dynamic Modulus Master Curve. *Airfield Highw. Pavements* **2017**, *2017*, 152–164. [CrossRef]
11. Birgisson, B.; Roque, R.; Page, G.C. Ultrasonic Pulse Wave Velocity Test for Monitoring Changes in Hot-Mix Asphalt Mixture Integrity from Exposure to Moisture. *Transp. Res. Rec. J. Transp. Res. Board* **2003**, *1832*, 173–181. [CrossRef]
12. Cheng, Y.-c.; Zhang, P.; Jiao, Y.-b.; Wang, Y.-d.; Tao, J.-l. Damage Simulation and Ultrasonic Detection of Asphalt Mixture under the Coupling Effects of Water-Temperature-Radiation. *Adv. Mater. Sci. Eng.* **2013**, *2013*, 838943. [CrossRef]

13. You, L.Y.; You, Z.P.; Dai, Q.L.; Guo, S.C.; Wang, J.Q.; Schultz, M. Characteristics of Water-Foamed Asphalt Mixture under Multiple Freeze-Thaw Cycles: Laboratory Evaluation. *J. Mater. Civil. Eng.* **2018**, *30*, 04018270. [CrossRef]

14. Pan, W.H.; Sun, X.D.; Wu, L.M.; Yang, K.K.; Tang, N. Damage Detection of Asphalt Concrete Using Piezo-Ultrasonic Wave Technology. *Materials* **2019**, *12*, 443. [CrossRef]

15. Franesqui, M.A.; Yepes, J.; Garcia-Gonzalez, C. Ultrasound data for laboratory calibration of an analytical model to calculate crack depth on asphalt pavements. *Data Brief* **2017**, *13*, 723–730. [CrossRef]

16. Ji, X.; Hou, Y.; Chen, Y.; Zhen, Y. Attenuation of acoustic wave excited by piezoelectric aggregate in asphalt pavement and its application to monitor concealed cracks. *Constr. Build. Mater.* **2019**, *216*, 58–67. [CrossRef]

17. Tigdemir, M.; Kalyoncuoglu, S.F.; Kalyoncuoglu, U.Y. Application of ultrasonic method in asphalt concrete testing for fatigue life estimation. *NDT E Int.* **2004**, *37*, 597–602. [CrossRef]

18. Wu, J.R. Fatigue Test of Asphalt Mixture and Ultrasonic Forecast. *Adv. Mater. Res.* **2010**, *168–170*, 488–491. [CrossRef]

19. Jiang, Z.Y.; Ponniah, J.; Cascante, G.; Haas, R. Nondestructive ultrasonic testing methodology for condition assessment of hot mix asphalt specimens. *Can. J. Civil. Eng.* **2011**, *38*, 751–761.

20. Xie, T.; Qiu, Y.J.; Jiang, Z.Z.; Lan, B. Compaction Quality of Bitumen Mixtures Based on Ultrasonic Methods. *Key Eng. Mater.* **2007**, *353–358*, 2341–2344. [CrossRef]

21. Xiong, R.; Chu, C.; Guan, B.; Sheng, Y. Performance damage characteristics of asphalt mixture suffered from the sulphate–water–temperature–load coupling action. *Int. J. Pavement Eng.* **2020**, *23*, 1368–1377. [CrossRef]

22. Grilli, A.; Bocci, E.; Graziani, A. Influence of reclaimed asphalt content on the mechanical behaviour of cement-treated mixtures. *Road Mater. Pavement* **2013**, *14*, 666–678. [CrossRef]

23. Zhang, J.; Fan, Z.; Wang, H.; Sun, W.; Pei, J.; Wang, D. Prediction of dynamic modulus of asphalt mixture using micromechanical method with radial distribution functions. *Mater. Struct.* **2019**, *52*, 49. [CrossRef]

24. Cheng, Y.; Li, L.; Zhou, P.; Zhang, Y.; Liu, H. Multi-objective optimization design and test of compound diatomite and basalt fiber asphalt mixture. *Materials* **2019**, *12*, 1461. [CrossRef] [PubMed]

25. Xiong, R.; Chu, C.; Qiao, N.; Wang, L.; Yang, F.; Sheng, Y.; Guan, B.; Niu, D.; Geng, J.; Chen, H. Performance evaluation of asphalt mixture exposed to dynamic water and chlorine salt erosion. *Constr. Build. Mater.* **2019**, *201*, 121–126. [CrossRef]

26. Norambuena-Contreras, J.; Castro-Fresno, D.; Vega-Zamanillo, A.; Celaya, M.; Lombillo-Vozmediano, I. Dynamic modulus of asphalt mixture by ultrasonic direct test. *NDT E Int.* **2010**, *43*, 629–634. [CrossRef]

27. Li, L.; Wu, C.; Cheng, Y.; Ai, Y.; Li, H.; Tan, X. Comparative Analysis of Viscoelastic Properties of Open Graded Friction Course under Dynamic and Static Loads. *Polymers* **2021**, *13*, 1250. [CrossRef] [PubMed]

materials

MDPI

Article

Multiscale Fatigue Performance Evaluation of Hydrated Lime and Basalt Fiber Modified Asphalt Mixture

Hang Diao [1,*], Tianqing Ling [1,*], Zhan Zhang [2], Bo Peng [3] and Qiang Huang [1]

[1]   School of Civil Engineering, Chongqing Jiaotong University, Chongqing 400074, China
[2]   Sichan Institute of Building Research, Chengdu 610081, China
[3]   School of Materials Science and Engineering, Chongqing Jiaotong University, Chongqing 400074, China
*   Correspondence: diaohang@mails.cqjtu.edu.cn (H.D.); lingtq@cqjtu.edu.cn (T.L.)

Abstract: Long-life pavement construction is an important research direction for sustainable road development. Fatigue cracking of aging asphalt pavement is one of the main reasons that affects its service life, and improving the fatigue resistance of aging asphalt pavement has become a key factor in promoting the development of long-life pavement. In order to enhance the fatigue resistance of aging asphalt pavement, hydrated lime and basalt fiber were selected to prepare a modified asphalt mixture. The resistance to fatigue is evaluated by the four-point bending fatigue test and self-healing compensation test, based on the energy method, the phenomenon-based approach, and other methods. The results of each method of evaluation were also compared and analyzed. The results indicate that the incorporation of hydrated lime can improve the adhesion of the asphalt binder, while the incorporation of basalt fiber can stabilize the internal structure. When incorporated alone, basalt fiber has no noticeable effect, while hydrated lime significantly improves the fatigue performance of the mixture after thermal aging. Mixing both ingredients produced the best improvement effect under various conditions, with a fatigue life improvement of 53%. In the multi-scale evaluation of fatigue performance, it was found that the initial stiffness modulus was unsuitable as a direct evaluation index of fatigue performance. Using the fatigue damage rate or the stable value of dissipated energy change rate as an evaluation index can clearly characterize the fatigue performance of the mixture before and after aging. The self-healing rate and self-healing decay index clearly reflected the fatigue damage healing process under repeated loading and could be used as relevant indices for evaluating the new-scale fatigue performance of asphalt mixtures.

Keywords: hydrated lime; basalt fiber; thermal aging; fatigue properties; self-healing performance

Citation: Diao, H.; Ling, T.; Zhang, Z.; Peng, B.; Huang, Q. Multiscale Fatigue Performance Evaluation of Hydrated Lime and Basalt Fiber Modified Asphalt Mixture. *Materials* **2023**, *16*, 3608. https://doi.org/10.3390/ma16103608

Academic Editor: Gilda Ferrotti

Received: 7 April 2023
Revised: 25 April 2023
Accepted: 28 April 2023
Published: 9 May 2023

1. Introduction

In recent years, with the increase in traffic volume on roads, the frequent occurrence of fatigue cracking in asphalt pavements has hindered normal passenger and cargo transportation, reduced the pavement's service life, increased road repair and maintenance costs, and severely affected both the economic benefits and the quality of traffic services [1–3]. The growth of vehicle load, as well as the emergence of the asphalt aging phenomenon, has resulted in the fatigue cracking of asphalt concrete pavements and reduced the service life of pavements; in particular, fatigue cracking and thermal oxygen aging are important reasons affecting the life of asphalt pavement [4–6]. Garbowski et al. [7,8] developed pavement damage identification tools to facilitate the timely detection and maintenance of pavements. This has become an important direction for the global development of a sustainable development strategy to improve the heat-resistant aging performance of asphalt pavement, decrease fatigue cracking, and increase the working life of roads, which is of great significance for transportation construction.

Existing research results have shown that fatigue damage of asphalt mixtures is mainly concentrated in the asphalt and asphalt mortar [9,10]. However, during mixing, paving, and use, asphalt and asphalt mortar will undergo thermal oxygen aging, severely affecting their fatigue performance. In engineering, additives are usually added to the mixture, as well as fibers, to improve their high-temperature resistance [11,12]. Among them, the incorporation of basalt fiber can suppress the occurrence and extension of fatigue cracks in asphalt pavement, improving its fatigue performance [13–15], with an especially significant improvement in crack resistance under high-temperature oxidation conditions [16]. The incorporation of hydrated lime, due to its good physical adsorption properties and high chemical activity, can inhibit the aging of asphalt and change the high and low-temperature properties of the asphalt mixture [17–20]. Das et al. [21] conducted experiments that show that replacing 20% of the mineral powder with hydrated lime in asphalt significantly improves its high-temperature aging resistance. Mwanza et al. [22] found that the softening point of asphalt gradually increases with the addition of hydrated lime, while the ductility and penetration decrease. Mohammad et al. [23] conducted a rheological performance analysis and concluded that the addition of hydrated lime enhances the permanent deformation properties and fatigue durability of asphalt mastic. Davar et al. [24] demonstrated that incorporating only basalt fibers into asphalt mixtures did not significantly change its fatigue life. Kathari et al. [25] showed that incorporating basalt fibers improved the high-temperature deformation resistance of asphalt mortar. Currently, many studies evaluate the fatigue properties of asphalt mixtures using a single theory, which makes comparisons difficult due to a lack of control over the results. In order to address this issue, numerous domestic and foreign scholars have conducted research. Amir Izadi et al. [26] accurately characterized the fatigue properties of asphalt mixtures at different aging states using the dissipated energy method, and the energy dissipated during repeated loading can be accurately characterized in terms of its fatigue behavior. Liu et al. [27] evaluated the fatigue characteristics of different aging states of asphalt concrete at various loading frequencies using the S-N fatigue equation. Lv et al. [28] found that aging had the greatest impact on the initial elastic modulus of asphalt mixtures, and there was a good correlation between fatigue life and the rate of dissipated energy change stabilization values. Sun et al. [29] proved that the self-healing performance of asphalt mixtures has a strong correlation with their fatigue resistance.

In summary, hydrated lime and basalt fibers were used as additives to prepare modified asphalt mixtures, and their fatigue properties were studied. Four-point bending fatigue tests and self-healing compensation tests were conducted, and the fatigue performance was evaluated by using phenomenological and energy-based methods at multiple scales. The applicability of different evaluation indicators was explored, and the self-healing index attenuation rate was introduced to evaluate the fatigue performance. This study aimed to provide a comprehensive evaluation of the fatigue properties of modified asphalt mixtures and to evaluate their fatigue resistance through the self-healing index attenuation rate.

2. Materials and Methods

2.1. Raw Materials

The asphalt used in this experiment is the base asphalt that meets the specified requirements, and its main technical specifications are shown in Table 1. Basalt fibers with a length of 6 mm were selected, and their basic performance parameters are shown in Table 2. The coarse and fine aggregates were composed of locally sourced limestone and stone chips. The filler used was hydrated lime powder and ordinary ground mineral powder of limestone. The specific parameters of the filler's specific surface area and particle size are shown in Table 3, and the basalt fiber and hydrated lime samples are shown in Figure 1.

Table 1. Basic performance of asphalt.

Project	Detection Value	Standard Requirements	Experiment Methods
Penetration (25 °C, 100 g, 5 s)/(0.1 mm)	67	60~80	T0604

Table 1. *Cont.*

	Project	Detection Value	Standard Requirements	Experiment Methods
	Ductility (15 °C, 5 cm/min)/(cm)	141	$\geq$100	T0605
	Ductility (10 °C, 5 cm/min)/(cm)	27	$\geq$20	T0605
	Softening point (°C)	48.5	$\geq$45	T0606
	Density (15 °C)/(g/cm^3)	1.01	Measured value	T0603
	Ductility (10 °C, 5 cm/min)/(cm)	27	$\geq$20	T0605
	Wax content (distillation method) (%)	1.8	$\leq$2.2	T0615
	Mass loss (%)	0.13	$\leq\pm$0.8	T0609
After RTFOT:	Penetration ratio (25 °C) (%)	71	$\geq$61	T0604
	Ductility (10 °C)/(cm)	8	$\geq$6	T0605

Table 2. Characteristics of basalt fiber.

Project	Index
Density/(g/cm^3)	2.65
Fiber diameter/μm	17
Average length/mm	6 ± 0.5
Tensile strength/MPa	1050
Elastic modulus/GPa	7.6
Fracture elongation/%	3
Operating temperature/°C	-260–650

Table 3. Particle size and specific surface area of mineral powder and hydrated lime powder.

Type	D90 (μm)	SSA (m^2/g)
limestone ore powder	73.8	1.13
Hydrated lime	43.2	2.48

(a)   (b)

Figure 1. Basalt fiber and hydrated lime samples. (**a**) basalt fiber; (**b**) hydrated lime.

2.2. Asphalt Mixture Ratio Design

In accordance with the preliminary experiment, the optimal oil-to-rock ratio was determined to be 5.1%, and the dense grading AC-13 was used, with the composition of the grading shown in Table 4.

Table 4. Aggregate gradation of AC-13.

Sieve size/mm	16.0	13.2	9.5	4.75	2.36	1.18	0.6	0.3	0.15	0.075
Passing/%	100	98	80	58	40	27.9	19.9	14.4	10.4	5.6

2.3. Preparation and Aging of Specimens

According to the previous experiment, we selected a 0.8 powder-to-rubber ratio; hydrated lime (HL) blending of 20% to replace the mineral powder (accounting for the mass of mineral powder); and basalt fiber (BF) blending of 0.3% (accounting for the total mass of asphalt mixture). For the preparation of a variety of asphalt mixtures, the specific types were divided as follows: not blended, blended with 20% HL, blended with 0.3% BF, 20% HL + 0.3% BF (recorded as CAM, HLAM, BFAM, and HLBFAM, respectively). The specific amount of blending and groupings are shown in Table 5.

Table 5. Group of mixture specimens.

Specimen Types	Powder to Glue Ratio	Thermal Aging	Hydrated Lime (%)	Limestone Powder (%)	Basalt Fiber (%)
CAM	0.8	No	0	100	0
HLAM	0.8	No	20	80	0
BFAM	0.8	No	0	100	0.3
HLBFAM	0.8	No	20	80	0.3
CAM-A	0.8	Yes	0	100	0
HLAM-A	0.8	Yes	20	80	0
BFAM-A	0.8	Yes	0	100	0.3
HLBFAM-A	0.8	Yes	20	80	0.3

In order to more accurately simulate the aging of the asphalt mixture during the transportation, paving, and use of the actual pavement, the tests in this thesis refer to the Chinese "Test Procedure for Asphalt and Asphalt Mixture for Highway Engineering" (JTG E20-2011) [30] and the American Association of Highway Workers AASHTO-PP2 specification; the specimens to be aged were subjected to short-term aging during the mixing process of the asphalt mixture and long-term aging after using a modified mixture forming the specimen. The steps to prepare the aged specimens were as follows: the mixed asphalt mixture was placed in an oven and stirred every hour under a temperature of 135 °C. After 4 h of forced-air heating, the short-term aging was complete. Then, the aged loose materials were formed into specimens. The specimens were placed in an oven with forced-air heating at 85 °C for 120 h to complete the long-term aging. The name of the aged asphalt mixture is the name of the first four asphalt mixtures followed by the code "-A" indicating that it is the aged version. For example, CAM-A represents the aged conventional asphalt mixture. Vehicle tire rutting test specimens were prepared according to the above process, cut into small beam specimens, and used for testing, as shown in Figure 2.

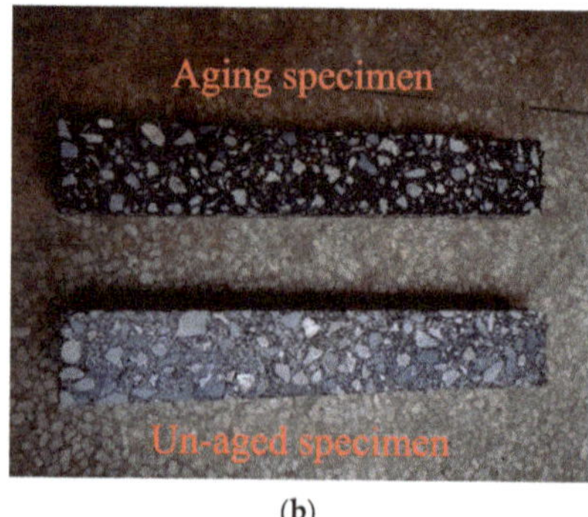

(a)   (b)

Figure 2. Fabrication and aging of small beam specimens: (**a**) mixture test plates; (**b**) comparison of aged and unaged specimens.

2.4. Methodology

In this study, the assessment of the fatigue performance of lime and basalt fiber modified asphalt mixes was carried out by applying various theories such as the phenomenological method and the energy method based on a four-point bending fatigue test and damage self-healing test. Figure 3 shows the test and analysis process.

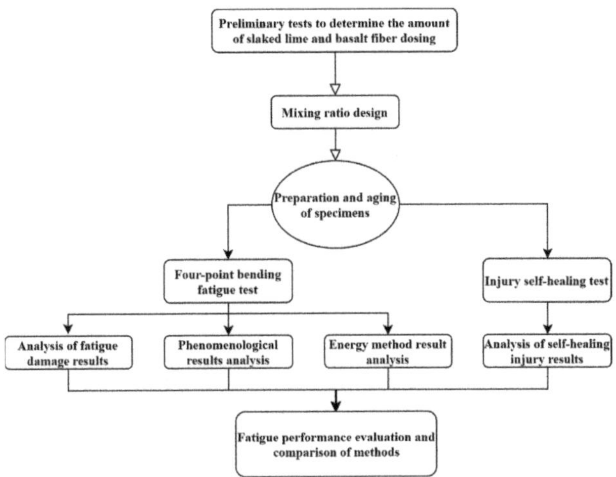

Figure 3. Test and analysis process.

2.4.1. Four-Point Bending Fatigue Test

In this experiment, the fatigue performance of various asphalt mixtures prepared in the previous step was tested through four-point bending tests. In order to improve the accuracy of the experiment, a temperature close to the equivalent temperature in Chongqing was chosen, and the test temperature was 17 °C. The constant strain control mode was used with a frequency of 10 Hz and continuous sine wave loading [31,32]. The strain levels ($\mu\varepsilon$) of 450, 550, and 650 were determined by pre-testing the specimens. According to relevant technical specifications and research experience, the test ended when the modulus of rigidity decreased to 50% of the initial value. At the 50th loading cycle, the current modulus of rigidity was taken as the initial modulus of rigidity [24,33].

2.4.2. Self-Healing Test for Injury

As the self-healing of asphalt pavements mainly takes place in the unloaded condition, a healing time was considered to be added to the two fatigue loadings, namely "fatigue–healing–fatigue" loading [34]. The asphalt mixture specimens underwent the first fatigue test to reach the corresponding damage state, and then they were placed in a specified environment for self-repair. After a certain time, a second fatigue test was performed to evaluate the self-healing performance under different conditions and use it as an indicator to assess fatigue performance.

The experiment used the four-point bending fatigue testing platform mentioned in the previous section, and the specimens were the same as before. Considering two influencing factors, temperature and initial damage level, the experiment was divided into three groups A, B, and C and conducted according to the following steps: (1) Set the test temperature at 17 °C, and all three groups use controlled strain level loading with a strain level of 650 $\mu\varepsilon$, a frequency of 10 Hz, and continuous sine wave loading. Groups A and B terminate the test when the modulus of rigidity decreases to 50% of the initial value, while Group C sets the termination criterion at 70%. (2) After completing the initial fatigue test, all three groups place their specimens in a constant temperature box for 5 h under different temperature conditions. Groups A and C are maintained at 17 °C, while Group B is maintained at 60 °C.

After 5 h of static storage, all specimens are placed in a constant temperature box at 17 °C for 3 h until they reach the test temperature for the next step. (3) The specimens after static storage undergo another round of fatigue testing, with the relevant parameters set as in step (1). The self-healing conditions in the experiment are shown in Table 6.

Table 6. Self-healing condition.

Group	Injury Degree (%)	Healing Temperature, Healing Time (°C, h)
A	50	17 °C, 6 h
B	50	(60 °C, 3 h) + (17 °C, 3 h)
C	70	17 °C, 6 h

3. Results and Discussion

3.1. Analysis Based on Fatigue Damage Results

The initial modulus of rigidity is an indicator of the original bending deformation resistance of an asphalt mixture. The initial bending modulus of rigidity under three different strain levels was tested, and the results are shown in Figure 4. It can be seen that a large amount of free state asphalt becomes stable after the addition of basalt fibers, reducing the stiffness of the asphalt mixture and lowering its modulus of rigidity. Meanwhile, the rheological properties of the asphalt change with the addition of lime due to neutralizing some polar molecules, which weakens the asphalt rheological properties and increases the modulus of rigidity. After thermal aging, the polarity of asphalt internal molecules increases, and a large number of functional groups are generated inside. At the same time, the volatilization of lightweight components in the asphalt reduces their rheological properties, so the modulus of rigidity of all four mixtures increases. When hydrated lime is added, it can inhibit the high-temperature oxidation of the asphalt mixture, thus reducing the negative effects of thermal aging and lowering the modulus of rigidity. When both basalt fiber and hydrated lime are added, the modulus of rigidity increases slightly, and the amount of free-state asphalt increases as well. If fatigue performance is evaluated based solely on the modulus of rigidity, the improvement in anti-fatigue performance is not significant compared to a conventional asphalt mixture. However, it is evident from the subsequent analysis that the simultaneous incorporation of the two materials improved the fatigue resistance of the aged asphalt mixes. Therefore, it can be concluded that there is no significant correlation between the modulus of rigidity and the fatigue performance of an asphalt mixture, indicating that it is not suitable as a direct indicator for evaluating fatigue performance.

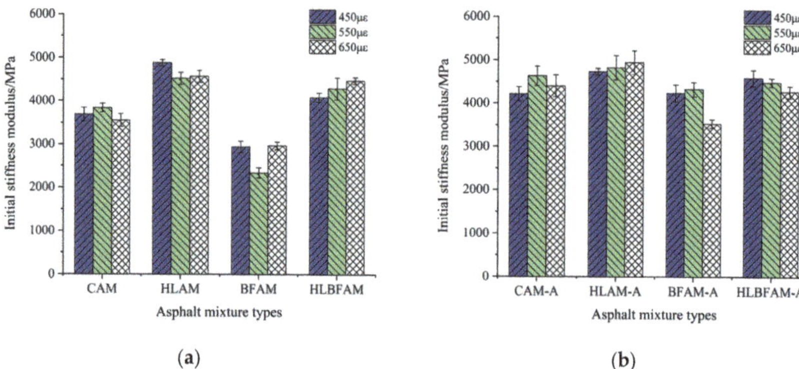

Figure 4. Initial stiffness modulus before and after aging of different types of asphalt mixtures: (**a**) unaged asphalt mixture; (**b**) aged asphalt mixture.

Existing research has indicated that the initial modulus of rigidity of an asphalt mixture merely represents its original bending deformation resistance, which is primarily determined by its own properties [35]. Considering that the process of fatigue failure is a long-term and complex one, it is observed that the fatigue performance and initial modulus of rigidity are not strongly correlated. Therefore, it cannot accurately reflect the fatigue damage process of asphalt mixtures. In this study, the fatigue damage rate D is introduced as an evaluation indicator of fatigue performance, which is expressed in Formula (1):

$$D = \frac{S_0 - S_i}{S_0} \tag{1}$$

where S_0 is the initial stiffness modulus; S_i is the stiffness modulus at the i-th loading.

The fatigue damage rate of each group of asphalt mixtures under different strains with the number of loading was plotted as shown in Figure 5.

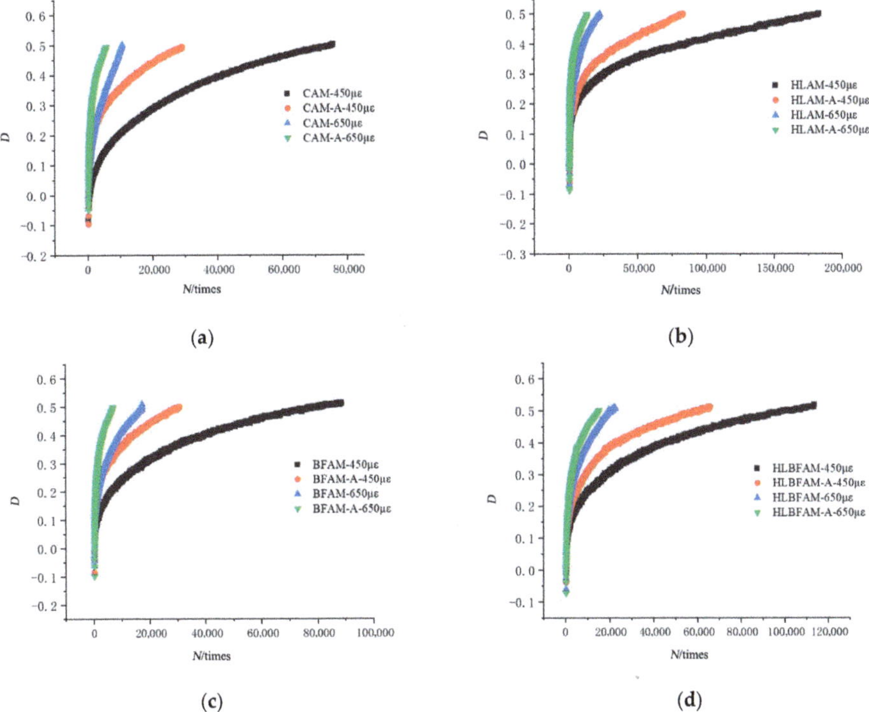

Figure 5. Variation trend of fatigue damage rate of various asphalt mixtures under different strain conditions: (**a**) ordinary mineral powder asphalt mixture; (**b**) hydrated lime asphalt mixture; (**c**) basalt fiber asphalt mixture; (**d**) basalt fiber + hydrated lime asphalt mixture.

As can be seen from Figure 5, in the early stage of loading, the fatigue damage rate of the asphalt mixture increases faster after aging. This can be explained by the hardening phenomenon that occurs within the aged asphalt mixture, which generates and develops microcracks more rapidly. However, the addition of hydrated lime effectively improves this condition. Comparing the fatigue damage curves of CAM-A and HLAM-A at the same strain level, it is found that the latter has a more gradual change trend and a longer loading cycle. This is because the addition of hydrated lime increases the viscosity of aged asphalt, enhances its adhesion to aggregates, and improves its fatigue life. In addition, the incorporation of basalt fiber into the asphalt binder can have a similar effect in asphalt mastic as reinforcement in reinforced concrete and can significantly delay the occurrence of internal fatigue cracking, as shown in Figure 5c. After thermal aging, the fatigue life of HLBFAM-A was improved by about 53% compared to CAM-A, indicating that the combined effects of both materials can delay the fatigue damage process of aged asphalt mixtures, thereby achieving a better fatigue performance, which is consistent with the conclusions made in [36].

For the purpose of analysis, the fatigue damage rate versus the number of loadings was separately plotted for 650 µε as shown in Figure 6.

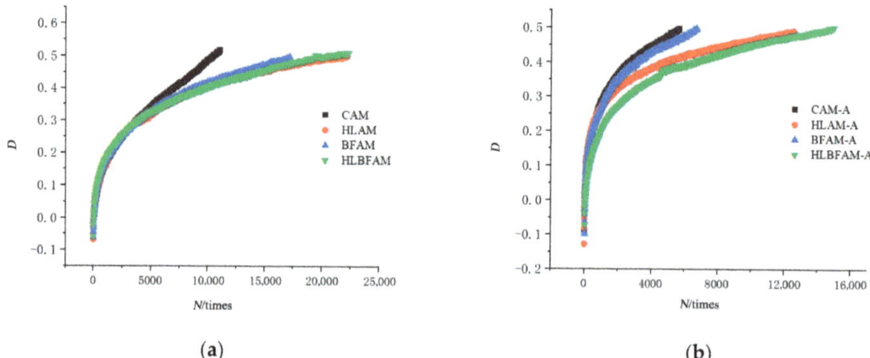

(a)   (b)

Figure 6. Variation trend of fatigue damage rate of various asphalt mixtures under 650 µε strain conditions: (**a**) unaged asphalt mixture; (**b**) aged asphalt mixture.

Analyzing the trend of fatigue damage rate before aging, it was found that at the same number of load cycles, the addition of basalt fibers and hydrated lime both reduced the fatigue damage rate of asphalt mixture with similar effects. Observing the change in curvature of the four curves in Figure 5b, it was found that the combined addition of hydrated lime and basalt fibers had the highest curvature, followed by the single addition of hydrated lime, the single addition of basalt fibers, and the control group. This indicates that the best anti-fatigue performance enhancement after aging was achieved with the combination of additives, and the addition of hydrated lime was more effective than basalt fibers as a single additive.

3.2. Analysis Based on Phenomenological Results

3.2.1. Fatigue Life Analysis

This study conducted fatigue life tests on asphalt mixtures, and the results are shown in Figure 7. The results indicate that the addition of hydrated lime produced the most significant improvement in fatigue life, which improved by about two times compared to the common asphalt mixture under 550 µε strain. The main reason for this is that hydrated lime has a large specific surface area and can adsorb the acid in the asphalt to improve its adhesion, thereby improving its fatigue characteristics. After thermal aging, the fatigue life of the asphalt mixture was reduced, and the reduction was minimal when hydrated lime was added alone, while the improvement was not significant when basalt fiber was added alone. This is because the adhesion of asphalt decreases after thermal aging, causing basalt fibers to become unable to stably embed into the asphalt, resulting in a loss of interlocking effect within the mixture. Under high strain (650 µε), the mixed addition of hydrated lime and basalt fibers showed the longest fatigue life among all asphalt mixtures, thanks to the fact that hydrated lime increased asphalt adhesion while basalt fibers formed a spatial mesh structure inside the mix [37], which significantly reduced particle slip in the mastic during the crack propagation stage, and also reduced the sensitivity of the strain level in fatigue cracking.

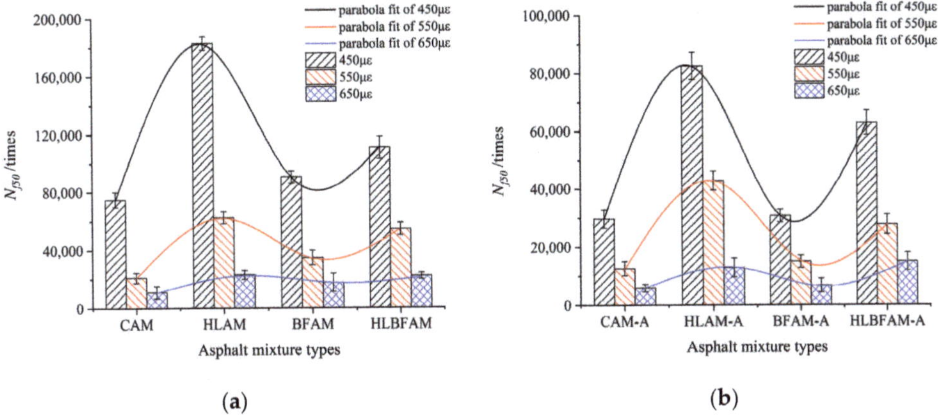

(a)

(b)

Figure 7. Fatigue life N_{f50} before and after aging of different types of asphalt mixtures: (**a**) unaged asphalt mixture; (**b**) aged asphalt mixture.

3.2.2. Analysis of Fatigue Equations

By using the results of the above experiments, the different asphalt mixes were fitted with the corresponding strain values to derive semi-logarithmic fatigue equations with a good correlation, and the fitting equations are shown in Equations (2) and (3). Figure 8 represents the model for the fatigue life prediction of asphalt mixes under different strains.

$$N_f = c\left(\frac{1}{\varepsilon}\right)^m \tag{2}$$

$$\ln(N_f) = a\varepsilon + b \tag{3}$$

where N_f is the fatigue life; ε is the strain amplitude; a, b, c, and m are constants.

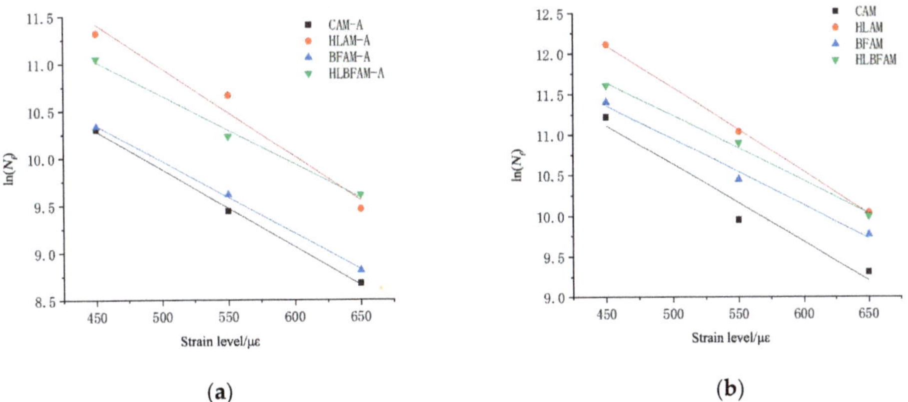

(a)

(b)

Figure 8. Fatigue life prediction model of asphalt mixtures under different strains: (**a**) unaged asphalt mixture; (**b**) aged asphalt mixture.

The fitted semi-logarithmic fatigue equations for different kinds of asphalt mixes are shown in Table 7.

Table 7. Different types of half logarithm fatigue equation of asphalt mixtures.

Type of Asphalt Mixture	Fatigue Equation	R^2
CAM	$\ln\left(N_f\right) = -0.00956\varepsilon + 15.418$	0.92991
HLAM	$\ln\left(N_f\right) = -0.01039\varepsilon + 16.779$	0.99926
BFAM	$\ln\left(N_f\right) = -0.00819\varepsilon + 15.049$	0.92748
HLBFAM	$\ln\left(N_f\right) = -0.00807\varepsilon + 15.278$	0.97301
CAM-A	$\ln\left(N_f\right) = -0.00813\varepsilon + 13.942$	0.99967
HLAM-A	$\ln\left(N_f\right) = -0.00928\varepsilon + 15.590$	0.94338
BFAM-A	$\ln\left(N_f\right) = -0.00759\varepsilon + 13.764$	0.99824
HLBFAM-A	$\ln\left(N_f\right) = -0.00738\varepsilon + 14.245$	0.98707

As observed in Table 7, the vast majority of asphalt mixtures have correlation coefficients close to one in each fitted result, indicating a strong correlation between fatigue life and the corresponding strain level of cyclic loading. However, different strain levels and loading frequencies produce differences in the internal stress change rate of asphalt mixtures, while the equations ignore the influence of such differences on the analysis results, which will produce variability when used as a fatigue performance assessment index.

3.3. Analysis of Results Based on Energy Method

3.3.1. Analysis of Dissipated Energy in a Single Loading Cycle

Previous studies have shown that asphalt mixtures generate energy (dissipated energy) to resist fatigue damage during the fatigue process, which is commonly used to reflect the development of fatigue damage. In this experiment, the loading method was a sine wave, and the waveform applied to the specimen was defined as $a\sin(\omega t)$. As asphalt mixtures are viscoelastic materials, there is a phase difference φ between the generation of strain and the action of stress. Therefore, each stress action on the specimen corresponds to a strain with phase lag $b\sin(\omega t + \varphi)$. Equations (4) and (5) are defined here:

$$x = a\sin(\omega t) \tag{4}$$

$$y = b\sin(\omega t + \varphi) \tag{5}$$

with Equation (6):

$$\cos(\omega t) = \frac{1}{\sin(\varphi)}\left(\frac{y}{b} - \frac{x}{a}\cos(\varphi)\right) \tag{6}$$

Combining (4), (5), and (6) yields the hysteresis equation:

$$\left(\frac{x}{a}\right)^2 - \left(\frac{y}{b}\right)^2 - \frac{2\cos(\varphi)}{ab}xy = \sin^2(\varphi) \tag{7}$$

Figure 9 shows the hysteresis curve in one loading cycle; it can be seen that the stress and strain form a closed loop curve, and the area inside the loop curve is the dissipation energy of this loading cycle, which is calculated by Equation (8). For fatigue tests conducted using the strain control mode, the x-axis represents a constant strain, while the y-axis represents stress, which decreases continuously as the test progresses and the specimen's fatigue damage accumulates. The circular curve will gradually shrink inward along the y-axis, showing that the dissipated energy of a single loading cycle is consequently reduced.

$$E_D = \int y\,dx = \int_0^{\frac{2\pi}{\omega}} ab\omega\sin(\omega t + \varphi)\cos(\omega t)\,dt = \pi ab\sin(\varphi) \tag{8}$$

Therefore, the dissipated energy is analyzed here with a single cycle of loading, and its calculation formula is shown in Equation (9):

$$E_{Di} = \pi \sigma_{ti} \varepsilon_{ti} \sin \varphi_i \qquad (9)$$

where E_{Di}, σ_{ti}, ε_t, and φ_i are the dissipation energy, maximum tensile stress, maximum tensile strain, and phase angle of cyclic loading up to i times, respectively.

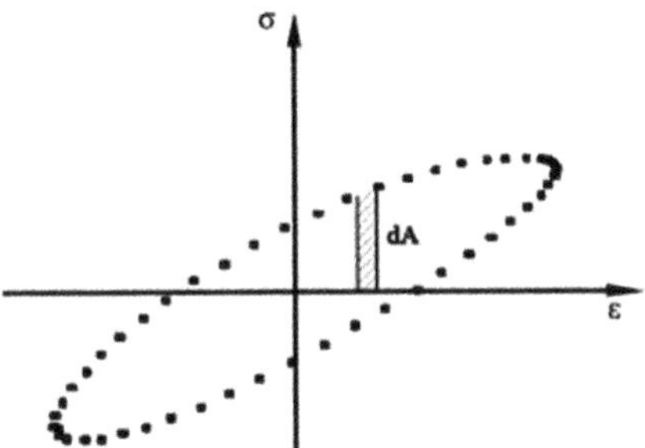

Figure 9. Hysteresis curve.

The relationship between the dissipated energy and loading cycles for various asphalt mixtures under different strain levels is shown in Figure 10. The results indicate that the asphalt mixture fatigue loading process can be classified into two phases: in the early stages of loading, dissipated energy rapidly decreases, and microcracks gradually form within the mixture; as the number of loading cycles increases, the sensitivity of the asphalt mixture to loading decreases, and the trend of dissipated energy changes slows down towards stability. Compared with common asphalt mixtures, the addition of basalt fiber can advance the change in dissipated energy towards a stable state, indicating its internal structure is relatively stable without obvious fatigue cracking. However, after thermal aging treatment, the viscoelastic properties of the asphalt deteriorate, and the internal anchoring structure is easily damaged (causing a decrease in the adhesive strength between asphalt and basalt fiber), such that the improvement effect of the single addition of basalt fiber is not significant. Observing Figure 10b,d, the aging asphalt mixture was found to have a significantly decreased ability to resist deformation, with the initial dissipation energy increasing, and the dissipation energy decreasing rapidly along with the number of loadings, while this effect was significantly weakened by adding hydrated lime. Since the addition of hydrated lime improved the asphalt viscosity, mixing with basalt fibers—which play a significant role in the internal stabilization of the mixture—greatly improved the fatigue resistance under the effect of thermal aging; the results also indicated that the longer the change in the stability period of the dissipation energy of the asphalt mixture, the better the fatigue performance.

3.3.2. Analysis of the Stable Value of the Rate of Change in Dissipated Energy

Previous studies have shown that some researchers calculate the rate of dissipated energy change for individual intervals (RDEC) evenly and use it as an evaluation index of the fatigue performance of asphalt mixtures [38]. When using RDEC to characterize the fatigue damage process of asphalt mixtures, which can be separated into a smooth phase and an accelerated damage phase, the stable value of the dissipated energy change rate (PV) represents the value of the dissipated energy change rate in the steady stage. The larger the PV value, the faster the microcracks expand inside the mixture, and the greater the fatigue damage. This paper introduces the calculation method of PV value proposed by Carpenter et al. [39]. The dissipated energy-loading cycle equation was fitted with a power function and the K value was calculated; the results are shown in Table 8. Then, through

Equation (10), the *PV* values of different asphalt mixtures under 550 με and 650 με strain conditions were calculated, and the results are shown in Figure 10.

$$PV = \frac{1 - \left(1 + \frac{100}{N_{f50}}\right)^{K}}{100} \tag{10}$$

where N_{f50} is the fatigue life; K is a constant.

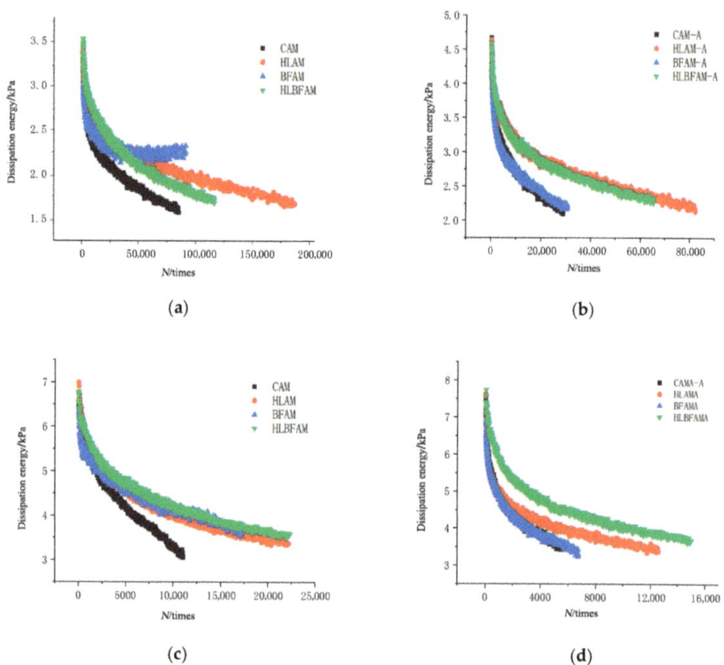

Figure 10. The dissipated energy varies with the number of cyclic loading: (**a**) strain 450 με, unaged; (**b**) strain 450 με, after aging; (**c**) strain 650 με, unaged; (**d**) strain 650 με, after aging.

Table 8. Dissipated energy fitting equation.

Strain Level (με)	Type of Asphalt Mixture	Fatigue Equation	K
	CAM	$y = 5.81x^{-0.103}$	−0.103
	HLAM	$y = 5.29x^{-0.082}$	−0.082
	BFAM	$y = 4.06x^{-0.055}$	−0.055
450 με	HLBFAM	$y = 6.93x^{-0.108}$	−0.108
	CAM-A	$y = 7.55x^{-0.110}$	−0.110
	HLAM-A	$y = 7.64x^{-0.098}$	−0.098
	BFAM-A	$y = 7.50x^{-0.111}$	−0.111
	HLBFAM-A	$y = 8.43x^{-0.109}$	−0.109
	CAM	$y = 13.09x^{-0.137}$	−0.137
	HLAM	$y = 12.93x^{-0.125}$	−0.125
	BFAM	$y = 9.47x^{-0.086}$	−0.086
650 με	HLBFAM	$y = 12.21x^{-0.115}$	−0.115
	CAM-A	$y = 11.87x^{-0.133}$	−0.133
	HLAM-A	$y = 9.72x^{-0.101}$	−0.101
	BFAM-A	$y = 10.76x^{-0.120}$	−0.120
	HLBFAM-A	$y = 13.82x^{-0.130}$	−0.130

From Figure 11, it can be seen that under the unaged condition, the PV values of asphalt mixtures containing basalt fiber and hydrated lime are lower because their energy consumption during the loading stage is relatively small, which prolongs the time of crack formation and development to a certain extent, thereby enhancing the fatigue performance. After thermal aging, the PV value of the mixture with added hydrated lime is only 1/3 of that of ordinary asphalt mixtures, which illustrates the promotion effect of hydrated lime on the fatigue resistance improvement of aging asphalt mixtures. Observing Figure 10a, we can see that under 650 με strain conditions, the asphalt mixture with only basalt fiber has the lowest PV value, which confirms the conclusion that the addition of BF in phenomenological methods will reduce the high strain sensitivity of asphalt mixtures.

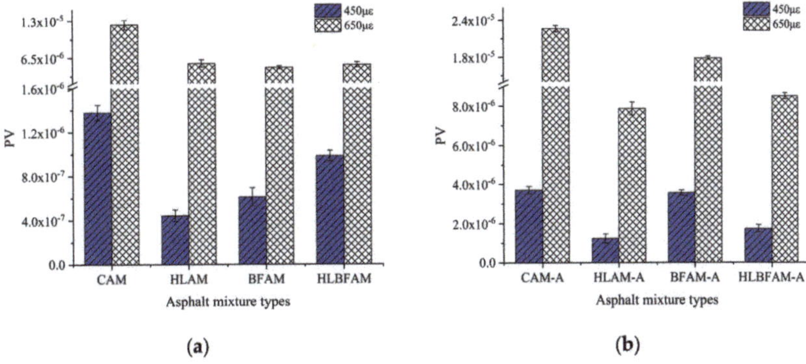

Figure 11. PV value of aged and unaged asphalt mixture: (**a**) unaged asphalt mixture; (**b**) aged asphalt mixture.

3.4. Analysis Based on Injury Self-Healing Results

After the self-healing test was completed, the fatigue life (N_{f1}), initial stiffness modulus (S_1), and stiffness modulus (S_{e1}) at the end of the test were recorded for the first cyclic loading, and the fatigue life (N_{f2}), initial stiffness modulus (S_2), and stiffness modulus (S_{e2}) at the end of the second cyclic loading, with the results presented in Table 9.

Table 9. Self-healing test results.

Specimen Types	Healing Conditions	Pre-Healing			Post-Healing		
		N_{f1}	S_1 (MPa)	S_{e1} (MPa)	N_{f2}	S_2 (MPa)	S_{e2} (MPa)
CAM	A	12,380	3946	1968	5071	3117	1543
	B	11,080	4129	2061	7719	3276	1599
	C	29,013	3516	1055	1581	2610	1303
HLAM	A	21,609	4993	2478	8634	3793	1862
	B	22,598	5219	2578	15,913	3913	1953
	C	54,908	5391	1617	2987	3792	1869
BFAM	A	18,213	3001	1497	6002	2381	1183
	B	15,435	3255	1601	9602	2481	1233
	C	42,193	3419	1026	4110	2019	1009
HLBFAM	A	20,011	4074	2009	7821	3276	1621
	B	19,374	4162	2079	14,821	3398	1701
	C	51,392	4510	1353	5992	2983	1490
CAM-A	A	6570	4501	2229	3080	4140	2053
	B	8314	4591	2302	4592	3917	1950
	C	18,920	4310	1293	774	2913	1456
HLAM-A	A	12,003	5173	2585	4910	4759	2374
	B	14,103	5617	2801	8913	4632	2302
	C	29,013	5201	1560	1102	3191	1590

Table 9. *Cont.*

Specimen Types	Healing Conditions	Pre-Healing			Post-Healing		
		N_{f1}	S_1 (MPa)	S_{e1} (MPa)	N_{f2}	S_2 (MPa)	S_{e2} (MPa)
BFAM-A	A	6819	3591	1795	3010	3329	1658
	B	7451	3699	1837	3910	3329	1649
	C	20,035	4013	1204	1620	2617	1353
HLBFAM-A	A	13,045	4310	2149	4310	4017	2004
	B	15,045	4263	2103	9310	3981	1974
	C	25,983	4983	1495	2410	3012	1503

Since damage self-healing is a long and complex process, its effects cannot be described by a single index [40]. Therefore, the self-healing rate (*HI*) expression introduced in this study was calculated and analyzed, as shown in Equation (11):

$$HI = \frac{D_1}{D_2} \tag{11}$$

where D_1 is the rate of decrease in the stiffness modulus at the first cyclic loading; D_2 is the rate of decrease in the stiffness modulus at the second cyclic loading.

The calculations of D_1 and D_2 in this test are shown in Equations (12) and (13):

$$D_1 = \frac{S_1 - S_{e1}}{N_{f1}} \tag{12}$$

$$D_2 = \frac{S_2 - S_{e2}}{N_{f2}} \tag{13}$$

Figure 12 shows the self-healing rates (*HI*) of four different types of asphalt mixtures under different temperatures and damage levels. Comparing the three healing conditions, it can be found that an excessive accumulation of fatigue damage will greatly weaken the self-healing ability of asphalt mixtures. After 70% damage accumulation, the self-healing performance decreased by 75% to 90%. In addition, under the same damage conditions, the higher the temperature, the higher the self-healing rate of the asphalt mixture.

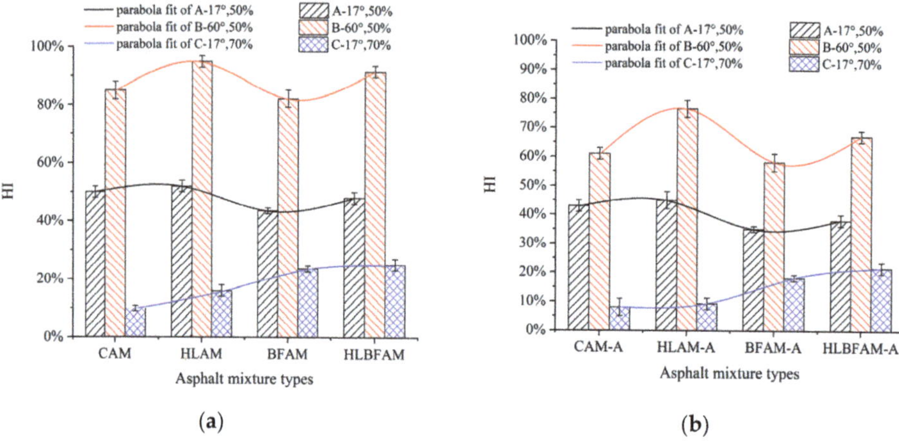

Figure 12. Self-healing rate of aged and unaged asphalt mixture: (**a**) unaged asphalt mixture; (**b**) aged asphalt mixture.

From Figure 12a, it can be seen that adding hydrated lime to the asphalt mixture has the best self-healing performance at 60 °C, with a self-healing rate of up to 96%, while there is no significant improvement at 17 °C. When the damage condition is 50%, the self-healing effect of the asphalt mixture containing only basalt fiber added is decreased. This is because the internal microcracks of the asphalt mixture are still in the stage of expansion, and the wetting and diffusion conditions between the asphalt interfaces are good at this time. The "bridging" effect of the fiber on the asphalt molecules does not have a significant promotion effect, and the uneven distribution of some fibers in the slurry will hinder their free movement and aggregation, thereby reducing their bonding performance with the aggregate. However, when the damage condition is 70%, the microcracks inside the mixture begin to expand, and the "bridging" effect of the fiber on the asphalt molecules begins to appear, producing better diffusion conditions and a better self-healing performance, with a doubled self-healing rate. Compared with the common mineral powder asphalt mixture, the asphalt mixture which incorporated both basalt fiber and hydrated lime had a better self-healing ability, and its self-healing rate increased by 6% to 15% under the larger initial damage and high temperature; in addition, the high temperature aging had no significant effect on its performance, and its self-healing property decreased the least compared with the unaged specimens in the above two cases, which had values of about 16% to 26%.

With the aging of the asphalt mixture, its self-healing performance has a certain degree of decline, justifying the introduction of a self-healing index decay rate (*HIDR*); the size of the *HIDR* can characterize the degree of self-healing performance decay after aging and reflect the effect of aging on the fatigue damage process. Thus, the fatigue performance analysis may use the *HIDR*, the definition of which is shown in Equation (14). The *HIDR* values of each asphalt mixture under the three healing conditions were calculated, and the results are shown in Figure 13.

$$HIDR = \frac{HI - HIA}{HI} \tag{14}$$

where *HIDR* is the self-healing index decay rate; *HI* is the self-healing rate; *HIA* is the self-healing rate after aging.

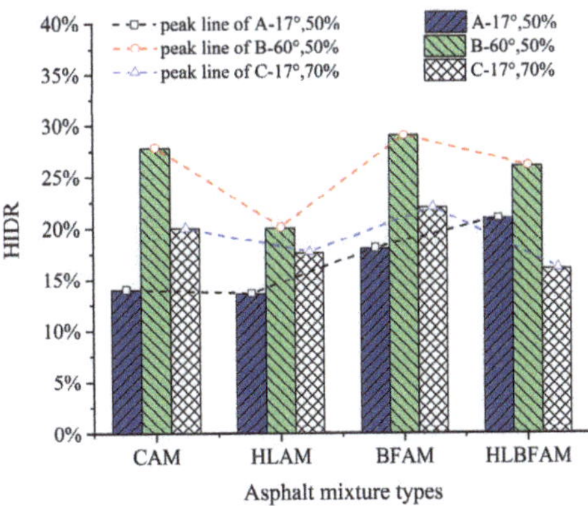

Figure 13. HI decay rate of different asphalt mixtures.

The results indicate that aging has a significant impact on the self-healing ability of asphalt mixtures under high temperature and high damage conditions. The main reason is that the light components in the asphalt decrease after aging, resulting in reduced fluidity, which severely affects the self-repair ability at high temperatures. Additionally, excessive accumulation of fatigue damage can cause the internal cracks of brittle aged asphalt materials to expand, making it difficult for asphalt molecules to flow back and diffuse for repair. Meanwhile, *HIDR* also confirms that adding basalt fiber and hydrated lime to asphalt mixtures during the thermal aging process can improve the fatigue self-healing performance of asphalt mixtures, which is consistent with the conclusion obtained previously.

4. Conclusions

This paper conducted four-point bending fatigue tests and damage self-healing tests to measure and record various parameters of four types of asphalt mixtures during the fatigue process before and after thermal aging. Based on changes in the modulus of rigidity, phenomenological methods, energy methods, and other approaches, the fatigue behaviors of the four asphalt mixtures before and after aging were analyzed, and the following conclusions were drawn:

1. For the unaged asphalt mixture, the adhesion strength between the asphalt binder is obviously improved by the addition of hydrated lime, and its fatigue characteristics are significantly improved under high temperature, and the fatigue life is increased significantly; the addition of basalt fiber plays an anchoring role in the mixture, which can enhance its sensitivity to changes in load and strain levels and make the internal structure more stable; the fatigue performance is improved well when the two are mixed.

2. For the asphalt mixture after thermal aging, under low strain conditions, single lime has a better fatigue resistance and its fatigue life is about three times that of the ordinary mix, but no significant effect was observed when blended with basalt fiber alone; here, we believe that the decline in fatigue performance was a result of asphalt aging due to reduced mobility, being more easily wrapped in the fiber surface, asphalt wetting, and a significantly reduced diffusion capacity. When mixed with hydrated lime, basalt fiber produced a good fatigue resistance under high strain conditions, its fatigue life and self-healing performance were about two times the ordinary asphalt mixture, and the overall enhancement produced the best effect.

3. The initial stiffness modulus reflects the original flexural capacity of the asphalt mixture which only represents the initial condition of the asphalt mixture; however, fatigue damage is a long-term procedure, making the initial stiffness modulus inappropriate for evaluating the fatigue properties of asphalt mixtures subjected to long-term loading. When analyzing the asphalt mixture based on the fatigue equation, the difference in the rate of change in the internal stress of the asphalt mixture under different strain levels and loading frequencies is not taken into account, which can lead to deviations in the analysis results.

4. The results of the fatigue damage rate and the steady value of the dissipation energy change rate analysis show the similarity, and both of these can reflect the fatigue resistance characteristics well. Under complex damage conditions, the fatigue performance of the asphalt mixture is evaluated by the self-healing rate and the decay rate of the self-healing index; both indicators can fully characterize the damage healing process under repeated loading, and have a good correlation with the fatigue characteristics, which can be used as fatigue performance evaluation indicators.

Author Contributions: Conceptualization, H.D. and Z.Z.; methodology, Z.Z. and T.L.; validation, H.D., T.L. and Z.Z.; formal analysis, H.D.; investigation, B.P. and Q.H.; data curation, H.D. and Z.Z.; writing—original draft preparation, H.D. and Z.Z.; writing—review and editing, H.D. and T.L.; visualization, B.P. and Q.H.; funding acquisition, T.L. All authors have read and agreed to the published version of the manuscript.

Funding: This research was funded by the Hebei Zunqin Expressway Science and Technology Demonstration Project, grant number (RP2021003700).

Institutional Review Board Statement: Not applicable.

Informed Consent Statement: Not applicable.

Data Availability Statement: Data sharing is not applicable to this article.

Conflicts of Interest: The authors declare no conflict of interest.

References

1. Pouranian, M.; Shishehbor, M. Sustainability Assessment of Green Asphalt Mixtures: A Review. *Environments* **2019**, *6*, 73. [CrossRef]
2. Mantalovas, K.; Di Mino, G. Integrating Circularity in the Sustainability Assessment of Asphalt Mixtures. *Sustainability* **2020**, *12*, 594. [CrossRef]
3. Wang, F.; Hoff, I.; Yang, F.; Wu, S.; Xie, J.; Li, N.; Zhang, L. Comparative assessments for environmental impacts from three advanced asphalt pavement construction cases. *J. Clean. Prod.* **2021**, *297*, 126659. [CrossRef]
4. Saleh, N.F.; Keshavarzi, B.; Yousefi Rad, F.; Mocelin, D.; Elwardany, M.; Castorena, C.; Underwood, B.S.; Kim, Y.R. Effects of aging on asphalt mixture and pavement performance. *Constr. Build. Mater.* **2020**, *258*, 120309. [CrossRef]
5. Zhang, H.; Soenen, H.; Carbonneau, X.; Lu, X.; Robertus, C.; Zhang, Y. Experimental and Statistical Analysis of Bitumen's Field Ageing in Asphalt Pavements. *Transp. Res. Rec.* **2022**, *2676*, 495–511. [CrossRef]
6. Alae, M.; Zhao, Y.; Zarei, S.; Fu, G.; Cao, D. Effects of layer interface conditions on top-down fatigue cracking of asphalt pavements. *Int. J. Pavement Eng.* **2020**, *21*, 280–288. [CrossRef]
7. Garbowski, T.; Pożarycki, A. Multi-level backcalculation algorithm for robust determination of pavement layers parameters. *Inverse Probl. Sci. Eng.* **2017**, *25*, 674–693. [CrossRef]
8. Garbowski, T.; Gajewski, T. Semi-automatic Inspection Tool of Pavement Condition from Three-dimensional Profile Scans. *Procedia Eng.* **2017**, *172*, 310–318. [CrossRef]
9. Wang, H.; Wang, J.; Chen, J. Micromechanical analysis of asphalt mixture fracture with adhesive and cohesive failure. *Eng. Fract. Mech.* **2014**, *132*, 104–119. [CrossRef]
10. Shen, S.; Chiu, H.-M.; Huang, H. Characterization of Fatigue and Healing in Asphalt Binders. *J. Mater. Civ. Eng.* **2010**, *22*, 846–852. [CrossRef]
11. Meng, F.; Gao, D.; Chen, F.; Huang, C. Fatigue Performance Test and Life Calculation of Fiber-Reinforced Asphalt Concrete. *ACSM* **2020**, *44*, 133–139. [CrossRef]
12. Brovelli, C.; Crispino, M.; Pais, J.C.; Pereira, P.A.A. Assessment of Fatigue Resistance of Additivated Asphalt Concrete Incorporating Fibers and Polymers. *J. Mater. Civ. Eng.* **2014**, *26*, 554–558. [CrossRef]
13. Lee, S.J.; Rust, J.P.; Hamouda, H.; Kim, Y.R.; Borden, R.H. Fatigue Cracking Resistance of Fiber-Reinforced Asphalt Concrete. *Text. Res. J.* **2005**, *75*, 123–128. [CrossRef]
14. Sim, J.; Park, C.; Moon, D.Y. Characteristics of basalt fiber as a strengthening material for concrete structures. *Compos. Part B* **2005**, *36*, 504–512. [CrossRef]
15. Wu, B.; Meng, W.; Xia, J.; Xiao, P. Influence of Basalt Fibers on the Crack Resistance of Asphalt Mixtures and Mechanism Analysis. *Materials* **2022**, *15*, 744. [CrossRef]
16. Morova, N. Investigation of usability of basalt fibers in hot mix asphalt concrete. *Constr. Build. Mater.* **2013**, *47*, 175–180. [CrossRef]
17. Wei, Q.; Ashaibi, A.A.; Wang, Y.; Albayati, A.; Haynes, J. Experimental study of temperature effect on the mechanical tensile fatigue of hydrated lime modified asphalt concrete and case application for the analysis of climatic effect on constructed pavement. *Case Stud. Constr. Mater.* **2022**, *17*, e01622. [CrossRef]
18. Mondal, A.; Islam, S.S.; Ransinchung, R.N.G.D. Synergistic Effect of Hydrated Lime and Warm Mix Asphalt Additive on Properties of Recycled Asphalt Mixture Subjected to Laboratory Ageing. *Int. J. Pavement Res. Technol.* **2022**, 1–15. [CrossRef]
19. Little, D.N.; Petersen, J.C. Unique Effects of Hydrated Lime Filler on the Performance-Related Properties of Asphalt Cements: Physical and Chemical Interactions Revisited. *J. Mater. Civ. Eng.* **2005**, *17*, 207–218. [CrossRef]
20. Das, A.K.; Singh, D. Evaluation of fatigue performance of asphalt mastics composed of nano hydrated lime filler. *Constr. Build. Mater.* **2021**, *269*, 121322. [CrossRef]
21. Das, A.K.; Singh, D. Investigation of rutting, fracture and thermal cracking behavior of asphalt mastic containing basalt and hydrated lime fillers. *Constr. Build. Mater.* **2017**, *141*, 442–452. [CrossRef]
22. Mwanza, A.D.; Hao, P.; Wang, H. Effects of Type and Content of Mineral Fillers on the Consistency Properties of Asphalt Mastic. *J. Test. Eval.* **2012**, *40*, 20120140. [CrossRef]
23. Mohammad, L.N.; Abadie, C.; Gokmen, R.; Puppala, A.J. Mechanistic Evaluation of Hydrated Lime in Hot-Mix Asphalt Mixtures. *Transp. Res. Rec.* **2000**, *1723*, 26–36. [CrossRef]
24. Davar, A.; Tanzadeh, J.; Fadaee, O. Experimental evaluation of the basalt fibers and diatomite powder compound on enhanced fatigue life and tensile strength of hot mix asphalt at low temperatures. *Constr. Build. Mater.* **2017**, *153*, 238–246. [CrossRef]
25. Kathari, P.M.; Sandra, A.K.; Sravana, P. Experimental investigation on the performance of asphalt binders reinforced with basalt fibers. *Innov. Infrastruct. Solut.* **2018**, *3*, 76. [CrossRef]
26. Izadi, A.; Motamedi, M.; Alimi, R.; Nafar, M. Effect of aging conditions on the fatigue behavior of hot and warm mix asphalt. *Constr. Build. Mater.* **2018**, *188*, 119–129. [CrossRef]
27. Liu, C.; Lv, S.; Peng, X.; Zheng, J.; Yu, M. Analysis and Comparison of Different Impacts of Aging and Loading Frequency on Fatigue Characterization of Asphalt Concrete. *J. Mater. Civ. Eng.* **2020**, *32*, 04020240. [CrossRef]
28. Lv, S.; Peng, X.; Liu, C.; Ge, D.; Tang, M.; Zheng, J. Laboratory investigation of fatigue parameters characteristics of aging asphalt mixtures: A dissipated energy approach. *Constr. Build. Mater.* **2020**, *230*, 116972. [CrossRef]
29. Sun, D.; Lin, T.; Zhu, X.; Cao, L. Calculation and evaluation of activation energy as a self-healing indication of asphalt mastic. *Constr. Build. Mater.* **2015**, *95*, 431–436. [CrossRef]

30. *JTG E20-2011*; Standard Test Methods of Bitumen and Bituminous Mixtures for Highway Engineering. China Communications Press: Beijing, China, 2011.
31. Zhang, J.; Wang, Y.D.; Su, Y. Fatigue damage evolution model of asphalt mixture considering influence of loading frequency. *Constr. Build. Mater.* **2019**, *218*, 712–720. [CrossRef]
32. Li, Y.; Liu, L.; Sun, L. Temperature predictions for asphalt pavement with thick asphalt layer. *Constr. Build. Mater.* **2018**, *160*, 802–809. [CrossRef]
33. Valdes-Vidal, G.; Calabi-Floody, A.; Sanchez-Alonso, E.; Miró, R. Effect of aggregate type on the fatigue durability of asphalt mixtures. *Constr. Build. Mater.* **2019**, *224*, 124–131. [CrossRef]
34. Qiu, X.; Cheng, W.; Xu, W.; Xiao, S.; Yang, Q. Fatigue evolution characteristic and self-healing behaviour of asphalt binders. *Int. J. Pavement Eng.* **2022**, *23*, 1459–1470. [CrossRef]
35. Rasouli, A.; Kavussi, A.; Qazizadeh, M.J.; Taghikhani, A.H. Evaluating the effect of laboratory aging on fatigue behavior of asphalt mixtures containing hydrated lime. *Constr. Build. Mater.* **2018**, *164*, 655–662. [CrossRef]
36. Zhang, Z. Effect of Slaking Lime and Basalt Fiber on Fatigue Performance of Asphalt Concrete under Thermal Aging. Master's Thesis, Chongqing Jiaotong University, Chongqing, China, 2021.
37. Wang, S.; Kang, A.; Xiao, P.; Li, B.; Fu, W. Investigating the Effects of Chopped Basalt Fiber on the Performance of Porous Asphalt Mixture. *Adv. Mater. Sci. Eng.* **2019**, *2019*, 2323761. [CrossRef]
38. Miao-Miao, Y.; Xiao-Ning, Z.; Wei-Qiang, C.; Shun-Xian, Z. Ratio of Dissipated Energy Change-based Failure Criteria of Asphalt Mixtures. *Res. J. Appl. Sci. Eng. Technol.* **2013**, *6*, 2514–2519. [CrossRef]
39. Carpenter, S.H.; Shen, S. Dissipated Energy Approach to Study Hot-Mix Asphalt Healing in Fatigue. *Transp. Res. Rec.* **2006**, *1970*, 178–185. [CrossRef]
40. Daniel, J.S.; Kim, Y.R. Laboratory Evaluation of Fatigue Damage and Healing of Asphalt Mixtures. *J. Mater. Civ. Eng.* **2001**, *13*, 434–440. [CrossRef]

Article

Influence of Fiber Type and Dosage on Tensile Property of Asphalt Mixture Using Direct Tensile Test

Shuyao Yang *, Zhigang Zhou and Kai Li

Key Laboratory of Road Structure and Material Ministry of Communication,
Changsha University of Science & Technology, Changsha 410114, China
* Correspondence: ysy@stu.csust.edu.cn; Tel.: +86-151-1137-0374

Abstract: In engineering practice, fiber addition is a frequently used method to improve the tensile property of asphalt mixture. However, the optimum fiber type and dosage have not been determined by direct tensile tests. In this paper, monotonic tensile tests were conducted on three kinds of stone mastic asphalt (SMA13) mixtures, that is, granular-lignin-fiber-reinforced SMA (GFSMA), flocculent-lignin-fiber-reinforced SMA (FFSMA), and basalt-fiber-reinforced SMA (BFSMA) at different fiber dosages to probe the influence of fiber dosage on their tensile mechanical indexes (tensile strength, ultimate strain, elastic modulus, and strain energy density) and to determine the optimum dosage of each kind of fiber. The results showed that with the elevation of fiber dosage, the tensile strength, elastic modulus, and strain energy density of all three kinds of asphalt mixtures increased first and then decreased, while the ultimate strain increased constantly. The optimum dosage was 0.50 wt%, 0.45 wt%, and 0.50 wt% for granular lignin fiber, flocculent lignin fiber, and basalt fiber, respectively. On this basis, strain-controlled direct tensile fatigue tests were conducted on the three kinds of asphalt mixtures at the corresponding optimum fiber dosage. The results indicated that asphalt mixture reinforced with 0.50 wt% granular lignin fiber exhibited ideal direct tensile fatigue performance with respect to fatigue life and accumulative dissipated energy. Therefore, granular lignin fiber is recommended as the favorable fiber type, and its optimum dosage is 0.50 wt%. Moreover, scanning electron microscopy (SEM) demonstrated that the essence of the impact of fiber dosage and type on the tensile property of SMA is whether the reinforcement effect on the mixture matrix outweighs the negative effect of the defects between fiber and mixture matrix, or whether the reverse applies.

Keywords: asphalt mixture; granular lignin fiber; flocculent lignin fiber; basalt fiber; fiber dosage; monotonic tensile tests; strain-controlled direct tensile fatigue tests; SEM

**check for
updates**

Citation: Yang, S.; Zhou, Z.; Li, K.
Influence of Fiber Type and Dosage
on Tensile Property of Asphalt
Mixture Using Direct Tensile Test.
Materials **2023**, *16*, 822. https://
doi.org/10.3390/ma16020822

Academic Editor: Yuri Ribakov

Received: 10 December 2022
Revised: 31 December 2022
Accepted: 4 January 2023
Published: 14 January 2023

1. Introduction

Asphalt concrete pavement has been widely used in highway construction by virtue of driving comfort, low noise level, and simple maintenance. Even though asphalt mixture in the surface course, which is the main structural layer of the highway, is mainly intended for bearing pressure, studies on the mechanical properties and failure mechanism of asphalt mixture in tension are still very necessary, which can be beneficial for getting a comprehensive understanding of the performance of asphalt mixture. Moreover, in engineering practice, many defects have occurred on asphalt concrete pavement resulting from tensile performance deficiency, among which cracking is the most typical one [1,2]. Once cracking occurs, it can induce stress concentration in asphalt mixture, which can not only accelerate the failure of the pavement structure, but also affect driving comfort and even driving safety [3]. Therefore, it is of great significance to improve the tensile properties of asphalt mixture.

Various kinds of additives have been used to improve the tensile properties of asphalt mixtures [4–6]. Among them, fibers have gained the most extensive research attention. It has been demonstrated that fibers such as lignin fiber, basalt fiber, carbon fiber [7], steel fiber [8], glass fiber [9], and polyester fiber [10] can markedly enhance the tensile properties of asphalt mixtures. Lignin fiber and basalt fiber are the most commonly used types [11], and a lot of research is focusing on the comparison of their improvement effect on tensile properties to determine the ideal one. Guan et al. [12], Zhao et al. [13], Eisa et al. [14], and Gupta et al. [15] found that both the tensile strength and tensile strain of asphalt mixture reinforced by basalt fiber were higher than those of asphalt mixture reinforced by lignin fiber. Furthermore, Wu et al. [16] took the aging factor into account and found that the mixture reinforced with lignin fiber exhibited poorer tensile properties than the mixture reinforced with basalt fiber, no matter what the aging degree was. It was worth noting that the above literature showed that basalt fiber performed better in tensile property improvement than lignin fiber. In contrast, Zhu et al. [17] found that lignin fiber could improve tensile performance more remarkably than basalt fiber. Moreover, Zhang et al. [18] found that lignin fiber enhanced the overall performance of asphalt mixture more markedly than basalt fiber.

Besides the ideal fiber type, the optimum fiber dosage should also be determined. If the fiber dosage is too low, the fiber can hardly exert its reinforcement effect and will become a superfluous inclusion or defect. If the fiber dosage is too high, matrix discontinuity may occur, which can retard load transfer and even degrade the performance of the fiber [19,20]. It was revealed that the anti-tensile performance of asphalt mixture mainly depends on the bonding strength between asphalt mortar and aggregate [21]. The bonding strength is reported to be closely associated with basic indexes of asphalt mortar, such as viscosity and dynamic shear modulus, which can vary after fiber incorporation [22–25]. Therefore, fiber dosage is an important issue that needs to be taken into consideration for a favorable tensile performance.

Indoor evaluation methods for the tensile performance of asphalt mixture mainly include the indirect tension test [26–28], bending beam test [29,30], semi-circular bending test [23,31], and direct tension test [32]. Research by direct tension test is scarce, and studies on fiber-reinforced asphalt mixture by direct tension test are even fewer. Additionally, it is believed that the regularity and accuracy of the strain and stiffness modulus measured by indirect tension tests are not satisfactory, and the stress mode of the specimen differs greatly from the actual stress state of the actual structure. The test results of bending beam tests lack stability and the manufacture craft of the specimen for semi-circular bending tests is complex [33]. For direct tensile tests, the specimen manufacture craft is easier, the loading mold is simpler, and the direct acquisition of stress–strain data is easier; in addition, the specimen is less discrete and more sensitive to the test conditions and material parameters and can better reflect the actual stress state of the structure. These advantages render the direct tensile test the ideal one among these evaluation methods [34].

In view of this, in this paper, stone mastic asphalt (SMA) mixtures modified with granular lignin fiber (GFSMA), flocculent lignin fiber (FFSMA), and basalt fiber (BFSMA) were selected as the research object. The monotonic tensile test was carried out on the asphalt mixtures at different fiber dosages to analyze the influence of fiber dosage on the stress–strain curves and fracture morphology of tensile specimens. Furthermore, the effects of dosage on the tensile mechanical indexes (tensile strength, ultimate strain, elastic modulus, and strain energy density) of the three kinds of fiber-reinforced asphalt mixtures were explored, and the optimum dosage for each kind of fiber was determined accordingly. Then, the strain-controlled direct tensile fatigue test was carried out on the three kinds of fiber-reinforced asphalt mixtures at the corresponding optimum fiber dosage. Finally, the influence mechanism of fiber type and dosage on the tensile properties of asphalt mixture was analyzed by scanning electron microscopy (SEM). The research results can provide a basis for fiber selection for mixture design in engineering practice.

2. Materials and Methods

2.1. Materials

2.1.1. Asphalt

Styrene-butadiene-styrene (SBS)-modified asphalt (PG 76-22) used in this research was provided by Shell (Xingyue) Co. Ltd. (Foshan, China). Their basic performance indexes are shown in Table 1.

Table 1. Basic performance index test results of the SBS-modified asphalt.

Index	Unit	Result	Method
Penetration (25 °C, 100 g, 5 s)	0.1 mm	43.3	ASTM D5-13
Ductility (5 °C, 5 cm/min)	cm	26.4	ASTM D113-17b
Softening point	°C	91.4	ASTM D36-14
Flash point	°C	>230	ASTM D92-12b
Density	g/cm^3	1.042	ASTM D70-17a
Residue after TFOT			
Mass loss	%	0.3	ASTM D6-11
Penetration ratio (25 °C)	%	74.3	ASTM D5-13
Ductility (5 °C, 5 cm/min)	cm	22	ASTM D113-17b

2.1.2. Fiber

Granular lignin fiber, flocculent lignin fiber, and basalt fiber used in this study were provided by Linxiang Building Materials Co. Ltd. (Changsha, China). Their basic performance indicators (provided by the manufacturer) are shown in Table 2, and the macro morphologies are displayed in Figure 1. It is stipulated in Technical Specification for Construction of Highway Asphalt Pavements (JTG F40-2004) that the dosage of lignin fiber should not be less than 0.30 wt% of the total weight of the mixture, and the dosage of basalt fiber should not be less than 0.40 wt%. Accordingly, the dosage of three kinds of fibers was set as follows: granular and flocculent lignin fibers (0.30 wt%, 0.40 wt%, 0.45 wt%, 0.50 wt%, and 0.55 wt%); basalt fiber (0.40 wt%, 0.45 wt%, 0.5 wt%, 0.55 wt%).

Table 2. Performance of fibers (offered by manufacturers).

		Fiber Type		
Index	Unit	Granular Lignin	Flocculent Lignin	Basalt
Length	mm	1.5	1	6
Diameter	mm	8	7	17
Oil absorption	times	6.09	7.14	3.20
pH value	/	7.40	7.30	7.60
Tensile strength	MPa	<300	<300	>2000
Density	g/cm^3	1.15	0.91	2.75

(a)

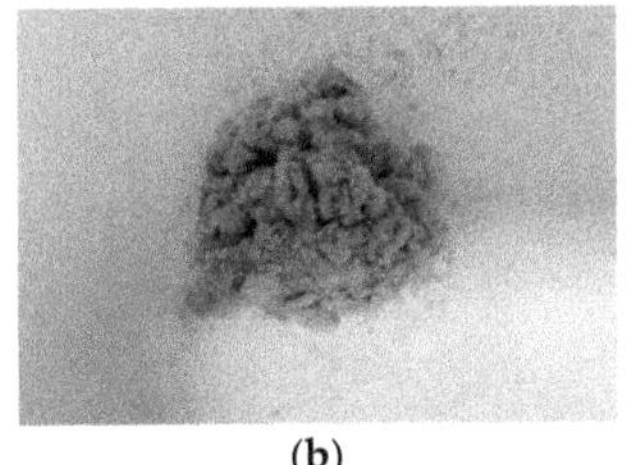
(b)

(c)

Figure 1. Macro morphology of fiber: (**a**) Granular lignin fiber; (**b**) Flocculent; (**c**) Basalt fiber.

2.1.3. Aggregates

Diabase was selected as coarse aggregate, limestone as fine aggregate, and limestone powder as mineral powder. Their basic technical performance indexes are presented in Table 3.

Table 3. Technical index test results of aggregates.

Index	Unit	Aggregate			Method
		Coarse	Fine	Filler	
Specific gravity	g/cm^3	2.756			ASTM C127-15
Specific gravity	g/cm^3		2.722		ASTM C128-15
Specific gravity	g/cm^3			2.715	ASTM D854-14
Los Angeles abrasion	%	11.6			ASTM C131-14
Flat and elongated particles	%	1.6			ASTM D4791-19
Fine aggregate angularity	%		37		AASHTO T304-17

2.1.4. Gradation

SMA13 gradation was used in this study and the gradation composition is presented in Table 4.

Table 4. SMA13 gradation.

Sieve Size/mm	16	13.2	9.5	4.75	2.36	1.18	0.6	0.3	0.15	0.075
Upper limit/%	100	100	75	34	26	24	20	16	15	12
Lower limit/%	100	90	50	20	15	14	12	10	9	8
Target/%	100	95	61	24.4	21	17.8	15.5	13.5	11.7	10.2

2.1.5. Optimum Asphalt Content (OAC)

The follow-up experiments were carried out on asphalt mixtures reinforced with three kinds of fibers at the corresponding optimum asphalt content (OAC), which had been obtained through the Marshall test, and the OAC results are displayed in Table 5. From Table 5, it could be found that OAC value increases with the adding of fiber dosage for each kind of fiber-reinforced asphalt mixture and the growth gradient all tended to be diminished with the elevation of fiber dosage. Among them, the growth gradient for GFSMA was the most notable, followed by BFSMA, and FFSMA was the last. Moreover, for mixtures at the same fiber dosage, the OAC value of FFSMA was the largest, followed by GFSMA, and BFSMA was the last. This might be attributed to the fact that lignin fiber has a larger specific surface area, thereby showing more excellence in absorbing asphalt than basalt fiber.

Table 5. Result of OAC.

Fiber Dosage (wt%)	OAC (%)		
	GFSMA	FFSMA	BFSMA
0.30	5.75	6.00	-
0.40	5.80	6.08	5.50
0.45	5.85	6.11	5.54
0.50	5.90	6.15	5.60
0.55	5.94	6.17	5.63

2.2. Specimen Preparation and Test Protocol

2.2.1. Mixing

The dry-mixing technique was adopted in this study to disperse the flocculent and bunched fibers, which was made easier by the friction of the aggregate [35]. The specific procedure is as follows: First, the fiber and aggregate were mixed together at 180 °C for

60 s, then asphalt was added and mixed for 90 s, and finally, the powder was added and mixed for 90 s.

2.2.2. Monotonic Tensile Test

The main purpose of monotonic tensile tests was to explore the influence of fiber dosage on the tensile property indexes of three kinds of fiber-reinforced asphalt mixtures so as to determine the optimum dosage for each kind of fiber.

Considering the excellent consistency with the field conditions, the wheel rolling compaction method was adopted in this paper to mold the rutting plates of asphalt mixture (300 mm × 300 mm × 50 mm) according to the Standard Test Method of Bitumen and Bituminous Mixtures for Highway Engineering (JTG E20-2011). Then the rutting plates were sawed into trabecular specimens (250 mm × 50 mm × 50 mm) along the wheel trace direction by the single-sided saw.

MTS 810 testing machine manufactured by MTS Corporation (Eden Prairie, MN, USA) was applied in the test. Each of the two ends of the trabecular specimens was stuck to a steel plate using epoxy resin steel glue. After 48 h, when the strength of the steel glue was built, specimens were placed in the assorted environment chamber with a target temperature of 20 °C for 4 h, and then they were taken out and clamped on the instrument for tests. In order to eliminate the eccentric tension in the loading process, pre-tensioning was carried out before each test. The loading speed was controlled at a displacement control mode by 5 mm/min [36]. The force value was acquired by the computer system of the loading system, and the strain value was obtained by a pair of extensometers symmetrically arranged on the two cutting sides of each trabecular along the longitudinal direction. In this test, four evaluation indexes were used, that is, the tensile strength, the ultimate strain, elastic modulus, and the strain energy density, and they were defined in Equations (1)–(4), respectively.

$$\sigma_{max} = \frac{F_{max}}{bd} \tag{1}$$

$$\varepsilon_{max} = \frac{\varepsilon_{1max} + \varepsilon_{2max}}{2} \tag{2}$$

$$S = \frac{\sum_{i=1}^{n}\left(\varepsilon_i - \frac{\sum_{i=1}^{n}\varepsilon_i}{n}\right)\left(\sigma_i - \frac{\sum_{i=1}^{n}\sigma_i}{n}\right)}{\left(\varepsilon_i - \frac{\sum_{i=1}^{n}\varepsilon_i}{n}\right)^2} \tag{3}$$

$$\omega = \int_0^{\varepsilon_{max}} \sigma d\varepsilon \tag{4}$$

where F_{max} = peak loading value in the course of tension; b = width of the specimen (50 mm); d = height of the specimen (50 mm); ε_{1max} = strain value measured by one of a pair of extensometers when loading attains the peak; ε_{2max} = strain value measured by the other of a pair of extensometers when loading attains the peak; (ε_i, σ_i) = data of the elastic stage of the stress − strain curves; $\int$ = integral of the stress-strain envelope area; σ = stress in the tension process; and ε = strain in the tension process.

2.2.3. Direct Tensile Fatigue Test

The direct tensile fatigue test was performed to determine an ideal fiber type that can most remarkably improve tensile fatigue. The procedure (preparation, clamping, and preservation) was the same as that for monotonic direct tensile tests. The strain-controlled mode, under which the stress–strain situation of asphalt mixtures was more consistent with that of actual pavements, was selected as the cyclic control mode. The procedure for determining the strain levels was as follows: firstly, the average of the ultimate strain of three kinds of fiber-reinforced asphalt mixtures at the corresponding optimum fiber dosage was calculated; then, the average value could be multiplied by 0.4, 0.5, 0.6, and 0.7 to acquire the target controlled strain. The half-sine waveform at a frequency of 10 Hz was applied. The test temperature was 20 °C. The fatigue life was defined as the cycle number at which the stiffness modulus drops to 50% of that of the 100th cycle (named as initial stiffness modulus).

The fatigue process of asphalt mixtures could be understood as the resistance of their internal structure to the impact of external action, in which energy would be continuously consumed for internal reconstruction. The accumulative dissipated energy was used as an index to evaluate the fatigue resistance besides the aforementioned index N, and it can be calculated by Equation (5):

$$W = \sum_{I=1}^{N} \pi \sigma_i \varepsilon_i sin(2\pi f \Delta t) \tag{5}$$

where W = accumulative dissipated energy; f = the applied loading frequency (Hz); and Δt = the time of strain peak lagging behind the stress peak (s).

2.2.4. SEM Test

For a better understanding of the impact mechanism of fiber dosage and type on the tensile properties of asphalt mixtures, samples sliced from the fracture section of monotonic tensile specimens were observed by scanning electron microscope. In the test, each sample was plated with a layer of 20 mm gold powder in a vacuum and then put into the electron microscope sample room for observation.

3. Discussion

3.1. Results of Monotonic Tensile Test

3.1.1. Tensile Stress–Strain Curves

The stress–strain curves of GFSMA, FFSMA, and BFSMA were depicted in Figure 2. On the whole, all of those stress curves could be divided into the following three stages: elastic stage, strain hardening stage, strain softening stage.

(1) Elastic stage: The stress–strain curve in this stage was close to a straight line. Specifically, when the strain was within 10–30% of the ultimate strain, the stress showed a synchronous linear upward trend with the elevation of strain.

(2) Strain hardening stage: The stress–strain curve in this stage presented an obvious deviation from its initial slope and was similar to an arc. During this stage, stress and strain both kept growing but the growth rate of the former was lower than the latter. By contrast, the growth rate of stress slowed down compared to that of stress with the previous stage.

(3) Strain softening stage: When stress reached the peak value, strain continued increasing while stress began to gradually decease. When the stress reaches the peak stress, the strain continues to increase while the stress begins to gradually decrease. This phenomenon might be explained as the following: When the stress and strain inside the specimen reached a certain extent, some fibers could be pulled out from the crack or even pulled off so that the resistance of the specimen to external loading was weakened and the tensile strength dropped eventually.

In addition, for all the three kinds of fiber-reinforced asphalt mixtures, the variation of fiber dosage had an impact on their own stress–stain curves to some degree. As fiber dosage increased, the wholeness of all of the curves tended to be plumper; the vertical span of the curves firstly went rising and then moved down; both the length and the slope of the first stage of the curves climbed up first and then declined; the arc coverage span of the second stage of the curves was first elongated and then shortened while the arc curvature exhibited the reverse variety; and the falling gradient of the third stage of the curves tended to be flattened.

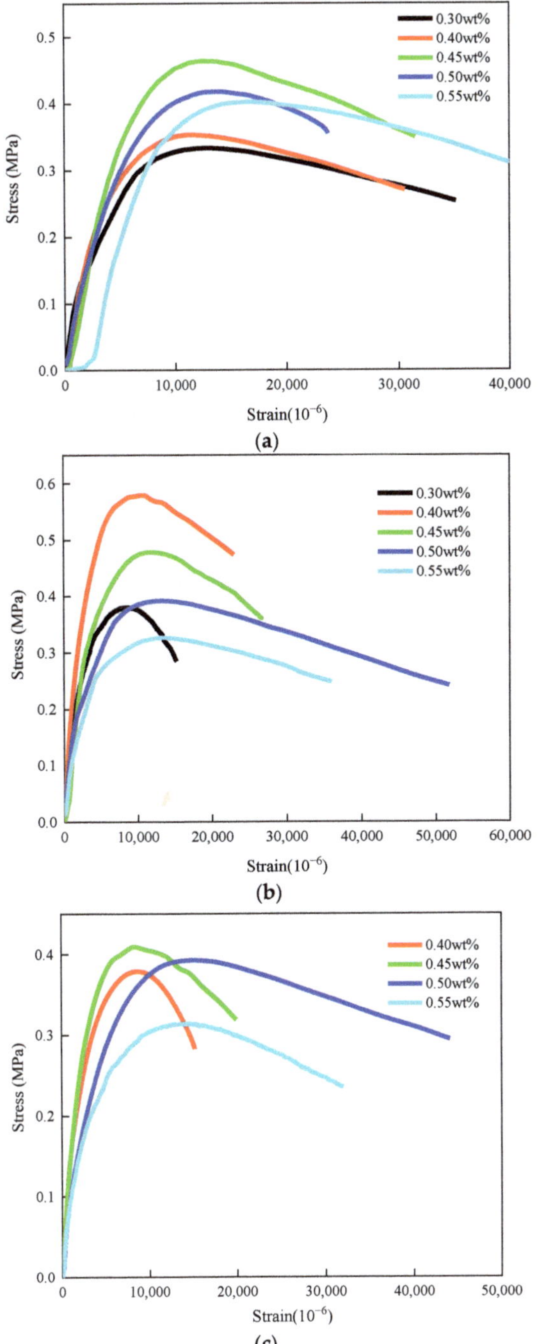

Figure 2. Stress–strain curves: (**a**) GFSMA; (**b**) FFSMA; (**c**) BFSMA.

3.1.2. Feature Analysis on Fracture Section

Given space constraints, the effects of fiber dosage on the crack morphology and fracture surface morphology of specimens were illustrated with GFSMA as an example. The results are presented in Figures 3 and 4, respectively.

  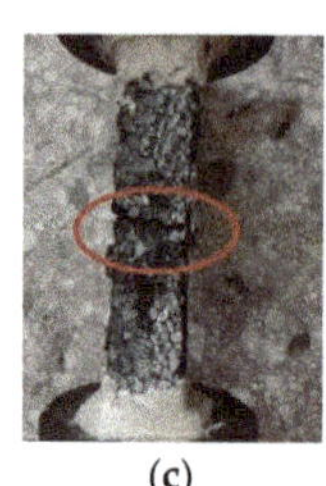  

(**a**) (**b**) (**c**) (**d**) (**e**)

Figure 3. Crack morphology of GFSMA specimens: (**a**) 0.3 wt; (**b**) 0.4 wt%; (**c**) 0.45 wt%; (**d**) 0.50 wt%; (**e**) 0.55 wt%.

(**a**) (**b**) (**c**)

(**d**) (**e**)

Figure 4. Fracture surface morphology of GFSMA specimens: (**a**) 0.3 wt%; (**b**) 0.4 wt%; (**c**) 0.45 wt%; (**d**) 0.50 wt%; (**e**) 0.55 wt%.

From Figure 3, it was observed that when the dosage was 0.3 wt%, the fractured section was generally parallel to the transverse section of the specimen. The cracks of the specimen were generally linear and did not yet extend longitudinally to the specimen. However, the edges of the section showed slightly uneven features, which were caused by the bridging effect of fibers. It could be observed from Figure 3 that there was some matrix debris on the fracture surface, and the fracture surface was slightly undulating. As the dosage continued climbing up, the fracture toughness of the specimen was significantly enhanced, and the whole crack or a segment thereof on the fracture surface was no longer flat. It could be easily noticed that cracks winded along the longitudinal direction of the specimen, and the crack width increased to a certain extent; in this case, the matrix may be torn. Some of the matrix was loosened and peeled off at the edge of the crack or even in the middle of the section during the test procedure. As a result, the fracture surface was very uneven, with matrix debris and fibers, and the two sections of the damaged specimens could not even be completely matched. However, when the dosage leaped to 0.55 wt%, the sinuousness of the macro cracks seemed to be less obvious than that of the mixture with 0.50 wt% fiber. Overall, from the perspective of the feature of the fracture section, the asphalt mixture at the fiber dosage of 0.50 wt% embodied higher fracture toughness than that at the other four dosages, which might cause it to have more excellent tensile properties.

In addition, comparing the five pictures in Figure 4, it was worth noting that there were more fibers on the fracture surface in Figure 3d compared with the other four pictures. In addition, as depicted by Figure 4d, some of the fibers had been pulled out of the matrix entirely and some had been broken. The two kinds of behaviors mentioned above both needed extra absorption of energy so that asphalt mixture exhibited more remarkable fracture toughness characteristics.

3.1.3. Results of Tensile Strength

The tensile strength results of GFSMA, FFSMA, and BFSMA at different fiber dosages are presented in Figure 5. It can be seen that for all three types of asphalt mixtures, the addition of an appropriate dosage of fiber could improve their tensile strength, but when the fiber dosage exceeded a certain critical value, the tensile strength would decrease with the growth of the fiber dosage. The reason for the above phenomenon may be as follows: At an appropriate elevation of fiber dosage, the average fiber distance could be shortened so that there could be more fibers participating in the adhesion with matrix, which would strengthen the interface between the matrix and fiber and eventually improve the tensile strength of mixture. However, the continuous rising of fiber dosage brought about the enlargement of coverage of the fiber and then the narrowness of the effective bonding interface between fiber and asphalt, which could lead to the decline of tensile strength of the mixture instead.

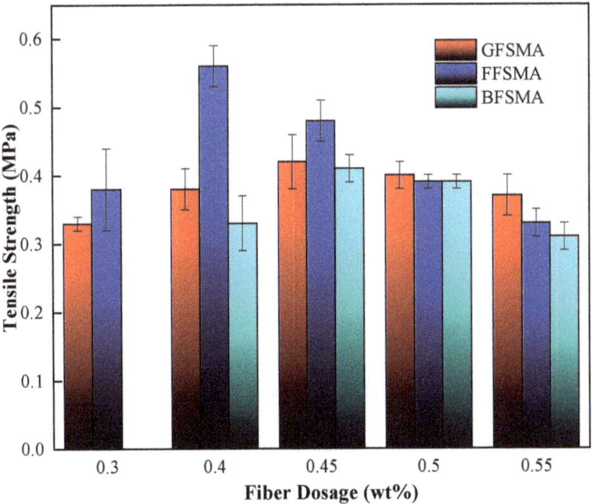

Figure 5. Tensile strength of mixtures.

The tensile strength of GFSMA at the fiber dosage of 0.40 wt% increased by 15% compared with that at the fiber dosage of 0.30 wt%. When the dosage reached up to 0.45 wt%, the tensile strength reached its peak value (about 0.42 MPa), which increased by 27% compared with that at the fiber dosage of 0.30 wt%. When the dosage continued increasing, the tensile strength began to decrease lightly but that at the fiber dosage of 0.55 wt% was still slightly higher than that at 0.30 wt% by 12%.

The tensile strength of FFSMA reached the peak value (about 0.55 MPa) at the fiber dosage of 0.40 wt%, which was about 47% higher than that at the fiber dosage of 0.30 wt%.

When the dosage increased from 0.40 wt% to 0.45 wt%, the tensile strength tended to descend dramatically but was still mildly higher than that at the fiber dosage of 0.30 wt% by nearly 26%. When the dosage reached 0.50 wt%, the tensile strength dropped to a value that was almost equal to that at 0.30 wt%. Then, the tensile strength declined to 0.33 MPa but the decrease gradient slowed down when the dosage ascended to 0.55 wt%.

For BFSMA, the tensile strength reached the peak value (about 0.41 MPa) when the dosage attained 0.45 wt%, which was remarkably higher than that at the fiber dosage of 0.4 wt% by about 24%. With the dosage further increasing, however, the tensile strength showed a declining tendency. When the fiber dosage attained 0.55 wt%, the tensile strength was even below that at the fiber dosage of 0.4 wt% by 6%.

The fractional function was imported to execute fitting between the tensile strength and fiber dosage for the three kinds of fiber-reinforced asphalt mixtures. The results are shown in Figure 6. It can be seen that the fitting correlation coefficients R^2 for the three kinds of fiber-reinforced asphalt mixtures ranged from 0.82 to 0.98, indicating that the fractional function could accurately describe the relationship between the tensile strength of the asphalt mixture and fiber dosage within 0.30–0.55 wt%.

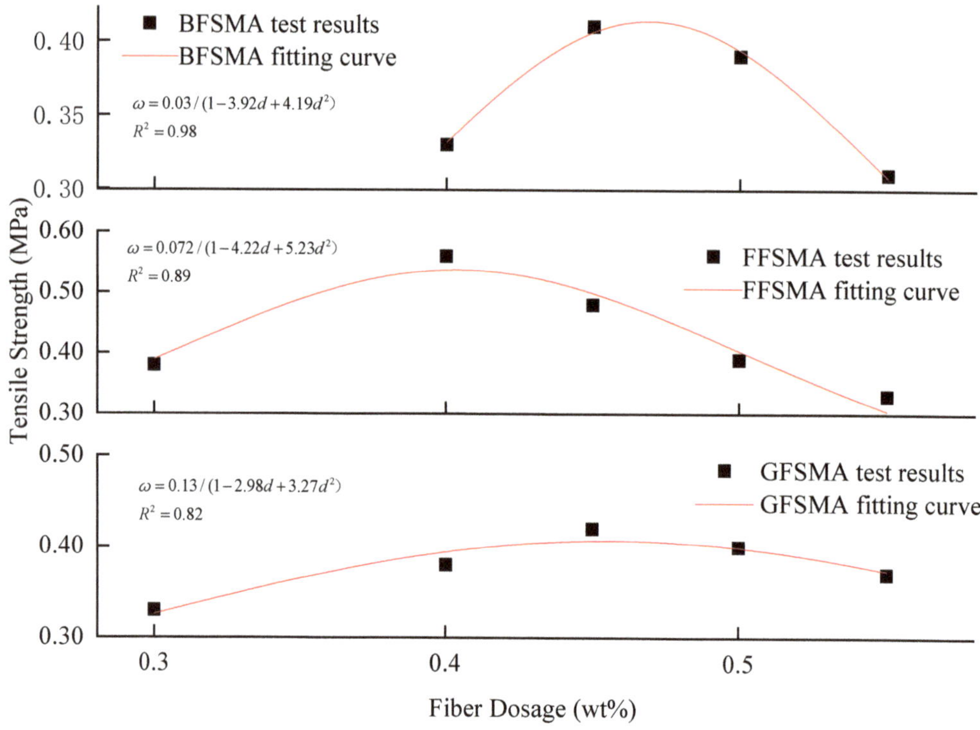

Figure 6. Fitting between fiber dosage and tensile strength.

3.1.4. Results of Ultimate Strain

The ultimate strain results of GFSMA, FFSMA, and BFSMA at different fiber dosages are shown in Figure 7. From Figure 7, it can be seen that the addition of an appropriate amount of fiber based on the dosage stipulated in the specification could enhance the ultimate strain of the asphalt mixtures. This can be explained as follows: On one hand, fibers mentioned in this research had excellent resistance to tensile deformation; on the other hand, they could function as bridge in the mixture, which would share and transfer deformation form matrix. The above two factors both contributed to the improvement on mixture's ultimate strain. In addition, with the increase of fiber dosage, more and more fibers were involved in the improvement issue, and then the ultimate stain exhibited a growing trend.

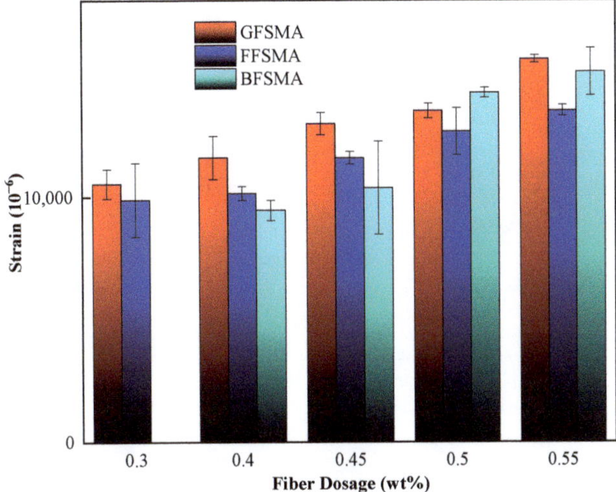

Figure 7. Ultimate strain of mixtures.

The ultimate strain of GFSMA at the fiber dosage of 0.4 wt%, 0.45 wt%, 0.50 wt%, and 0.55 wt% increased by 10%, 23%, 28%, and 48%, respectively, when compared with that at the fiber dosage of 0.30 wt%. In addition, the increment gradient at 0.55 wt% was the most significant.

For FFSMA, when the dosage was 0.40 wt%, the increment of the ultimate strain was very limited (only 3%) compared to the ultimate strain at 0.30 wt%, but with the continuous climbing of the dosage, the fiber improvement impact on the ultimate strain was more and more significant. Compared to the ultimate strain at 0.30 wt%, the ultimate strain increased by 17%, 28%, and 37%, respectively, for 0.45 wt%, 0.50 wt%, and 0.55 wt%. In addition, it was not difficult to find that with the increase in fiber dosage, the growth gradient of the ultimate strain tended to be flatter.

For BFSMA, as the fiber dosage went up from 0.40 wt% to 0.45 wt%, the enhancement effect of basalt fiber on the ultimate strain was not remarkable, with an increment of only 10%. When the dosage increased to 0.50 wt%, the ultimate strain increased sharply and was more than half of that at 0.40 wt%. When the dosage reached 0.55 wt%, the ultimate strain increased by 60% compared with that at 0.4 wt%. When the dosage increased to 0.50 wt%, the ultimate strain increased dramatically and was more than half of that at 0.4 wt%. When the dosage reached 0.55 wt%, the ultimate strain increased by 60% compared with that at 0.40 wt%, but the increase slowed down.

The ultimate strain and fiber dosage of the three kinds of fiber-reinforced asphalt mixtures were fitted, and the fitting curves are shown in Figure 8. The fitting results indicated that the first-order function could better describe the relationship between the tensile strength and the fiber dosage with the fitting correlation coefficients R^2 in the range of 0.90–0.93 for all three kinds of fiber-reinforced asphalt mixtures.

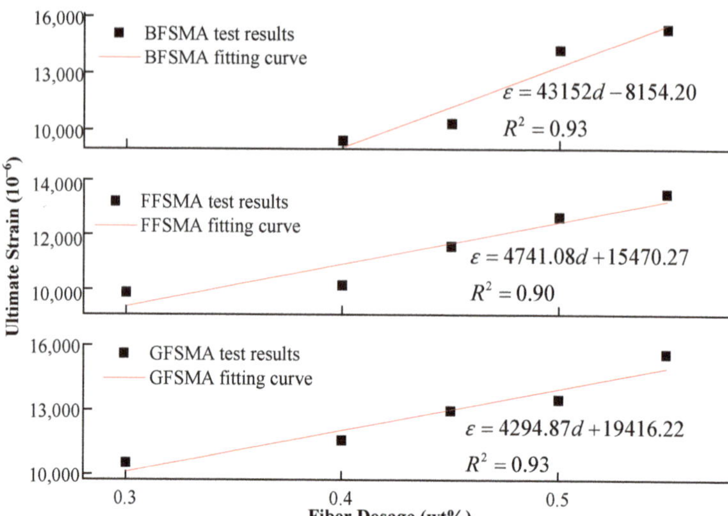

Figure 8. Fitting between fiber dosage and ultimate strain.

3.1.5. Results of Elastic Modulus

The elastic modulus of GFSMA, FFSMA, and BFSMA at different fiber dosages are presented in Figure 9. It can be seen that for GFSMA and FFSMA, the addition of an appropriate dosage of fiber could improve their elastic modulus, but when the fiber dosage exceeded a certain critical value, it would decrease with the increase of the fiber dosage. However, for BFSMA, the elastic modulus kept a gradual downward trend with the elevation of fiber dosage. The phenomenon might be explained by the difference between the variation gradient of the tensile strength and the ultimate strain.

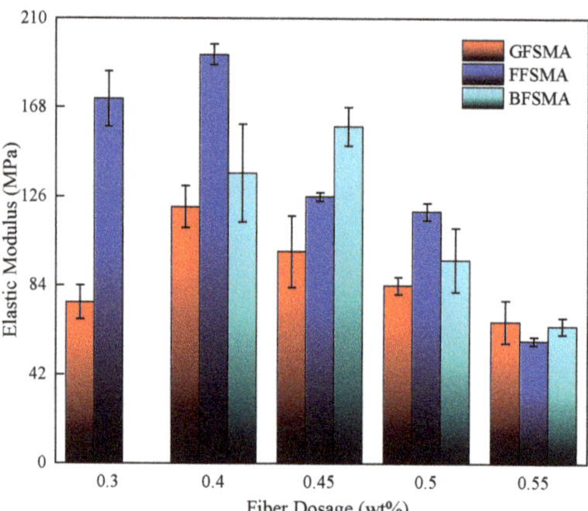

Figure 9. Elastic modulus of mixtures.

For GFSMA, there was a marked growth of nearly 60% in the elastic modulus at the fiber dosage of 0.40 wt% compared with that at the fiber dosage of 0.30 wt%. Then, the elastic modulus kept a falling tendency at a stable descending gradient. However, it was worth noting that the elastic modulus was not below that at the fiber dosage of 0.3 wt% until the fiber dosage attained 0.55 wt%.

Similarly, the elastic modulus of FFSMA showed a slight increase by about 12% at the fiber dosage of 0.40 wt% than that at the fiber dosage of 0.30 wt%. When the dosage increased from 0.40 wt% to 0.45 wt%, the elastic modulus tended to descend. When the dosage reached 0.55 wt%, the elastic modulus dropped more sharply with a reduction rate of 66% compared with that at the fiber dosage of 0.3 wt%.

For BFSMA, as the fiber dosage increased from 0.40 wt% to 0.45 wt%, there was a growth by about 16% likewise. When the fiber dosage exceeded 0.45 wt%, the elastic modulus began to decline steadily and the elastic modulus at the fiber dosage of 0.50 wt% and 0.55 wt% descended by about 29% and 52%, respectively, compared with that at 0.4 wt%.

The fractional function was also imported to execute fitting between the elastic modulus and fiber dosage for the three kinds of fiber-reinforced asphalt mixtures. The results are shown in Figure 10. It can be seen that the fitting correlation coefficients R^2 for the three kinds of fiber-reinforced asphalt mixtures ranged from 0.86 to 0.94, indicating that the fractional function could accurately describe the relationship between the elastic modulus of asphalt mixture and fiber dosage within 0.3 wt%–0.5 wt%.

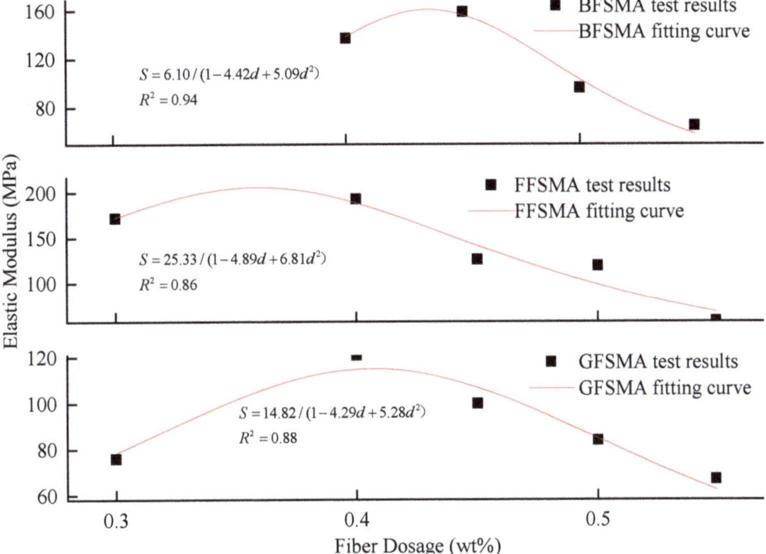

Figure 10. Fitting between fiber dosage and elastic modulus.

3.1.6. Results of Strain Energy Density

From the above analysis of tensile strength and ultimate strain, it could be summarized that the variation regulation for the tensile strength and the ultimate strain with the variation of fiber dosage for each kind of fiber-reinforced asphalt mixture showed great difference. In addition, the two indexes reached their own peak value at different fiber dosages, which could subsequently result in an inconsistency between the optimum fiber dosage determined, respectively, based on the two indicators. Therefore, a comprehensive indicator called strain energy density, which can reflect both the tensile strength and ultimate strain [37], was chosen to evaluate the tensile properties of the asphalt mixtures.

The results of the strain energy density of GFSMA, FFSMA, and BFSMA at different dosages are displayed in Figure 11. It can be seen that the strain energy density of all three types of fiber-reinforced asphalt mixtures ascended first and then descended, which was similar to the trend of the tensile strength. It can be illustrated as follows: Fibers were distributed irregularly and overlapped with each other to form a three-dimensional network, which could have "reinforcement" and "anti-cracking" effects. Moreover, the realization of the two effects mentioned above needed strain energy absorbed by energy. Therefore, with the increase of fiber dosage, there were supposed to be more strain energy in the mixture. However, at the same time, the continuous addition of fiber could lead to uneven distribution and even agglomeration, both of which could diminish the effect of "reinforcement" and "anti-cracking" of fiber and result in the drop of strain energy density in the end.

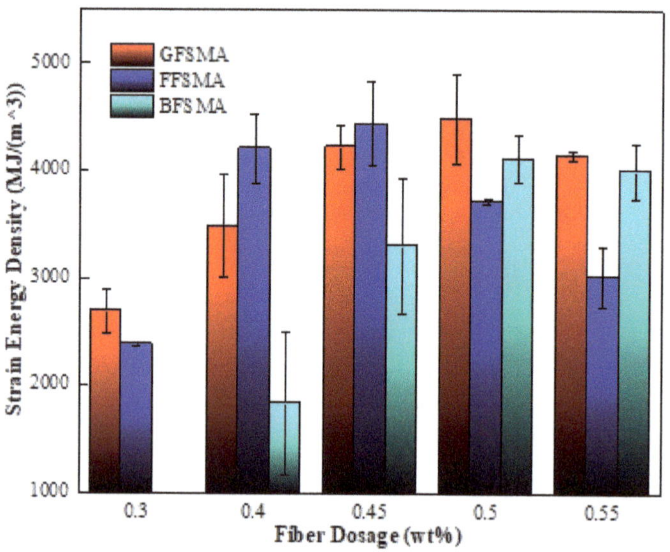

Figure 11. Strain energy density of mixtures.

The strain energy density of GFSMA at 0.40 wt% and 0.45 wt% increased by 29% and 57%, respectively, compared with that at 0.30 wt%. When the dosage attained 0.50 wt%, the strain energy density reached the peak value (about 4481 MJ/m^3), which was 67% higher than that at 0.30 wt%. When the dosage increased up to 0.55 wt%, the strain energy density was slightly lower than that at 0.50 wt% but was still more than half of that at 0.30 wt%.

The strain energy density of FFSMA at 0.4 wt% significantly increased by 76% compared with that at 0.3 wt%. When the dosage reached 0.45 wt%, the strain energy density reached the peak value (4439 MJ/m^3), which was 87% higher than that at 0.30 wt%. With the continuous increase of dosage to 0.50 wt% and 0.55 wt%, although the strain energy density of FFSMA began to show a nearly uniform downward trend, it was approximately 56% and 27% higher than that at 0.30 wt%, respectively.

The strain energy density of BFSMA at 0.45 wt% leaped by 81% compared with that at 0.40 wt%. When the dosage went up to 0.50 wt%, the strain energy density reached the peak value (4112 MJ/m^3), which was more than twice that at 0.40 wt%. When the dosage was 0.55 wt%, the strain energy density was nearly 1.2 times higher than that at 0.40 wt% but declined compared with that at 0.50 wt%.

The strain energy density of the three kinds of fiber-reinforced asphalt mixtures was fitted with fiber dosage, and the fitting image is displayed in Figure 12. The fitting results showed that the fractional function could describe the relationship between the strain energy density of all three kinds of asphalt mixtures and fiber dosage, and the fitting correlation coefficients R^2 were in the range of 0.96–0.99.

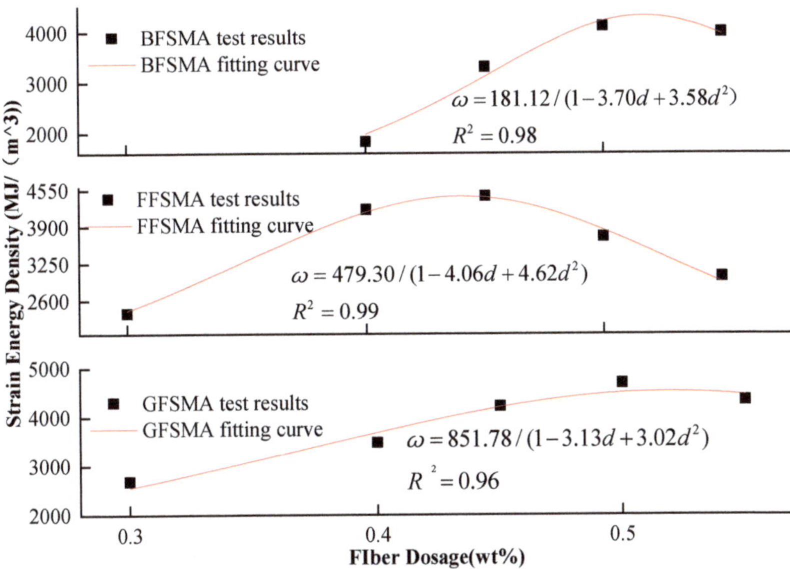

$$\omega = 181.12 / (1 - 3.70d + 3.58d^2)$$
$$R^2 = 0.98$$

$$\omega = 479.30 / (1 - 4.06d + 4.62d^2)$$
$$R^2 = 0.99$$

$$\omega = 851.78 / (1 - 3.13d + 3.02d^2)$$
$$R^2 = 0.96$$

Figure 12. Fitting between fiber dosage and strain energy density.

Furthermore, it could be summarized from the above analysis results that within the given dosage range, the optimal dosage of granular lignin fiber, flocculent lignin fiber, and basalt fiber was 0.50 wt%, 0.45 wt%, and 0.5 wt%, respectively, based on the criteria of maximal strain energy density.

3.2. Results of Direct Tensile Fatigue Test

3.2.1. Fatigue Life

The strain-controlled direct tensile fatigue tests were carried out for GFSMA, FFSMA, and BFSMA at the corresponding optimum dosage. The four controlled strain were 5048με, 6310με, 7572με, and 8834με. Five replicates were fabricated for each kind of fiber-reinforced asphalt mixture at each controlled strain. The fatigue life results are shown in Table 6. From Table 6, it could be concluded that within the given range of strain, the fatigue life of the three kinds of fiber-reinforced asphalt mixtures at the same strain was ranked as GFSMA > BFSMA > FFSMA. The fatigue life of GFSMA was 3.7 times and 2.2 times that of FFSMA and BFSMA, respectively, under the strain of 5048με; 4.6 times and 2.1 times longer under the strain of 6310με; 4.7 times and 1.7 times longer under the strain of 7572με; and 3.6 times and 1.8 times longer under the strain of 8834με.

Table 6. Result of fatigue life.

Strain	Specimen Number	GFSMA		FFSMA		BFSMA	
		Fatigue Life (Times)	Average (Times)	Fatigue Life (Times)	Average (Times)	Fatigue Life (Times)	Average (Times)
5048	1	48,824		11,654		23,161	
	2	42,361		13,502		17,238	
	3	43,894	45,209	12,404	12,200	17,921	20,513
	4	46,267		13,795		18,286	
	5	44,699		9644		15,960	

Table 6. *Cont.*

Strain	Specimen Number	GFSMA		FFSMA		BFSMA	
		Fatigue Life (Times)	Average (Times)	Fatigue Life (Times)	Average (Times)	Fatigue Life (Times)	Average (Times)
6310	1	28,708		6481		15,478	
	2	27,491		5301		11,117	
	3	29,905	28,671	7499	6273	13,873	13,755
	4	31,250		6158		10,990	
	5	26,000		5925		17,316	
7572	1	12,452		2630		8165	
	2	13,500		3307		10,940	
	3	17,018	13,950	3357	2954	6771	7990
	4	11,617		2562		9804	
	5	15,164		2914		4269	
8834	1	6836		2125		3404	
	2	5764		1345		1847	
	3	6375	6339	1319	1742	3163	3555
	4	5922		1844		5085	
	5	6757		2077		4280	

In addition, it could also be seen that the fatigue life of the three types of fiber-reinforced asphalt mixtures exhibited varying degrees of decrease with the growth of strain level. Equation (6) was used to depict the relationship between fatigue life and strain level of three kinds of asphalt mixtures in double logarithmic co-ordinates. The fitting results are shown in Figure 13.

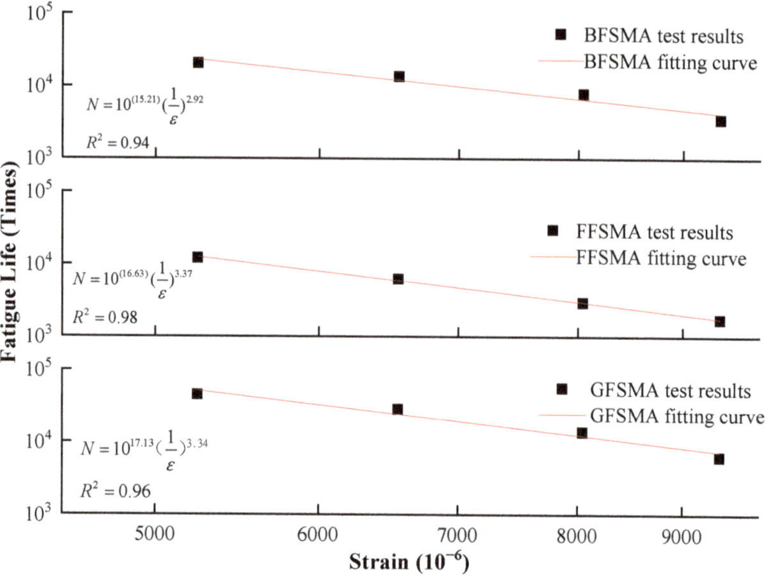

Figure 13. Fitting between strain level and fatigue life.

It could be found from Figure 13 that in terms of the *k* value, the three fiber-reinforced asphalt mixtures were ranked as BFSMA < FFSMA < GFSMA; in terms of the *c* value, the three asphalt mixtures were ranked as BFSMS < GFSMA < FFSMA. The above results showed that GFSMA had moderate sensitivity to strain and ideal fatigue resistance with the highest fatigue line position. By contrast, BFSMA had the lowest fatigue resistance and sensitivity to strain.

It could be found from the figure that the *k* value of the three fiber asphalt mixtures were in order as follows: BFSMA < FFSMA < GFSMA; the *c* value of the three fiber asphalt mixtures were in order as follows: BFSMS < GFSMA < FFSMA. The above results showed that GFSMA had moderate sensitivity to strain as well as the best fatigue resistance with the highest fatigue line position. By contrast, BFSMA was observed to be of the poorest fatigue resistance and sensitivity to strain.

$$N = k\left(\frac{1}{\varepsilon}\right)^{c} \tag{6}$$

where N = fatigue life; ε = strain; c = sensitivity of fatigue life to the variation of strain (the slope of fitting curve); and k = fatigue resistance (the intercept of fitting curve).

3.2.2. Cumulative Dissipated Energy

The cumulative dissipated energy results of three types of fiber-reinforced asphalt mixtures are shown in Figure 14. From Figure 14, it can be seen that in terms of the cumulative dissipated energy at the same strain, the three kinds of asphalt mixtures were ranked as GFSMA > BFSMA > FFSMA, which is consistent with the ranking in terms of fatigue life. Specifically speaking, the cumulative dissipated energy of GFSMA was 1.7 times and 1.2 times higher than that of FFSMA and BFSMA, respectively, under the strain of 5048με; 2.6 times and 1.5 times higher under the strain of 6310με; 2.5 times and 1.9 times higher under the strain of 7572με; and 2.4 times and 1.6 times higher under the strain of 8834με. It could be summarized that the tensile fatigue performance of GFSMA at 0.5 wt% is superior to that of FFSMA at 0.45 wt% or BFSMA at 0.50 wt% in terms of fatigue life and cumulative dissipative energy.

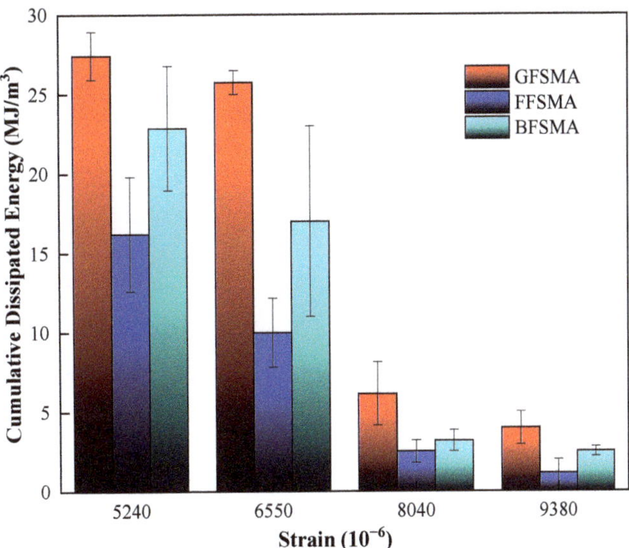

Figure 14. Cumulative dissipated energy results of mixtures.

Based on the data of cumulative dissipated energy and fatigue life of three types of fiber-reinforced asphalt mixtures under each strain, fitting was performed in the double logarithmic co-ordinates using Equation (7) of the dissipated energy model proposed by Dijk and Visser. The fitting results (by 95% confidence limits) are shown in Table 7.

$$W = AN^Z \qquad (7)$$

where W = accumulative dissipated energy (MPa); N = fatigue life; and A and z = fitting parameter.

Table 7. Fitting Results of Parameters.

Strain (10^{-6})	lg$A$	Z	R^2
5240	-0.35	0.39	0.94
6550	-1.36	0.62	0.99
8040	-2.49	0.78	0.96
9380	-3.08	0.97	0.99

The relationship between each of the above two fitting parameters and strain can be approximately expressed by a linear function, and the fitting results are shown in Figure 15.

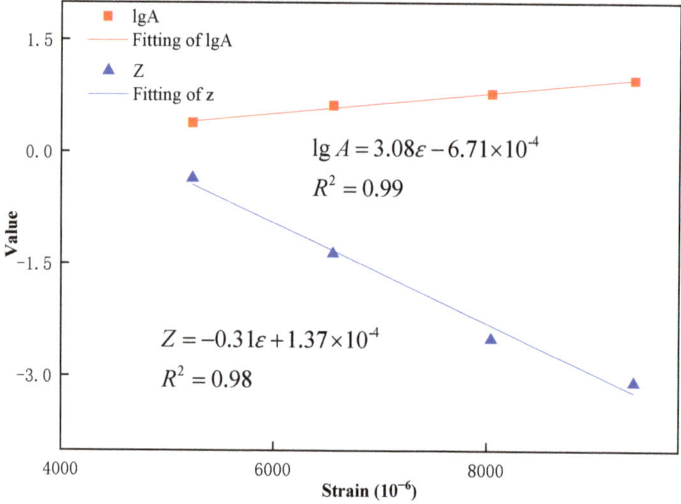

Figure 15. Fitting results of parameters.

Finally, the relationship between the strain and the cumulative dissipative energy of the three fiber-reinforced asphalt mixtures could be obtained by plugging the relationship formula displayed in Figure 13 into Equation (7), and the derived formula is shown in Equation (8).

$$W = 10^{3.08\varepsilon - 0.000671} N^{-0.31\varepsilon + 0.000137} \qquad (8)$$

3.3. Analysis of SEM Test Results

The influence mechanism of fiber dosage on the tensile performance of asphalt mixture was explained by taking the cross-section morphology of a tensile specimen of GFSMA at five dosages as examples; the influence mechanism of fiber type on the tensile properties of asphalt mixture was explained by taking the cross-section morphology of tensile specimens of GFSMA, FFSMA, and BFSMA at the corresponding optimum dosage as examples. The SEM images acquired are shown in Figure 16.

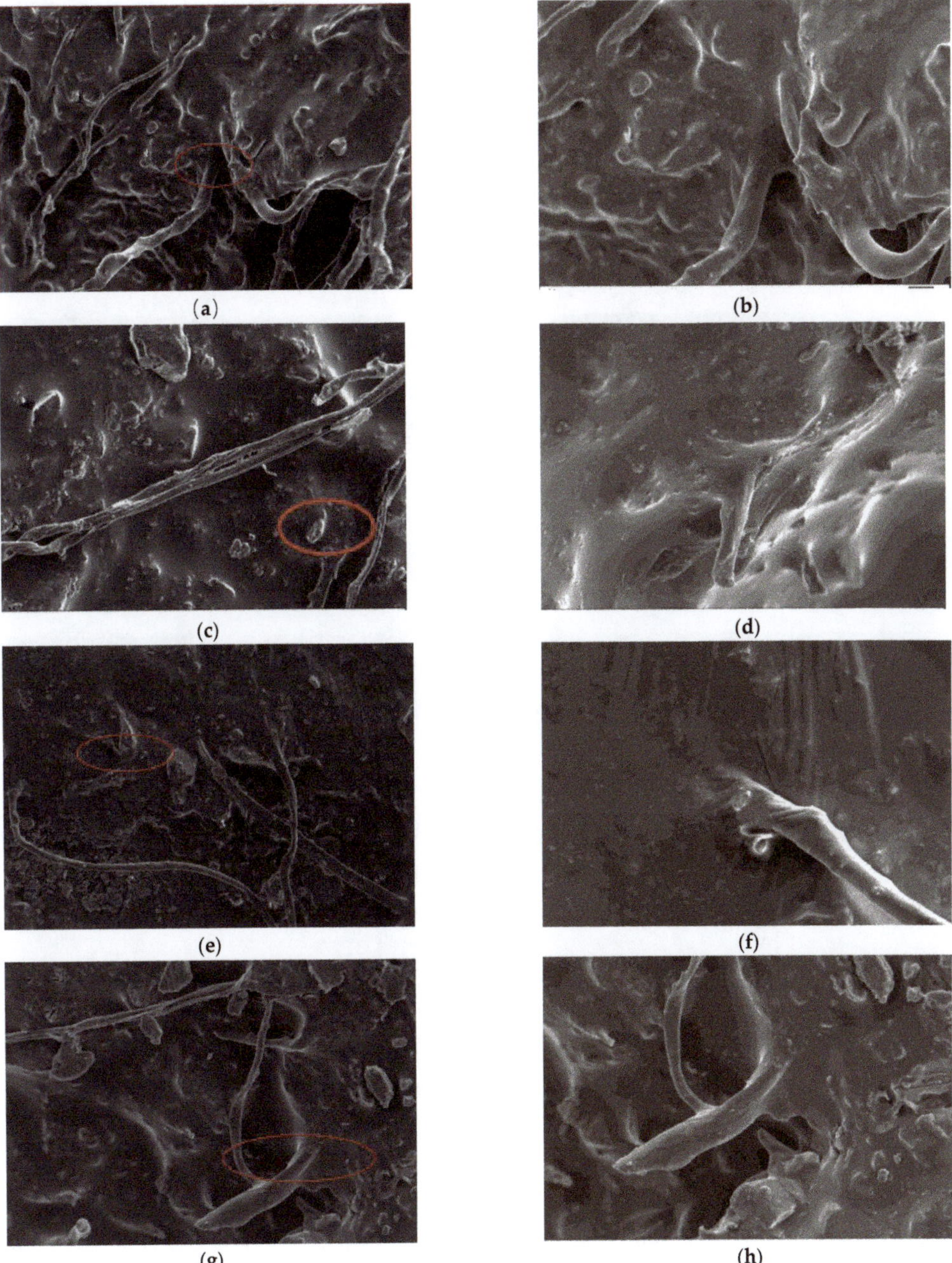

Figure 16. *Cont.*

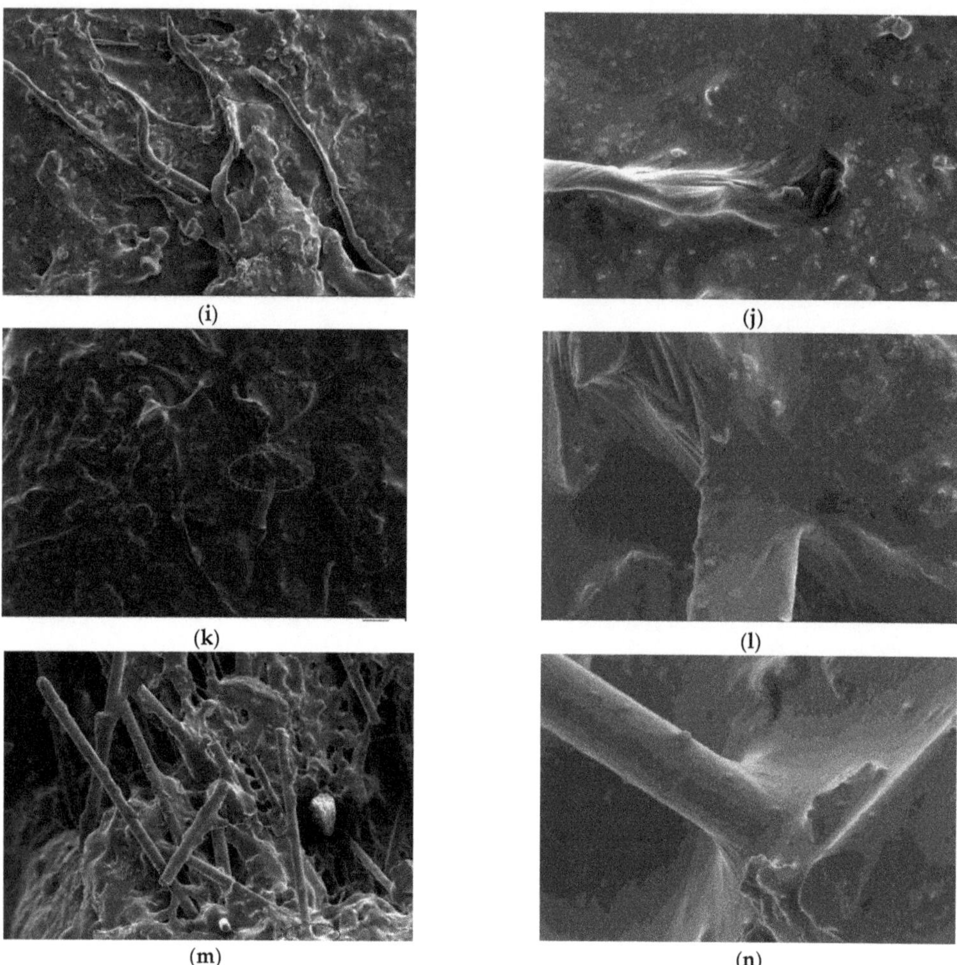

Figure 16. Microscopic Morphology: (**a**) 0.30 wt% GFSMA; (**b**) local zoom of 0.30 wt% GFSMA; (**c**) 0.40 wt% GFSMA; (**d**) local zoom of 0.40 wt% GFSMA; (**e**) 0.45 wt% GFSMA; (**f**) local zoom of 0.45 wt% GFSMA; (**g**) 0.5 wt% GFSMA; (**h**) local zoom of 0.5 wt% GFSMA; (**i**) 0.55 wt% GFSMA; (**j**) local zoom of 0.55 wt% GFSMA; (**k**) 0.45 wt% FFSMA; (**l**) local zoom of 0.45 wt% FFSMA; (**m**) 0.50 wt% BFSMA; (**n**) local zoom of 0.50 wt% BFSMA.

Fibers are dispersed in the mortar in a disorderly overlapping manner and crisscross with the asphalt to form a three-dimensional network structure. The external load is generally directly applied to the matrix, and then the stress can be transferred to the fiber through the fiber end and the interface between the fiber and the matrix. The stress imposed on the fiber is rapidly diffused, which can prevent the crack from propagation. That is the very embodiment of the bridging effect of fiber.

Fibers can also absorb the asphalt to its surface. The roots of fibers are well-bonded with the asphalt to exert an anchoring effect, and the end forms an antenna with the asphalt to exert an interlocking effect [18]. With the addition of fiber to the mixture, the asphalt mortar film wrapped on the aggregate surface becomes thicker, impeding the mutual sliding between asphalt and aggregate, which will enhance the overall ability of the structure to resist external loads and then improve tensile properties. This reflects the reinforcement effect of fiber on the mixture matrix. Negative effects can be brought about once fiber dosage exceeds a certain limit.

Moreover, unevenness in the following aspects can be observed in Figure 16i,j. First, fibers are dispersed in the mixture unevenly. Second, fibers are mechanically twisted and knotted with each other in the local region. Third, the asphalt adsorbed on the fibers has an uneven "flocculent" distribution.

Furthermore, with the increase in the dosage, the wrinkle at the fiber anchorage becomes more and more prominent, which indicates the trend of stripping and pull-out from the "root". From Figure 16j, it can be observed that there are pulled-out fibers, which can result in holes on the interface. The unevenness and pull-out of fiber can lead to undesirable negative effects, which will increase the internal defects of the mixture, degrade the internal continuity of the asphalt mixture, and reduce the compactness.

As long as the dosage of fiber is appropriate, the positive effect, namely, the bridging effect and adhesion effect, can be boosted. Nevertheless, when the dosage attains a certain value, the negative effect brought about by uneven dispersion and pull-out of fiber may outweigh the positive effect, which may thereby degrade the overall tensile property and crack the resistance of asphalt mixtures.

Based on the comparison among Figure 16g,h and Figure 16k–n, it can be found that basalt fiber has the best dispersion uniformity, followed by granular lignin fiber, and the dispersion uniformity of flocculent lignin fiber is the worst. This can be explained by the difference in the diameter of the three types of fibers. Specifically, a larger fiber diameter indicates a smaller number of fibers and a smaller amount of entanglement in the composited material [38]. Furthermore, basalt fiber has higher tensile strength and length–diameter ratio than the other two types of fibers, which contributes to its better performance in larger stress transfer and bridging. From this point of view, BFSMA is supposed to have better anti-tensile performance. However, what is noteworthy is that the asphalt film wrapped on the flocculent lignin fiber is the thickest, followed by the film wrapped on the granular lignin fiber, and the asphalt film wrapped on basalt fiber is the thinnest. The reason for this difference is that the lignin fiber has the largest specific surface area, as reflected by the largest number of convex structures observed on the surface, making it easier to absorb more and thicker asphalt. The superior tensile properties of GFSMA can be attributed to the situation where the positive effect brought about by the excellent adhesion properties outweighs its weakness in evenness and bridging.

4. Conclusions

In this research, the effects of fiber dosage and type on the tensile properties of three kinds of fiber-reinforced asphalt mixtures were investigated, respectively. In addition, the SEM test was performed to explore the influence mechanism of fiber dosage. According to the test results, the corresponding conclusions are summarized as follows:

The results of the monotonic direct tensile tests showed that when the fiber dosage keeps increasing within the given range, the tensile strength, elastic modulus, and strain energy density of the three kinds of fiber-reinforced asphalt mixtures first went up and then went down, while the ultimate strain showed a continuous rising trend.

In light of the criterion of maximal strain energy density, the optimum fiber dosage of GFSMA, FFSMA, and BFSMA was determined to be 0.50 wt%, 0.45 wt%, and 0.50 wt%, respectively. In addition, in terms of the strain energy density of three kinds of fiber-reinforced asphalt mixtures with the corresponding optimum dosage, the performance of GFSMA was the best, followed by FFSMA, and the performance of BFSMA was the worst.

The strain-controlled direct tensile fatigue tests showed that direct tensile fatigue of GFSMA at the fiber dosage of 0.50 wt% was superior to that of FFSMA at 0.45 wt% or BFSMA at 0.50 wt% in terms of fatigue life and cumulative dissipative energy. Therefore, it was concluded that granular fiber could more remarkably improve the tensile properties of asphalt mixtures, and its optimum dosage was 0.50 wt%.

The analysis results of the SEM test suggested that the essence of the influence of fiber dosage and type on the asphalt mixture is whether the reinforcement effect of the fiber on the mixture matrix outweighs the negative effect of interface defects between fiber and mixture matrix or the reverse applies.

Author Contributions: Conceptualization, S.Y. and Z.Z.; Data curation, S.Y. and K.L.; Formal analysis, S.Y.; Funding acquisition, Z.Z.; Investigation, S.Y.; Methodology, S.Y.; Resources, S.Y. and K.L.; Supervision, Z.Z.; Validation, S.Y. and K.L.; Writing—original draft, S.Y.; Writing—review & editing, S.Y. and Z.Z. All authors have read and agreed to the published version of the manuscript.

Funding: This research was funded by the Science and Technology Planning Project of Transportation of Guangdong Province, grant number 1912-0002.

Institutional Review Board Statement: Not applicable.

Informed Consent Statement: Not applicable.

Data Availability Statement: Not applicable.

Conflicts of Interest: The authors declare no conflict of interest.

References

1. Álvarez, A.E.; Walubita, L.F.; Sánchez, F. Using fracture energy to characterize the hot mix asphalt cracking resistance based on the direct-tensile test. *Rev. Fac. Ing.* **2012**, *20*, 126–137.
2. Xu, W.Z. Research on Cracking Behavior of High Dosage Hot Reclaimed Asphalt Mixture Based on Digital Speckle Method. Master's Thesis, Shenyang University of Civil Engineering and Architecture, Shenyang, China, 2021.
3. Alfalah, A.; Offenbacker, D.; Ali, A.; Decarlo, C.; Lein, W.; Mehta, Y.; Elshaer, M. Assessment of the impact of fiber types on the performance of fiber-reinforced hot mix asphalt. *Transp. Res. Rec.* **2020**, *2674*, 337–347. [CrossRef]
4. Sheng, Y.P.; Zhang, B.; Yan, Y.; Li, H.B.; Chen, Z.J.; Chen, H.X. Laboratory Investigation on the Use of Bamboo Fiber in Asphalt Mixtures for Enhanced Performance. *Arab. J. Sci. Eng.* **2019**, *44*, 4629–4638. [CrossRef]
5. Rathore, M.; Haritonovs, V.; Zaumanis, M. Performance evaluation of warm asphalt mixtures containing chemical additive and effect of incorporating high reclaimed asphalt content. *Materials* **2021**, *14*, 3793. [CrossRef] [PubMed]
6. Zheng, D.; Qian, Z.D.; Li, P.; Wang, L.B. Performance evaluation of high-elasticity asphalt mixture containing inorganic nano-titanium dioxide for applications in high altitude regions. *Constr. Build. Mater.* **2019**, *199*, 594–600. [CrossRef]
7. Zhang, K.; Liu, Y.; Nassiri, S.; Li, H.; Englund, K. Performance evaluation of porous asphalt mixture enhanced with high dosages of cured carbon fiber composite materials. *Constr. Build. Mater.* **2021**, *274*, 122066. [CrossRef]
8. Guo, Q.; Wang, H.; Gao, Y.; Jiao, Y.; Dong, Z. Investigation of the low-temperature properties and cracking resistance of fiber-reinforced asphalt concrete using the DIC technique. *Eng. Fract. Mech.* **2020**, *229*, 106951. [CrossRef]
9. Luo, D.; Khater, A.; Yue, Y.; Abdelsalam, M.; Zhang, Z. The performance of asphalt mixtures modified with lignin fiber and glass fiber: A review. *Constr. Build. Mater.* **2019**, *209*, 377–387. [CrossRef]
10. Alfalah, A.; Offenbacker, D.; Ali, A.; Mehta, Y. Evaluating the impact of fiber type and dosage rate on laboratory performance of Fiber-Reinforced asphalt mixtures. *Constr. Build. Mater.* **2021**, *310*, 125217. [CrossRef]
11. Slebi-Acevedo, C.J.; Lastra-González, P.; Pascual-Muñoz, P.; and Castro-Fresno, D. Mechanical performance of fibers in hot mix asphalt: A review. *Constr. Build. Mater.* **2019**, *200*, 756–769. [CrossRef]
12. Guan, B.; Chen, S.; Xiong, R. Low temperature anti-cracking performance of brucite-fiber-reinforced asphalt concrete. *Adv. Mater. Res.* **2011**, *228*, 23–28. [CrossRef]
13. Zhao, H.; Guan, B.; Xiong, R.; Zhang, A. Investigation of the performance of basalt fiber reinforced asphalt mixture. *Appl. Sci.* **2020**, *10*, 1561. [CrossRef]
14. Eisa, M.S.; Basiouny, M.E.; Daloob, M.I. Effect of adding glass fiber on the properties of asphalt mix. *Int. J. Pavement Res. Technol.* **2021**, *14*, 403–409. [CrossRef]
15. Gupta, A.; Castro-Fresno, D.; Lastra-Gonzalez, P.; Rodriguez-Hernandez, J. Selection of fibers to improve porous asphalt mixtures using multi-criteria analysis. *Constr. Build. Mater.* **2021**, *266*, 121198. [CrossRef]
16. Wu, B.; Wu, X.; Xiao, P.; Chen, C.; Xia, J.; Lou, K. Evaluation of the long-term performances of SMA-13 containing different fibers. *Appl. Sci.* **2021**, *11*, 5145. [CrossRef]
17. Zhu, Y.; Li, Y.; Si, C.; Shi, X.; Qiao, Y.; Li, H. Laboratory evaluation on performance of fiber-modified asphalt mixtures containing high percentage of RAP. *Adv. Civ. Eng.* **2020**, *4*, 1–9. [CrossRef]
18. Zhang, J.; Huang, W.; Zhang, Y.; Lv, Q.; Yan, C. Evaluating four typical fibers used for OGFC mixture modification regarding drainage, raveling, rutting and fatigue resistance. *Constr. Build. Mater.* **2020**, *253*, 119131. [CrossRef]
19. Du, Y.; Ma, Y. Study on tensile properties of carbon fiber epoxy resin composites in humid and hot environment. *J. Northwestern Polytech. Univ.* **2022**, *40*, 33–39. [CrossRef]

20. Li, L.; Luan, Y.; Wu, J.; Du, X.; Wu, W. Test study on dynamic tensile behaviors of steel Grid-polyethylene fiber reinforced engineered cementitious composites at low and medium loading rates. *Mater. Rep.* **2022**, *36*, 49–54.

21. Guo, Q.; Li, G.; Gao, Y.; Wang, K.; Dong, Z.; Liu, F.; Zhu, H. Experimental investigation on bonding property of asphalt-aggregate interface under the actions of salt immersion and freeze-thaw cycles. *Constr. Build. Mater.* **2019**, *206*, 590–599. [CrossRef]

22. Luo, Y.; Zhang, K.; Li, P.; Yang, J.; Xie, X. Performance evaluation of stone mastic asphalt mixture with different high viscosity modified asphalt based on laboratory tests. *Constr. Build. Mater.* **2019**, *225*, 214–222. [CrossRef]

23. Wu, B.; Meng, W.; Xia, j.; Xiao, P. Influence of basalt fibers on the crack resistance of asphalt Mixtures and mechanism analysis. *Materials* **2022**, *15*, 744. [CrossRef]

24. Wan, J.; Wu, S.; Xiao, Y.; Liu, Q.; Schlangen, E. Characteristics of ceramic fiber modified asphalt mortar. *Materials* **2016**, *9*, 788. [CrossRef]

25. Fu, Z.; Tang, Y.; Ma, F.; Wang, Y.; Shi, K.; Dai, J.; Hou, Y.; Li, J. Rheological properties of asphalt binder modified by nano-TiO_2/ZnO and basalt fiber. *Constr. Build. Mater.* **2022**, *320*, 126323. [CrossRef]

26. Phan, T.M.; Nguyen, S.N.; Seo, C.B.; Park, D.W. Effect of treated fibers on performance of asphalt mixture. *Constr. Build. Mater.* **2021**, *274*, 122051. [CrossRef]

27. Du, Y.; Wang, J.; Chen, J. Cooling asphalt pavement by increasing thermal conductivity of steel fiber asphalt mixture. *Sol. Energy* **2021**, *217*, 308–316.

28. Pei, Z.; Lou, K.; Kong, H.; Wu, B.; Wu, X.; Xiao, P.; Qi, Y. Effects of fiber diameter on crack resistance of asphalt mixtures reinforced by basalt fibers based on digital image correlation technology. *Materials* **2021**, *14*, 7426. [CrossRef] [PubMed]

29. Xu, C.; Zhang, Z.; Liu, F. Improving the low-temperature performance of RET modified asphalt mixture with different modifiers. *Coatings* **2020**, *10*, 1070. [CrossRef]

30. Xia, C.; Wu, C.; Liu, K.; Jiang, K. Study on the Durability of bamboo fiber asphalt mixture. *Materials* **2021**, *14*, 1667. [CrossRef]

31. Xiong, A.; Huang, W.; Lv, Q.; Guan, W. Selection of low temperature performance index of asphalt mixture evaluated by semi-circular bending tensile test. *Pet. Asph.* **2020**, *34*, 33–39.

32. Li, X.; Gibson, N.; Youtcheff, J. Evaluation of asphalt mixture cracking performance using monotonic direct tension test in the AMPT. *Asph. Paving Technol.* **2017**, *86*, 603–627. [CrossRef]

33. Wu, S.; He, R.; Chen, H.; Luo, Y. Low temperature characteristics of asphalt mixture based on the semi-circular bend and thermal stress restrained specimen test in alpine cold regions. *Constr. Build. Mater.* **2021**, *311*, 125300. [CrossRef]

34. Zhou, Z.; Yuan, X.; Tan, H. Fatiuge damage of asphalt mixture specimen with pre-made gaps in direct tensio test analysis. *Chin. J. Highw. Transport.* **2013**, *26*, 8.

35. Zhang, C.; Qian, Z.D.; Liu, Y.; Wang, X. Evaluation and influence factors of dispersion property of cellulose fiber pellets in SMA mixture. *J. Southeast Univ. Nat. Sci. Ed.* **2018**, *48*, 713–718.

36. Wei, H.; Zhang, H.; Li, J.; Zheng, J.; Ren, J. Effect of loading rate on failure characteristics of asphalt mixtures using acoustic emission technique. *Constr. Build. Mater.* **2023**, *364*, 129835. [CrossRef]

37. Xue, Y.C.; Qian, Z.D.; Xia, R.H. Study on rubber granule epoxy asphalt mixture based on low temperature performance. *J. Hunan Univ. Nat. Sci. Ed.* **2016**, *43*, 120–128.

38. Zhu, S.; Xu, Z.; Qin, X.; Liao, M. Fiber-reinforcing effect in the mechanical and road performance of Cement-emulsified asphalt mixtures. *Materials* **2021**, *14*, 2779. [CrossRef]

materials

Article

Evaluation of the Tensile and Puncture Properties of Geotextiles Influenced by Soil Moisture under Freezing Conditions

Lanjun Liu [1], Haiku Zhang [2], Jinhuan Zhu [1], Shixin Lv [3] and Lulu Liu [3,4,*]

[1] Shangqiu Institute of Technology, Shangqiu 476000, China; lljtmzy@163.com (L.L.); z2434837747@163.com (J.Z.)
[2] Fuyuan Ding Installation Engineering Company Limited, Shangqiu 476000, China; 15729203788@163.com
[3] School of Highway, Chang'an University, Xi'an 710064, China; lsx1052332819@163.com
[4] School of Mechanics and Civil Engineering, China University of Mining and Technology, Xuzhou 221116, China
* Correspondence: liululu@cumt.edu.cn

Abstract: Freezing conditions under different humidity will influence the mechanical properties of geotextiles, leading to the gradual fracture of geotextiles. It brings hidden danger to the whole isolation, reinforcement and protection of rock and soil. It is particularly important to study the tensile and puncture properties of geotextiles considering low temperature and moisture content. In this paper, a series of tensile and puncture tests of geotextiles are performed under different low temperatures (0, −3, −6, −9, and −12 °C) and at different moisture content levels (0, 5, 10, 30, 50, and 80%). From the microscopic perspective, the failure mechanism considering the low temperature and moisture content was explained comprehensively. Experimental results indicate that with a decrease in freezing temperature, the tensile strength of geotextiles increases as a parabolic function while the elongation at failure decreases as an exponential function. Additionally, the puncture strength of geotextiles presented a parabolic increase with the decreasing temperature. Under the freezing temperature environment, the higher moisture content of geotextiles can generate a higher puncture strength increment. This research contributes to a more comprehensive understanding of the tensile and puncture properties of geotextile materials considering low temperature and moisture content. It can provide important guidance for the design of slopes, the reinforcement of earthen dams, and roadbed reinforcement with geotextiles in cold regions.

Keywords: geotextiles; tensile and puncture properties; failure mechanism; low temperature; moisture content

Citation: Liu, L.; Zhang, H.; Zhu, J.; Lv, S.; Liu, L. Evaluation of the Tensile and Puncture Properties of Geotextiles Influenced by Soil Moisture under Freezing Conditions. *Materials* **2024**, *17*, 376. https://doi.org/10.3390/ma17020376

Academic Editors: Changho Lee and Andrea Sorrentino

Received: 15 November 2023
Revised: 27 December 2023
Accepted: 29 December 2023
Published: 11 January 2024

1. Introduction

Geotextile materials have emerged as a cornerstone in the realm of civil engineering, particularly revered for their multifaceted utility in projects requiring enduring resilience and adaptability [1–3]. Their applications span an impressive range of engineering challenges, from fortifying roadbeds and buttressing slopes to facilitating highway upkeep and use in the undergirding of tunnel construction. The widespread adoption of geotextiles is attributable to a unique combination of physical properties: their lightness simplifies handling and transportation; excellent permeability allows for efficient water drainage while averting waterlogging; remarkable tensile strength ensures durability under strain; and resistance to high temperatures and corrosion guarantees longevity in harsh environments [4]. These characteristics are not merely theoretical assertions but are well documented in a number of scientific studies [5–7]. In the specific context of geotechnical engineering, geotextiles serve a pivotal role in safeguarding the integrity of geological substrates. When applied, they are subjected to a biaxial tension state, a direct consequence of the load distribution from above and lateral constraints. However, this state of stress renders them vulnerable to potential damage. For instance, when encased materials shift, generating stresses that exceed the fabric's tensile and penetration thresholds, or when they encounter sub-surface sharp objects like gravel or stones, the integrity of the geotextiles can be compromised [8,9]. The occurrence of such punctures not only undermines the material's protective capabilities but also poses a latent risk to the structural stability of the engineering project. Given these risks, the study of geotextile durability—specifically, tensile and puncture resistance—is of paramount importance [10]. Delving into the properties that govern their performance under stress can yield valuable insights, leading to the development of more robust geotextile variants. This knowledge is essential for advancing the reliability of geotextile applications in critical infrastructure, thereby ensuring that these materials continue to perform their intended protective functions throughout the lifespan of the engineering projects they support [11].

Geotextiles, when deployed as a layer embedded within the soil, are highly susceptible to the prevailing environmental conditions. This susceptibility becomes notably significant in colder climates where sub-zero temperatures are common. In such environments, the drop in temperature leads to the formation of ice within the soil matrix, which can severely impact the geotextiles by altering their structural characteristics. The presence of ice can drastically affect tensile strength, which is the capacity of the geotextile to withstand pulling forces, and puncture resistance, which is the ability to resist breaking or tearing upon the application of a sharp force [12]. Considering the pivotal role that geotextiles play as a reinforcing and stabilizing element in construction, it becomes imperative to conduct an in-depth analysis of their performance. This analysis must focus on understanding how low-temperature scenarios, especially when coupled with moisture, can influence the material properties and integrity of geotextiles. By doing so, it is possible to ensure the reliability and durability of construction projects that incorporate these materials in challenging weather conditions.

As for geotextiles, the investigations in the literature mainly focused on evaluating tensile properties [13,14], puncture properties, creep behavior [15], interfacial behavior [16,17], and microstructure [18,19] under ambient temperature conditions.

The effects of high temperature on the behavior of geotextiles have been investigated by several researchers. For example, Kongkitkul et al. indicated that the tensile rupture strength of polypropylene (PP), high-density polyethylene (HDPE), and polyester (PET) geogrids decreased by 9.2%, 26.7%, and 4.5%, respectively, as the temperature increased from 30 to 50 °C [20]. Similarly, Chantachot et al. also revealed that an increasing ambient temperature could reduce the rupture tensile strength and elastic stiffness of geogrids while improving creep strain during sustained loading, which was consistent with the previous results [21,22].

So many studies have been conducted on the performance of geotextiles working under and above ambient conditions, while limited knowledge is available on the behavior of geotextiles working in freezing temperatures. Allen (1983) reported that a freezing temperature (-12 °C) could reduce the strain at failure value of needle-punched polyester by 59–86% compared with an ambient temperature, while the failure modulus and elongation could reach 60% and 88% of the ambient temperature, respectively. Nevertheless, there is a lack of in-depth understanding regarding the tensile and puncture properties of geotextiles considering low-temperature and wet conditions, especially the failure mechanism of geotextiles due to the formation of ice under freezing conditions.

In general, considering the influence of low temperature and moisture content, the tensile and puncture properties and failure mechanism of geotextiles need further study. In light of this, a series of tensile and puncture tests were performed in this study. The effect of different freezing temperatures and moisture content on tensile strength, elongations at failure, and puncture strength are analyzed in detail. Furthermore, the failure mechanism considering low temperature and moisture content are explained comprehensively.

2. Experimental Content

2.1. Materials

In this comprehensive study, the geotextile samples under examination were meticulously sourced from an active construction zone located along the Bao-Han highway, specifically pinpointed at a marker of K36 + 300 within the Baoji region of Shaanxi Province, China. The production process of these geotextiles involves a mechanical interlacing technique that intricately combines multiple layers of diverse fibers, all aligned in the direction of the machinery to ensure uniformity and structural integrity. The resulting geotextile product boasts an average thickness measurement of 2.8 mm, which contributes to its substantial feel and effectiveness in application. Additionally, it presents a significant mass per unit area, registering at 400 g/m^2, which is indicative of its robustness and suitability for the demanding tasks it is designed for in civil engineering projects.

In this study, we have chosen polypropylene and polyester fibers as the primary materials for the geotextile. After careful consideration, we decided on a 50/50 ratio of polypropylene to polyester fibers. This ratio balances the mechanical strength and durability of the materials, making them suitable for a variety of civil engineering applications. This decision is based on a comprehensive evaluation of the properties of both fibers, aimed at ensuring the geotextile exhibits optimal performance characteristics.

2.2. Experimental Apparatus

A tensile testing machine was used to test the tensile strength and elongations. The machine has a variety of stress and strain control modes, with a load rate of 0.5–500 mm/min, a maximum load level of 100 kN, and load control precision of ±0.1% [23].

A puncture testing machine was applied to measure puncture strength with a strength test range of 0–20 kN, a puncture down speed of 100 mm/min, and an effective maximum dynamic range of 90 mm [24].

A freeze–thaw box was used to simulate the ambient and freezing temperatures for geotextile specimens. The adjustable temperature ranged from 80 to −20 °C and control accuracy was ±0.1 °C.

In the experimental section of our paper, particularly in the subsection discussing SEM images and characterization techniques, the following paragraph is added: 'To comprehensively characterize the geotextile samples, Scanning Electron Microscopy (SEM) was employed for microstructural analysis. The specific SEM equipment utilized was the JEOL JSM-7800F Prime, manufactured by JEOL Ltd., Tokyo, Japan. This high-resolution field emission SEM is renowned for its exceptional imaging capabilities, with technical specifications including magnification up to 200,000× and superior resolution. The sample preparation process involved coating and appropriate drying of the geotextile samples, ensuring high-quality images for SEM analysis. These detailed equipment specifications and sample preparation steps are crucial for understanding the reproducibility and accuracy of the experimental results.

In this study, we employed the Instron Universal Testing Machine, model 5969, produced at Nanjing Instrument Factory in Jiangsu Province, China, for the testing of tensile strength and elongation. The Instron 5969 is widely recognized and used in the field of material testing due to its high precision in control and its extensive range in measuring stress and strain. Its broad range of stress and strain control modes, along with a maximum load level of 100 kN and load control precision of ±0.1%, make it an ideal choice for our study. The high accuracy and reliability of this equipment are crucial for ensuring the precision of our geotextile material performance tests.

The tensile tests were conducted in accordance with the ASTM D4595 [23] standard, while the puncture tests followed the guidelines of ASTM D6241 [24].

2.3. Specimen Preparation

In preparation for a comprehensive study on the effects of extreme cold on geotextiles, a vital material in construction, specimens were meticulously prepared to mimic the harsh winter conditions of Shaanxi Province. During Shaanxi's coldest months, temperatures can plummet to as low as -10 °C, posing a significant challenge to the durability and performance of construction materials. In order to understand how these cold temperatures affect the mechanical properties of geotextiles, researchers conducted a series of humidity tests on the samples, particularly at 0%, 5%, 15%, 30%, 50%, and 80% humidity. The methodology was meticulous, with each geotextile first equilibrated to its specific moisture content by being placed in a controlled curing chamber maintained at a constant 20 ± 2 °C at 95% relative humidity for 72 h. Following this, they were then exposed to a freeze–thaw cycle in a box set at varying freezing temperatures of -3, -6, -9, and -12 °C for 12 h to simulate the harsh winter conditions. Upon removal from these severe freezing conditions, the geotextile samples were immediately subjected to mechanical testing, specifically tensile and puncture resistance tests. This quick transition from the freeze–thaw box to testing was critical to minimize any potential alterations in the samples due to temperature changes, which could compromise the integrity of the data. The production of the geotextile employed a mechanical interlacing technique, effectively combining polypropylene and polyester fibers in a layered structure, ensuring product uniformity and structural integrity. Additionally, vinylon was incorporated into the fiber composition to enhance the water and heat resistance of the geotextile in specific environments. The inclusion of vinylon is based on its unique performance attributes, allowing the geotextile to be suitable for a wider range of applications, especially in projects demanding high water and heat resistance. In addition, five parallel specimens were prepared for each test scheme. Under the given moisture content and test temperature, five parallel specimens were tested, and the average values were taken as test results to reduce fluctuation errors in test data.

3. Results and Discussion

3.1. Tensile Properties

3.1.1. Tensile Strength

Figure 1a illustrates the relationship between tensile strength and temperature across various levels of moisture content within geotextile materials. From the graph, a clear pattern emerges where the tensile strength appears to elevate in a parabolic manner as the temperature drops below the freezing point. This trend is consistent across the range of temperatures studied, suggesting a robust link between colder conditions and the mechanical properties of the materials in question. It is noteworthy that, at any specific freezing temperature, the tensile strength is positively correlated with the moisture content; in other words, the more moisture present within the geotextile fibers, the higher the tensile strength recorded. This enhancement of tensile strength in wetter conditions, particularly under sub-zero temperatures, implies that the presence of water molecules may be playing a crucial role in reinforcing the structural integrity of the geotextiles when they are subjected to freezing conditions. The data may suggest that the formation of ice crystals within the fibers could be contributing to this strengthening effect, as the freezing process possibly facilitates additional bonding within the material structure, leading to an increase in tensile resistance. This phenomenon highlights a potential avenue for improving the performance of geotextile materials in cold and wet environments by optimizing moisture content to harness this strength-augmenting effect.

In the freezing temperature environment, the molecular activity inside the geotextile is hindered, and the intermolecular force increases. Therefore, a greater external load is required to overcome the increased interaction between molecular bonds and cause the directional arrangement between molecular bonds to move [18]. For a given temperature, the change rate of tensile strength under different moisture content (5–80%) is between 13.0 and 14.67% compared with that under dry conditions. For a given moisture content, the change rate of tensile strength at different freezing temperatures (-3–12 °C) is 11.71–13.87% compared with that under the temperature of 0 °C. Therefore, the effect of freezing temperature on tensile strength is less than that of moisture content.

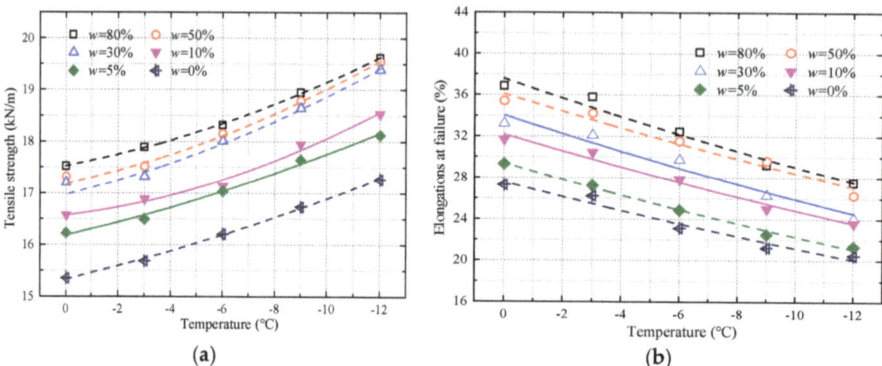

Figure 1. Tensile strength and elongations at failure versus temperature under different moisture content in geotextiles: (**a**) tensile strength versus temperature; (**b**) elongations at failure versus temperature.

Three function models (the exponential, linear, and parabolic functions) were used to fit the test data. The parabolic function can present the best fitted effect (Table 1), and the fitted equation of the tensile strength and freezing temperature of geotextiles is as follows:

$$Pt = aT^2 + bT + c \tag{1}$$

where Pt represents the tensile strength, kN/m; T represents the temperature, °C; and a, b, and c represent the fitted coefficients.

Table 1. Fitted results of temperature and the tensile strength of geotextiles.

Moisture Content	Fitted Equations	Correlation Coefficient
0%	$Pt = -0.0033T^2 - 0.123T + 15.339$	0.996
5%	$Pt = -0.0039T^2 - 0.183T + 16.19$	0.987
15%	$Pt = -0.0087T^2 - 0.062T + 16.575$	0.977
30%	$Pt = 0.0068T^2 - 0.122T + 16.983$	0.993
50%	$Pt = 0.007^2 - 0.112T + 17.184$	0.996
80%	$Pt = 0.0064T^2 - 0.099T + 17.522$	0.999

3.1.2. Elongations at Failure

From 0 to -12 °C, the change in elongations at failure for geotextiles with a moisture content of 30% was largest, reaching 34.22%. When moisture content was less than 5%, the influence of moisture content on elongations at failure was decreased gradually compared with higher moisture content. It was noteworthy that, at the same freezing temperature, the wet geotextiles that generated more ice exhibited more brittle behavior than the dry geotextiles, which generated lower elongations at failure. In Figure 1b, a comprehensive analysis showcases the impact of temperature on the stretching capability of geotextiles at the point of rupture when subjected to varying levels of moisture content. The study indicates a clear correlation between the decrease in freezing temperature and the reduced capacity of the geotextiles to elongate upon reaching their failure threshold. This reduction is not arbitrary but follows an exponential relationship with the descending temperatures, highlighting a fundamental change in material behavior in colder climates.

While there is an observable increase in the tensile strength of geotextiles as the temperature drops, it comes at a significant cost to the material's overall ductility. In environments characterized by a specific moisture content paired with freezing temperatures, the toughness of the geotextile material, or the ability to absorb energy prior to failure, was adversely affected. This transition results in a compromised yield capacity, where the material's molecular chains relax more swiftly, thus precipitating a marked decrease in the ability of the geotextiles to stretch to their limits without breaking.

This phenomenon was particularly evident when the geotextiles were exposed to a temperature range from 0 to $-12\,^\circ$C with 30% moisture content. Under these conditions, the elongations at failure observed a significant decline, with a recorded decrease of 34.22%. Interestingly, when the moisture content was maintained below 5%, the impact on the elongations at failure began to taper off, especially when compared to geotextiles with higher moisture content. This suggests that moisture content plays a critical role in the material's performance, particularly under freezing conditions. A notable observation was that geotextiles with higher moisture content, which consequently formed more ice within their structure, tended to exhibit a more brittle nature. This brittleness translated into a lower threshold for elongation at failure compared to their drier counterparts. Such a difference underscores the importance of considering both temperature and moisture content when evaluating the durability and suitability of geotextile materials for use in cold environments, where ice formation can drastically alter their mechanical properties and performance.

Similar to tensile strength, the exponential function model (Table 2) can describe the change tendency of elongations at failure with temperature, and the fitted equation of elongations at failure under freezing temperatures of geotextiles is as follows:

$$L = de^{fT} \tag{2}$$

where L represents the elongations at failure, %, and d, e, and f represent the fitted coefficient.

Table 2. Fitted results of temperature and the elongations of geotextiles at failure.

Moisture Content	Fitted Equations	Correlation Coefficient
0%	L = 27.58e0.126T	0.964
5%	L = 29.4e0.028T	0.992
15%	L = 32.18e0.026T	0.972
30%	L = 34.05e0.027T	0.951
50%	L = 36.09e0.024T	0.954
80%	L = 37.56e0.026T	0.965

3.2. Puncture Properties

3.2.1. Puncture Strength

Figure 2a provides a graphical representation of the relationship between the puncture strength of geotextiles and temperature across various moisture levels. From the chart, it becomes evident that as the moisture content in the geotextiles escalates from a completely dry state to a saturation level of 80%, there is only a minor decline in puncture strength, specifically a reduction of 0.09 kilonewtons (kN) at room temperature. This marginal change indicates noteworthy resistance in the material's puncture strength in relation to moisture alteration. The underlying reason for this resilience can be attributed to the intermolecular friction among the polymer chains constituting the geotextiles, which only moderately diminishes as the moisture content rises. Despite the presence of moisture, the fibers maintain a considerable amount of their interaction, hence preserving the structural integrity of the geotextile.

However, an interesting phenomenon occurs when the geotextiles are subjected to sub-zero temperatures. As the temperature drops below the freezing point, to $-3\,^\circ$C, $-6\,^\circ$C, $-9\,^\circ$C, and then to $-12\,^\circ$C, there is a noticeable incremental increase in the material's puncture strength, with increases of 0.88 kN, 0.91 kN, 0.94 kN, 1.04 kN, and 1.16 kN, respectively. This trend clearly demonstrates that freezing temperatures act to enhance the puncture strength of the geotextile fabric. This is likely due to the formation of ice within the fabric, which binds the polymer molecules more tightly together, essentially reinforcing the material's structure against puncture forces.

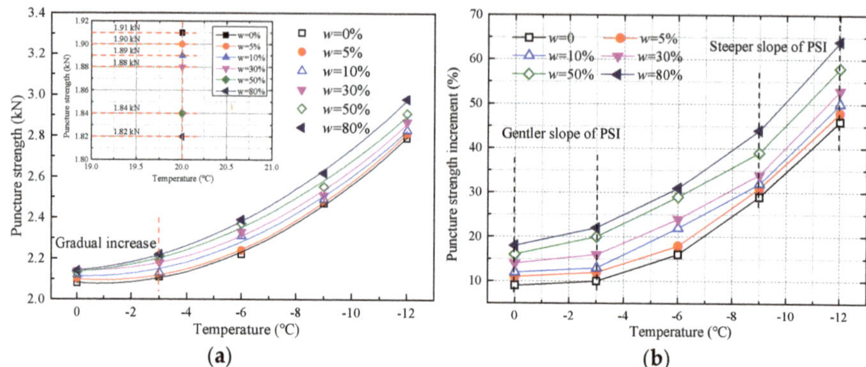

Figure 2. Puncture strength and puncture strength increment versus temperature at different moisture content levels: (**a**) puncture strength versus temperature; (**b**) puncture strength increment versus temperature.

Notably, the rate at which the puncture strength increases with the decrease in temperature is not constant. In the temperature range from 0 °C to −3 °C, the rate of increase in puncture strength is relatively slower. In contrast, from −3 °C to −12 °C, the rate of increase is more pronounced. The explanation for this might be associated with the stage at which the geotextiles initially have to disrupt the ice's bond strength before contending with the intermolecular bonding strength of the polymers. As the temperature plunges, the bonding force between the polymer particles intensifies due to decreased molecular activity. Consequently, a greater amount of external force is required to disrupt these bonds, resulting in the amplification of puncture strength. This suggests that geotextiles exhibit an enhanced performance under freezing conditions, making them particularly effective for applications in cold environments where they are subjected to sharp objects or conditions that could potentially cause punctures [25].

In addition, the parabolic function (Table 3) with the best fitted effect for puncture strength and freezing temperature was obtained:

$$P_p = gT^2 + hT + i \tag{3}$$

where Pp represents the puncture strength, kN; T represents the temperature, °C; and g, h, and i represent fitted coefficients.

Table 3. Fitted results of temperature and the puncture strength of geotextiles.

Moisture Content/%	Fitted Equation	R2
0%	Pp = 0.0057T² − 0.0029T + 2.0809	0.9979
5%	Pp = 0.0059T² − 0.0011T + 2.097	0.9982
10%	Pp = 0.0053T² − 0.0041T + 2.114	0.9945
30%	Pp = 0.0053T² − 0.0041T + 2.144	0.9558
50%	Pp = 0.0046T² − 0.0081T + 2.137	0.9894
80%	Pp = 0.0049T² − 0.0103T + 2.143	0.9842

3.2.2. Puncture Strength Increment

To describe the effect of low temperature on the puncture strength of geotextiles under different moisture content levels more comprehensively, the concept of the puncture strength increment (PSI) of geotextiles was proposed:

$$PSI = 100 \times (Pp\text{-}Pp\text{−}20\,°C)/Pp\text{−}20\,°C \tag{4}$$

where PSI denotes the puncture strength increment, %; Pp denotes the puncture strength at different freezing temperature, kN; and Pp−20 °C denotes the puncture strength at an ambient temperature (20 °C), kN.

The relationship between the puncture strength increment and temperature at different moisture content levels is plotted in Figure 2b. At temperatures less than −3 °C, there is a gentler slope in PSI, while the steeper slope in PSI occurred when the temperature exceeded −9 °C. The obvious variation in PSI under different moisture content levels can be noted. For example, at a given temperature (−6 °C), the PSI at the moisture content level of 80% was 1.78 times that of the PSI at a moisture content level of 5%. The lower the temperature, the greater the puncture strength increment was. Additionally, under the freezing temperature environment, higher moisture content in geotextiles can generate greater puncture strength increments.

3.3. Failure Mechanism Analysis Based on Low Temperature and Moisture Content

The main components of staple needle-punched geotextiles are polyester, polypropylene, and vinylon, each of which has porous spaces [18]. The 'geotextile matrix' refers to the intricate network of fibers composing the geotextile fabric. Within this matrix, water molecules primarily situate in the interstices between fibers.

As temperatures drop below freezing, these water molecules freeze, forming ice within these interstitial spaces. This ice formation occurs on a macroscopic scale within the voids between fibers, rather than within the fibers themselves. It is important to note that the ice does not create a continuous matrix akin to that in composite materials but rather fills the gaps between fibers, altering the fabric's overall structure.

The comprehensive analysis of Figures 3 and 4 reveals a critical aspect of material science concerning geotextiles subjected to tensile stresses under freezing conditions. Through polymer molecular bonding, the internal mechanism change chart of geotextiles under freezing temperatures and ambient temperatures is established in Figures 3 and 4, respectively. Among them, e_{ef} and e_{pf} represent the elastic and plastic deformation of geotextiles at freezing temperatures. e_{eu} and e_{pu} denote the elastic and plastic deformation of geotextiles at room temperature.

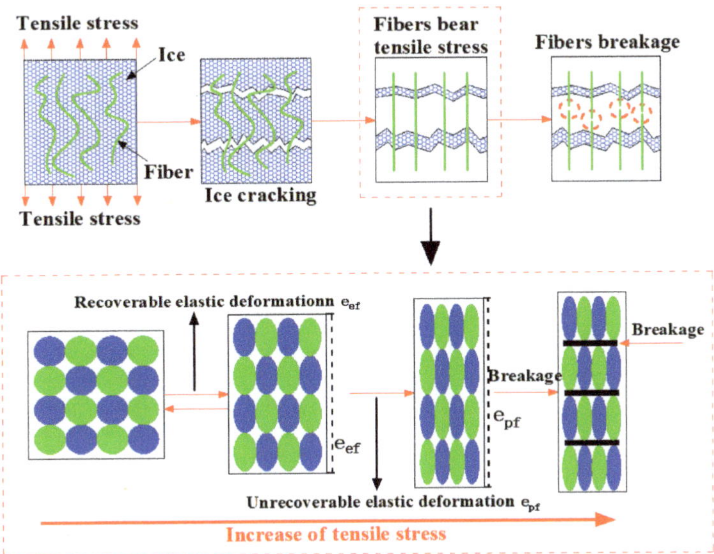

Figure 3. Analysis of the failure mechanism in wet geotextiles at freezing temperatures.

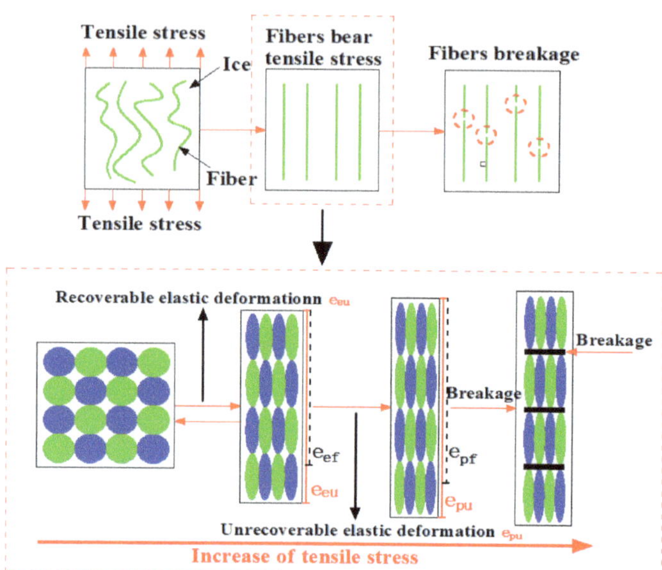

Figure 4. Analysis of the failure mechanism in dry geotextiles at normal temperatures.

The formation of ice impacts the mechanical properties of the geotextile by introducing an additional phase within the matrix, which changes the way the material responds to mechanical forces. When subjected to tensile stress, the geotextile now contains both the original fibrous material and the interspersed ice, leading to a composite-like behavior where both components contribute to the overall response of the material to stress. This unique interaction at the macroscopic level, between the geotextile fibers and the ice, is critical to understanding the failure mechanism of geotextiles under freezing conditions. From the failure progress of geotextiles, the puncture failure can be regarded as the tensile failure.

This phenomenon has implications for the mechanical properties of geotextiles. The inherent bonding strength between the polymer chains that make up the geotextile fibers, and that between the ice molecules, forms a dual barrier to material failure. Initially, as the tensile load is applied, it is the ice that resists, with its molecular bonds cleaving under stress. Following this, the geotextile fibers, now interlaced with fragments of shattered ice, begin to bear the load. These ice fragments act almost like a secondary reinforcing agent, creating an additional bonding force that enhances the tensile capacity of the geotextile fibers. This interaction continues until the ultimate tensile strength of the polymer chains is reached and they succumb to the applied force, leading to failure.

To further elaborate on the directional nature of the measurements, our tensile tests were primarily conducted along the principal fiber orientation of the geotextiles, aligning with their structural and manufacturing design. This direction, typically parallel or perpendicular to the roll direction, is crucial for assessing tensile behavior under freezing conditions. The formation of ice in the geotextile matrix and its subsequent impact on tensile strength and failure were thoroughly evaluated in these specific directions. This approach provides insights into how the ice influences the geotextile's mechanical response when subjected to stress in these key orientations.

Furthermore, Figure 5a–f provide additional insights into the behavior of geotextiles under these conditions. These figures depict the directional nature of fiber failure in wet geotextiles as compared to those in dry conditions. The ice plays a significant role in this directional failure, suggesting that the presence of frozen water within the fabric influences the alignment and subsequent breakage pattern of the fibers. The SEM analysis presented in these figures, particularly between Figure 5c,e, illustrates a clear correlation between temperature and the orderly nature of the failure. The colder the temperature, the more pronounced and orderly the direction of geotextile failure, underscoring the role of temperature in the mechanical failure process of geotextile fibers.

In conclusion, the interactive effects of temperature, ice formation, and polymer bonding within geotextiles present a complex mechanical behavior under tensile stress. The study presented in these figures contributes significantly to our understanding of the performance and resilience of geotextile materials in harsh, freezing environments, offering valuable insights for the design and application of geotextiles in cold regions.

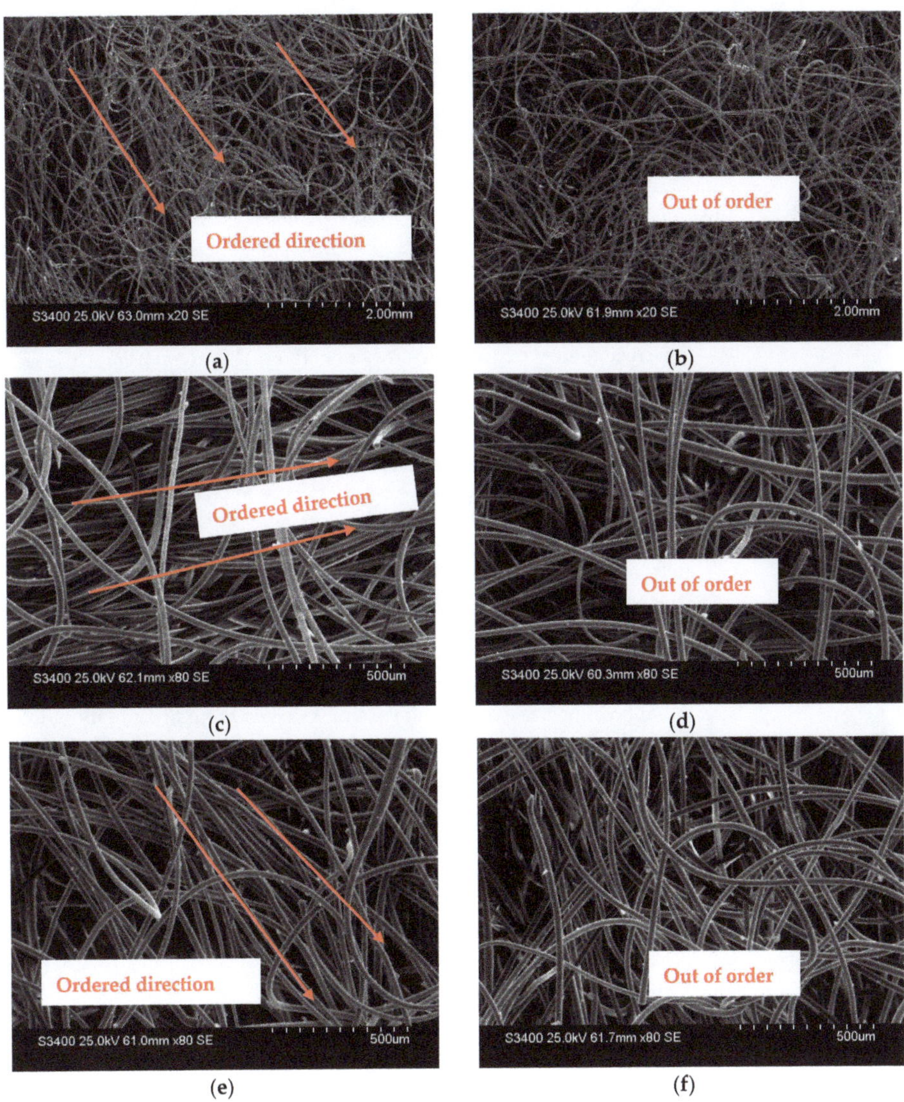

Figure 5. SEM images of geotextiles after tensile failure in dry and wet conditions. (**a**) T = −5 °C, w = 80%, 20 times. (**b**) T = −5 °C, w = 0%, 20 times. (**c**) T = −3 °C, w = 80%, 80 times. (**d**) T = −3 °C, w = 0%, 80 times. (**e**) T = −6 °C, w = 80%, 80 times. (**f**) T = −6 °C, w = 0%, 80 times.

When the geotextile fiber begins to bear forces, due to the effect of freezing temperature on the geotextile fiber, the elastic deformation (e_{ef}) at freezing temperatures is less than (e_{eu}) at ambient temperatures, and the plastic deformation (e_{pf}) at freezing temperatures is less than (e_{pu}) at ambient temperatures. In the elastic deformation stage, only recoverable tensile deformation occurs for polymer molecules due to the shorter time period. In the plastic deformation stage, the polymer molecule exhibits irreversible deformation, and only part of the deformation can be restored [25]. In comparison with the ambient temperature, although the tensile strength increases under the freezing temperature, the overall toughness decreases, which enhances brittleness and decreases the yield ability of the geotextile fiber. Consequently, the activity of polymer molecular chains decreases under the influence of low temperature, and the active region gradually shrinks, shortening the elastic deformation and plastic deformation phases [26,27].

4. Conclusions and Suggestions

In this study, a series of laboratory tests were conducted to evaluate the tensile and puncture properties of geotextiles considering the effect of various freezing temperatures and moisture content levels. The main conclusions are drawn as follows:

1. The parabolic and exponential function model of tensile strength and elongations at failure with the decreasing temperature of geotextiles was proposed considering different moisture content levels. In addition, the tensile strength and elongations at failure reached the maximum values of 14.67% and 34.22%, respectively, at the moisture content of 30% when the temperature decreased from 0 to $-12\,^{\circ}\mathrm{C}$.
2. The puncture strength of geotextiles presents a parabolic increase with decreasing temperature. Moreover, under the freezing temperature environment, the higher the moisture content of geotextiles, the greater the puncture strength increment is.
3. Under the freezing temperature, the appearance of broken ice bodies surrounding the geotextile fibers provides a bonding force to resist tensile failure and enhances the tensile strength of the geotextiles. At low temperature, the activity of polymer molecular chains decreases and the active zone gradually shrinks, thus shortening the deformation of geotextiles.

Our methodology in this study primarily involves the application of freeze–thaw cycles to assess the durability and performance of geotextiles. These cycles simulate real-world environmental conditions, providing valuable data on the materials' resilience. Moving forward, we plan to expand our research to include a broader range of environmental simulations, such as exposure to varying levels of humidity and ultraviolet radiation. This comprehensive approach will enable a deeper understanding of the long-term performance of geotextiles in diverse conditions, contributing significantly to the field of civil engineering materials.

Author Contributions: Writing—original draft, L.L. (Lanjun Liu); methodology, H.Z. and J.Z.; writing—review and editing, L.L. (Lulu Liu) and S.L. All authors have read and agreed to the published version of the manuscript.

Funding: The authors are grateful to the National Natural Science Foundation of China (Grant No. 42302320), the Youth Program of the Natural Science Foundation of Jiangsu Province (Grant No. BK20221136), the Open Fund of the National Engineering Research Center of Highway Maintenance Technology (Changsha University of Science and Technology) (Grant No. kfj220104), the Jiangsu Excellent Postdoctoral Program (Grant No. 2022ZB529), the China Postdoctoral Science Foundation (Grant No. 2023M733746), the Young Elite Scientist Sponsorship Program by Cast (Grant No. YESS20220076), and the Key Research and Development Program of Shaanxi (Program No. 2021SF-459).

Institutional Review Board Statement: Not applicable.

Informed Consent Statement: Not applicable.

Data Availability Statement: Data are contained within the article.

Acknowledgments: Authors are grateful to the above mentioned funders.

Conflicts of Interest: The authors declare no conflicts of interest. Haiku Zhang was employed by the Fuyuan Ding Installation Engineering Company Limited Shangqiu. The remaining authors declare that the research was conducted in the absence of any commercial or financial relationships that could be construed as a potential conflict of interest.

References

1. Li, J.; Zhang, J.; Yang, X.; Zhang, A.; Yu, M. Monte Carlo simulations of deformation behaviour of unbound granular materials based on a real aggregate library. *Int. J. Pavement Eng.* **2023**, *24*, 2165650. [CrossRef]
2. Wiewel, B.V.; Lamoree, M. Geotextile composition, application and ecotoxicology—A review. *J. Hazard. Mater.* **2016**, *317*, 640–655. [CrossRef] [PubMed]
3. Liu, L.; Cai, G.; Zhang, J.; Liu, X.; Liu, K. Evaluation of engineering properties and environmental effect of recycled waste tire-sand/soil in geotechnical engineering: A compressive review. *Renew. Sustain. Energy Rev.* **2020**, *126*, 109831. [CrossRef]
4. Prambauer, M.; Wendeler, C.; Weitzenböck, J.; Burgstaller, C. Biodegradable geotextiles—An overview of existing and potential materials. *Geotext. Geomembr.* **2019**, *47*, 48–59. [CrossRef]
5. Correia, N.d.S.; Bueno, B.D.S. Effect of bituminous impregnation on nonwoven geotextiles tensile and permeability properties. *Geotext. Geomembranes* **2011**, *29*, 92–101. [CrossRef]
6. Wu, H.; Yao, C.; Li, C.; Miao, M.; Zhong, Y.; Lu, Y.; Liu, T. Review of application and innovation of geotextiles in geotechnical engineering. *Materials* **2020**, *13*, 1774. [CrossRef]
7. Sudarsanan, N.; Karpurapu, R.; Amrithalingam, V. An investigation on the interface bond strength of geosynthet-ic-reinforced asphalt concrete using Leutner shear test. *Constr. Build. Mater.* **2018**, *186*, 423–437. [CrossRef]
8. Liu, X.; Congress SS, C.; Cai, G.; Liu, L.; Liu, S.; Puppala, A.J.; Zhang, W. Development and validation of a method to predict the soil thermal conductivity using thermal piezocone penetration testing (T-CPTU). *Can. Geotech. J.* **2022**, *59*, 510–525. [CrossRef]
9. Liu, X.; Congress SS, C.; Cai, G.; Liu, L.; Puppala, A.J. Evaluating the thermal performance of unsaturated bentonite–sand–graphite as buffer material for waste repository using an improved prediction model. *Can. Geotech. J.* **2022**, *60*, 301–320. [CrossRef]
10. Saathoff, F.; Oumeraci, H.; Restall, S. Australian and German experiences on the use of geotextile containers. *Geotext. Geomembranes* **2007**, *25*, 251–263. [CrossRef]
11. Moo-Young, H.K.; A Gaffney, D.; Mo, X. Testing procedures to assess the viability of dewatering with geotextile tubes. *Geotext. Geomembranes* **2002**, *20*, 289–303. [CrossRef]
12. Kutay, M.E.; Aydilek, A.H. Retention performance of geotextile containers confining geomaterials. *Geosynth. Ternational* **2004**, *11*, 100–113. [CrossRef]
13. Hakimelahi, N.; Bayat, M.; Ajalloeian, R.; Nadi, B. Effect of woven geotextile reinforcement on mechanical behavior of calcareous sands. *Case Stud. Const. Mat.* **2003**, *18*, e02014. [CrossRef]
14. Rawal, A.; Sayeed, M.A.; Saraswat, H.; Shah, T. A comparison of wide-width tensile strength to its axi-symmetric tensile strength of hybrid needlepunched nonwoven geotextiles. *Geotext. Geomembranes* **2013**, *36*, 66–70. [CrossRef]
15. Sawicki, A.; Kazimierowicz-Frankowska, K. Creep behaviour of geosynthetics. *Geotext. Geomembr.* **1998**, *16*, 365–382. [CrossRef]
16. Bacas, B.M.; Cañizal, J.; Konietzky, H. Shear strength behavior of geotextile/geomembrane interfaces. *J. Rock Mech. Geotech. Eng.* **2015**, *7*, 638–645. [CrossRef]
17. Pinho-Lopes, M.; De Lurdes Lopes, M. Influence of mechanical damage induced in laboratory on the soil-geosynthetic interaction in inclined-plane shear. *Constr. Build. Mater.* **2018**, *185*, 468–480. [CrossRef]
18. Gautier, K.B.; Kocher, C.W.; Drean, J.-Y. Anisotropic mechanical behavior of nonwoven geotextiles stressed by uniaxial tension. *Text. Res. J.* **2007**, *77*, 20–28. [CrossRef]
19. Punetha, P.; Mohanty, P.; Samanta, M. Microstructural investigation on mechanical behavior of soil-geosynthetic interface in direct shear test. *Geotext. Geomembr.* **2017**, *45*, 197–210. [CrossRef]
20. Kongkitkul, W.; Tabsombut, W.; Jaturapitakkul, C.; Tatsuoka, F. Effects of temperature on the rupture strength and elastic stiffness of geogrids. *Geosynth. Int.* **2012**, *19*, 106–123. [CrossRef]
21. Chantachot, T.; Kongkitkul, W.; Tatsuoka, F. Effects of temperature rise on load-strain-time behaviour of geogrids and simulations. *Geosynth. Int.* **2018**, *25*, 287–303. [CrossRef]
22. Zornberg, J.G.; Byler, B.R.; Knudsen, J.W. Creep of geotextiles using time–Temperature superposition methods. *J. Geotech. Geoenvironmental Eng.* **2004**, *130*, 1158–1168. [CrossRef]
23. *ASTM D4595*; Standard Test Method for Tensile Properties of Geotextiles by the Wide-Width Strip Method. ASTM: West Conshohocken, PA, USA, 2011.
24. *ASTM D6241*; Standard Test Method for the Static Puncture Strength of Geotextiles and Geotextile-Related Products Using a 50-mm Probe. ASTM: West Conshohocken, PA, USA, 2014.
25. Hsieh, J.-C.; Li, J.-H.; Huang, C.-H.; Lou, C.-W.; Lin, J.-H. Statistical analyses for tensile properties of nonwoven geotextiles at different ambient environmental temperatures. *J. Ind. Text.* **2017**, *47*, 331–347. [CrossRef]

26. Liu, T.; Yang, X.; Zhang, Y. A Review of Gassy Sediments: Mechanical Property, Disaster Simulation and In-Situ Test. *Front. Earth Sci.* **2022**, *10*, 915735. [CrossRef]
27. Liu, L.; Cai, G.; Liu, S. Compression properties and micro-mechanisms of rubber-sand particle mixtures considering grain breakage. *Constr. Build. Mater.* **2018**, *187*, 1061–1072. [CrossRef]

 *materials*

Article

Effects of Aging and Immersion on the Healing Property of Asphalt–Aggregate Interface and Relationship to the Healing Potential of Asphalt Mixture

Haimei [1], Lili Li [2], Qinglin Guo [2,*], Tongmao Zhao [2], Pan Zuo [2] and Fengming E [2]

1. Inner Mongolia Vocational and Technical College of Communications, Chifeng 024005, China
2. School of Civil Engineering, Hebei University of Engineering, Handan 056038, China
* Correspondence: guoql@hebeu.edu.cn

Abstract: The self-healing ability of asphalt–aggregate bonding interfaces can maintain the mechanical properties of asphalt mixtures. However, the interface's healing ability will also be affected by moisture and aging. In order to clarify the influence of moisture and aging on the healing ability of a bonding interface, the effects of healing period and temperature on the self-healing level of interfacial strength were measured. The healing master curve of the strength was established. Thereafter, the effects of soaking time, salt solution concentration, and thermal aging on the healing degree of interfacial strength were measured. Based on digital image processing technology and the meso-finite element method, the influence of the interface on the healing performance of the mixture was simulated and analyzed, which was then verified by the beam bend healing test. The results show that the healing index of bonding strength increases with the ascent of healing temperature and period. Healing index gradually decreases with the extension of soaking period, and the higher the concentration of salt in the solution, the worse the healing performance of interfacial strength. After asphalt aging, the healing potential of the interface is weakened. There is a good linear relationship between the healing level of an asphalt–aggregate interface and the level of strength and fracture energy of the mixture. However, the actual healing level of an asphalt mixture is obviously lower than that of the interface, due to the addition of mineral filler. This paper provides a method for predicting the recovery performance of asphalt pavement.

Keywords: asphalt–aggregate interface; self-healing level; moisture and aging; healing potential of asphalt mixture; finite element simulation

Citation: Haimei; Li, L.; Guo, Q.; Zhao, T.; Zuo, P.; E, F. Effects of Aging and Immersion on the Healing Property of Asphalt–Aggregate Interface and Relationship to the Healing Potential of Asphalt Mixture. *Materials* **2023**, *16*, 3574. https://doi.org/10.3390/ma16093574

Academic Editor: Giovanni Polacco

Received: 20 March 2023
Revised: 23 April 2023
Accepted: 3 May 2023
Published: 6 May 2023

1. Introduction

The self-healing ability of materials entails that the materials can heal the damage and cracks when micro-cracks and local damage occur inside of the materials, so that the material's performance can be partially or fully recovered [1–5]. For asphalt pavement, cracks, pits, and other diseases may reduce the service life of the pavement and increase the maintenance cost. These diseases are mainly caused by the decay of asphalt adhesion, which is caused by asphalt aging, moisture damage, and other factors. Zhu et al. [1] found that the self-healing of asphalt at medium and high temperatures made the failed asphalt–aggregate interface adhere again, showing a self-healing characteristic of the interface, which makes an asphalt mixture with micro-cracks heal to a certain extent and recover its strength, and then prevents the cracks from expanding. Guo et al. [2] indicated that the strength and pre-peak failure energy of an asphalt mixture can be recovered to more than 80% after healing for 15 min at 90 °C. Therefore, the self-healing ability of asphalt concrete is helpful to improve the durability of pavement and prolong its service life [6–11].

In order to conveniently evaluate the adhesive strength between the asphalt and aggregate, the bitumen bond strength (BBS) test is gradually introduced to measure the adhesive property of an asphalt–aggregate interface [12]. Johannes et al. [13] carried out the bonding performance test of asphalt materials through a BBS experiment and evaluated the moisture damage resistance of asphalt in chip seal by using the bonding strength. Copeland et al. [14] investigated the influence of aging on asphalt adhesion through the BBS test. The correlation between the bonding strength of bitumen and rutting test results was determined. Hoki et al. [15] evaluated the effect of film thickness and soaking time on the bonding strength of interfaces using the BBS test. They indicated that the bonding strength declines with the increase in film thickness and soaking time. Guo et al. [16] studied the influence of water soaking, salt solution soaking, and freeze–thaw cycles on the bonding strength of the asphalt–aggregate interface by the BBS test, and established a coupled damage model considering soaking time, solution concentration, and freeze–thaw cycles to predict the bonding strength of an interface. The results showed that the BBS test can accurately measure the bonding strength between the asphalt and aggregate quickly, and reflect the bonding level between asphalt and the aggregate of an asphalt mixture.

In recent years, Hu et al. [17] studied the tensile strength of the bonding interface and self-healing ability of high-viscosity modified asphalt. Their results showed that the modifier reduced the self-healing ability of asphalt, and high-temperature immersion would lead to the decrease in adhesion and healing ability. A high-viscosity modifier enhances the polarity of modified asphalt and increases the thickness of structural asphalt by absorbing the light components, thus improving the bonding performance. Huang et al. [18] indicated that the moisture damage mechanism of asphalt pavement was mainly induced by the cohesion and adhesion damage of the asphalt–aggregate interface.

Sun et al. [7] indicated that the healing of asphalt–aggregate interface includes the cohesion healing and adhesion healing. Recently, Xu et al. [19] pointed out that asphalt aging caused the oxidation reaction, and the index of carbonyl and sulfoxide functional groups increased, which eventually led to a decrease in the self-healing performance of the binder. Zhou et al. [20] investigated the self-healing performance of five kinds of modified asphalt using the BBS test, and analyzed the influence of moisture and modifier on the interface's healing ability. The results showed that the self-healing ability of different asphalt types was also different, and the void at the aggregate's surface had a direct influence on the molecular selectivity and the healing ability of the interface. The self-healing process of the bonding interface was scanned using CT technology. Higher temperatures and dry conditions are beneficial to the self-healing of asphalt, while moisture is not conducive to the short-term or long-term healing of asphalt, but it can improve the healing rate at the middle stage [21,22]. In 2021, Huang et al. [23] carried out a bonding interface healing test and a four-point bending fatigue test, evaluated the healing performance of asphalt–aggregate bonding interfaces, and analyzed the correlation between the healing level of the interface and the fatigue healing performance of the mixture. Their results showed that there was a good linear correlation between the healing level of the bonding interface and the healing potential of the mixture.

Based on the above analysis and experimental investigation, it can be inferred that aging and moisture would affect the healing ability of the asphalt–aggregate interface, and the healing ability of the interface would influence the healing performance of the asphalt mixture. In fact, the pavement bears the actions of rainfall, high temperature, ultraviolet rays, and so on. The bonding strength and healing ability of the asphalt–aggregate interface will gradually decrease, which directly affects the self-healing ability of the mixture. However, the healing process of the asphalt mixture includes the adhesion healing of the interface and the cohesion healing inside of the asphalt mortar [24]. Although the current research has demonstrated the influence of single factors on the self-healing property of the interface, an understanding of the relationship between the healing level of the interface and the healing potential of the mixture was necessary to improve the mixture durability. It is difficult to quantitatively analyze this relationship solely through experiments. Fortunately, finite element simulation technology provides a convenient way to explore this relationship.

The research goal of this work is to determine the influence law of soaking and aging on the healing ability of the asphalt–aggregate interface, and to explore the relationship between the interface's healing ability and the recovery degree of the mixture's performance. In view of this goal, the relationships among healing time, healing temperature, and healing index of the asphalt–aggregate interface were measured first in this paper. Based on the principle of time–temperature equivalence, the self-healing master curve of the interfacial strength is established, and the effects of different factors, including soaking time, salt solution concentration, and aging degree, on the healing characteristics of the asphalt–aggregate interface are investigated. The finite element simulation was carried out in order to evaluate the healing performance of the asphalt mixture. The influence of the healing ability of the asphalt–aggregate interface on the self-healing potential of the asphalt mixture was analyzed and verified. The research algorithm is described in Figure 1.

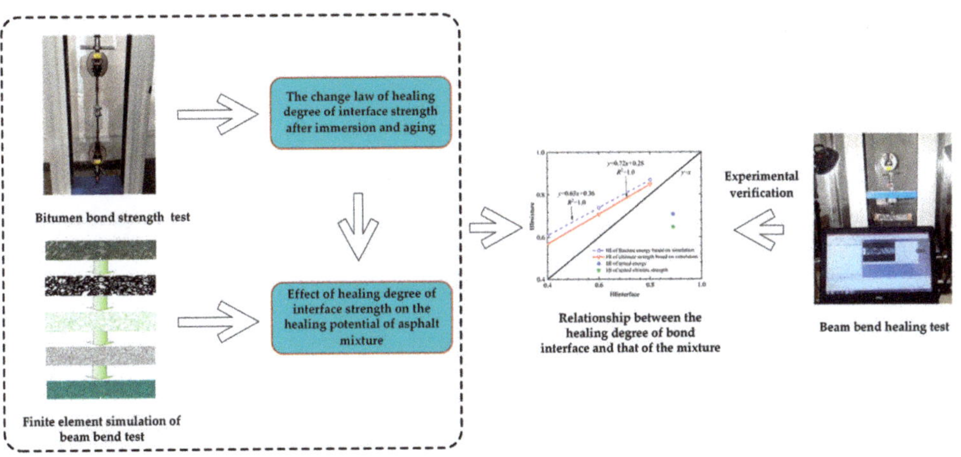

Figure 1. Research algorithm.

2. Materials and Methods

2.1. Materials

The asphalt used in this experiment was AH-70# petroleum asphalt, which has a penetration grade of 70. Its basic properties are presented in Table 1. Basalt aggregates were selected to prepare the asphalt mixture, and the apparent gravity and moisture uptake ratio of the aggregates are listed in Table 2. The gradation named AC-13, which is recommended by China technical specifications JTG D50-2017 [25], was selected for the experiment, as presented in Figure 2. According to the Marshall test results, the optimal asphalt content of the mixture was 5.0%, the apparent gravity of the mixture was 2.447 g/cm^3, and the air void in the asphalt mixture was 4.1%. A plane specimen with a size of 30 cm × 30 cm × 5 cm was prepared to cut into the beams of 30 cm × 5 cm × 5 cm, and a notch of 5 mm in depth and 3 mm in width was cut at the bottom of the mid-span of the beam.

Table 1. Properties of AH-70# asphalt.

Items	Values	Test Methods
Penetration (25 °C, 0.1 mm)	75	ASTM D5
Softening point (°C)	47.8	ASTM D36
Ductility (15 °C, cm)	>100	ASTM D113
Flashing point (°C)	285	ASTM D92

Table 2. Properties of aggregates.

Sieve Size (mm)	16	13.2	9.5	4.75	2.36	1.18	0.6	0.3	0.15	0.075
Apparent gravity (g/cm^3)	2.716	2.723	2.681	2.679	2.694	2.708	2.713	2.713	2.712	2.712
Water absorption (%)	0.49	0.57	0.63	0.52	-	-	-	-	-	-

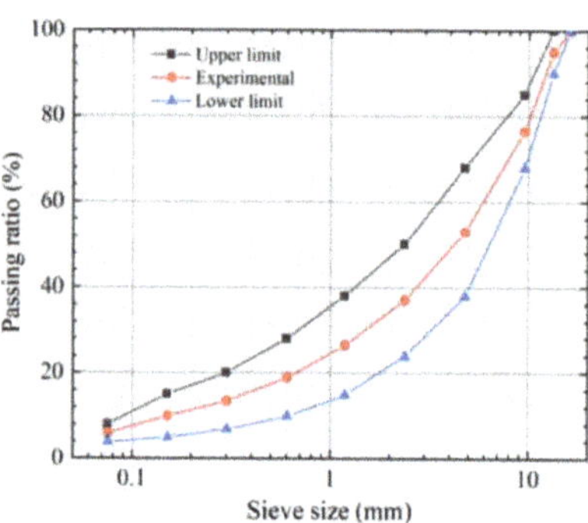

Figure 2. Experimental gradation.

2.2. Healing Test of Asphalt–Aggregate Bonding Interface

A disc specimen, with a diameter of 4.2 cm, was prepared by coring the stone plate with a water drill, and its surface was polished with sandpaper to ensure that the roughness of the interface was consistent. The moisture inside of the disc specimen was removed by drying at 105 °C for 12 h. Thereafter, the disc specimen was placed in an oven at 135 °C for 4 h, taken out, and asphalt was dropped onto the center of the disc specimen, covered with another disc specimen, and compacted. The asphalt was required to evenly coat the specimen interface, and the thickness of the asphalt film was controlled at 0.1 mm. Based on the specific test method of reference [16], the "aggregate–asphalt–aggregate" bonded sample was finally made, as presented in Figure 3. Thereafter, the sample was cooled at room temperature, and the tensile test was carried out after it was kept at 20 °C for 6 h. In this paper, the electronic universal testing machine from cangzhou city of hebei province was used for the pull-off test, with a loading rate of 1 mm/min, and the test was stopped when the specimen was destroyed. Three parallel samples were set for each group test. The specific process is presented in Figure 4.

Figure 3. Sandwich specimen of the asphalt–aggregate bonding interface.

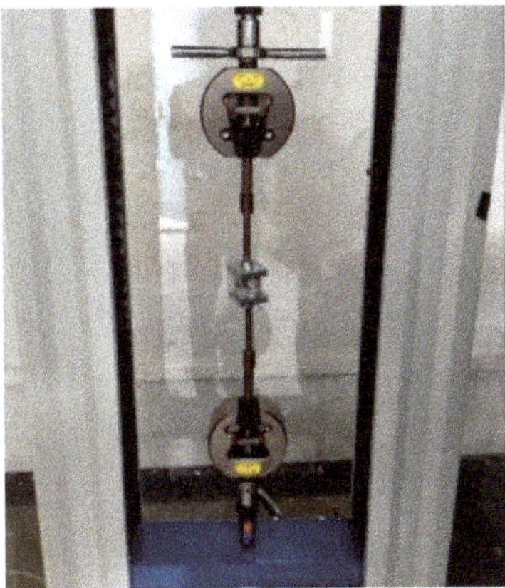

Figure 4. Bitumen bond strength (BBS) test.

After the test, the damaged specimens were stacked and healed at 10 °C, 40 °C, 60 °C, and 80 °C. After the healing, the second pull-off test was carried out, and four parallel specimens were prepared for each group. The healing index of the interfacial strength was obtained in order to evaluate the healing performance of the interface. The healing index can be calculated using the following equation.

$$HI = \frac{\sigma_2}{\sigma_1} \tag{1}$$

where, *HI* is the healing index of the interfacial strength; σ_1 and σ_2 are the tensile strengths before and after healing, respectively, MPa.

In seasonally frozen areas, snow-melting agents are often used to remove the snow. In order to investigate the influence of salt solutions on the self-healing performance of the interface, a salt solution was prepared by using $Cacl_2$ to simulate the influence of salt corrosion on the healing performance of the interface. The asphalt was also aged using the rolling thin film oven test (RTFOT) method, and the influence of aging degree on the interface's healing performance was discussed.

2.3. Beam Bend Healing Test of Asphalt Mixture

The healing of the bonding interface leads to the mechanical performance recovery of the asphalt mixture. In order to determine the self-healing ability of the asphalt mixture, the healing performance of the asphalt mixture in an outdoor, high-temperature environment was measured using the beam bend healing test. The bending test was carried out using the aforementioned beam specimen, and the support space was 200 mm ± 0.5 mm. The beam specimen was kept at 10 °C for more than 4 h, and then the bending failure test was carried out. The vertical loading rate was 1 mm/min. During the test, the deformation and load on the bottom of the mid-span were recorded in real time. The experimental procedure is presented in Figures 5 and 6. In the first bend test, the test was stopped when the mid-span vertical deformation reached 2 mm. Subsequently, the damaged beam was placed outdoors for 48 h of healing. The healing temperature of the beam specimen was between 23 °C and 45 °C, with an average temperature of 28 °C. The typical temperature field on the beam's surface is presented in Figure 7. After 48 h of healing, the second bending test was carried out after beam conditioning at 10 °C for 4 h, and the test procedure was consistent with the first bending test. Three parallel samples were set for each group test.

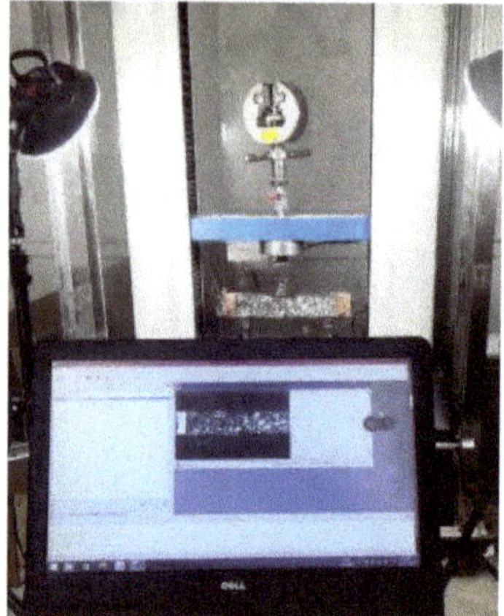

Figure 5. Three-point bending test.

Figure 6. Bending failure of the beam specimen.

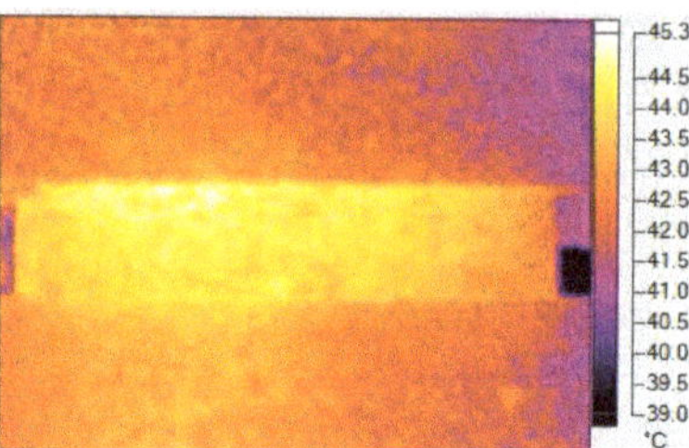

Figure 7. Temperature field of the beam specimen at 14:00 on 25 June 2019.

After finishing the test, the healing degree of the mixture was evaluated using the fracture strength and fracture energy. It can be determined according to the following equations.

$$\sigma(t) = \frac{3LF(t)}{2bh^2} \tag{2}$$

$$\varepsilon(t) = \frac{6td(t)}{L^2} \tag{3}$$

$$G_F = \int_0^{\varepsilon f} \sigma \varepsilon d\varepsilon \tag{4}$$

$$HI_S = \frac{R_{healed}}{R_0} \tag{5}$$

$$HI_{GF} = \frac{G_{Fhealed}}{G_{F0}} \tag{6}$$

where, $\sigma(t)$ represents the tensile stress at the bottom of the mid-span, MPa. $\varepsilon(t)$ is the tensile strain at the bottom of the mid-span. ε_f is the failure strain corresponding to 2 mm vertical deformation, $F(t)$ is the applied load, N; L represents the support space, mm. b and h are the width and height of the beam, respectively, mm. $d(t)$ is the deformation at the mid-span of the beam, mm. HI_s is the strength recovery index, R_0 and R_{healed} are the failure strength before and after healing, respectively, MPa. G_0 and $G_{Fhealed}$ are the failure energy before and after healing, respectively, J. HI_{GF} indicates the healing index of failure energy.

2.4. Mesoscopic Finite Element Simulation for Asphalt Mixture

The healing of the asphalt–aggregate interface will directly affect the healing performance of the mixture. However, the beam bend healing test can only analyze the self-healing ability of the mixture from the macroscopic perspective, and cannot establish the relationship between the interface healing level and the self-healing potential of the mixture. Herein, the meso-finite element method was adopted in order to reveal this relationship. A finite element (FE) beam model of the asphalt mixture was established, based on digital image processing technology, in order to consider the realistic mesostructure of the asphalt mixture. The modeling procedure is presented in Figure 8.

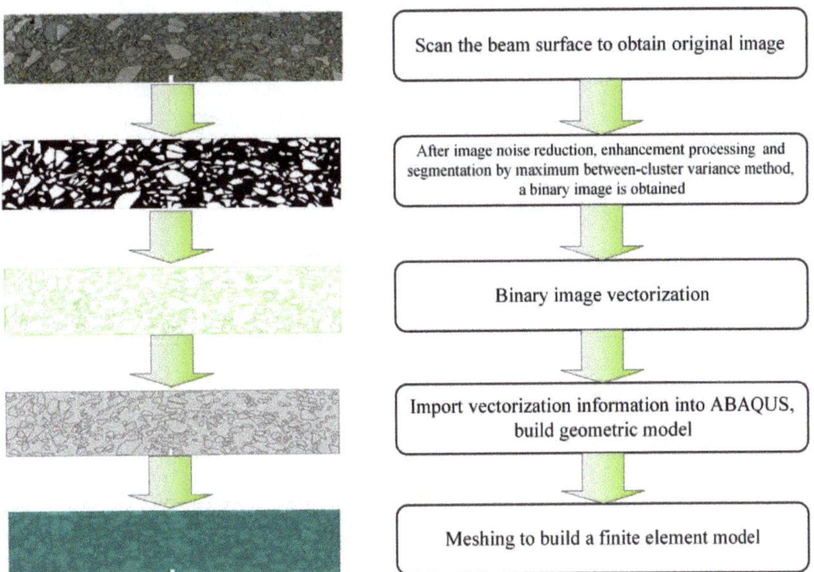

Figure 8. Reconstruction process of the 2D finite element model for the asphalt mixture.

The cohesive zone model (CZM) has been widely used to investigate the interface debonding process in the material fracture field. Therefore, the bilinear cohesive zone model was selected in order to describe the constitutive relation of the bonding interface. The tractor-separation curve is presented in Figure 9.

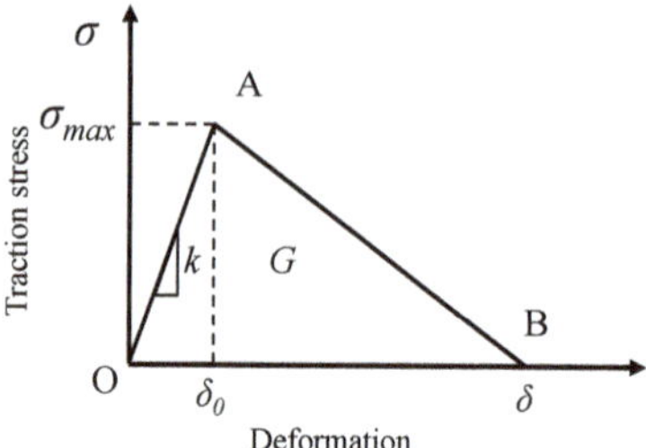

Figure 9. Tractor-separation curve of the CZM. Where σ_{max} represents the maximum stress, MPa; δ_0 is the deformation of the CZM element when the ultimate strength is reached. k represents the stiffness, MPa; G represents the failure energy of the CZM.

When the traction stress satisfies the following relationship, the cohesive element begins to be damaged.

$$\left\{\frac{\langle T_n\rangle}{T_n^0}\right\}^2 + \left\{\frac{T_s}{T_s^0}\right\}^2 = 1 \tag{7}$$

where < and > are Macaulay brackets, which indicates that the compression stress will not cause damage to the cohesive element. T_n^0 and T_s^0 are traction stresses in normal direction and shear direction, respectively. T_n and T_s are pure normal traction and pure shear traction, respectively, at the beginning of damage. Once the failure starts, the material goes into a softening state. This process is quantified by defining a damage variable D, as presented in Equation (8).

$$D = \frac{\delta_f(\delta_{max} - \delta_0)}{\delta_{max}(\delta_f - \delta_0)} \tag{8}$$

where δ_0, δ_{max}, and δ_f are the effective displacement at the beginning of damage, the maximum effective displacement obtained during loading, and the effective displacement at the point of complete failure, respectively. In this paper, the power-law failure criterion based on energy is used to describe the fracture evolution in mixed mode:

$$\left\{\frac{G_n}{G_n^c}\right\}^2 + \left\{\frac{G_t}{G_t^c}\right\}^2 = 1 \tag{9}$$

where G_n^c and G_t^c are the fracture energy in normal and shear directions, respectively. G_n and G_t are the dissipated energy generated in the normal and shear direction, respectively, during the loading procedure. In the meshing stage, the three-node plane stress element, called CPS3, was applied to mesh the asphalt mortar, and the four-node quadrilateral linear plane stress element, called CPS4R, was used to mesh the coarse aggregate. The FE model was discretized by free mesh, and the side length of the element was 0.5 mm. A self-compiled program developed by MATLAB was adopted in order to edit the initial 'inp file' generated by ABAQUS. In this process, the zero-thickness CZM element was inserted into the mortar and the interface, one after the other. The completed CZM model is presented in Figures 10 and 11.

Figure 10. CZM elements between the aggregate and asphalt mortar.

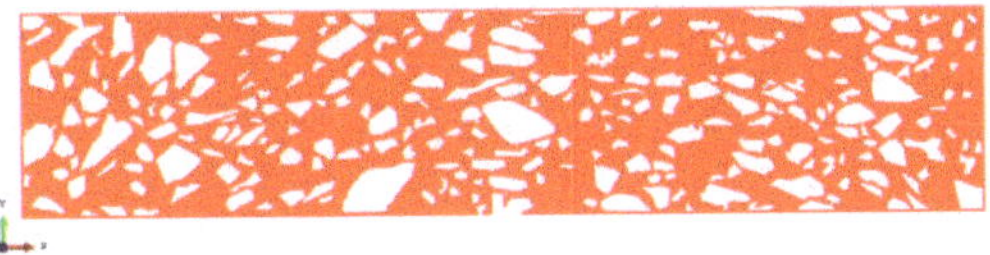

Figure 11. CZM elements inside of the asphalt mortar.

The boundary and loading points of the beam are presented in Figure 12. The support spacing of the finite element model was 20 cm. The parameters of the CZM element, asphalt mortar, and aggregate are listed in Tables 3–5. The parameters in Tables 3 and 4, from the results of Liu [26], were adopted as the initial values, and then the trial calculation was conducted. The final values were determined when the difference between the experimental and the simulated load-displacement curve before the peak was at its minimum. The parameters in Table 5 are consistent with the results of E [27].

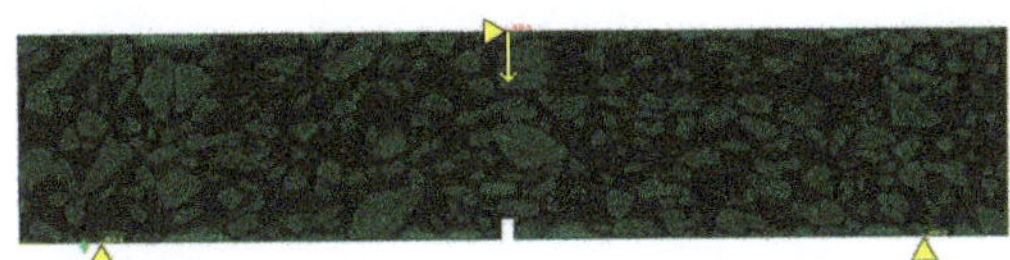

Figure 12. Model boundary and loading conditions.

Table 3. Elastic parameters for the aggregate and asphalt mortar.

Material	Elastic Modulus (MPa)	Poisson's Ratio
Aggregate	20,000	0.2
Asphalt mortar	800	0.5

Table 4. Model parameters for cohesion before healing.

Interface Type	Fracture Strength (MPa)		Fracture Energy (J·m^2)	
	Normal Direction	Tangent Direction	Normal Direction	Tangent Direction
Interface of aggregate–mortar	0.6	1	500	500
Interface of mortar	1	2	500	500

Table 5. Parameter table for Prony series viscoelastic coefficients of the asphalt mortar.

i	1	2	3	4	5
g_i	0.542	0.166	0.1	0.098	0.034
T_i	0.048	0.631	6.711	48.78	613.497

3. Results and Discussion

3.1. Effects of Healing Period and Temperature on Healing Level of Interface Strength

The healing level of the interface was measured using the bitumen bond strength test at different temperatures and healing periods, and the healing index of the bonding interface was calculated based on the tensile strength, as presented in Figure 13.

As presented in Figure 13, the healing index of the bonding interface gradually increases with the extension of healing time. The higher the healing temperature is, the shorter the recovery period of the strength will be. For the same healing period, the higher the healing temperature is, the better the recovery of the strength is. The self-healing degree of the interface strength between the asphalt and aggregate obviously depends on the healing time and temperature. In order to obtain a good healing effect, it is feasible to prolong the healing time or increase the temperature. This is mainly because asphalt is a typical viscoelastic material. The higher the temperature is, the lower the viscosity of the asphalt, and the more active its molecular movement is. The accelerated molecular movement is helpful to enhance the healing of the asphalt film at the interface.

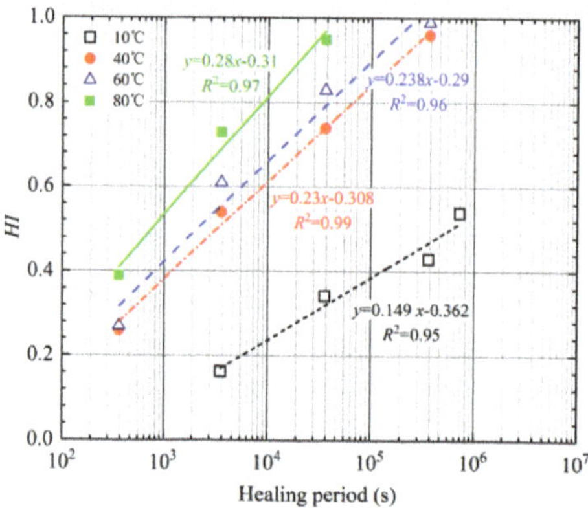

Figure 13. Healing-index of the bonding interface of asphalt–aggregate at different temperatures.

3.2. Master Curve of Healing Index

In this work, the Compertz model, as presented in Equation (10), was selected to describe the trend of the healing index, and the parameters of the healing model can be determined through the nonlinear least square method on the data in Figure 13.

$$HI(t) \; = \; a \, \exp(b \, \exp(c \times lgt)) \tag{10}$$

where a, b, and c are model parameters. Referring to the change in HI, the parameters should follow the conditions: $b < 0$, $c < 0$; the parameters of the HI model are listed in Table 6.

Table 6. Parameters of the HI model.

Parameters	10 °C	40 °C	60 °C	80 °C
a	1.000	1.000	1.000	1.000
b	−9.047	−10.140	−20.500	−12.050
c	−0.460	−0.792	−1.079	−0.994
SSE	0.003	0.006	0.003	0.000
R^2	0.969	0.976	0.991	1.000
RMSE	0.057	0.056	0.036	0.006

It can be observed from Table 6 that the correlation coefficients (R^2) of nonlinear fitting are all above 0.96, which indicates that the Compertz model has good applicability, and it can accurately describe the change law of the healing index. A low RMSE indicates that there is little difference between the predicted value and the measured value. According to the development trend of the healing index in Figure 12, the healing index is influenced by both temperature and healing time. When the temperature is below 60 °C, the healing period is in the range of 100 s–1,000,000 s, while when the temperature reaches 80 °C, the healing time is sharply shortened, and when the healing time is 30,000 s, the healing index is close to 1. Thus, both b and c first increase and then decrease, which is caused by the cross influence of temperature and healing time.

In order to evaluate the interface healing ability at different temperatures, the temperature shift factor is calculated based on the reference temperature of 60 °C. A trial method is proposed to determine the initial value of the temperature shift factor considering the prediction accuracy of the master curve. On this basis, the glass transition temperature T_g and the reference temperature T_S can be determined by Equation (11).

$$lg\alpha_T = \frac{-C_1 \times (T - T_S)}{C_2 + T - T_S} \tag{11}$$

where $C_1 = 8.86$ and $C_2 = 101.6$.

There is a conversion relationship between the reference temperature and the glass transition temperature, as calculated by Equation (12).

$$T_S = T_g + 50 \tag{12}$$

Generally, the temperature selected in the experiment is fixed. These temperatures may not cover T_S. T_S is unknown. Herein, a certain temperature T_0 in the actual test sequence is often selected as the reference temperature, and the healing index at other temperatures can be moved to the curve of T_0 after shifting from the test temperature to the reference temperature. The temperature shift factor can be determined using Equation (13).

$$lg\alpha_{T0} = \frac{-8.86 \times (T_0 - T_S)}{101.6 + T_0 - T_S} \tag{13}$$

The shift factor of any temperature to the reference temperature, defined as $lg\alpha'_{T0}$, can be obtained by using Equations (11) and (13).

$$lg\alpha'_{T0} = lg\alpha_T - lg\alpha_{T0} = -C_1 C_2 \frac{-(T - T_0)}{(C_2 + T - T_S)(C_2 + T_0 - T_S)} \tag{14}$$

According to the measured data of different temperatures, $lg\alpha'_{T0}$ can be determined by moving to the reference temperature, with the fixed reference temperature T_0 at 60 °C, so that the initial value of the temperature shift factor relative to the reference temperature can be calculated by Equation (14) when $T_0 = T_S$. Thereafter, the temperature shift factor is optimized by trial calculation, and SSE, R^2, and RMSE are used to determine the optimal temperature shift factor with a minimum error. The results are presented in Table 7.

Table 7. $lg\alpha'_{T0}$ with 60 °C as the reference temperature.

$lg\alpha'_{T0}$	10 °C	40 °C	80 °C
Initial value	8.58	2.172	−1.457
Optimal value	6.65	1.80	−1.20

Based on the optimal value of $lg\alpha'_{T0}$ in Table 7, T_S is 323.288 K and T_g is 273.288 K. The temperature shift factor $lg\alpha_T$ can be determined according to Equation (11), which is listed in Table 8. Subsequently, the correlation coefficients R^2 and SSE are used as control indexes to optimize $lg\alpha_T$ through the trial calculation. In the optimization process, the change in SSE is presented in Figure 14. The closer SSE is to zero, the smaller the difference between the predicted and the experimental HI is. The optimized results are also presented in Table 8.

Table 8. $lg\alpha_T$ for different temperatures.

$lg\alpha_T$	10 °C	40 °C	60 °C	80 °C
Initial value	5.79	0.98	−0.78	−2.01
Optimal value	2.32	0.20	−0.08	−0.20
R^2	0.991	0.987	0.997	1.000
SSE	0.035	0.008	0.010	0.001

The measured healing indexes of different temperatures were nonlinearly fitted based on temperature shift factor, as presented in Figure 15. Parameters of the master curve of the healing index are presented in Table 9.

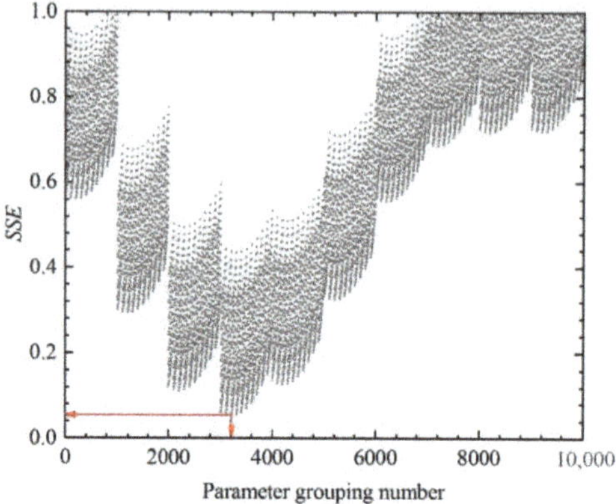

Figure 14. Variation of SSE in temperature shift factor optimization.

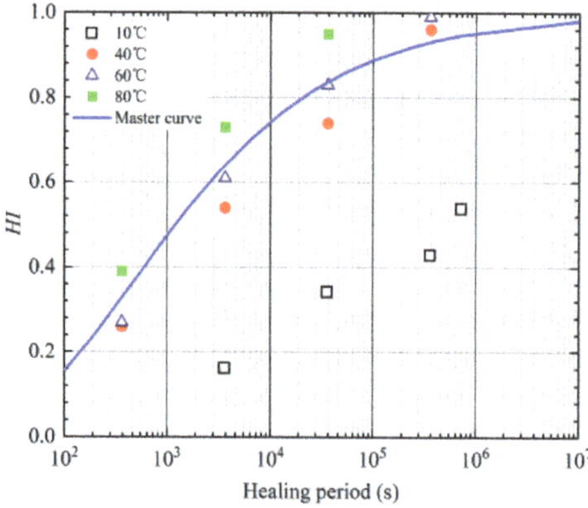

Figure 15. Master curve of *HI* based on the measured healing index.

Table 9. Parameter values of the optimal *HI* principal curve model.

Parameters	a	b	c	R^2	Equation of Master Curve
Value	1.00	−11.70	−0.92	0.988	$HI(t) = \exp(-11.7\exp(-0.92lgt/lg\alpha))$

3.3. Effects of Immersion and Aging on the Healing Level of Interface

Asphalt pavement will be affected by environmental factors, such as rainfall and aging, in the service stage. The change in asphalt molecular structure will inevitably change the healing level of the asphalt–aggregate interface. After soaking in water, the healing index of the bonding strength of the interface is presented in Figure 16.

As can be observed from Figure 16, the healing index of interfacial bonding strength is higher than those of untreated specimens when the soaking period is shorter than 16.8 h. This is because the immersion time is short, and a small amount of water enters the bonding area of the interface [8]. In addition, the specimen is soaked at 20 °C, which accelerates the heat transfer and promotes the healing of interface bonding. Therefore, the healing degree of bonding strength after short-term soaking is higher than those of untreated ones. After soaking for more than 6 h, the healing index of the strength decreases gradually. After 48 h of soaking, the healing index decreased by 34%. This proved that long-term soaking was unfavorable to the healing potential of the interface. The longer the soaking time is, the worse the self-healing performance of the interface is. With the extension of soaking time, water permeates into the interface to form a water film, which hinders the diffusion and re-bonding of asphalt molecules on the surface of the aggregate, resulting in a decline in self-healing level. It can be inferred that moisture has a significant effect on the healing ability of the asphalt–aggregate bonding interface. Zhou et al. [22] stated that moisture is not conducive to the short-term and long-term healing of asphalt, but it can improve the healing rate at the middle stage. The findings in this paper are consistent with those of Zhou. The middle stage is between approximately 6 and 12 h. Long-term water intrusion will reduce the self-healing potential of asphalt pavement, and eventually accelerate damage and cracking.

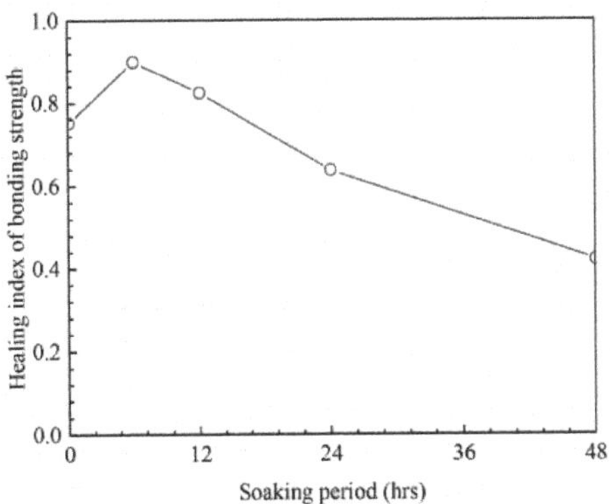

Figure 16. Influence of soaking period on the healing index of bonding strength.

In this test, the soaking time was set at 6 h. The change in the healing index is presented in Figure 17.

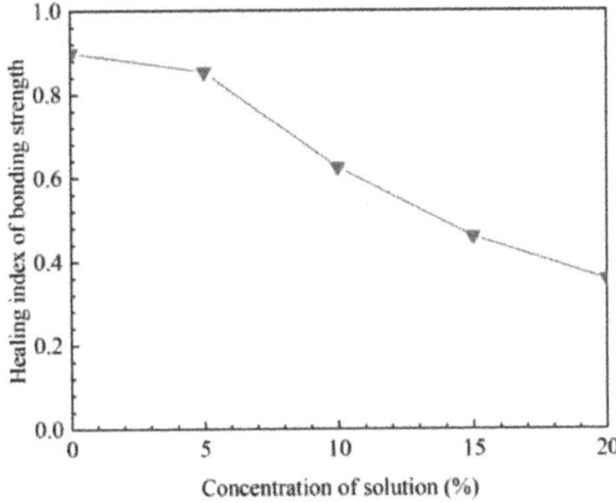

Figure 17. Influence of salt solution on the healing index of bonding strength after 6 h soaking.

According to the changing trend in Figure 17, the healing index of interfacial bonding strength decreases with the increase in solution concentration. The self-healing property declines slowly when the concentration is no more than 5%. Afterwards, the healing index decreases linearly with the increase in the concentration. Therefore, a high concentration solution has a negative effect on the healing ability of the interfacial strength. Chlorine is the main component of snow-melting agents. It is necessary to control the dosage of chlorine salt when spraying snow-melting agents in winter, which is beneficial to reduce the concentration of salt solution and the influence of chlorine salt on the healing performance of asphalt pavement.

In order to explore the influence of asphalt aging on the healing performance of the asphalt–aggregate interface, the asphalt was aged using the RTFOT method in a laboratory setting. The aged asphalt was adopted to make sandwich bonding specimens. The self-healing index of the asphalt–aggregate interface was then tested, as presented in Figure 18.

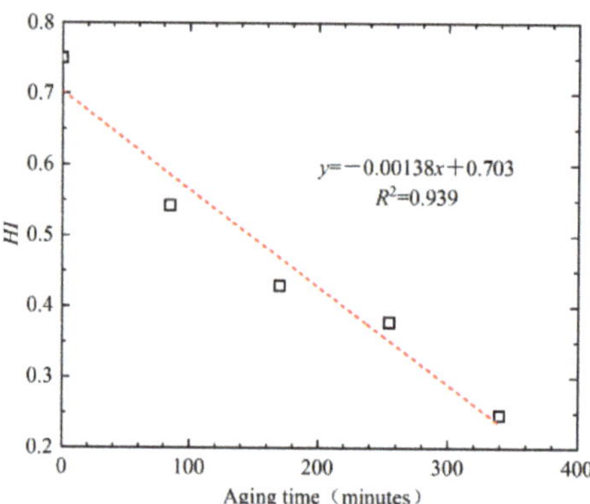

Figure 18. Influence of aging on the healing index.

It can be observed from Figure 18 that with the prolongation of aging time, the healing index decreases linearly. The longer the time over which the asphalt is aged, the lower its healing index is, and the worse its healing ability is. This is because asphalt includes saturate, aromatic, resin, and asphaltene components. During the aging process, with the continuous input of oxygen, light components are gradually oxidized and the small molecules condense into asphaltene in asphalt, while aromatic and saturated components decrease, resulting in the change in asphalt colloid structure from a solution gel to a gel structure. On the other hand, the molecular chain of asphalt is destroyed and reorganized in the process of oxidation, which increases its relative molecular weight, as well as the movement resistance of asphalt molecules, and slows down the diffusion rate of asphalt molecules. Finally, the interface's healing ability decreases. In addition, the penetration, ductility, softening point, and viscosity of asphalt before and after aging were also tested, and the correlation between healing index and these indexes was analyzed, as presented in Figure 19.

It can be observed from Figure 19a,b that the healing index rises with the increase in penetration. The healing index first rises rapidly with the ascent of ductility, and then the growth gradually slows down. It can be observed from Figure 19c,d that the healing index decreases linearly with the increase in softening point and viscosity. As a result, the aging of asphalt leads to the decrease in viscosity, the increase in softening point, and the deterioration of fluidity. However, there is a close relationship between the asphalt's fluidity and the healing ability of the interface. Asphalt with a high penetration, low softening point, and high ductility has excellent healing ability. The changes in these indexes have macroscopic effects, which are caused by the change in asphalt components and colloid structure. For this reason, these common indexes may be used as an indirect basis on which to evaluate the healing potential of the bonding strength at the interface. It must be indicated that polymer-modified asphalts still require more data to support this statement in their case.

3.4. Relationship between the Interface Healing Level and Healing Potential of Mixture

To reveal the relationship between the healing index of the interface and the healing performance of the asphalt mixture, the self-healing performance of the asphalt mixture was simulated using the meso-finite element method in this paper, and the parameters for the cohesive zone model of the bonding element were designated to different healing levels, as presented in Table 10. The load-displacement curve of the bending beam for the asphalt mixture was obtained after finite element simulation and presented in Figure 19. The measured load-displacement curves are also demonstrated in Figure 20.

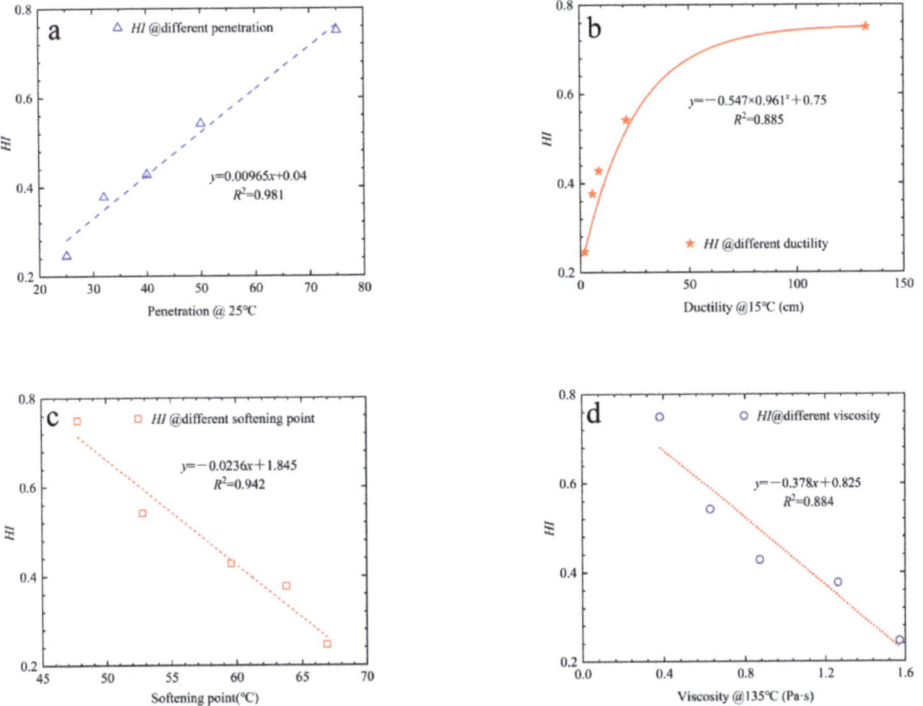

Figure 19. Relationship between the indexes of aged asphalt and the healing index. (**a**) *HI* @ different penetration, (**b**) *HI* @ different ductility, (**c**) *HI* @ different softening point, (**d**) *HI* @ different viscosity.

Table 10. Model parameters for the CZM element at different healing levels.

Interface Type	Healing Index	Fracture Strength (MPa)		Fracture Energy (J·m²)	
		Normal Direction	Tangent Direction	Normal Direction	Tangent Direction
Interface of aggregate–mortar	0.8	0.48	0.8	400	400
	0.6	0.36	0.6	300	300
	0.4	0.24	0.4	200	200
Interface of asphalt mortar	0.8	0.8	1.6	400	400
	0.6	0.6	1.2	300	300
	0.4	0.4	0.8	200	200

As described in Figure 20, for the first loading condition, the flexural bearing capacity of the beam is the highest, and the FE simulated and measured values before the peak are consistent. However, the simulated values begin to deviate from the measured values, and the load of the FE model degrades faster, while the actual asphalt mixture has good post-peak bearing capacity. This may be due to the fact that the CZM element did not consider the influence of asphalt viscosity. Although there are some differences between the experimental and simulated values, the meso-finite element method is still feasible in the simulation of the mechanical properties of the asphalt mixture. The finite element simulation results also show that the flexural bearing capacity of the healed asphalt mixture gradually declines with the decrease in the healing index of the interface. Comparing the measured and simulated results of the asphalt mixture, it can be observed that the flexural bearing capacity of the asphalt mixture is very close to the condition that the healing index of the interface is 0.6 when the beam specimens of the asphalt mixture were healed at 23–45 °C (the average temperature was 28 °C) for 2 days. It is helpful to save the waiting time for the healing process, speed up the analysis procedure, and save the test consumption.

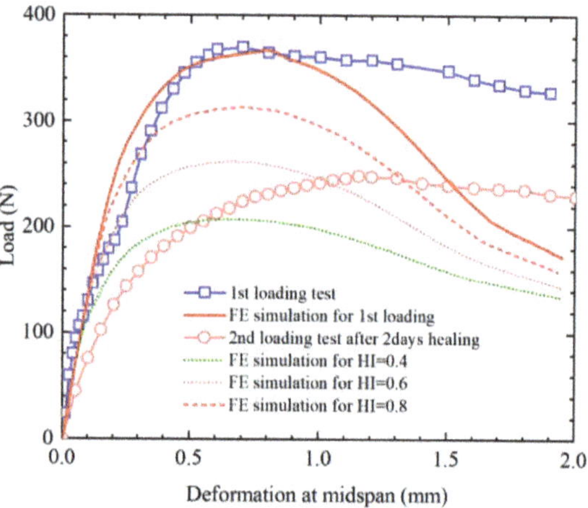

Figure 20. Load-displacement curves of the beam before and after healing.

In order to clarify the relationship between the healing performance of the interface and the healing potential of the mixture, the healing performance of the interface and the healing degree of fracture energy or ultimate strength of the mixture were compared, as presented in Figure 21. In this paper, the average healing temperature of the mixture is 28 °C, and the *HI* of the interface after healing can be calculated by using the master curve.

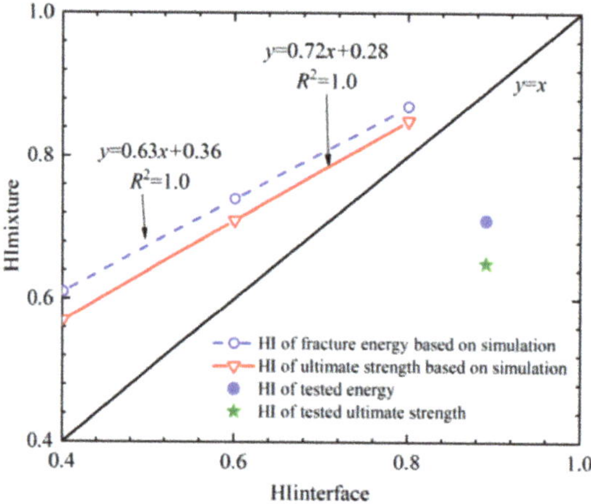

Figure 21. Healing level between the interface and the mixture.

As presented in Figure 21, it can be observed that there is an obvious linear correlation between the healing index of the bonding strength of the asphalt–aggregate interface and the self-healing performance of the mixture. It can be observed from this relationship that the self-healing ability of the asphalt–aggregate interface has a direct impact on the performance of the asphalt mixture. On the basis of the simulated and the measured results, the energy recovery of the asphalt mixture after self-healing is the fastest, and the self-healing amplitude of the strength is slightly lower than the fracture energy. It is proved that there are significant differences among different indexes for evaluating the healing potential of the asphalt mixture. The simulation results show that the healing index of the asphalt mixture will be higher than that of the interface, while the measured results show that the healing level of the mixture (0.65–0.71) is obviously lower than that of the interface (0.89). This is mainly because the interface studied in this paper only includes asphalt and stone, without considering the influence of mineral powder and fillers. In addition, the bilinear failure criterion of the CZM element and 2D FE simulation also result in a difference between the test value and the simulation value. Cheng et al. [28] and Varma et al. [29] stated that mineral powder would increase asphalt viscosity. According to the relationship in Figure 19d, the increase in viscosity leads to a decrease in the self-healing ability of the interface. Therefore, the actual healing level of the asphalt mixture is lower than that of the interface.

4. Conclusions

In this paper, the influence law of time and temperature on the self-healing index of the asphalt–aggregate bonding interface was measured, the influence degrees of aging, immersion, and other factors on the self-healing ability of interface strength were discussed, and the relationship between the self-healing degree of the asphalt–aggregate interface and the self-healing ability of the mixture was analyzed. Through this research, the following conclusions can be drawn.

A higher healing temperature and longer healing time can effectively improve the healing level of the asphalt–aggregate interface.

The Compertz model can accurately describe the change in the healing index over time. Based on the time–temperature equivalence principle, the main healing curve of bonding strength can be established.

Immersion, salting solutions, and aging all have adverse effects on the healing properties of the bonding interface. The longer the immersion time, the worse the healing index of the interface. With the increase in solution concentration, the healing property of the interface between the asphalt and aggregate gradually decreases. With the extension of aging time, the self-healing ability of interface strength decreases linearly. There is a good correlation between common asphalt indexes and its healing ability.

The healing level of the asphalt–aggregate interface is linearly correlated with the healing level of the strength and fracture energy of the mixture. The actual healing level of the asphalt mixture is obviously lower than that of the interface.

It is feasible to simulate and analyze the self-healing of the asphalt mixture using the finite element method, which is not only green and sustainable, but also convenient. It should also be noted that more studies are required to verify the accuracy of the FE method. More experiments on various asphalt mixtures should also be conducted to ensure the robustness and accuracy of the model proposed in this paper.

Author Contributions: Conceptualization Q.G., H. and F.E.; methodology, F.E., T.Z. and P.Z. Software, L.L. and H.; Investigation, T.Z., P.Z. and H.; writing—original draft preparation, H., L.L. and Q.G.; funding acquisition, L.L. and Q.G. All authors have read and agreed to the published version of the manuscript.

Funding: The authors express their appreciation for the financial support of the National Natural Science Foundation of China under Grant No. 51508150; 52178266.

Institutional Review Board Statement: Not applicable.

Informed Consent Statement: Not applicable.

Data Availability Statement: Not applicable.

Conflicts of Interest: The authors declare no conflict of interest.

References

1. Zhu, X.Y.; Lu, C.H.; Dai, Z.W.; Li, F. Application of self-healing engineering materials: Mechanical problems and research progress. *China Sci. Bull.* **2021**, *66*, 2802–2819. (In Chinese) [CrossRef]
2. Guo, Q.; Liu, Q.; Wu, C.; Li, L.; Li, Y.; Liu, F. Local temperature field and healing level of crack in conductive asphalt and mixture. *J. Jilin Univ. (Eng. Technol. Ed.)* **2022**, *52*, 1386–1393.
3. Guo, Q.; Chen, Z.; Liu, P.; Li, Y.; Hu, J.; Gao, Y.; Li, X. Influence of basalt fiber on mode I and II fracture properties of asphalt mixture at medium and low temperatures. *Theor. Appl. Fract. Mech.* **2021**, *112*, 102884. [CrossRef]
4. Guo, Q.; Bian, Y.; Li, L.; Jiao, Y.; Tao, J.; Xiang, C. Stereological estimation of aggregate gradation using digital image of asphalt mixture. *Constr. Build. Mater.* **2015**, *94*, 458–466. [CrossRef]
5. Jiao, Y.; Zhang, L.; Guo, Q.; Guo, M.; Zhang, Y. Acoustic Emission-Based Reinforcement Evaluation of Basalt and Steel Fibers on Low-Temperature Fracture Resistance of Asphalt Concrete. *J. Mater. Civ. Eng.* **2020**, *32*, 04020104. [CrossRef]
6. Bo, L.; Fang, L.; Kai, S.; Qian, G.; Liu, Z.; Zheng, J. Review on the self-healing of asphalt materials: Mechanism, affecting factors, assessments and improvements. *Constr. Build. Mater.* **2021**, *266*, 120453.
7. Sun, D.; Sun, G.; Zhu, X.; Guarin, A.; Li, B.; Dai, Z.; Ling, J. A comprehensive review on self-healing of asphalt materials: Mechanism, model, characterization and enhancement. *Adv. Colloid Interface Sci.* **2018**, *256*, 65–93. [CrossRef]
8. Guo, Q.; Liu, Q.; Zhang, P.; Gao, Y.; Jiao, Y.; Yang, H.; Xu, A. Temperature and pressure dependent behaviors of moisture diffusion in dense asphalt mixture. *Constr. Build. Mater.* **2020**, *246*, 118500. [CrossRef]
9. Guo, Q.; Li, L.; Cheng, Y.; Jiao, Y.; Xu, C. Laboratory evaluation on performance of diatomite and glass fiber compound modified asphalt mixture. *Mater. Des.* **2015**, *66*, 51–59. [CrossRef]
10. Wei, H.; Li, J.; Wang, F.; Zheng, J.; Tao, Y.; Zhang, Y. Numerical investigation on fracture evolution of asphalt mixture compared with acoustic emission. *Int. J. Pavement Eng.* **2021**, *23*, 3481–3491. [CrossRef]
11. Margaritis, A.; Hasheminejad, N.; Pipintakos, G.; Jacobs, G.; Blom, J.; Bergh, W.V.D. The impact of reclaimed asphalt rate on the healing potential of bituminous mortars and mixtures. *Int. J. Pavement Eng.* **2022**, *23*, 4664–4674. [CrossRef]
12. Huang, W.; Zhou, L. Evaluation of Adhesion Properties of Modified Asphalt Binders with Use of Binder Bond Strength Test. *Transp. Res. Rec.* **2017**, *2632*, 88–98. [CrossRef]
13. Johannes, P. The use of the binder bond strength (BBS) test in quantifying moisture damage resistance of bituminous binders used in chip seals. In Proceedings of the Conference on Asphalt Pavements for Southern Africa, Sun City, South Africa, 16–19 August 2015; Southern African Bitumen Association: Cape Town, South Africa, 2015; pp. 9–19.
14. Copeland, A.; Youtcheff, J.; Shenoy, A. Moisture Sensitivity of Modified Asphalt Binders: Factors Influencing Bond Strength. *Transp. Res. Rec.* **1998**, *1*, 18–28. [CrossRef]
15. Ban, H.; Kim, Y.-R.; Pinto, I. Integrated Experimental–Numerical Approach for Estimating Material-Specific Moisture Damage Characteristics of Binder–Aggregate Interface. *Transp. Res. Rec.* **2011**, *2209*, 9–17. [CrossRef]
16. Guo, Q.; Li, G.; Gao, Y.; Wang, K.; Dong, Z.; Liu, F.; Zhu, H. Experimental investigation on bonding property of asphalt-aggregate interface under the actions of salt immersion and freeze-thaw cycles. *Constr. Build. Mater.* **2019**, *206*, 590–599. [CrossRef]
17. Hu, M.; Sun, D.; Lu, T.; Ma, J.; Yu, F. Laboratory Investigation of the Adhesion and Self-Healing Properties of High-Viscosity Modified Asphalt Binders. *Transp. Res. Rec. J. Transp. Res. Board* **2020**, *2674*, 307–318. [CrossRef]
18. Huang, W.; Lv, Q.; Xiao, F. Investigation of using binder bond strength test to evaluate adhesion and self-healing properties of modified asphalt binders. *Constr. Build. Mater.* **2016**, *113*, 49–56. [CrossRef]
19. Xu, H.; Zhang, W.; He, Z.; Kong, L.; Huang, J. Influence of Asphalt Aging and Regeneration on Self-healing Performance. *J. Build. Mater.* **2022**, *25*, 1070–1076.
20. Zhou, L.; Huang, W.; Lv, Q. Evaluation and mechanism analysis of asphalt self-healing performance under dry and wet conditions. *J. Build. Mater.* **2021**, *24*, 137–145. (In Chinese)
21. Zhou, L.; Huang, W.; Sun, L.; Lv, Q.; Zhang, X. Influence factors of asphalt self-healing performance and micro-analysis by CT scanning. *J. Harbin Inst. Technol.* **2021**, *9*, 1–9.
22. Lv, Q.; Huang, W.; Zheng, M.; Hao, G.; Yan, C.; Sun, L. Investigating the asphalt binder/mastic bonding healing behavior using bitumen bonding strength test and X-ray Computed Tomography scan. *Constr. Build. Mater.* **2020**, *257*, 119504. [CrossRef]
23. Huang, W.; Zhou, L.; Lv, Q.; Guan, W. Effects of Multiple Modifiers on Adhesive and Self-Healing Prop-erties of Asphalt Based on Bitumen Bond Strength Test. *J. Tongji Univ. (Nat. Sci.)* **2021**, *49*, 670–679.
24. Wang, C.; Chen, Y.; Gong, G. Cohesive and adhesive healing evaluation of asphalt binders by means of the LASH and BBSH tests. *Constr. Build. Mater.* **2021**, *282*, 122684. [CrossRef]
25. JTG D50-2017; Specifications for Design of Highway Asphalt Pavement. China Communications Press: Beijing, China, 2017.
26. Liu, Q. Micromechanics Analysis on Moisture Damage of Asphalt Mixture under the Action of Dry-Wet Cycles. Master's Thesis, Hebei University of Engineering, Handan, China, 2021.
27. E, F. Self-Healing Law of Aggregate Asphalt Interface Strength and Its Influence on Healing Property of Mixture. Master's Thesis, Hebei University of Engineering, Handan, China, 2020.

28. Cheng, Y.; Tao, J.; Jiao, Y.; Tan, G.; Guo, Q.; Wang, S.; Ni, P. Influence of the properties of filler on high and medium temperature performances of asphalt mastic. *Constr. Build. Mater.* **2016**, *118*, 268–275. [CrossRef]
29. Varma, R.; Balieu, R.; Kringos, N. A state-of-the-art review on self-healing in asphalt materials: Mechanical testing and analysis approaches. *Constr. Build. Mater.* **2021**, *310*, 125197. [CrossRef]

 *materials*

Article

Seismic Performance of Drained Piles in Layered Soils

Yaohui Yang [1,*], Gongfeng Xin [1], Yumin Chen [2,3], Armin W. Stuedlein [4] and Chao Wang [5]

1   Shandong Hi-Speed Group Innovation Research Institute, Jinan 250014, China
2   Key Laboratory of Ministry of Education for Geomechanics and Embankment Engineering, Hohai University, Nanjing 210098, China
3   College of Civil and Transportation Engineering, Hohai University, Nanjing 210098, China
4   School of Civil and Construction Engineering, Oregon State University, Corvallis, OR 97331, USA
5   School of Traffic & Transportation Engineering, Changsha University of Science & Technology, Changsha 410114, China
*   Correspondence: yangyaohui1905@163.com

Abstract: The provision of drains to geotechnical elements subjected to strong ground motion can reduce the magnitude of shaking-induced excess pore pressure and the corresponding loss of soil stiffness and strength. A series of shaking table tests were conducted within layered soil models to investigate the effectiveness of drained piles to reduce the liquefaction hazard in and near pile-improved ground. The effect of the number of drains per pile and the orientation of the drains relative to the direction of shaking were evaluated in consideration of the volume of porewater discharged, the magnitude of excess pore pressure generated, and the amount of de-amplification in the ground's motion. The following main conclusions can be drawn from this study. Single, isolated piles and a group of drained piles were tested in three series of shake table tests. Relative to conventional piles, the drained piles exhibited improved performance with regard to the generation and dissipation of excess pore pressure and stiffness of the surrounding soil, with increases in performance correlated with increases in the discharge capacity of the drained pile. The acceleration time histories observed within the pile-improved soil indicated a coupling of the rate and magnitude of porewater discharge, excess pore pressure generated, and de-amplification of strong ground motion. The amount of de-amplification reduced with increases in the number of drains per pile and corresponding reductions in excess pore pressure. The improved performance should prove helpful in the presence of sloping ground characterized with low-permeability soil layers that inhibit the dissipation of pore pressure and have demonstrated the significant potential for post-shaking slope deformation.

Keywords: drainage piles; shake table tests; liquefiable soils; excess pore pressure; discharge flow

Citation: Yang, Y.; Xin, G.; Chen, Y.; Stuedlein, A.W.; Wang, C. Seismic Performance of Drained Piles in Layered Soils. *Materials* **2023**, *16*, 5868. https://doi.org/10.3390/ma16175868

Academic Editor: René de Borst

Received: 27 July 2023
Revised: 16 August 2023
Accepted: 17 August 2023
Published: 27 August 2023

1. Introduction

The consequences of earthquake-induced soil liquefaction has produced a significant amount of damage to civil infrastructure, with the manifestation of excessive settlement, tilt, and lateral movement of buildings, bridges, lifelines, and waterfront structures [1–5] The 1964 Niigata and Good Friday, Anchorage, earthquakes provided engineers with the first modern opportunities to document the devastating effects of liquefaction. Hamada [6] describes observations of ruptured concrete piles discovered twenty years following the Niigata earthquake, with lateral deformations exceeding 0.6 m at their heads and clear indication of plastic hinging some 2 m below the pile connection. The numerous earthquakes since then have continued to form the basis for the evaluation of liquefaction susceptibility, triggering, and its consequences, and researchers have put significant effort into improving the understanding of soil–pile interaction in liquefying soil [7–13].

Numerous ground-improvement methodologies have been developed and refined to address the need to prevent damaging magnitudes of lateral deformation. Ground improvements include densification via dynamic compaction [14], vibro-compaction [15], drilled and driven displacement piles [16–19], and sand compaction piles [20]; reinforcement using vibro-replacement and drilled piles; seismic isolation using damping materials [21,22], deep-soil mixed columns and panels, and jet grout columns [23–29]; and drainage [30–32]. While some methods (e.g., displacement-type ground improvements) provide densification and reinforcement, vibro-compaction (i.e., stone columns) offers some efficiency by providing densification, reinforcement, and drainage, depending on the degree of mixing with native fines [25,26]. In the presence of significant silty fines, the installation of stone columns following improvement with pre-fabricated vertical drains (PVDs) can lead to improvements in the magnitude of densification [33,34]. However, the use of two or more different ground-improvement technologies can lead to significant costs associated with mobilization, scheduling, and coordination. Therefore, efforts to develop multi-function ground-improvement methodologies that can lead to further efficiencies in mitigating the consequences of liquefaction is of interest to the profession [18].

The inclusion of drainage within liquefying soils is one technology that continues to hold substantial merit for restraining large lateral deformations. Seed and Booker [35] proposed a methodology for incorporating drainage into the design of stone columns, which spurred significant research interest and investigation. The authors of [36–42] describe steel pipes and sheet piles fitted with drains to serve as a liquefaction countermeasure; shake table tests showed that the provision of drainage led to sufficient dissipation of excess pore pressure so as to prevent large deformations. Further recent developments have considered the range of drainage from the improvement in densification of silty sands with drained piles [18], the mitigation of liquefaction below embankments using plastic board PVDs [43], the direct in situ evaluation of the cyclic response of PVD-improved ground [44], and centrifuge testing [45]. When installed with deep foundations, drains have been shown to reduce the magnitude of excess pore pressure as well as the shaking-induced bending moments [46].

This paper demonstrates the performance of a new drained pile that combines the provision of vertical drains along the longitudinal axis of pre-cast piles. Shaking table tests were carried out on various model configurations to investigate the performance of drainage piles with a view of the discharge volumes of porewater, the generation and dissipation of shaking-induced excess pore pressure, and the de-amplification of acceleration. The experimental investigation focused on the orientation of drains relative to the direction of shaking, the number of drains per pile, and the response of single piles and groups of piles. The volume of porewater discharged, the excess pore pressures, and the acceleration response of the soil and piles were strongly affected by the number of drains and the orientation of the drain relative to the predominant direction of shaking. These results indicate that drained piles can serve to improve the seismic response of pile-improved ground and pile-supported structures without the use of separate construction equipment.

2. Drainage Piles: Concept and Prototypes

2.1. General Concept

The drainage pile evaluated in this paper combines the advantages of the flexural stiffness provided by a prefabricated (i.e., pre-cast) concrete pile with the ability to dissipate excess pore pressure via vertical drainage. Grooves are cast along the full length of the pile along its longitudinal axis, facilitating the placement of pre-fabricated vertical drains. The shape and size of the grooves can be modified to accommodate various drainage elements. For example, if cast in a circular arrangement, prefabricated, flexible steel-wire drain pipes can be readily installed to provide filtration and drainage. If the grooves are quadrate, rectangular PVDs can be used. The use of a geotextile filter fabric to cover the drain prevents clogging during installation and while in service. Owing to the small surface area of the drain along the pile relative to the remainder of the rough concrete surface, there is no significant impact on the resistance of axial loads through shaft resistance. Similar to the drainage piles tested by Stuedlein et al. [18], drainage of driving-induced excess pore pressure through contractive, silty sands can lead to increases in densification around the pile during pile installation, depending on the pile spacing, providing an additional benefit of the drainage pile. During shaking, the provision of drains with sufficient discharge capacity allows for the reduction in the loss of the mean effective stress surrounding the pile shaft and a corresponding reduction in lateral deformations arising from soil liquefaction. The first application of this kind of drainage pile was in Jiangyin, Jiangsu Province of China [47–49], using 23 m long, 0.3 m square precast concrete piles with a single rectangular drain running the length of the pile. The structure was supported on 203 of these piles distributed under the footprint area of 1100 m^2.

2.2. Model-Scale Drainage Piles

The drainage piles evaluated in this study are shown in Figure 1. The overall model pile cross-section is 5 cm × 5 cm and is 60 cm long, and it is constructed of micro-concrete, which consists of cement, sand, and water using a ratio of 1:2:0.5, respectively. Considering the selected scale factor of 1:10 described below, these model piles represent 0.5 m square prototype piles, 6 m in length. Various drain configurations were investigated in the model tests, including a conventional pile, singly and doubly drained piles (i.e., with one and two drains, respectively), and a configuration with four drains (i.e., one drain on each side of the pile). The drain consisted of a rectangular plastic drainage board-type PVD 15 mm in width and 10 mm in depth and fitted with a geotextile to prevent clogging by the surrounding sand. The permeability coefficient of the geotextile was 0.162 cm/s.

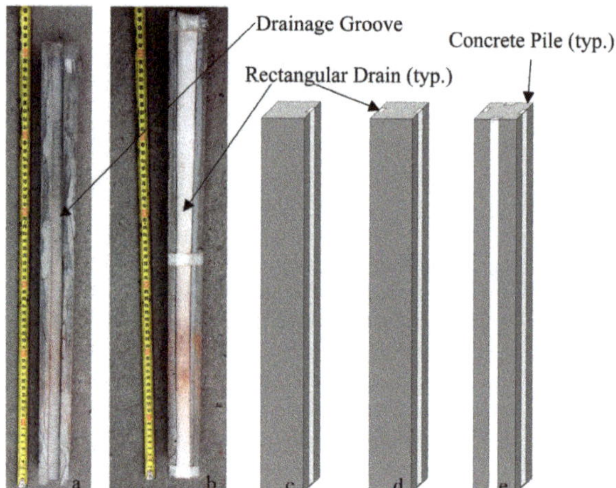

Figure 1. Drainage pile geometry investigated here: (**a**) model micro-concrete pile showing groove for drain, (**b**) model pile fitted with drain, schematic indicating typical square concrete pile with drain, (**c**) single drain, (**d**) two drains, and (**e**) four drains.

2.3. Shaking Table Tests

Significant improvements in the understanding of the mechanisms of soil liquefaction and soil–pile interaction have been achieved using 1 g shake table testing [10,18,50–53]. The study described here was conducted using the shake table facility at Chongqing University, capable of simultaneous horizontal and vertical shaking with a maximum base excitation and frequency of 2 g and 50 Hz, respectively. Shaking-table tests were conducted within a laminar soil box with inner dimensions of 950 mm in length, 850 mm in width, and 600 mm in height. The selection of model component dimensions considered established 1 g scaling laws [54], as described below.

2.3.1. Model Geometry and Materials

Given the potential for large deformations of liquefiable soil capped with an impermeable crust [55], the shake table tests were conducted using a two-layer system consisting of an upper clay layer overlying loose, liquefiable sand. Test specimens were constructed by pluviating $7^{\#}$ silica sand ($G_s = 2.64$, $D_{50} = 0.17$ mm, $\rho d_{,max} = 1.65$ g/cm^3, $\rho d_{,min} = 1.34$ g/cm^3), of similar gradation to Toyoura sand (Figure 2), with water from a controlled height to produce uniform relative densities of 45% within the 40 cm thick liquefiable sand layer. Pluviation was paused intermittently to facilitate placement of various instruments, described subsequently, at the target locations. The piles were fixed to the base of the laminar box container such that the sand was pluviated around the piles. A 10 cm thick capping layer of clay was placed on top of the sand to prevent the vertical dissipation of excess pore pressure and complete the shake table specimen. The clay cap material was sourced from Southwest China, and is characterized by a specific gravity, G_s, of 2.72; an in-place void ratio, e, equal to 0.91; a dry density, ρd, of 1.73 g/m^3; liquid and plastic limits, ω_L and ω_p, respectively, of 71 and 42; and hydraulic conductivity of $k = 4.2 \times 10^{-7}$ cm/s. Figure 2 presents the particle size distribution of clay cap material. Following pluviation of the sand, the clay cap was hand-placed on the sand layer and around the piles, allowing the PVDs to discharge porewater through the cap for collection, and trimmed to produce a thickness of 10 cm.

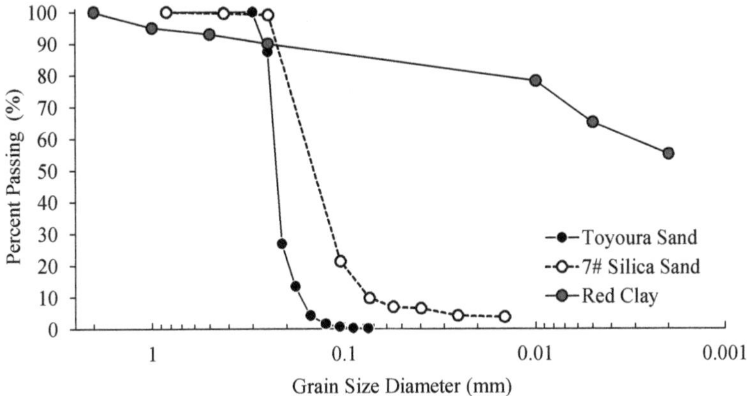

Figure 2. Particle size distributions of materials used in this study and comparison of 7# silica sand to Toyoura sand.

2.3.2. Instrumentation

Instruments were deployed to observe the performance of the drainage pile-improved ground, including the shaking-induced excess pore pressure, acceleration, and the flow of water discharged from the drains. The pore pressure transducers (PPTs) used were characterized with a capacity of 20 kPa and accuracy of 0.1% full-scale (i.e., 0.02 kPa). Accelerometers were characterized with a range of ± 10 g and sensitivity of 500 mV/g over the frequencies of interest. The PPTs and accelerometers were placed 20 cm from the base of the model container and 5 cm and 10 cm (i.e., one and two pile diameters) from the piles. In some shake tests, several piles were instrumented with accelerometers to capture the motion at the pile heads. The discharge flow from the PVDs was measured using flow transducers and pumps, as shown in Figure 3, and were arranged to capture and measure water exiting the top of the model pile without applying pump suction to water below the clay cap material.

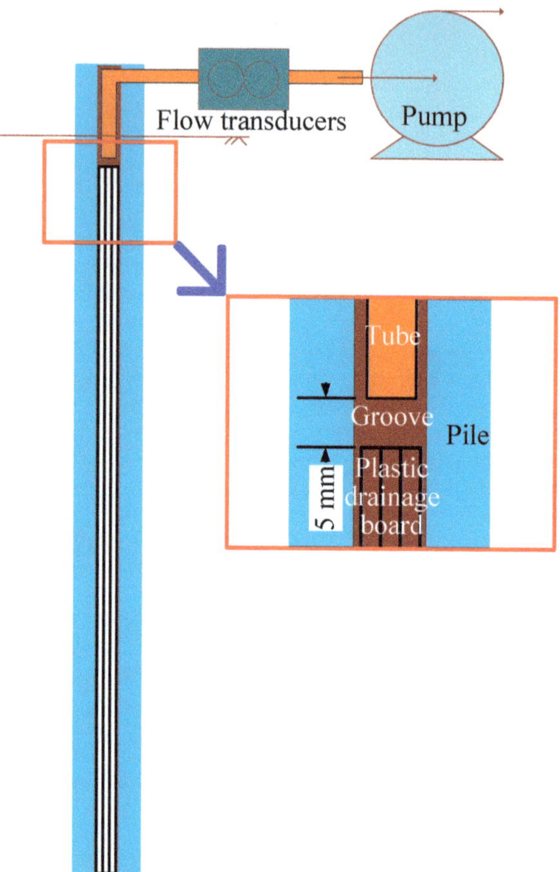

Figure 3. Schematic indicating provisions for measuring the volume of porewater discharged.

2.3.3. Test Configurations

Figure 4 presents the experimental configuration of each test specimen, with corresponding test and pile designations shown in Table 1. The first shake table test (designated T1) included one conventional pile (i.e., a pile with no drainage provision; designated T1ND) and three piles with a single drain orientated parallel (T1SD0), perpendicular (T1SD90), and 45° (T1SD45) to the direction of shaking (Figure 4). The piles were located 35 and 20 cm from the edges of the laminar box and placed to act as isolated piles, enforced using an impermeable membrane to prevent flow to adjacent piles. The second shake table test (designated T2) also evaluated four piles (at the same locations as T1) but considered piles with two drains (designated T2DD0 and T2DD90) and with four drains (designated T2QD090 and T2QD45), which allowed further investigation into the most efficient orientation of drainage piles. The third shake table test designated T3 was conducted on a 3×3 pile group spaced at three pile widths (i.e., 15 cm) from one another and with single drains oriented parallel to the direction of shaking (designated T3GSD0) in order to study the response of a group of free-headed drainage piles.

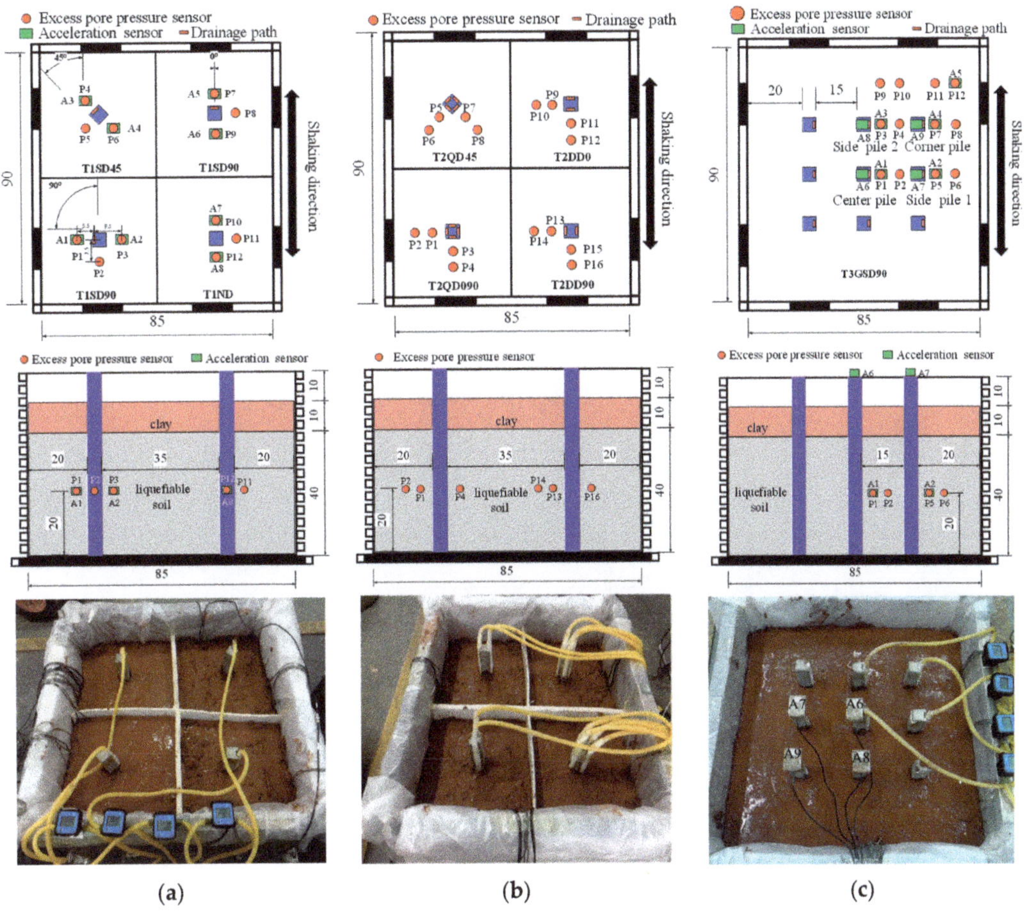

Figure 4. Plan, elevation, and instrumentation arrays used for shake table tests: (**a**) test series T1 with single piles, including T1ND, T1SD0, T1SD90, and T1SD45; and the blue square is flow transducer; the white line is impermeable foam board; the yellow line is tube; and (**b**) test series T2 with single piles with T2DD0, T2DD90, T2QD090, and T2QD45; and (**c**) test series T3 with a group of piles fitted with a single drain, T3GSD90 (Unit: cm).

Table 1. Shake table test configurations.

Shake Table Test Number	Pile Designation	Pile Configuration	Drainage Configuration	Drainage Area (cm²)	Orientation of Drain with Shaking Direction (°)
T1	T1ND	Single pile	No drain	0	N/a
T1	T1SD0	Single pile	Single drain	90	0
T1	T1SD90	Single pile	Single drain	90	90
T1	T1SD45	Single pile	Single drain	90	45
T2	T2DD0	Single pile	Double drain	180	0
T2	T2DD90	Single pile	Double drain	180	90
T2	T2QD090	Single pile	Quad drain	360	0, 90
T2	T2QD45	Single pile	Quad drain	360	±45
T3	T3GSD90	Group of piles	Single drain	810	90

2.3.4. Ground Motion and Scaling

Uniaxial shaking was accomplished with a 10 sec sinusoidal motion with 5 Hz frequency that increased to its peak amplitude of 0.2 g over a duration of two seconds, remained constant for 6 s, and then reduced to zero over a two-second duration (Figure 5). In order to distinctly analyse the results, sinusoidal motion was adopted in the test. This motion allows a smooth ramp-up and -down of the shaking table and has been found appropriate for liquefaction testing [13]. The selected scale factor for these model tests was 1:10; as such, the ground motion acceleration amplitude is 1:1, whereas the frequency scales to $1/n^{0.5}$ (Table 2; [56]). Thus, an equivalent experiment at full scale would experience ten seconds of shaking with peak amplitude of 0.2 g and frequency of 2.2 Hz.

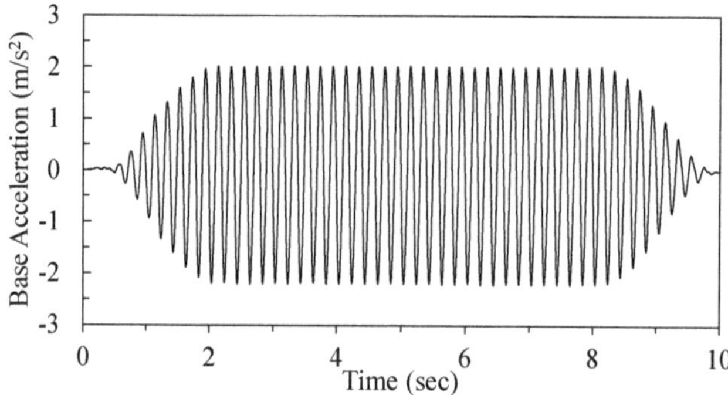

Figure 5. Acceleration time history applied by shake table for each experiment.

Table 2. Similitude laws for 1 g shaking table tests (after Iai, [56]).

Items	Model	Prototype
Scaling factor	1	n
Length	$1/n$	1
Density	1	1
Displacement	$1/n$	1
Stress	1	1
Frequency	$1/n^{0.5}$	1
Acceleration	1	1

3. Shake Table Results and Analysis

3.1. Comparison of Shaking-Induced Discharge Flow Volumes

The drainage of porewater during strong ground motion to relieve excess pore pressure represents the key advantage associated with the use of the drainage pile. Figure 6 presents the discharge flow time histories for shake table tests conducted with the drainage piles. Figure 6a presents the discharge time histories for isolated, single piles fitted with one drain oriented at 0°, 45°, and 90° (i.e., T1SD0, T1SD45, and T1SD90, respectively) from the direction of shaking. During shaking, the drained piles exhibit different rates of shaking-induced discharge: the pile with the drain oriented 90° from the direction of shaking produced the greatest volume of porewater, whereas pile T1SD0 produced the least volume. Following the end of shaking, the rate of discharge (or discharge flow) gradually slowed over the first 20 s to slow to a near-zero flow thereafter. The total volume of pore water removed from the liquefiable soil equalled 201 and 448 mL for the T1SD0 and T1SD90, respectively. The isolated pile with a single drain oriented at 45° performed nearly the same as T1SD90, yielding a total porewater discharge volume of 393 mL (approximately 88% of T1SD90).

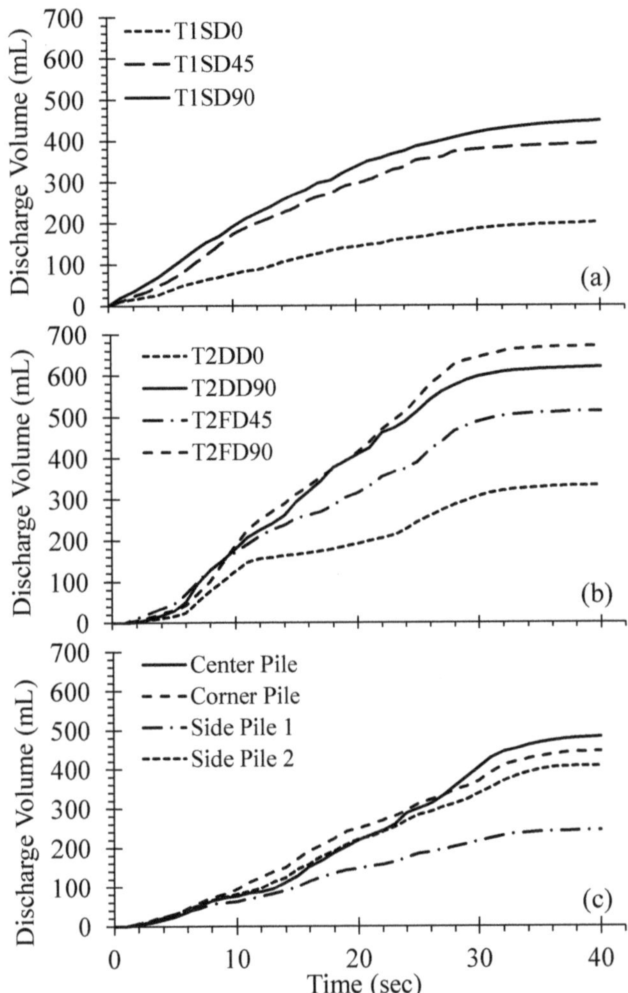

Figure 6. Comparison of discharge volume time histories for (**a**) test series T1 on single piles with one drain per pile, (**b**) test series T2 on single piles with two and four drains per pile, and (**c**) test series T3 on a group of singly-drained piles.

Figure 6b presents the discharge flow time histories for the isolated, single piles with two and four drains observed during the second shake table test (T2 series; Table 1). In the case of two drainage paths, the piles with two drains oriented parallel to the direction of shaking (i.e., T2DD0) produced greater discharge volumes than the comparable, singly-drained pile (T1SD0), with a total discharge volume of 333 mL. However, pile T2DD90 with drains oriented perpendicular to the direction of shaking removed nearly twice the porewater volume as T2DD0 at the end of shaking (i.e., 620 mL). Piles with four drainage paths oriented at 45° and 90° to the direction of shaking, designated T2QD45 and T2QD90, discharged 513 and 671 mL of porewater at the end of shaking. Thus, it appears that piles with two drains oriented perpendicular to the direction of shaking exhibit greater efficiency than piles with four drains if the orientation of the drains is rotated toward the predominant direction of shaking. However, given that earthquake motions exhibit significant variations in direction, arbitrary directionality seems to be best mitigated through the use of four drains per pile.

Figure 6c presents the discharge flow time histories for drained piles set within a pile group evaluated in the third shake table test (i.e., T3, Table 1). A single drain was fitted to each pile in this test series and all piles were oriented at 90° or perpendicular to the direction of shaking. The shaking-induced flow was monitored for four of the nine piles, located at the centre and one corner of the pile group, and a side pile leading the centre pile and one trailing the corner piles. The instrumented piles exhibited similar magnitudes of discharge flow during shaking; however, following shaking, the centre, corner, and the second side pile exhibited a greater amount of discharge flow, producing discharge volumes of 483, 445, and 408 mL, respectively. Side pile 1, trailing the corner piles, produced the smallest magnitude of post-shaking discharge volume of 245 mL.

Figure 7 compares the maximum discharge volume of isolated, single piles with one, two, and four drains and with drain orientations relative to the direction of shaking. Doubling the drainage capacity from one to two drains results in a significant increase in discharge volume for the orientations investigated. While doubling the drainage capacity from two to four drains results in further increases in discharge volume, the incremental increase in volume is smaller than that when doubling the capacity from one to two drains. This suggests that there may be limited benefit in providing four drainage paths to a drained pile; however, the predominant direction of shaking is usually not known with significant accuracy as noted earlier.

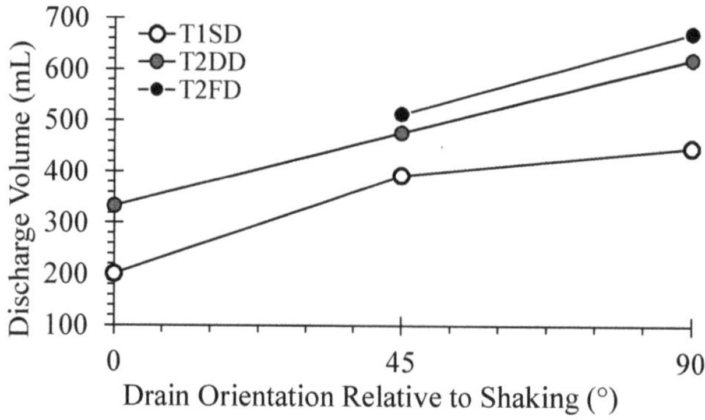

Figure 7. Comparison of discharge volumes observed for single piles with various number of drains and drain orientations relative to shaking.

3.2. Comparison of Excess Pore Pressure Generation and Dissipation

Drained piles have the potential to mitigate the consequences of liquefaction through the accelerated dissipation of excess pore pressure during and immediately following strong ground motion. Figure 8 demonstrates the rate of excess pore pressure generation and dissipation during the shake table tests of the single, isolated piles in terms of the excess pore pressure ratio, r_u, along with the distribution of PPTs for each pile. PPTs located 180° from the drains (i.e., P3 and P6 for T1SD90 and T1SD45, respectively) consistently produced the largest excess pore pressures, with peak r_u of about 0.88 (n. b., P9 did not function during the test, but likely would have produced a similar r_u response). On the other hand, PPTs directly in front of the drains consistently exhibited the smallest generation of excess pore pressure, with peak r_u ranging from 0.53 to 0.58. For these PPTs, the lowest peak r_u (i.e., 0.53) was observed for the drain arranged perpendicular to shaking (T1SD90; Figure 8a), whereas the highest peak r_u (i.e., 0.58) was generated for the case with the single drain oriented perpendicular to the direction of shaking (T1SD0; Figure 8c). Excess pore pressures generated along the sides of the pile and observed using PPTs P2, P5, and P8 ranged from 0.60 to 0.64; this indicates, along with the PPTs, that the drains produce an azimuthal variation in r_u and corresponding hydraulic gradient driving flow from the region away from the drain towards the drain. This implies that a preferred orientation of singly drained piles placed in sloping ground may exist in order to restrain ratchet-type movements downslope. Furthermore, the excess pore pressures observed for most of the PPTs dissipated to near-zero within 10 s following shaking; this represents a significant advantage of slopes with crust caps that can retard the dissipation of migrating porewater and lead to delayed failure [54,55].

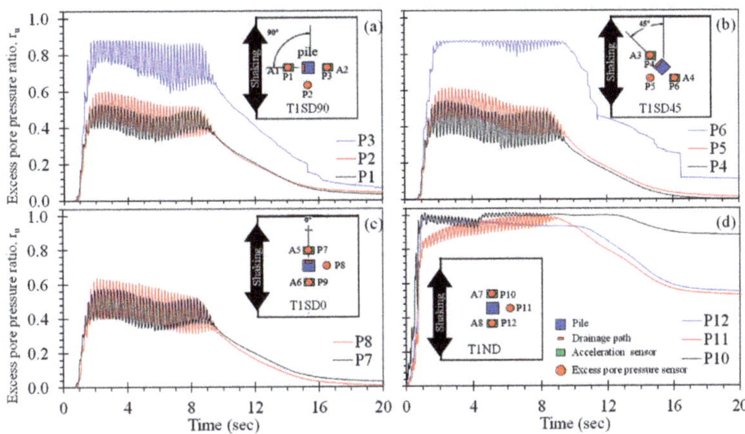

Figure 8. Excess pore pressure time histories for test series T1: (**a**) pile T1SD90, (**b**) pile T1SD45, (**c**) pile T1SD0, and (**d**) conventional pile T1ND.

In comparison, the single, isolated, conventional pile (T1ND, Figure 8d) tested to compare the r_u response in the absence of a drain produced significantly larger peak r_u values equal to 1.0 at all locations. For this pile, the duration to full liquefaction varied with location: P10 and P12, placed in front and behind the pile and parallel to the ground motion, exhibited liquefaction near-simultaneously at about 1.0 s into the ground motion, whereas P11, placed along the side of the pile in the direction perpendicular to the direction of shaking (relative to the pile), indicated the onset of liquefaction at 7.5 s of shaking. The rate of excess pore pressure dissipation following shaking observed in all PPTs was significantly lower than for the case of the drained piles. In general, the presence of a single drain on the model piles significantly reduced the rate and magnitude of excess pore pressure generated as compared to the conventional pile.

Figure 9 presents the rate of excess pore pressure generation and dissipation during the second shake table test series with piles fitted with two and four drains along with the distribution of PPTs for each pile. In general, the near-pile r_u time histories (observed using P1 and P3, P5 and P7, P9 and P11, and P13 and P15), indicate slightly lower magnitudes than the singly drained piles (compare to Figure 8). Comparison of T2QD090 (Figure 9a) to T2DD0 and T2DD90 (Figure 9c,d) indicate that the use of four drains results in reduced excess pore pressures, particularly in the region further away from the piles. The largest magnitude of excess pore pressure was observed using P12 for the doubly drained pile T2DD0, with peak r_u of about 0.86. The remainder of the PPTs exhibited peak r_u values of 0.59 or smaller: the average peak r_u for piles with two drains was 0.58, whereas the average maximum r_u for piles with four drains was 0.46. Thus, use of a greater number of drains will produce a greater volume of stabilized soil surrounding the pile, resulting in smaller permanent displacements if installed in sloping ground.

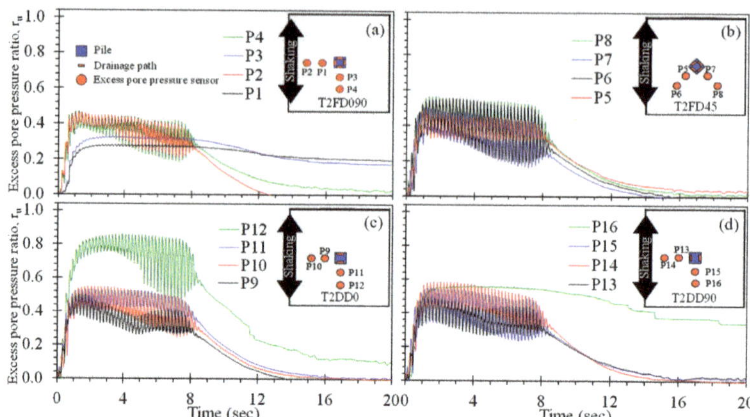

Figure 9. Excess pore pressure time histories for test series T2: (**a**) pile T2QD090, (**b**) pile T2QD45, (**c**) pile T2DD0, and (**d**) T2DD90.

The excess pore pressure ratio time histories for the T3 test series of a group of singly-drained piles is shown in Figure 10. Comparison of the pore pressure response of the group of singly drained piles to single, isolated singly drained piles in Figure 8 indicates that the pile group serves to drain a larger volume of soil. The pore pressure response within or near the pile group (PPTs P1 through P8) exhibited an average peak r_u of 0.5, comparable to the case of single piles with four drainage paths. This indicates that singly-drained piles may be more cost-effective than multi-drained piles when installed in groups. Away from the pile group, the peak r_u ranged from 0.68 to 1.00 (PPTs P9 through P12), with largest excess pore pressure response for the PPTs furthest from the corner pile (i.e., P12) and characterized by the longest drainage path.

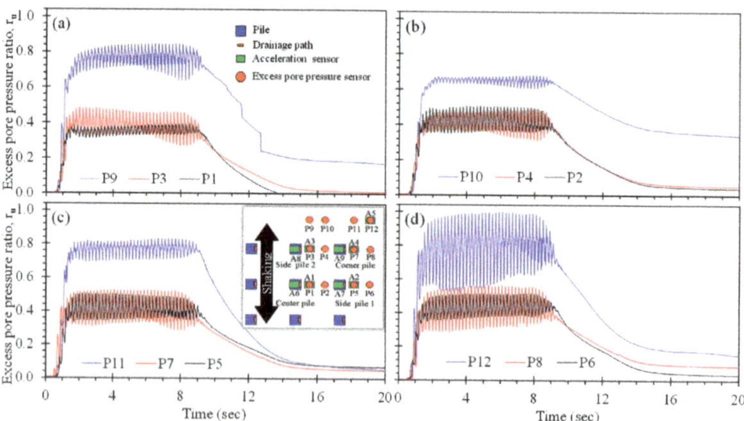

Figure 10. Excess pore pressure time histories for test series T3: (**a**) pore pressures observed along the centre of the pile group, (**b**) pore pressures observed between the centre and outer column of piles, (**c**) pore pressures observed immediately adjacent to the outer column of piles, and (**d**) pore pressures observed away from the outer column of piles.

Comparison of the three test series indicates that the volume of porewater discharged varied with the number of drains per pile, the orientation of the drains relative to the direction of shaking, and the position of a drained pile within a group (Figures 6–10). The presence of a drain serves to reduce r_u in the liquefiable soil as a function of the radial distance from the drain, as shown in Figure 11a for the T1 test series. Here, the excess pore pressure averaged over the period of the ground motion with constant amplitude (i.e., strong shaking from 2 to 8 s) is plotted with radial distance, indicating a sharp reduction in r_u to a near-constant magnitude as the proximity to the drain increases. The excess pore pressure field during shaking implied by Figure 11a does not appear significantly sensitive to the drain orientation; however, the total volume of porewater discharged (during and following shaking) was shown to vary with drain orientation relative to the direction of shaking (Figure 7). Figure 11b presents the variation in porewater volume discharged at time t = 10 s (at the end of shaking) and 40 s (representing total porewater discharged) with the average hydraulic gradient computed using the PPTs nearest and farthest from the drain for the T1 and T2 test series and during the period of strong, constant shaking. The hydraulic gradients increase with the decrease in the number of drains per pile and with increasing orientation away from the direction of shaking. Furthermore, at the end of shaking (t = 10 s), the effect of the average hydraulic gradient on the discharge volume increases with decreasing number of drains per pile. However, as time elapses following the end of shaking and excess pore pressures dissipate, the effect of increasing drain orientation away from the direction of shaking on discharge volume increases. These results suggest that the average hydraulic gradient developed during shaking controls the magnitude of porewater discharged until sufficient discharge capacity is provided (by the increased number of drains per pile), whereupon the volume of discharged porewater becomes controlled by the hydraulic conductivity of the soil. Head losses associated with porewater entering the drain increase with the increasing discharge velocities [57], hence increasing hydraulic gradients. These test results suggest that the ability of drained piles to quickly reduce excess pore pressures during and following shaking is more strongly controlled by the hydraulic gradient when discharge capacity is small (i.e., singly drained piles) and more strongly controlled by the orientation of the drained pile when discharge capacity is large.

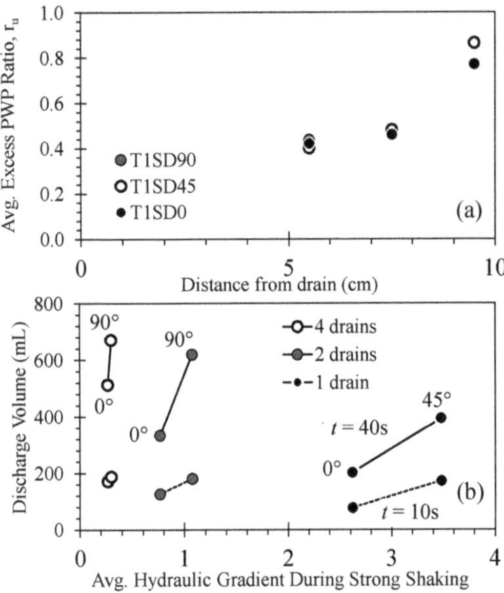

Figure 11. Summary of excess pore pressure response and volume of porewater discharged for test series T1 and T2: (**a**) variation in average excess pore pressure during strong shaking with radial distance from the drain and (**b**) variation in discharge volume at time t = 10 and 40 s with average hydraulic gradient during strong shaking.

3.3. Comparison of Acceleration Time Histories

The acceleration time histories observed in the soil adjacent to the piles, presented in Figure 12, demonstrated the role of drain orientation on the propagation of seismic energy through the soil–pile model. Note that the accelerometers were placed in different physical locations for each drained pile, but along the same orientation relative to the front and back of a given drained pile; therefore, some differences in the response may be attributed to the spatial distribution of shaking within the box and soil–pile–soil interaction relative to the direction of shaking. Pile T1SD90 (Figure 12a) exhibited the highest magnitude of sustained acceleration, as well as the largest peak horizontal acceleration (PHA) of 2.04 m/s², as compared to T1SD45 (Figure 12b) and T1SD0 (Figure 12c). The acceleration time histories for T1SD90 indicate that the magnitude of shaking varies with distance from the drain, with higher magnitudes of acceleration corresponding to the smaller distances. This observation holds for drained piles oriented 45 and 0° from the direction of shaking.

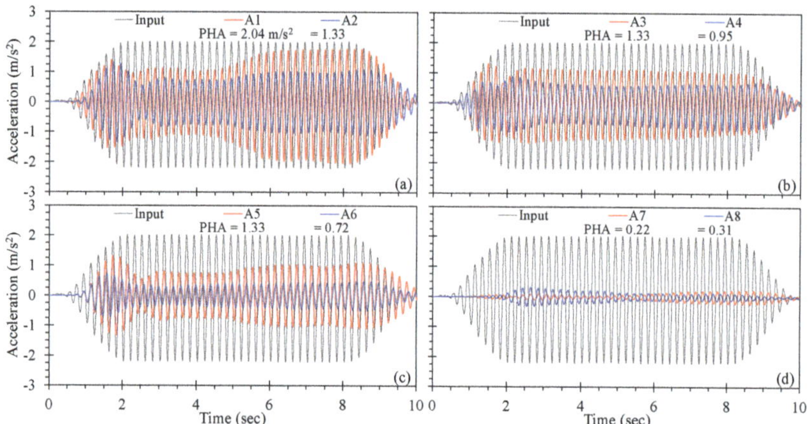

Figure 12. Acceleration time histories for test series T1: (**a**) pile T1SD90, (**b**) pile T1SD45, (**c**) pile T1SD0, and (**d**) conventional pile T1ND.

Excess pore pressures observed for the single, isolated drained piles reduced with increasing proximity to the drain. Thus, the reduction in excess pore pressure is associated with increased soil stiffness, allowing for greater propagation of seismic energy through the soil mass at that location. This may be confirmed in Figure 12d, which shows that the full liquefaction produced for pile T1ND, with no provision for drainage, resulted in significant de-amplification of the input ground motion. The T1 test series also shows that as the volume of porewater discharged from the drains increases (Figures 6a and 7), the magnitude of peak and sustained acceleration decrease. Thus, the direction of shaking relative to the drain serves to reduce the liquefaction hazard through a reduction in acceleration, which appears to lead to a reduction in excess pore pressure and increase discharge volume.

Figure 13 provides the acceleration time histories for the pile group observed in the soil adjacent to and top of selected piles (refer to Figure 4). Similar to the T1 series, the accelerometers were placed 10 cm above the base of the model and indicate the attenuation of the input ground motion to varying extents. In general, the comparison of Figures 10 and 13 shows that accelerations measured closer to the centre of the pile group exhibited the least amount of de-amplification and was associated with the least magnitude of excess pore pressure. The acceleration time histories observed using A1 through A4, placed in or near the pile group, indicated similar responses and acceleration magnitudes. Acceleration magnitudes increased until the excess pore pressures reached relatively stable plateaus (at approximately 2 s of shaking; Figure 10), and then reduced with continuing shaking followed by a gradual rise (at about 6 s) as the drains began to discharge porewater

(Figure 6c). The PHA for accelerometers A1 though A4 were observed towards the end of shaking and ranged from 1.78 to 1.95 m/s^2. Accelerations measured furthest from the drained pile group (i.e., A5) produced the greatest reduction in magnitude, which was associated with greatest r_u, as shown in Figure 10d.

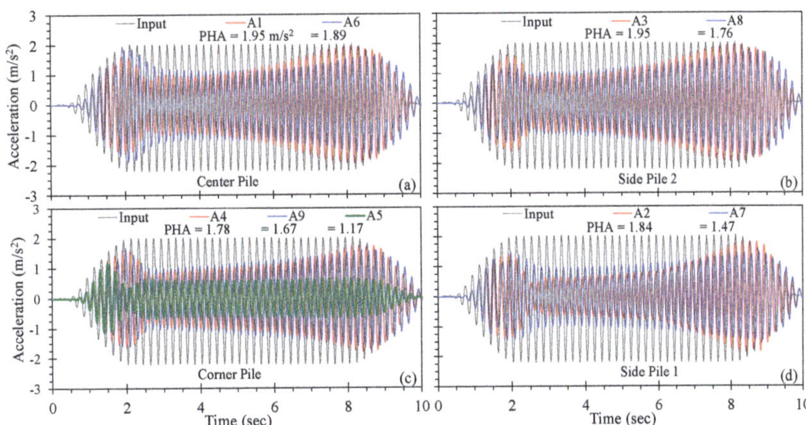

Figure 13. Acceleration time histories for test series T3: (**a**) adjacent to and on top of the centre pile, (**b**) adjacent to and on top of the second side pile, (**c**) adjacent to and on top of the corner pile, and (**d**) adjacent to and on top of the first side pile.

Acceleration time histories measured at the top of the selected drained piles were largely similar those measured immediately adjacent to a given pile, indicating negligible damping associated with soil–pile–soil interaction. This may have resulted from the use of free-headed, unloaded piles in the pile group appropriate for mitigation of lateral spreading in the free field [13]. Pile groups constructed with a relatively rigid pile cap and subjected to inertial loading would likely have exhibited differential motion due to soil–pile–soil interactions (e.g., gapping and drag) that would likely result in a different response. The acceleration time histories varied slightly with pile position in the group: the maximum acceleration of a pile top was observed in the centre of the group, whereas the minimum acceleration was observed at Side Pile 1. These observations show that the discharge capacity of drained piles control the magnitude of excess pore pressures and de-amplification, and therefore, the strength and stiffness of the pile-improved ground, and that the interaction between discharge capacity, excess pore pressure, and acceleration can be inferred from accepted soil dynamics principles.

4. Conclusions

A series of shake table tests were conducted to investigate the effectiveness of drained piles in reducing the liquefaction hazard in and near pile-improved ground within a layered subsurface profile. Various configurations were evaluated, included piles with differing numbers of drains and orientations relative to the direction of shaking. The response of the drained pile-improved soil was evaluated using the discharged volume of porewater, the generation and dissipation of excess pore pressure, and the attenuation of acceleration. The following main conclusions can be drawn from this study:

1. The volume of porewater discharged was dependent on the orientation of the drains relative to the predominant direction of shaking and the number of drains per pile. The efficiency of the drains increased as the orientation of the drains increased from 0° to 90° relative to the direction of shaking. The provision of two drains per pile resulted in a significant increase in the volume of porewater discharged as compared

to the singly drained pile, but the incremental increase in discharge volume reduced when doubling the drains again to four per pile.

2. Drained piles demonstrated the ability to significantly reduce the magnitude of shaking-induced excess pore pressures in proximity to the drain. The excess pore pressure ratio reduced slightly with an increase in the number of drains per pile; however, the extent of soil with reduced excess pressure increased with the increase in the discharge capacity or number of drains. The orientation of the drain relative to the direction of shaking also influenced the magnitude of excess pore pressure.

3. The provision of drains to the model piles resulted in a sharper reduction in post-shaking excess pore pressure, a critical feature for layered soils that include a low-permeability crust, which can inhibit the dissipation of pore pressure and which may result in large post-shaking deformations when the crust is accompanied by sloping ground conditions.

4. The amount of porewater discharged and excess pore pressure generated within and near the group of drained piles depended on the position of the pile within the group; however, the soil within the group exhibited relatively low variation in the distribution of excess pore pressure.

5. The acceleration time histories observed within the pile-improved soil indicated a coupling of the rate and magnitude of porewater discharge, excess pore pressure generated, and de-amplification of strong ground motion. The amount of de-amplification reduced with increases in the number of drains per pile and corresponding reductions in excess pore pressure. Therefore, removal of excess pore pressure-driven porewater maintains the integrity of the ground motion due to prevention of stiffness degradation associated with liquefaction.

Author Contributions: Conceptualization, Y.Y. and G.X.; methodology, Y.Y. and G.X.; formal analysis, A.W.S. and Y.C.; investigation, Y.Y.; resources, Y.Y.; data curation, Y.C.; writing—original draft preparation, Y.Y. and C.W.; project administration, G.X.; funding acquisition, Y.C. All authors have read and agreed to the published version of the manuscript.

Funding: This research is supported by the National Natural Science Foundation of China (No. 51879090).

Institutional Review Board Statement: "Not applicable" for studies not involving humans or animals.

Informed Consent Statement: Informed consent was obtained from all subjects involved in the study.

Data Availability Statement: Data available on request due to restrictions e.g., privacy or ethical. The data presented in this study are available on request from the corresponding author. The data are not publicly available because they contain proprietary information or trade secrets.

Conflicts of Interest: The authors declare no conflict of interest.

References

1. Hamada, M. Large ground deformations and their effects on lifelines: 1964 Niigata earthquake. In *Case Studies of Liquefaction and Lifeline Performance during Past Earthquakes. Vol. 1: Japanese Case Studies*; Technical Report NCEER-92-0001; Hamada, M., O'Rourke, T.D., Eds.; National Center for Earthquake Engineering Research: Buffalo, NY, USA, 1992; Volume 1.
2. Youd, T.L. *Liquefaction-Induced Damage to Bridges*; Transportation Research Record, Transportation Research Board and the National Research Council: Washington, DC, USA, 1993; Volume 1411, pp. 35–41.
3. Tokimatsu, K.; Kojima, H.; Kuwayama, S.; Abe, A.; Midorikawa, S. Liquefaction-induced damage to buildings I 1990 Luzon Earthquake. *J. Geotech. Eng.* **1994**, *120*, 290–307. [CrossRef]
4. Bray, J.D.; Dashti, S. Liquefaction induced building movement. *Bull. Earthq. Eng.* **2014**, *2*, 1129–1156. [CrossRef]
5. Cubrinovski, M.; Jacka, M.E.; Bray, J.D.; Ballegooy, S.V.; Malan, P.; O'Rourke, T.D.; Crawford, S.A. Assessment of liquefaction-induced land damage for residential christchurch. *Earthq. Spectra* **2014**, *30*, 31–55.
6. Hamada, M. Performance of Foundations against Liquefaction-induced Permanent Ground Displacements. In Proceedings of the 12th World Conference on Earthquake Engineering, Auckland, New Zealand, 30 January–4 February 2000; p. 8.
7. Miwa, S.; Ikeda, T.; Sato, T. Damage process of pile foundation in liquefied ground during strong ground motion. *Soil Dyn. Earthq. Eng.* **2006**, *26*, 325–336. [CrossRef]

8. Bhattacharya, S.; Madabhushi, S.P.G.; Bolton, M.D. An alternative mechanism of pile failure in liquefiable deposits during earthquakes. *Géotechnique* **2004**, *54*, 203–213. [CrossRef]

9. Wilson, D.W.; Boulanger, R.W.; Kutter, B.L. Observed seismic lateral resistance of liquefying sand. *J. Geotech. Geoenviron. Eng.* **2000**, *126*, 898–906. [CrossRef]

10. Tokimatsu, K.; Suzuki, H. Porewater Pressure Response around Pile and its Effects on p-y Behavior during Soil Liquefaction. *Soils Found.* **2004**, *44*, 101–110. [CrossRef] [PubMed]

11. Rollins, K.M.; Gerber, T.M.; Ashford, S.A. Lateral resistance of a full-scale pile group in liquefied sand. *J. Geotech. Geoenviron. Eng.* **2005**, *131*, 115–125. [CrossRef]

12. Rollins, K.M.; Hales, L.J.; Ashford, S.A.; Camp, W.M. P-y curves for large diameter shafts in liquefied sand from blast liquefaction tests. In *Seismic Performance and Simulation of Pile Foundations*; ASCE: Reston, VA, USA, 2005; pp. 11–23.

13. Li, W.; Chen, Y.; Stuedlein, A.W.; Liu, H.; Zhang, X.; Yang, Y. Performance of X-shaped and Circular Pile-Improved Ground Subject to Liquefaction-induced Lateral Spreading. *Soil Dyn. Earthq. Eng.* **2018**, *109*, 273–281. [CrossRef]

14. Kumar, S. Reducing liquefaction potential using dynamic compaction and construction of stone columns. *Geotech. Geol. Eng.* **2001**, *19*, 169–182. [CrossRef]

15. Adalier, K.; Elgamal, A. Mitigation of liquefaction and associated ground deformations by stone columns. *Eng. Geol.* **2004**, *72*, 275–291. [CrossRef]

16. Mitchell, J.K. In Place Treatment of Foundation Soils. *J. Soil Mech. Found. Div. ASCE* **1968**, *96*, 73–110. [CrossRef]

17. Siegel, T.C.; NeSmith, W.M.; NeSmith, W.M.; Cargill, P.E. Ground Improvement Resulting from Installation of Drilled Displacement Piles. In Proceedings of the 32nd Annual DFI Conference; Deep Foundations Institute: Hawthorne, NJ, USA, 2007; p. 10.

18. Stuedlein, A.W.; Gianella, T.N.; Canivan, G.J. Densification of Granular Soils using Conventional and Drained Timber Displacement Piles. *J. Geotech. Geoenviron. Eng.* **2016**, *142*, 04016075. [CrossRef]

19. Gianella, T.N.; Stuedlein, A.W. Performance of Driven Displacement Pile-Improved Ground in Controlled Blasting Field Tests. *J. Geotech. Geoenviron. Eng.* **2017**, *143*, 04017047. [CrossRef]

20. Kitazume, M. *The Sand Compaction Pile Method*; Taylor and Francis Group, PLC.: London, UK, 2005; p. 232.

21. Aloisio, A.; Contento, A.; Xue, J.; Fu, R.; Fragiacomo, M.; Briseghella, B. Probabilistic formulation for the q-factor of piles with damping pre-hole. *Bull. Earthq. Eng.* **2023**, *21*, 3749–3775. [CrossRef]

22. Panah, A.K.; Khoshay, A.H. A new seismic isolation system: Sleeved-pile with soil-rubber mixture. *Int. J. Civ. Eng.* **2015**, *13*, 124–132.

23. Dobson, T. Case Histories of the Vibro Systems to Minimize the Risk of Liquefaction. In *Soil Improvement: A Ten Year Update*; ASCE Geotechnical Special Publication: Atlantic, NJ, USA, 1987; pp. 167–183.

24. Mitchell, J.K.; Baxter, C.D.P.; Munson, T.C. *Performance of Improved Ground during Earthquakes, Improvement for Earthquake Hazard Mitigation*; ASCE Geotechnical Special Publication: Atlantic, NJ, USA, 1995; pp. 1–36.

25. Baez, J.I. A Design Model for the Reduction of Soil Liquefaction by Vibro-stone Columns. Ph.D. Thesis, University of Southern California, Los Angeles, CA, USA, 1995; p. 222.

26. Boulanger, R.W.; Idriss, I.M.; Stewart, D.; Hasbash, Y.; Schmidt, B. Drainage Capacity of Stone Columns or Gravel Drains for Mitigating Liquefaction. *Geotech. Spec. Publ.* **1998**, *75*, 678–690.

27. Filz, G.; Sloan, J.; McGuire, M.; Collin, J.; Smith, M. Column-Supported Embankments: Settlement and Load Transfer. In *Geotechnical Engineering State of the Art and Practice*; ASCE: Oakland, CA, USA, 2012; pp. 54–77.

28. Rayamajhi, D.; Nguyen, T.V.; Ashford, S.A.; Boulanger, R.W.; Lu, J.; Elgamal, A.; Shao, L. Numerical study of shear stress distribution for discrete columns in liquefiable soils. *J. Geotech. Geoenviron. Eng.* **2014**, *140*, 04013034. [CrossRef]

29. Rayamajhi, D.; Tamura, S.; Khosravi, M.; Boulanger, R.W.; Wilson, D.W.; Ashford, S.A.; Olgun, C.G. Dynamic Centrifuge Tests to Evaluate Reinforcing Mechanisms of Soil-Cement Columns in Liquefiable Sand. *J. Geotech. Geoenviron. Eng.* **2015**, *141*, 04015015. [CrossRef]

30. Rollins, K.M.; Anderson, J.; McCain, A.; Goughnour, R. Vertical composite drains for mitigating liquefaction hazard. In Proceedings of the 13th International Offshore and Polar Engineering Conference, Honolulu, HI, USA, 25–30 May 2003; pp. 498–505.

31. Howell, R.; Rathje, E.; Kamai, R.; Boulanger, R. Centrifuge Modeling of Prefabricated Vertical Drains for Liquefaction Remediation. *J. Geotech. Geoenviron. Eng.* **2012**, *138*, 262–327. [CrossRef]

32. Vytiniotis, A.; Whittle, A. Analysis of PVDrains for Mitigation of Seismically Induced Ground Deformations in Sand Slopes. *J. Geotech. Geoenviron. Eng.* **2017**, *143*, 9. [CrossRef]

33. Rollins, K.M.; Price, B.E.; Dibb, E.; Higbee, J.B. Liquefaction mitigation of silty sands in Utah using stone columns with wick drains. In *Ground Modification and Seismic Mitigation*; ASCE: Oakland, CA, USA, 2006; pp. 343–348.

34. Rollins, K.M.; Quimby, M.; Johnson, S.R.; Price, B. Effectiveness of stone columns for liquefaction mitigation of silty sands with and without wick drains. In *Advances in Ground Improvement: Research to Practice in the United States and China*; ASCE: Oakland, CA, USA, 2009; pp. 160–169.

35. Seed, H.B.; Booker, J.R. Stabilization of Potentially Liquefaction Sand Deposits Using Gravel Drains. *J. Geotech. Eng. Div.* **1977**, *103*, 756–768.

36. Sasaki, Y.; Taniguchi, E. Shaking table tests on gravel drains to prevent liquefaction of sand deposits. *Soils Found.* **1982**, *22*, 1–14. [CrossRef] [PubMed]

37. Matsubara, K. Analysis of gravel drain against liquefaction and its application to design. In Proceedings of the Ninth Conference on Earthquake Engineering, Tokyo and Kyoto, Japan, 2–9 August 1988; pp. 249–254.

38. Adalier, K.; Elgamal, A.; Meneses, J.; Baez, J.I. Stone columns as liquefaction countermeasure in non-plastic silty soils. *Soil Dyn. Earthq. Eng.* **2003**, *23*, 571–584. [CrossRef]

39. Sadrekarimi, A.; Ghalandarzadeh, A. Evaluation of gravel drains and compacted sand piles in mitigating liquefaction. *Ground Improv.* **2005**, *9*, 91–104. [CrossRef]

40. Kita, H.; Iida, T.; Nishitani, M. Experimental study on countermeasures for liquefaction by steel piles with drain. In Proceedings of the Tenth World Conference on Earthquake Engineering, Madrid, Spain, 19–24 July 1992; pp. 1701–1706.

41. Pestana, J.M.; Hunt, C.E.; Goughnour, R.R.; Kammerer, A.M. Effect of Storage Capacity on Vertical Drain Performance in Liquefiable Sand Deposits. In Proceedings of the 2nd International Conference on Ground Improvement Techniques, Singapore, 8–9 October 1998; pp. 373–380.

42. Iai, S.; Koizume, K. Estimation of Earthquake Induced Excess Pore Water Pressure for Gravel Drains. In Proceedings of the 7th Japan Earthquake Engineering Symposium; Kyoto University: Kyoto, Japan, 1986; pp. 679–684.

43. Tanaka, Y.; Mizoguchi, Y.; Asada, T.; Nomura, T.; Higashi, S.; Masashi, J.; Iwamoto, J. Group Installation of Plastic Board Drains At Embankment Toes for Liquefaction and Lateral Flow Countermeasures. In Proceedings of the 4th International Conference on Earthquake Geotechnical Engineering, Thessaloniki, Greece, 25–28 June 2007; pp. 1519–1527.

44. Chang, W.J.; Rathje, E.M.; Stokoe, K.H.; Cox, B.R. Direct evaluation of effectiveness of prefabricated vertical drains in liquefiable sand. *Soil Dyn. Earthq. Eng.* **2004**, *24*, 723–731. [CrossRef]

45. Brennan, A.J.; Madabhushi, S.P.G. Liquefaction remediation by vertical drains with varying penetration depths. *Soil Dyn. Earthq. Eng.* **2006**, *26*, 469–475. [CrossRef]

46. Harada, N.; Towhata, I.; Takatsu, T.; Tsunoda, S.; Sesov, V. Development of new drain method for protection of existing pile foundations from liquefaction effects. *Soil Dyn. Earthq. Eng.* **2006**, *26*, 297–312. [CrossRef]

47. Xiang-Ying, W.; Han-Long, L.; Qiang, J.; Yu-Min, C. Field tests on response of excess pore water pressures of liquefaction resistant rigid-drainage pile. *Chin. J. Geotech. Eng.* **2017**, *39*, 645–651. (In Chinese)

48. Xiang-Ying, W.; Yu-Min, C.; Qiang, J.; Han-Long, L. Soil pressures of the anti-liquefaction rigid-drainage pile during pile driving. *Rock Soil Mech.* **2018**, *39*, 2184–2192. (In Chinese)

49. Qiang, J.; Yumin, C.; Xiangying, W.; Hanlong, L. Field tests on squeezing effects of the rigid-drainage pile. *China Civ. Eng. J.* **2018**, *51*, 87–94. (In Chinese)

50. Cubrinovski, M.; Kokusho, T.; Ishihara, K. Interpretation from large-scale shake table tests on piles undergoing lateral spreading in liquefied soils. *Soil Dyn. Earthq. Eng.* **2006**, *26*, 275–286. [CrossRef]

51. Dungca, J.R.; Kuwano, J.; Takahashi, A.; Saruwatari, T.; Izawa, J.; Suzuki, H.; Tokimatsu, K. Shaking table tests on the lateral response of a pile buried in liquefied sand. *Soil Dyn. Earthq. Eng.* **2006**, *26*, 287–295. [CrossRef]

52. Haeri, S.M.; Kavand, A.; Rahmani, I.; Torabi, H. Response of a group of piles to liquefaction-induced lateral spreading by large scale shake table testing. *Soil Dyn. Earthq. Eng.* **2012**, *38*, 25–45. [CrossRef]

53. Motamed, R.; Towhata, R. Shaking table model tests on pile groups behind quay walls subjected to lateral spreading. *J. Geotech. Geoenviron. Eng.* **2010**, *136*, 477–489. [CrossRef]

54. Malvick, E.J.; Kutter, B.L.; Boulanger, R.W. Postshaking Shear Strain Localization in a Centrifuge Model of a Saturated Sand Slope. *J. Geotech. Geoenviron. Eng.* **2008**, *134*, 164–174. [CrossRef]

55. Kokusho, T. Water film in liquefied sand its effect on lateral spread. *J. Geotech. Geoenviron. Eng.* **1999**, *125*, 817–826. [CrossRef]

56. Iai, S. Similitude for shaking table tests on soil-structure-fluid model in 1g gravitational field. *Int. J. Rock Mech. Min. Sci. Geomech.* **1989**, *27*, 3–24. [CrossRef]

57. Pestana, J.M.; Hunt, C.E.; Goughnour, R.R. *FEQDrain: A Finite Element Computer Program for the Analysis of the Earthquake Generation and Dissipation of Pore Water Pressure in Layered Sand Deposits with Vertical Drains*; Report No. UCB/EERC-97/15 1997; University of California: Berkeley, CA, USA, 1997; p. 91.

materials

Article

Research on the Flexural Performance of Steel Pipe Steel Slag Powder Ultra-High-Performance Concrete Components

Xianyuan Tang [1,2], Chenzhuo Feng [1,2], Jin Chang [3,4,*], Jieling Ma [1] and Xiansong Hu [1]

1   School of Construction and Traffic Engineering, Guilin University of Electronic Science and Technology, Guilin 541004, China; thy1188@126.com (X.T.); fcz1272071476@163.com (C.F.); majielingzc1128@163.com (J.M.); 18484635906@163.com (X.H.)
2   Guangxi Key Laboratory of Intelligent Transportation, Guilin University of Electronic Science and Technology, Guilin 541004, China
3   College of Civil Engineering, Changsha University, Changsha 410022, China
4   Hunan Engineering Research Center for Intelligent Construction of Fabricated Retaining Structures, Changsha University, Changsha 410022, China
*   Correspondence: changjin1906@126.com

Abstract: In order to study the flexural performance of the combined structure of steel-pipe and steel slag powder ultra-high-performance concrete (UHPC), nine round steel-pipe beams filled with steel slag powder UHPC of different types were fabricated according to the orthogonal test method with the steel pipe type, coarse aggregate content, steel fiber admixture, and curing system as parameters. The broken ring morphology, deformation characteristics, deflection distribution, and flexural bearing capacity of the steel-pipe–UHPC beams were analyzed via a pure bending test and a finite element simulation. The results show that the damage morphology of the round steel-tube–UHPC beams prepared by using steel slag powder UHPC as the inner filling material was "bow damage" under the pure bending load, and the load capacity was higher. When the cross-sectional deflection reached $L/30$, the external load was still not reduced, and the steel-tube–steel-slag powder-UHPC beam had a better plastic deformation capacity and a later flexural bearing capacity. The type of steel tube had a significant influence on the flexural bearing capacity of the steel-tube–UHPC beam, and the larger the diameter of the steel tube section and the thicker the tube wall, the higher its flexural bearing capacity. The calculated ultimate flexural bearing capacity by the finite element software and the test results had a stable error between 5.6% and 11.2%, which indicates that the model was reasonably established. The research results can provide a reference for the application of steel pipe UHPC engineering.

Keywords: ultra-high-performance concrete; steel slag powder; grouting steel-pipe; flexural performance

Citation: Tang, X.; Feng, C.; Chang, J.; Ma, J.; Hu, X. Research on the Flexural Performance of Steel Pipe Steel Slag Powder Ultra-High-Performance Concrete Components. *Materials* **2023**, *16*, 5960. https://doi.org/10.3390/ma16175960

Academic Editor: Dolores Eliche Quesada

Received: 21 May 2023
Revised: 6 July 2023
Accepted: 12 July 2023
Published: 30 August 2023

1. Introduction

Ultra-high-performance concrete (UHPC) is a new type of cementitious composite material with excellent mechanical properties and durability that has a promising application in large-span bridge structures [1–3]. The application of steel-pipe–concrete arch bridges in China is progressing very fast, and the construction level is at the forefront of the world [4–6]. However, as the span diameter of steel-pipe–concrete arch bridges increases, the cross section continues to increase, resulting in a large increase in material usage, poor economy, and more prominent stability and construction safety problems. In order to solve the above technical bottlenecks, practice the concept of green development, further improve the load-bearing capacity of the components, and reduce the weight of the structure, high-strength steel tubes and ultra-high-performance concrete are increasingly used in the arch bridge structure [7–11].

The combined mechanical properties of steel tubes and UHPC are the theoretical basis for their application to arch bridges, and several researchers have carried out relevant studies on them. Fu [12] prepared a new micro-expansion steel-tube UHPC from the composition and shrinkage characteristics of UHPC and studied the design of the expansion stress and the compressive performance of a steel-tube–UHPC short column. Wang et al. [13], based on the mechanical theory and deformation coordination relationship, derived the expressions for the steel-tube discount factor and the reinforcement factor of the in-filled concrete and analyzed the axial compression load capacity and deformation capacity of round steel-tube–UHPC short columns. Lu et al. [14] investigated the damage morphology and stress–strain curves of a round steel-tube UHPC with different steel-tube thicknesses by conducting axial compression tests on the steel-tube-restrained UHPC specimens. Tang et al. [15] conducted an experimental study on the axial compressive load capacity of short columns of round steel tubes containing coarse aggregate UHPC, which showed that the outer diameter of the steel tubes had the greatest influence on the axial compressive load capacity of the short columns, followed by the steel strength and tube wall thickness, and the concrete strength had less influence; Zhou et al. [16] analyzed the influence of different steel contents and steel strengths on the axial compressive strength of steel-tube UHPC based on the axial compressive test of nine steel-tube–UHPC short columns. The effect of the different steel contents and steel strengths on the axial compressive strength of the steel-tube UHPC was analyzed, indicating that the higher the steel content and steel strength, the better the mechanical properties of the steel-tube UHPC, and it is recommended to use a high-strength steel to improve the ultimate load capacity of steel-tube–UHPC short columns. Zhang et al. [17] studied the load carrying capacity of a high-strength steel-tube–UHPC combined arch and found that the stable load carrying capacity of the steel-tube–concrete arch specimens increased with the increase in the strength of the in-filled concrete. Wei et al. [18] conducted pure bending tests on round steel-tube–UHPC beams and investigated the effects of different steel tube strengths and hoop coefficients on the flexural load carrying capacity of the round steel-tube–UHPC beams. Deng et al. [19] studied the damage form of square high-strength steel-tube–UHPC beams under pure bending load and the effects of the steel content and steel yield strength on the flexural bearing capacity and flexural stiffness of the beams via tests, which showed that the higher the steel content and the thicker the steel tube wall, the greater the flexural bearing capacity and flexural stiffness.

In summary, the research has focused on the short-column axial compression bearing capacity and deformation capacity, and the member bending resistance, compression bending, and other performance research is relatively small. A pure bending test is the basis for the study of the steel-pipe–UHPC composite force, and in the actual project of a steel-pipe–concrete arch bridge, the steel-pipe–concrete structure is subject to a bending moment, which is not uncommon. Therefore, an in-depth study of the bending performance of steel-pipe UHPC is necessary. In this paper, an ecological UHPC was developed by using solid-waste steel slag from the steel industry processed into micronized powder instead of quartz powder, and the ecological UHPC was used as the inner filling material of the steel pipe via orthogonal tests in order to study the flexural performance of the combined members. It is significant for reducing the production cost and saving resources, and it can also provide a reference for the application research of ecological inner-fill concrete material in a steel-pipe–concrete structure and provide technical support for the design of a larger-span and green steel-pipe–concrete arch bridge to promote the development of such bridges.

2. Experiment Overview

2.1. Experimental Design

In order to comprehensively analyze the flexural performance of steel-pipe–steel-slag-micronized-UHPC members, this real test was designed with nine different types of steel-pipe–steel-slag-micronized-UHPC beams by using four factors and three levels of an L9 (3^4) orthogonal experimental scheme with the steel pipe type, coarse aggregate content, steel fiber volume dosing, and maintenance system as parameters. Among them, the steel pipe was selected as 50 × 4 (steel pipe diameter D × wall thickness t) mm, 89 × 6 mm, and

108×. All the types of steel pipes were 1350 mm in length, and the coarse aggregate content was 0, 20%, and 40%. For the steel-fiber volume dosing, 0, 1.0%, and 1.5%, and 3 doping were selected [20]. First, the specimens were left in the room with a temperature of $20 \pm 5\,°C$ and a relative humidity >50% for 1 d; then, they underwent high temperature maintenance at $90 \pm 5\,°C$ and a relative humidity >90% for different lengths; and finally, they underwent standard maintenance until the age of the test period (28 d). The maintenance condition A was high temperature maintenance for 1 d, B was high temperature maintenance for 2 d, and C was high temperature maintenance for 3 d.

The test factors and level settings are shown in Table 1, and the specimens are numbered according to the specimen type, steel-pipe diameter, and in-fill steel-slag micro-powder UHPC ratio: Group C (coarse aggregate) represents the specimen with the in-fill steel-slag micro-powder UHPC containing only coarse aggregate without steel fiber; group S (steel fiber) represents the specimen with the in-fill steel-slag micro-powder UHPC containing only steel fiber without coarse aggregate; group CS represents the specimen with the in-fill steel-slag micro-powder UHPC containing both coarse aggregate and steel fiber; and group T represents the specimen with the in-fill steel-slag micro-powder UHPC containing neither coarse aggregate nor steel fiber.

Table 1. Table of test factors and level parameters of the specimens.

Test Piece Number	Steel Pipe Type $D \times t$ /mm	Steel Content α	Yield Strength of Steel Pipe f_y/MPa	Conservation Conditions	Hoop System Number ξ
S-50-1	50 × 4	0.417	278	A	1.18
CS-50-2	50 × 4	0.417	278	B	1.02
C-50-3	50 × 4	0.417	278	C	1.38
S-89-4	89 × 6	0.336	275	C	0.86
C-89-5	89 × 6	0.336	275	A	1.25
CS-89-6	89 × 6	0.336	275	B	0.93
T-108-7	108 × 8	0.378	273	B	1.13
CS-108-8	108 × 8	0.378	273	C	1.02
CS-108-9	108 × 8	0.378	273	A	0.93

Here, the steel content rate $\alpha = \frac{A_s}{A_c}$, A_s is the cross-sectional area of the steel pipe, and A_c is the cross-sectional area of the UHPC filled with steel-slag-micronized powder; ζ is the steel-pipe–steel-slag-micronized-UHPC beam tightening factor; $\zeta = \alpha \frac{f_y}{f_{ck}}$, f_y is the yield strength of the steel pipe; and f_{ck} is the axial compressive strength of the steel-slag-powder UHPC.

2.2. Test Piece Fabrication

After mixing the UHPC mix filled with steel slag, it was immediately poured into the steel pipe, and the steel pipe was tilted and hammered on the outside of the steel pipe to ensure the concrete was dense. After pouring, the excess UHPC mix was scraped off, smoothed, and covered with plastic film when the mix was close to the initial set, left to stand for 1 d in a test chamber at a temperature of $20 \pm 5\,°C$ and a relative humidity of >50%, and then maintained until the end of the test period according to the maintenance system.

The steel pipes for the steel–UHPC beams were three diameter types of Q235 hot-rolled seamless round steel pipes of 50, 89, and 108 mm. The steel strength was determined by tensile tests; three standard specimens were used as one group; and the material properties were tested according to the requirements of the standard [21]. The measured yield strength of the steel (f_y) was 278 MPa, and the yield strain (ε_y) was 1.5×103 $\mu\varepsilon$. The ultimate tensile strength (f_u) was 438.7 MPa, the modulus of elasticity (E_s) was 2.06×10^5 MPa, and the Poisson's ratio (μ_s) was 0.283.

For the in-fill concrete, the steel-slag-micronized UHPC was made according to the literature [22], and the coarse aggregate was selected from 4.75–20 mm limestone crushed rock produced in northern Guangxi, with parent rock strength of 100–120 MPa, crushing value ≤10%, mud content ≤0.5%, and needle and flake particle content ≤8% [23]. The other materials were consistent with the previous steel-slag-micronized UHPC raw materials, and the sieving curves of the various materials are shown in Figure 1.

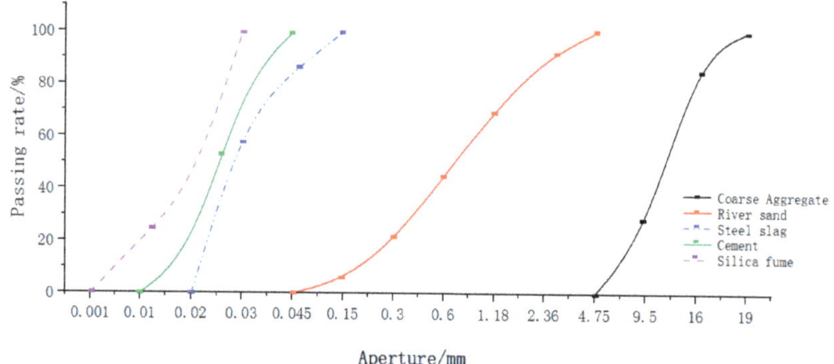

Figure 1. Cumulative screening curve of raw materials.

In the process of making steel-pipe–steel-slag-micronized-UHPC specimens, cubic compressive test blocks, axial compressive test blocks, flexural test blocks, and static compressive modulus of elasticity test blocks were prepared simultaneously, and the main working properties and mechanical properties of concrete were tested in strict accordance with the relevant requirements in the standards [24,25]. The results of the concrete main performance tests are detailed in Table 2.

Table 2. Steel-slag micronized-powder UHPC main performance test.

Matching Ratio Group	Working Performance/mm		28 d Mechanical Properties/MPa			
	Slump	Extensibility	Cubic Compressive Strength	Axial Compressive Strength	Flexural Strength	Modulus of Elasticity
1	265	610	117.4	98.1	16.26	46,400
2	245	590	137.1	112.7	15.32	55,200
3	230	420	112.4	83.9	10.47	50,300
4	260	610	133.6	107.8	20.83	48,800
5	260	690	108.1	73.8	9.01	49,500
6	235	410	128.4	99.5	14.83	47,200
7	270	640	120.1	91.1	8.61	43,100
8	255	580	139.2	102.4	15.38	53,700
9	210	320	139.7	110.5	14.21	56,800

2.3. Measurement Point Arrangement and Loading System

The traditional four-point loading method was used in the test. The 200-ton hydraulic jack was used to load the test piece, and the loading force was measured by the pressure sensor. The length of the pure bending section of the steel-pipe–steel-slag-powder-UHPC beam (i.e., the distance between the two supporting points during the test, L) was 1200 mm. A resistance strain gauge was used to measure the longitudinal strain of the steel tube. The strain gauges were pasted at the middle part of the positive span of the test piece, and eight strain gauges were pasted in total. The pasting position is shown in Figure 2.

Resistance displacement gauges were arranged in the span of the specimen and at the three-quarter point position to measure its vertical deflection. One displacement gauge was set at each of the two end-support point positions to measure the deformation, and the displacement gauge located in the middle of the span was pre-set to a larger range downward [26]. Before loading, the member was strictly aligned according to the loading schematic, and the test loading schematic is shown in Figure 3.

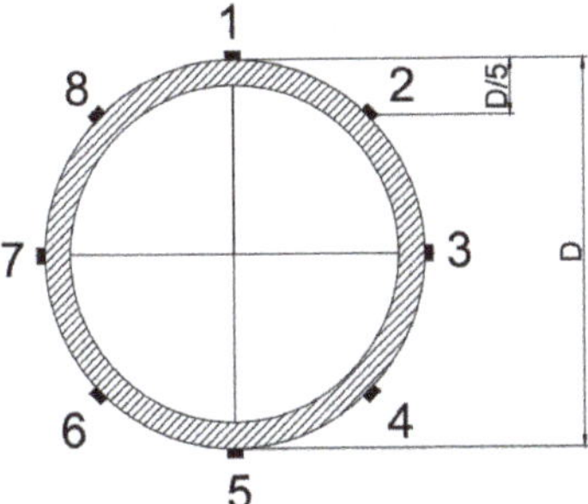

Figure 2. Location of the strain gauges in the span center.

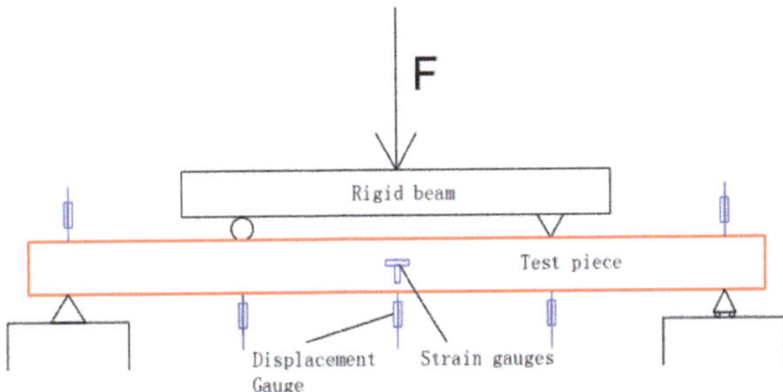

Figure 3. Schematic diagram of the pure bending test loading.

The test was loaded by graded loading. At the beginning of loading, each level of load was 1/10 of the expected ultimate load; when the steel started to yield, each level of load was 1/15 of the expected ultimate load, and the holding time of each level of load was 2 min, so that the deformation of the specimen was fully developed. The loading was stopped when the deformation of the specimen was large.

3. Experimental Results and Analysis

3.1. Destruction Phenomenon

During the loading process, the damage morphology of the nine different types of the steel-pipe–steel-slag-powder-UHPC beam was basically similar, and the damage form of its typical specimen (CS-108-8) is shown in Figure 4, which shows that the brittle damage of the steel-pipe–steel-slag-powder-UHPC beams did not occur during the loading process, and the ductility was good.

Figure 4. Typical damage form of the steel-pipe–steel-slag-micronized-UHPC beam.

At the initial stage of loading, there was no obvious vertical deflection in each specimen; with the increasing load, there was a crisp sound from the in-filled steel-slag powder UHPC, and the beam started to obviously deform when the loading was close to its yield load. The vertical deflection in the span of the specimen increased rapidly, while the load was almost unchanged. The loading was stopped when the vertical deflection in the span was too large.

During the loading process, the surface of each specimen was continuous and smooth, and none of them showed the local convexity of ordinary steel-pipe–concrete beams, indicating that the steel pipe and the steel-slag micro-powder-filled UHPC jointly withstood the external load and provided better play to the advantages of both.

A typical specimen (CS-108-8) was selected to have its middle steel tube cut open, as shown in Figure 5. Figure 5 shows that the in-fill UHPC and steel pipe joint were tight, and the cracks of the core slag-powder UHPC were relatively uniformly distributed between the quarter point and the span; when the specimen was damaged, only a few cracks of the slag-powder UHPC extended to the compression zone of the section, and no crush damage occurred in the compression zone.

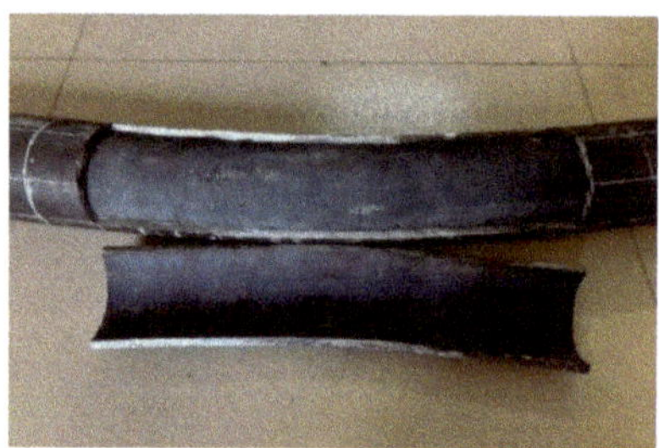

Figure 5. Typical damage pattern of the in-filled UHPC.

3.2. Load Deflection Curve

The spanwise moment–strain relationships of the nine steel-pipe–steel-slag-powder-UHPC beams were similar, and Figure 6 shows the measured spanwise moment–strain curves of a typical specimen (CS-89-6), where the tensile strain was taken as positive and the compressive strain as negative.

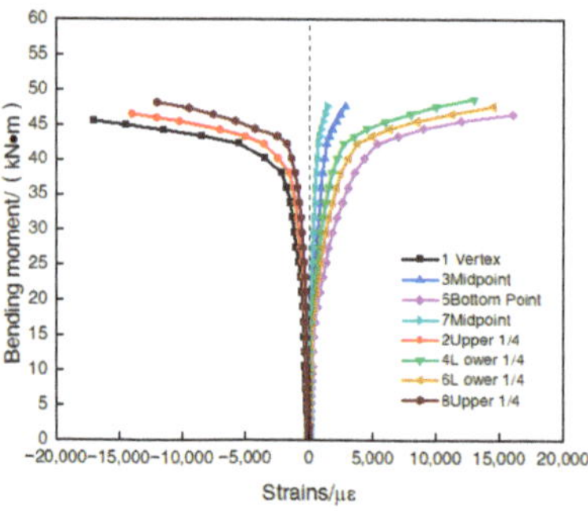

Figure 6. Bending-moment–strain relationship curve of a typical specimen.

Figure 6 shows that at the beginning of loading, the strains in the compression zone and tension zone of the specimen were generally the same, and the strains in the upper and lower corresponding parts of the steel tube were symmetrical to the neutral axis distribution, indicating that the neutral axis coincided with the shaped axis of the section; as the load increased, the steel-slag micro-powder UHPC in the tension zone gradually cracked, making the growth rate of strains in the tension zone of the section faster than in the compression zone, and the steel tube in the tension zone entered the plastic stage before the steel tube in the compression zone and the neutral axis. The strain gauges pasted in the center of the steel tube were analyzed. The analysis of the strain gauges pasted in the center of the steel tube (#3 and #7) showed that the amount of change in both slowly increased and the value was positive, which also indicated that the neutral axis of the steel-slag-powder UHPC beam slowly moved up during the loading process.

3.3. Deflection Distribution Curve

Four typical steel-pipe–steel-slag-powder-UHPC beams were selected, and their distribution curves were drawn along the length of the specimen according to the vertical deflection under all levels of loading, as shown in Figure 7, where the horizontal coordinates are the distance of each measurement point from the left end hinge support, i.e., the length of the pure bending section (L), and the vertical coordinates are the deflection values at the mid-span and quarter-point positions during the specimen loading. To further analyze the distribution of the beam deflection along the span direction, a symmetrically distributed sinusoidal half-wave curve is plotted with a black dashed line.

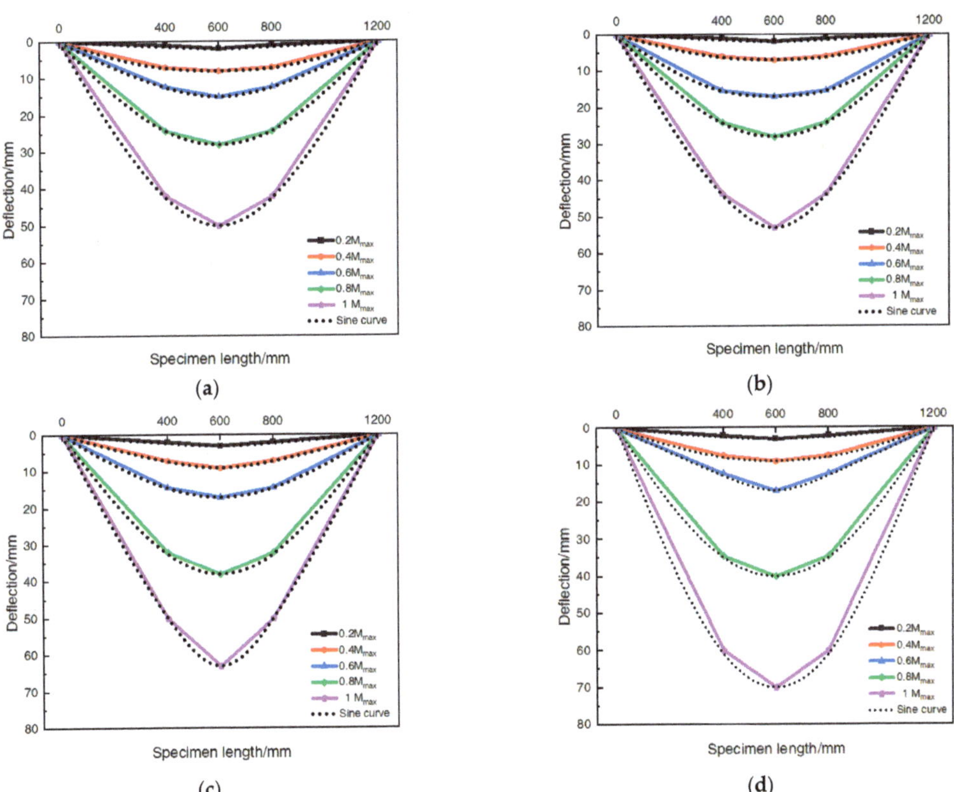

Figure 7. Deflection distribution curve. (**a**) CS-50-2; (**b**) C-50-3; (**c**) CS-89-6; (**d**) CS-108-8.

The analysis of Figure 7 shows that the steel-tube–UHPC beam was deformed by "bow damage" during the loading process, each beam was in good agreement with the corresponding sinusoidal half-wave curve, and the deflection distribution curve generally conformed to the law of change of the sinusoidal half-wave function under the same steel tube type (tube diameter/times). The change in the deflection distribution curve was generally the same for the same steel tube type (tube diameter (wall thickness)), and there was no obvious change due to the change in the in-filled steel-slag micro-powder UHPC ratio, which indicated that the change in the deflection distribution of the member was less affected by the type of inner filling material.

3.4. Bending Moment–Deflection Curve

The spanwise moment–deflection curves of each of the nine steel-pipe–steel-slag microfabrication-UHPC beams were classified according to the steel-pipe diameters and plotted, as shown in Figure 8.

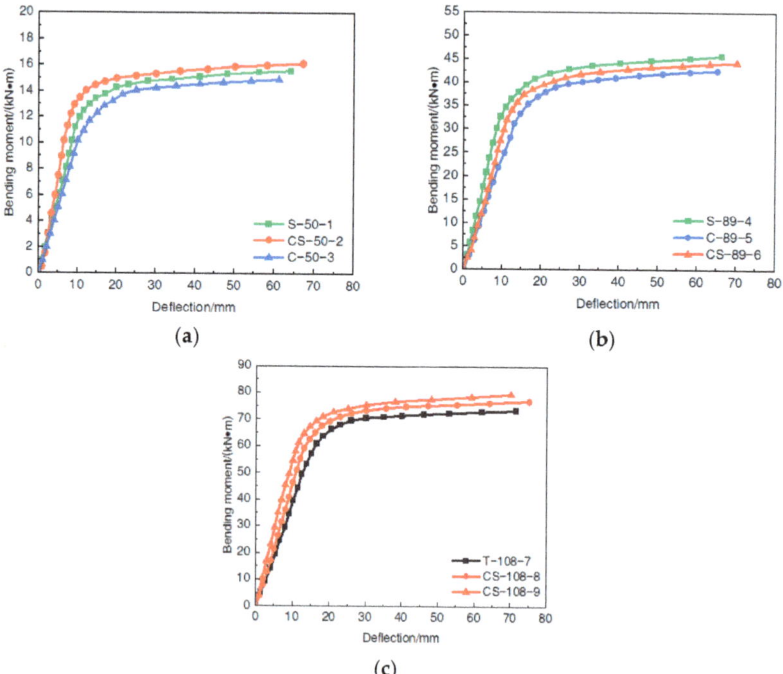

Figure 8. Span moment–deflection curve. (**a**) 50 mm diameter steel pipe; (**b**) 89 mm diameter steel pipe; (**c**) 108 mm diameter steel pipe.

Comparison analysis of Figure 8 shows that the shape of the moment–deflection relationship curves of the nine steel-pipe–steel-slag-powder-UHPC beams in the process of the pure bending loading was basically similar; all could be divided into an elastic phase, an elasto-plastic phase, and a strengthening phase. (1) In the elastic phase, the interaction between the steel pipe and the in-filled steel-slag-powder UHPC was small; both bore the load alone; the span deflection changed very slowly; and the specimen as a whole had no obvious deformation. The growth rate of the bending moment was significantly faster than the growth rate of the deflection, and the two showed a good linear growth relationship. (2) In the elastic–plastic stage, with the increase in the bending moment, the steel-pipe fibers at the outermost edge of the tensile zone reached the proportional limit, the growth rate of the bending moment decreased, the growth rate of the deflection gradually increased, the beam underwent bending deformation, and the bending moment–deflection curve turned into a nonlinear growth state. (3) In the strengthening stage, after the steel-tube–UHPC beam yielded, the plasticity of the span cross section developed continuously, the role of the steel tube sleeve and the close fitting support of the in-filled steel-slag-powder UHPC were brought into play, and its bending moment still had a small rising trend.

From the test results, it can be seen that the steel-tube–steel-slag powder-UHPC beams had pure bending action even when the span cross-sectional deflection reached L/30 and the external load acting on the specimen was still not reduced, indicating that the steel-tube–UHPC beam had good plastic deformation capacity and later flexural load-bearing capacity with good ductility.

4. Finite Element Analysis

4.1. Model Building

The finite element analysis software was used to simulate the pure bending test of the steel-pipe–steel-slag-powder-UHPC beam. In order to ensure the accuracy of the finite element simulation experiment of the steel-pipe–steel-slag-powder-UHPC beam, the intrinsic structure relationship models of its constituent steel pipe and inner-filled steel-slag-powder UHPC needed to be clarified separately.

4.1.1. Steel Principal Structure Relationship Model

The steel pipe used in this test was a Q235 hot-rolled seamless round steel pipe, whose steel was low-carbon soft steel; the stress–strain relationship curve of the steel can be simplified into five stages [27]. The simplified stress–strain relationship curve is shown in Figure 9, and the present structure relationship model expression is shown in Equation (1), where: f_p is the proportional limit of the steel, f_y is the yield strength of the steel, and f_u is the ultimate tensile strength of the steel.

$$\sigma_s = \begin{cases} E_s \times \varepsilon_s & (\varepsilon_s \leq \varepsilon_e) \\ -A\varepsilon_s^2 + B\varepsilon_s + C & (\varepsilon_e \leq \varepsilon_s \leq \varepsilon_{e1}) \\ f_y & (\varepsilon_{e1} \leq \varepsilon_s \leq \varepsilon_{e2}) \\ f_y \times \left[1 + 0.6 \times \frac{\varepsilon_s - \varepsilon_{e2}}{\varepsilon_{e3} - \varepsilon_{e2}}\right] & (\varepsilon_{e1} \leq \varepsilon_s \leq \varepsilon_{e2}) \\ 1.6 f_y & (\varepsilon_s > \varepsilon_{e3}) \end{cases} \tag{1}$$

where

E_s is the modulus of elasticity of the steel;

$\varepsilon_e = 0.8 f_y / E_s,\ \varepsilon_{e1} = 1.5\varepsilon_e,\ \varepsilon_{e2} = 10\varepsilon_{e1},\ \varepsilon_{e3} = 100\varepsilon_{e1};$

$A = 0.2 f_y / (\varepsilon_{e1} - \varepsilon_e)^2,\ B = 2A\varepsilon_{e1},\ C = 0.8 f_y + A\varepsilon_e^2 - B\varepsilon_e.$

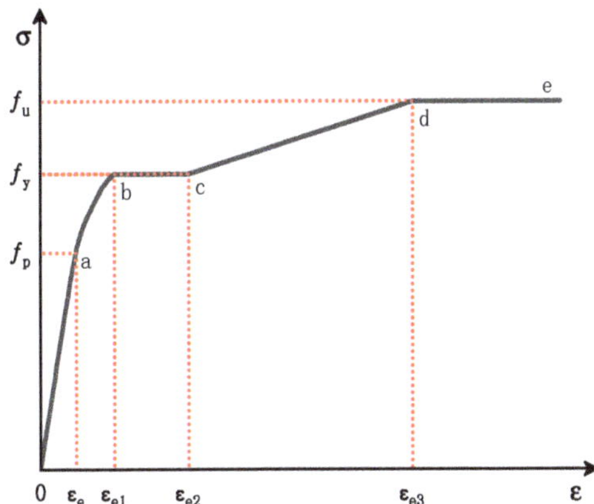

Figure 9. Stress–strain relationship curve for steel.

4.1.2. Steel-Pipe–Steel-Slag-Powder-UHPC Intrinsic Structure Relationship Model

During the pure bending test of the steel-pipe–steel-slag-powder-UHPC beam, its upper UHPC was damaged in compression, and the lower UHPC was cracked in tension. Therefore, the stress–strain relationship between the compressive and tensile zones of the UHPC needed to be determined separately.

For the stress–strain relationship of filled concrete in steel-pipe–concrete members, much research has been carried out by scholars. Zhong et al. [28] proposed a stress–strain relationship model in 2013, which had a more accurate simulation effect by fully considering the high-strength concrete-filled material. Therefore, in this paper, the stress–strain relationship of the steel-slag-powder UHPC in the compressed zone was simulated by this model. Equation (2) provides the expression of the model,

$$\sigma = \begin{cases} f'_c \times \frac{ax+bx^2}{1+(a-2)\times x+(b+1)\times x^2} & (0 < \varepsilon \le \varepsilon_{c0}) \\ f'_c & (\varepsilon_{c0} < \varepsilon \le \varepsilon_{cc}) \\ f_r + (f'_c - f_r) \times \exp\left[-\left(\frac{\varepsilon-\varepsilon_{cc}}{\alpha}\right)^\beta\right] & (\varepsilon_{cc} < \varepsilon) \end{cases}. \tag{2}$$

where:

σ—stress variables; ε—strain variables;

ε_{c0}—uniaxial ultimate strains in the concrete; ε_{cc}—peak strain in the restrained concrete;

f'_c—compressive strength of the concrete cylinders;

f_B—tightening stresses; f_r—residual stress; ζ—tightening coefficient;

$x = \varepsilon/\varepsilon_{c0}$; $\varepsilon_{c0} = 0.00076 + \sqrt{(0.626f'_c - 4.33) \times 10^{-7}}$;

$a = \frac{E_c \times \varepsilon_{c0}}{f'_c}$; $b = \frac{(a-1)^2}{0.55} - 1$;

$\varepsilon_{cc} = \varepsilon_{c0} \times e^k$; $k = (2.9224 - 0.00367f'_c) \times \left(\frac{f_B}{f'_c}\right)^{0.3124+0.0002f'_c}$;

$f_B = \frac{(1+0.027f_y) \times e^{-0.02\frac{D}{t}}}{1+1.6e^{-10}(f'_c)^{4.8}}$;

$f_r = 0.7(1 - e^{-1.38\zeta}) \times f'_c \le 0.25f'_c$;

$\alpha = 0.04 - \frac{0.036}{1+e^{6.08\zeta-3.49}}$; $\beta = 1.2$.

The tensile-zone steel-slag-powder UHPC intrinsic structure relationship was calculated using the energy damage criterion [28], which takes into account the tensile softening properties of concrete and its corresponding fracture energy G_F, as shown in Equation (3)

$$G_F = (0.0469d_{max}^2 - 0.5d_{max} + 26) \times \left(\frac{f'_c}{10}\right)^{0.7}. \tag{3}$$

where d_{max} is the maximum aggregate particle size of concrete, and the concrete cracking stress is σ_{to}. The concrete tensile strength, shown in Equation (4), is from ref. [29]:

$$\sigma_{to} = 0.26 \times \left(1.25f'_c\right)^{\frac{2}{3}}. \tag{4}$$

4.2. Model Validation

4.2.1. Destruction Form

In accordance with the above method, finite element modeling analysis was performed on the nine tested steel-pipe–steel-slag-powder-UHPC beams, the damage morphology of the simulated members was compared with that of the tested members, and a typical specimen (CS-108-8) was selected for comparison as shown in Figure 10. Figure 10 shows that the final damage patterns of the simulated members and the corresponding test members were similar in appearance and that the degree of agreement between them was good, indicating that it is feasible to use the finite element method to simulate and analyze the steel-pipe–steel-slag-powder-UHPC beam.

Figure 10. Comparison of the damage patterns CS-108-8.

4.2.2. Bending Moment–Deflection Curve

The moment–deflection relationship curve can reflect the bending capacity of the member more clearly; the moment–deflection curve obtained from the finite element simulation was compared with the test curve, and the comparison is shown in Figure 11. Figure 11 shows that the moment–deflection relationship curve of the steel-pipe–steel-slag-powder-UHPC beam calculated by the finite element simulation was slightly lower than the test curve, but the curve trends of both were basically the same, and the overall agreement was good, indicating that the cell selection, mesh division, interface model, and boundary conditions of the model were reasonably set, and the finite element method can be used to analyze the effect of the change in the relevant parameters of the steel-pipe–steel-slag-powder-UHPC beam on the bending capacity. The simulation method can be used to analyze the effect of changes in parameters on the flexural performance of the UHPC beam.

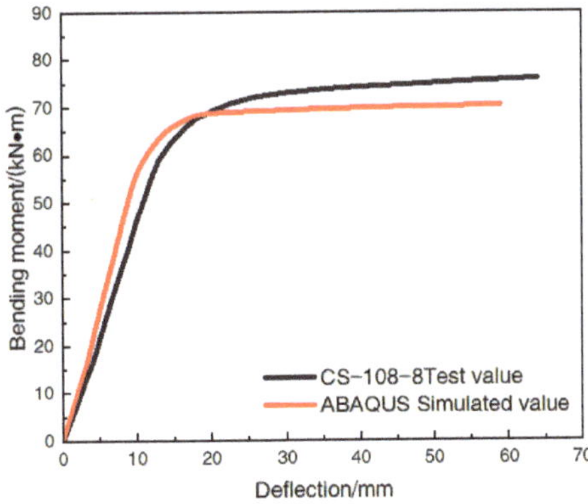

Figure 11. Comparison of the bending moment–deflection curves for each specimen CS-108-8.

4.2.3. Flexural Bearing Capacity

The maximum flexural load capacity of each member was calculated via a numerical simulation of the steel-pipe–steel-slag-powder-UHPC beam using the finite element model, which was compared with the test value. The results of the comparative analysis of each specimen are shown in Table 3.

Table 3. Comparison between the test value of the ultimate flexural bearing capacity and the finite element simulation value.

Specimen Number	Test Value/ (kN·m)	Analog Value/ (kN·m)	Error/ %
S-50-1	14.1	12.8	9.2
CS-50-2	14.4	13.6	5.6
C-50-3	13.3	12.1	9.0
S-89-4	43.2	40.2	6.9
C-89-5	40.4	38.0	5.9
CS-89-6	42.4	39.7	6.4
T-108-7	71.2	66.6	6.5
CS-108-8	73.6	67.8	7.9
CS-108-9	76.8	68.2	11.2
Error Mean μ			7.6

The analysis of Table 3 shows that the error between the finite element simulation and the test value was between 5.6% and 11.2%, with an average error of 7.6%. However, the error was within an acceptable range, indicating that it is feasible to use this finite element model to simulate the pure bending test of the steel-pipe–steel-slag-powder-UHPC beam.

4.3. Analysis of the Simulation Parameters

4.3.1. Change in the Steel Content

Table 4 shows the model parameters of each simulated member under different steel content rates. The ultimate flexural bearing capacity obtained by the finite element calculation of each member is summarized in Table 5. The effect of the steel content rate change on the moment–deflection curve in the span of the steel-pipe–steel-slag-powder-UHPC beam is shown in Figure 12.

Table 4. Parameters of the model members at different steel content rates.

Model Number	Steel Pipe Type D × t × L/mm	Steel Content α	Steel Tube Yielding Strength f_y/MPa	Cubic Compressive Strength f_{cu}/MPa	Axial Compression Resistance Strength f_{ck}/MPa
TA-89-4-0.207	89 × 4 × 1350	0.207	275	120	93
TA-89-4.5-0.237	89 × 4.5 × 1350	0.237	275	120	93
TA-89-5-0.269	89 × 5 × 1350	0.269	275	120	93
TA-89-5.5-0.302	89 × 5.5 × 1350	0.302	275	120	93
TA-89-6-0.336	89 × 6 × 1350	0.336	275	120	93

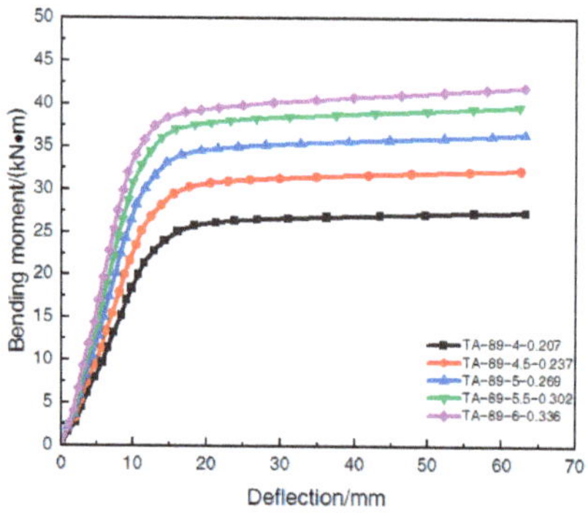

Figure 12. Effect of the steel content on the moment–deflection curve.

Table 5. Effect of the steel content on the flexural load carrying capacity.

Model Number	Ultimate Flexural Load Capacity /(kN·m)	Improvement Rate /(%)
TA-89-4-0.207	27.3	-
TA-89-4.5-0.237	32.8	20.1
TA-89-5-0.269	36.9	35.4
TA-89-5.5-0.302	39.9	46.3
TA-89-6-0.336	41.6	52.4

The analysis of Table 5 shows that when the steel content of the steel-pipe–steel-slag-powder-UHPC beam increased from 0.207 to 0.237, 0.269, 0.302, and 0.336, respectively, its ultimate flexural bearing capacity increased by 20.1%, 35.4%, 46.3%, and 52.4%, respectively, which indicates that the increase in steel content caused a significant increase in the flexural bearing capacity of the member. However, the rate of increase in the flexural load capacity decreased with the increase in the steel content. The growth rate of the flexural load capacity of the member was 20.1% when the steel content increased from 0.207 to 0.237, 12.8% when the steel content increased from 0.237 to 0.269, 8.1% when the steel content increased from 0.269 to 0.302, and only 4.2% when the steel content increased to 0.336.

Figure 12 shows that in the elastic phase, the increase in the steel content caused a significant increase in the initial flexural stiffness of the member. Similarly, the ultimate flexural load capacity of the member increased with the increase in the steel content.

In summary, the change in the steel content has a significant effect on the initial flexural stiffness and ultimate flexural bearing capacity of the steel-pipe–steel-slag-powder-UHPC beams, both of which increased with the increase in steel content.

4.3.2. Change in the Yield Strength of Steel

Table 6 shows the model parameters of each simulated member under different steel yield strengths. The finite element simulation values of the ultimate flexural bearing capacity of each member are shown in Table 7. The effect of the steel yield strength variation on the moment–deflection curve in the span of a steel-pipe–steel-slag-micro-powder-UHPC beam is shown in Figure 13.

Table 6. Parameters of the model members at different steel yield strengths.

Model Number	Steel Pipe Type $D \times t \times L$/mm	Steel Conten α	Steel Tube Yielding Strength f_y/MPa	Cubic Compressive Strength f_{cy}/MPa	Axial Compression Resistance Strength f_{ck}/MPa
TB-89-4.5-235	89 × 4.5 × 1350	0.237	235	120	93
TB-89-4.5-275	89 × 4.5 × 1350	0.237	275	120	93
TB-89-4.5-345	89 × 4.5 × 1350	0.237	345	120	93
TB-89-4.5-390	89 × 4.5 × 1350	0.237	390	120	93
TB-89-4.5-420	89 × 4.5 × 1350	0.237	420	120	93

Table 7. Effect of the yield strength of the steel on the flexural load carrying capacity.

Model Number	Ultimate Flexural Bearing Capacity /(kN·m)	Improvement Rate /(%)
TB-89-4.5-235	32.8	-
TB-89-4.5-275	37.4	14.1
TB-89-4.5-345	41.2	25.8
TB-89-4.5-390	44.5	36.0
TB-89-4.5-420	47.2	44.0

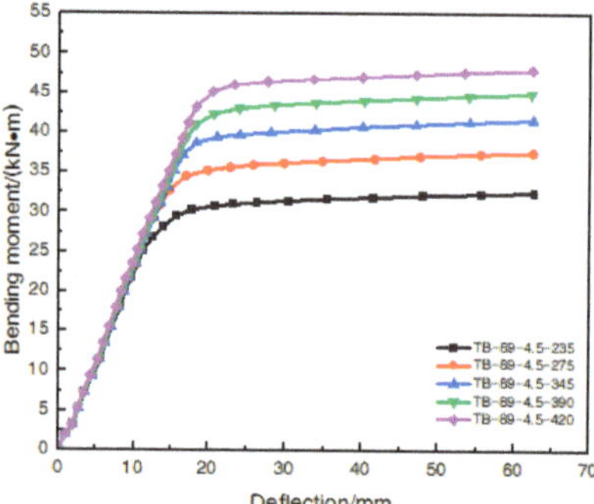

Figure 13. Effect of the yield strength of steel on the bending moment–deflection curve.

The analysis of Table 7 shows that when the steel yield strength of the steel-tube-steel-slag-powder-UHPC beams increased from 235 MPa to 275 MPa, 345 MPa, 390 MPa, and 420 MPa, respectively, the ultimate flexural load capacity of the member increased by 14.1%, 25.8%, 36.0%, and 44.0%, respectively, and the increase in the steel yield strength caused a significant increase in the flexural load capacity of the member.

However, with the gradual increase in the yield strength of the steel-tube-steel-slag-powder-UHPC beam, its ultimate flexural bearing capacity increase rate slowly decreased. When the yield strength of the steel increased from 235 MPa to 275 MPa, the member flexural load capacity increased by 14.1%; when the yield strength of the steel increased from 275 MPa to 345 MPa, the growth rate of flexural load capacity decreased to 10.2%; when the yield strength of the steel increased from 345 MPa to 390 MPa, the growth rate decreased to 8.1%; and when the yield strength of the steel increased to 420 MPa, the growth rate was only 5.9%. When the yield strength of the steel increased to 420 MPa, the growth rate was only 5.9%.

As shown in Figure 13, when the bending moment was less than 25 kN/m, there was no significant difference in the change in the deflection for the members with different steel yield strengths, indicating that the yield strength of the steel pipe had a small effect on the flexural stiffness of the steel-tube-steel-slag-powder-UHPC beams at the initial and use stages; when the member was subjected to a gradually increasing bending moment, its ultimate flexural bearing capacity increased with the increase in the steel yield strength.

In summary, the change in the steel yield strength had a small effect on the flexural stiffness of the steel-tube-steel-slag-powder-UHPC beams in the initial and service stages but had a significant effect on the ultimate flexural load capacity.

4.3.3. Strength Variation in the UHPC with the Inside-Filling Steel-Slag Powder

Table 8 shows the model parameters of each simulated member under different steel-slag-powder UHPC strengths; in order to study the effect of the variation in the strength of the in-filled steel-slag powder UHPC on the beam, the ultimate flexural bearing capacity of each member obtained by finite element simulation is shown in Table 9, and the effect of the variation in the strength of the in-filled steel-slag-powder UHPC on the moment–deflection curve in the span of the steel-pipe–UHPC beam is shown in Figure 14.

The analysis of Table 9 shows that when the strength of the UHPC filled with steel-slag micro powder increased from 120 MPa to 130 MPa, 140 MPa, 150 MPa, and 160 MPa, respectively, the ultimate flexural bearing capacity of the member increased by 1.6%, 3.2%, 4.9%, and 6.6%, respectively, and the rate of the increase in the flexural bearing capacity was less than 10%, which was not significant, indicating that the increase in the strength of the UHPC filled with steel-slag micro powder had less effect on the flexural bearing capacity of the member. The improvement in UHPC strength had a small effect on the flexural bearing capacity of the members.

Table 8. Model parameters at different intensities.

Model Number	Steel Pipe Type D × t × L/mm	Steel Content α	Steel Tube Yielding Strength f_y/MPa	Cubic Compressive Strength f_{cu}/MPa	Axial Compression Resistance Strength f_{ck}/MPa
TC-89-4.5-120	89 × 4.5 × 1350	0.237	275	120	93
TC-89-4.5-130	89 × 4.5 × 1350	0.237	275	130	101
TC-89-4.5-140	89 × 4.5 × 1350	0.237	275	140	108
TC-89-4.5-150	89 × 4.5 × 1350	0.237	275	150	116
TC-89-4.5-160	89 × 4.5 × 1350	0.237	275	160	124

Table 9. Influence of the strength of the in-fill UHPC on the flexural load carrying capacity.

Model Number	Ultimate Flexural Bearing /(kN·m)	Improvement Rate /(%)
TC-89-4.5-120	32.8	-
TC-89-4.5-130	33.3	1.6
TC-89-4.5-140	33.8	3.2
TC-89-4.5-150	34.4	4.9
TC-89-4.5-160	34.9	6.6

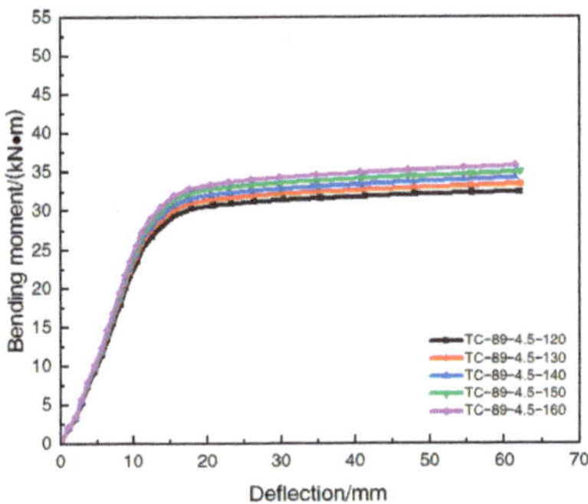

Figure 14. Effect of the strength of the in-filled UHPC on the bending moment–deflection curve.

As shown in Figure 14, the moment–deflection curves of the members in the elastic phase were not significantly different under different steel-slag-powder UHPC strengths. The different strengths of the UHPC mainly played a role after the elasto-plastic phase of the specimen, and the increase in the material strength had little effect on the ultimate flexural load capacity enhancement of the specimen.

In summary, the change in the strength of the in-filled UHPC material had a small effect on the flexural bearing capacity of the steel-tube-steel-slag-powder-UHPC beams, and the improvement in the material strength had a weak effect on the enhancement of the ultimate flexural bearing capacity of the member. Therefore, in an actual engineering application, it is possible to appropriately reduce the strength of the in-fill UHPC to reduce the cost.

5. Conclusions

(1) All types of the steel-tube-steel-slag-powder-UHPC beams were subject to "bow damage" during loading, and the deflection distribution curve basically conformed to the variation law of the sinusoidal half-wave function, showing good ductility. The deflection distribution curves of the same cross-sectional members were basically the same, and there was no significant change due to the change of the in-fill steel-slag-powder–UHPC ratio, which indicates that the influence of the in-fill material type on the deflection distribution of the members is relatively small.

(2) The shapes of the moment–deflection curves of the steel-tube-steel-slag-powder-UHPC beams were basically similar, and they could all be divided into an elastic phase, an elasto-plastic phase, and a strengthening phase. When the cross-sectional deflection reached L/30, the external load acting on the specimen continued to increase, indicating that the steel-tube-steel-slag-powder-UHPC beams had a good plastic deformation capacity with later flexural bearing capacity.

(3) The type of steel pipe had a significant effect on the flexural bearing capacity of the steel-tube-steel-slag-powder-UHPC beams; the larger the diameter of the steel pipe section and the thicker the wall, the higher the flexural bearing capacity. The amount of steel fiber admixture also had a certain degree of influence on the flexural load-bearing capacity. The admixture of steel fiber played a role in hindering the cracking and crack development of the UHPC, which effectively improved the flexural load-bearing capacity of the steel-pipe–UHPC beam. The amount of coarse aggregate and the length of the high-temperature maintenance had less influence on the flexural load-bearing capacity of the steel-pipe–UHPC beams, and the ecological steel-pipe–UHPC with coarse aggregate can be prepared when the proportion of coarse aggregate dosing is low.

(4) The steel-tube-steel-slag-powder-UHPC beams established by the finite element software matched well with the bending moment–deflection curves of the corresponding test members, and the calculated ultimate flexural bearing capacity and the test results had a stable error between 5.6% and 11.2%, which indicates that the model was reasonably established; so, the finite element method can be used for simulation analysis under the restricted test conditions.

(5) The finite element calculation analysis showed that the change in the steel content rate had a significant effect on the initial flexural stiffness and ultimate flexural bearing capacity of the beam; when the steel content rate increased, the initial flexural stiffness and ultimate flexural bearing capacity of the member increased significantly; the yield strength of steel had a small effect on the flexural stiffness of the steel-tube–steel-slag powder-UHPC beams in the initial and use stages but had a significant effect on the ultimate flexural bearing capacity; and the change in the in-fill UHPC strength had a small effect on the flexural bearing capacity of the beam. The change in the UHPC strength had less effect on the flexural load-bearing capacity of the beam, and the appropriate reduction in the strength of the in-fill UHPC can be considered in an actual project to reduce the construction cost.

(6) Compared with traditional steel pipe concrete, steel-tube-steel-slag-powder-UHPC beam has high toughness, high elasticity, low shrinkage, and other excellent performance, which can effectively reduce the impact of the difference between the performance of steel and concrete materials. At the same time, steel slag micronized powder UHPC for industrial solid waste reuse, adding coarse aggregate, can reduce its preparation costs, provide economic and environmental protection, and save green energy. The next step should be to deepen the steel-tube-steel-slag-powder-UHPC beam material's structural integration research in order to promote the steel pipe concrete material's lightweight, high performance, and green sustainable development and improve its application performance and use range.

Author Contributions: X.T.: Conceptualization and writing—original draft. C.F.: Data curation and formal analysis. J.C.: Software. J.M. and X.H.: Validation. All authors have read and agreed to the published version of the manuscript.

Funding: This work is supported by the National Natural Science Foundation of China (42067044; 42107166), Guangxi Postgraduate Education Innovation Program (YCSW2021171), the Hunan Provincial Department of Education Project (22B08253), and the Open Fund of Hunan Engineering Research Center for Intelligent Construction of Fabricated Retaining Structures (22K06).

Institutional Review Board Statement: This article is not be involved in ethical issues.

Informed Consent Statement: Informed consent was obtained from all subjects involved in the study.

Data Availability Statement: All data, models, and code generated or used during the study are available from the corresponding author by request.

Conflicts of Interest: The authors declare no conflict of interest.

References

1. Chen, B.C.; Ji, T.; Huang, Q.W.; Wu, H.Z.; Ding, Q.J.; Zhan, Y.W. A review of ultra-high performance concrete research. *J. Build. Sci. Eng.* **2014**, *31*, 1–24.
2. Shao, X.D.; Qiu, M.H.; Yan, B.F.; Luo, J. Progress of research and application of ultra-high performance concrete in bridge engineering at home and abroad. *Mater. Guide* **2017**, *31*, 33–43.
3. Azreen, N.M.; Raizal, R.; Mugahed, A.; Voo, Y.L.; Haniza, M.; Hairie, M.; Rayed, A.; Hisham, A. Simulation of ultra-high-performance concrete mixed with hematite and barite aggregates using Monte Carlo for dry cask storage. *Constr. Build. Mater.* **2020**, *263*, 120161. [CrossRef]
4. Mou, T.M.; Fan, B.K.; Zhao, Y.C.; Li, S. Application and development of steel pipe concrete bridges in China. *Highway* **2017**, *62*, 161–165.
5. Chang, J.; Xu, Y.F.; Xiao, J.; Wang, L.; Jiang, J.Q.; Guo, J.X. Influence of acid rain climate environment on deterioration of shear strength parameters of Natural residual expansive soil. *Transp. Geotech.* **2023**, *40*, 101017. [CrossRef]
6. Liu, Y.J.; Sun, L.P.; Zhou, X.H.; Peng, J.P.; Zhang, N.; Li, H. Advances in engineering applications and research of steel pipe concrete bridge towers. *Chin. J. Highw.* **2022**, *35*, 1–21.
7. Tu, C.L.; Shi, Y.J.; Liu, D. Calculation method for axial compression load capacity of high-strength steel pipe concrete columns. *Adv. Build. Steel Struct.* **2020**, *22*, 99–107.
8. Chen, B.C.; Wei, J.G.; Su, J.Z.; Huang, W.; Chen, Y.C.; Huang, Q.W.; Chen, Z.H. Advances in the application of ultra-high performance concrete. *J. Build. Sci. Eng.* **2019**, *36*, 10–20.
9. Hu, H.W.; Vizzari, D.; Zha, X.D.; Mantalovas, K. A comparison of solar and conventional pavements via life cycle assessment. *Transp. Res. Part D Transp. Environ.* **2023**, *119*, 103750. [CrossRef]
10. Chang, J.; Li, J.; Hu, H.W.; Qian, J.; Yu, M. Numerical investigation of aggregate segregation of superpave gyratory compaction and its influence on mechanical properties of asphalt mixtures. *J. Mater. Civ. Eng.* **2023**, *35*, 04022453. [CrossRef]
11. Li, J.; Zhang, J.; Yang, X.; Zhang, A.; Yu, M. Monte Carlo simulations of deformation behaviour of unbound granular materials based on a real aggregate library. *Int. J. Pavement Eng.* **2023**, *24*, 2165650. [CrossRef]
12. Fu, Z.D.; Lv, L.N.; Xiao, J.; He, Y.J.; Shen, P.L. The effect of CSA expansion agent on the properties of ultra-high performance concrete. *J. Mater. Sci. Eng.* **2019**, *37*, 559–564.
13. Wang, Z.; Wang, J.Q.; Liu, T.X.; Xiu, H.L. Calculation model of axial compression bearing capacity and deformation capacity of short UHPC columns with round steel tubes. *J. Cent. South Univ.* **2019**, *50*, 428–436.
14. Lu, Q.R.; Xu, L.H.; Ji, Y.; Yu, M.; Yang, Y.X. A compressive principal structure model for steel tube-confined ultra-high performance concrete. *J. Silic.* **2020**, *48*, 1201–1211.
15. Tang, J.J.; Wang, T.; Li, J.Y. Orthogonal analysis of axial compression load capacity of short columns of round steel pipes with ultra-high performance concrete containing coarse aggregates. *People's Pearl River* **2021**, *42*, 52–56.
16. Zhou, X.J.; Mou, T.M.; Tang, H.Y.; Fan, B.K. Experimental Study on Ultrahigh Strength Concrete Filled Steel Tube Short Columns under Axial Load. *Adv. Mater. Sci. Eng.* **2017**, *2017*, 8410895. [CrossRef]
17. Zhang, Y.X.; Liu, A.R.; Zeng, X.B.; Fu, J.Y.; Huang, Y.H. Study on the load-bearing capacity of high-strength steel tube-ultra-high-performance concrete arch. *J. Build. Struct.* **2021**, *42*, 365–372.
18. Wei, J.G.; Luo, X.; Chen, B.C.; Lv, J.Y. Study on the flexural performance of round high-strength steel tube UHPC beams. *Eng. Mech.* **2021**, *38*, 183–194.
19. Deng, Z.C.; Sun, T.; Li, J.Y. Study on the flexural performance of high-strength steel tube restrained ultra-high performance concrete beams. *J. Tianjin Univ.* **2021**, *54*, 1111–1120.
20. Huang, W.R.; Yang, Y.Z.; Cui, T.; Liu, Y.J. Study on shrinkage deformation performance of ultra-high performance concrete containing coarse aggregates. *Concrete* **2021**, *8*, 99–103.
21. GB/T 228.1-2021; Tensile Test of Metallic Materials Part 1: Room Temperature Test Method. Standardization Administration of the People's Republic of China: Beijing, China, 2021.
22. Tang, X.Y.; Ma, J.L.; Luo, J.; He, B.B.; Lu, C.J. Analysis of factors influencing mechanical properties of steel slag powder ecological type ultra-high performance concrete. *Silic. Bull.* **2023**, *42*, 607–617.
23. JTG/T 3650-2020; Technical Specification for Construction of Highway Bridges and Culverts. People's Communication Press: Beijing, China, 2020.
24. GB/T 50080-2016; Standard for Test Methods for the Performance of Ordinary Concrete Mixes. China Academy of Building Research: Beijing, China, 2016.
25. GB/T 50081-2019; Standard for Test Methods of Physical and Mechanical Properties of Concrete. Ministry of Housing and Urban-Rural Development: Beijing, China, 2019.

26. Hu, H.W.; Zha, X.D.; Li, Z.H.; Lv, R.D. Preparation and performance study of solar pavement panel based on transparent Resin-Concrete. *Sustain. Energy Technol. Assess.* **2022**, *52*, 102169. [CrossRef]

27. Xiang, Y.; Leroy, G. Stress-strain curves for hot-rolled steels. *J. Constr. Steel Res.* **2017**, *133*, 36–46.

28. Zhong, T.; Zhi, B.W.; Qing, Y. Finite element modelling of concrete-filled steel stub columns under axial compression. *J. Constr. Steel Res.* **2013**, *89*, 121–131.

29. Rao, Y.L.; Zhang, C.C.; Li, Y.; Huang, Y.S.; Li, D. Experimental study on the axial compression bearing capacity of high-strength cold-formed rectangular steel tube concrete short columns. *J. Huaqiao Univ.* **2019**, *40*, 338–343.

Article

Study on the Strength and Failure Characteristics of Silty Mudstone Using Different Unloading Paths

Jijing Wang [1,*], Hualin Zhang [2,3,*], Shuangxing Qi [1], Hanbing Bian [2,3], Biao Long [1] and Xinbo Duan [1]

[1] School of Traffic & Transportation Engineering, Changsha University of Science & Technology, Changsha 410114, China; 18002010031@stu.csust.edu.cn (S.Q.); lxy215512@163.com (B.L.); dyw12323@163.com (X.D.)
[2] IMT Nord Europe, University Lille, F-59000 Lille, France; hanbing.bian@univ-lille.fr
[3] ULR 4515-LGCgE Laboratoire de Génie Civil et Géo-Environnement, JUNIA, University Artois, F-59000 Lille, France
* Correspondence: wangjijing@csust.edu.cn (J.W.); hualin.zhang.etu@univ-lille.fr (H.Z.)

Abstract: To investigate the strength and failure characteristics of silty mudstone using different stress paths, silt-like mudstone specimens were subjected to triaxial unloading tests. The results indicate the following. (1) When subjected to equivalent initial deviator stress levels and differing confining pressures, the peak stress, residual stress, and elastic modulus, exhibited during unloading, increased concordantly with greater initial confining pressure. Both the peak strain and residual strain increased with rising initial confining pressure. The increase in peak strain and residual strain initially decelerated, then noticeably increased, before ultimately decreasing again. Additionally, the unloading failure time and strain rate demonstrated a negative correlation as the confining pressure increased. (2) Under different initial deviatoric stress conditions, the peak stress, residual stress, and residual strain, under unloading confining pressure conditions, decreased as the initial deviatoric stress levels elevated. Conversely, the peak strain and elastic modulus initially increased, then decreased under increasing initial deviatoric stress conditions. The unloading failure time and strain rate were both observed to decrease as the initial deviatoric stress levels increased. (3) Utilizing the Mohr stress circle enabled the characterization of the shear strength variation in the specimens during the unloading process. The cohesion and internal friction angle remained relatively consistent across the different unloading stress paths appraised, with cohesion being greater in path I versus path II, whereas the internal friction angle exhibited an inverse relationship. (4) The specimen failed during unloading due to lateral expansion caused by unloading confining pressure and collapse failure. The failure fracture surfaces predominantly manifested shear failure morphologies.

Keywords: silty mudstone; unloading path; mechanical properties; shear strength; failure morphology

Citation: Wang, J.; Zhang, H.; Qi, S.; Bian, H.; Long, B.; Duan, X. Study on the Strength and Failure Characteristics of Silty Mudstone Using Different Unloading Paths. *Materials* **2023**, *16*, 5155. https://doi.org/10.3390/ma16145155

Academic Editor: Dimitrios Papoulis

Received: 22 June 2023
Revised: 12 July 2023
Accepted: 17 July 2023
Published: 21 July 2023

1. Introduction

Silty mudstone is a type of soft rock which is widely distributed in the south of China. With the expansion of infrastructure in south-west China, this type of rock is frequently encountered in engineering practices. For instance, during highway construction, as the rock mass is continuously excavated, the initial in situ stresses are "unloaded", which may lead to the redistribution of stresses. This triggers unloading deformation and potential damage, including micro or macro cracks within the rock mass [1–4]; these consequences are similar to the well-known Excavation Damaged/Distributed Zone (EDZ) found in tunnel engineering. Over the past decade, there have been numerous landslides triggered by the failure of soft rock in China, such as the Anlesi landslide and the Dalixi landslide, the landslide that occurred in the Three Gorges Reservoir area, and the landslide that occurred along the Jinchuan–Xiaojin highway in Sichuan Province; these are typical examples of unloading deformation caused by mudstone cut slope excavation [5–7]. To ensure the safety of human beings and infrastructures, it is of great importance to investigate the strengths and failures of silty mudstone under unloading conditions to provide fundamental strength parameters for construction practitioners.

Thus far, numerous researchers have directed their attention towards studying the strength, deformation, and expansion characteristics of rock specimens under unloading confining pressure conditions. These studies have primarily been conducted using experimental approaches, yielding valuable research findings. Through a comparison of conventional and unloading triaxial tests, Wang et al. [8] discovered that lower peak deviatoric stress and axial strain levels were exhibited during the unloading triaxial test. Conversely, other mechanical parameters, such as the internal friction angle and triaxial compressive strength, were found to exhibit higher levels than those obtained during the conventional triaxial test. Similarly, Zhang et al. [9] examined the impact of different stress paths on coal specimen mechanics, and they found that greater unloading rates heightened damage risk. Comparing the triaxial unloading test with the conventional triaxial compression test, a reduction in the coal specimen's cohesive force and an increase in internal friction were noted. Moreover, Wang et al. [10] analyzed how the confining pressure and unloading rate impacted crack propagation; the results indicated that under greater confining pressure, crack formation post-failure, and the rate of crack expansion, increased with higher unloading rates. Takeda et al. [11] examined the semi-permeability evolution of Wakkanai mudstones during cyclic loading and unloading, emphasizing the importance of considering the applied stresses when estimating argillite semi-permeability. In response to stress evolution induced by activities such as excavation and mining, Zhang et al. [12] executed a multi-level axial stress triaxial unloading test that defined the alteration law of mechanical parameters (friction, cohesion, and dilatancy angle) using an unloading factor representing rock specimen damage mechanisms. Current research indicates that during a triaxial confining pressure unloading test, the failure mode, crack propagation, and strength characteristics largely hinge on factors such as initial confining pressure, unloading rate, and unloading path. Furthermore, different rock types exhibit unique mechanical behaviors under triaxial unloading confining pressure conditions. Shale displays significant brittleness, which escalates as the unloading rate and confining pressure increases. As the loading rate increases, sandstone shows an increasing stress brittleness reduction coefficient, indicating that its surface brittle failure decreases gradually. With the increasing unloading rate, the rock's brittle failure initially strengthens and subsequently weakens; abnormal brittle failure is detected when the unloading rate is high [13,14].

In addition, significant progress has also been made in the investigation concerning the failure characteristics of rocks under cyclic loading in the early and subsequent unloading stages. Fan et al. [15] conducted comprehensive unloading tests on sandstone using the true triaxial system, which combined a damage control cyclic loading path with an unloading confining pressure path. The results suggested that the prior cyclic loading damage had an effect on strength and deformation characteristics, energy conversion, and failure modes. Huang et al. [16] demonstrated that the size and unloading rate of the initial confining pressure had a significant impact on the failure mode and strain energy conversion (accumulation, dissipation, and release). Zhao et al. [17], through the experiment, obtained mechanical variables (instantaneous elastic strain, viscoelastic strain, transient plastic strain, viscoplastic strain), indicating a positive relationship between the instantaneous deformation modulus and creep stress. These studies highlight how the multi-stage loading and unloading process during a cycle enhances a soft rock's resistance to transient elastic and plastic deformation. Chen et al. [18] conducted a triaxial test on sandstone under confining pressure unloading conditions, measuring permeability and acoustic emission signals before peak stress was reached. The results demonstrate a strong correlation between the stress–strain curve, permeability–axial strain curve, and the acoustic emission activity mode under unloading conditions. Gupta et al. [19] examined shale deformation using various stress paths, highlighting the significant influence of stress level and bedding plane orientation on microcrack geometry. Additionally, Dai and Zhang et al. [20,21] showed that the strain increment is significantly influenced by the unloading path, whereas the unloading rate of the confining pressure is comparatively less affected.

Due to the relatively low strength, and the substantial sensitivity to water, silty mudstone has long been considered an inappropriate engineering material. Few investigations on its mechanical strength have been reported, particularly during "unloading" process. Most existing research has primarily focused on the mechanical properties and failure morphology of hard rocks using different loading and unloading paths, with fewer studies considering the deformation and comprehensive effects of different loading and unloading paths on soft rocks [22–24]. Therefore, in this paper, silt-like mudstone specimens, composed of gypsum, barite powder, and nanomaterials, were used to conduct triaxial tests using different unloading paths so that we could investigate the influence of different initial confining pressures and deviatoric stresses on the strength and failure characteristics of silty mudstone. The goal of this paper is to provide information on the safe operation and slope stability of silty mudstone areas, providing theoretical references for understanding rock performance under complex loading and unloading conditions.

2. Materials

2.1. Specimens Preparation

The silty mudstone used in this paper was collected from the cutting slope of the Longlang Expressway, Hunan Province. The rock was rigorously characterized prior to testing. The major chemical components included SiO_2, Al_2O_3, Fe_2O_3, K_2O, MgO, TiO_2, and Na_2O. Notably, the collective mass fraction of SiO_2 and Al_2O_3 constituted approximately 80% of all the chemical components. The granular composition consisted mainly of quartz, feldspar, clay minerals, and a small amount of carbonate minerals.

The inherent defects and low homogeneity of natural silty mudstone have a substantial impact on test results. To mitigate the impact of initial states on results, similar material specimens exhibiting excellent homogeneity were prepared. Silt-like specimens, composed of Nano-TiO_2, Nano-Al_2O_3, NTi, Nano-bentonite, gypsum, and barite powder mixed with water, bore a close resemblance to silty mudstone in terms of its physical and mechanical characteristics, as reported in previous studies [25–27]. The silt-like mudstone specimens were tested using a controlled moisture content of 17.5%, which is within the range found in original rock (14.6–21.5%). The specimens were of a standard size (Φ50 × H100 mm).

To validate the initial homogeneity of similar materials, a RSM-SY5(T) non-metallic ultrasonic tester was used on the processed silt-like mudstone specimens, in accordance with the ASTM D2845-standard [28], as displayed in Figure 1a. Two transducers were set on each specimen while Vaseline was applied for enhanced contact, and the sensor was connected to transducers. The travel time of the ultrasonic waves was measured, and the longitudinal wave velocity was calculated using the specimen length. Figure 1b illustrated the longitudinal wave velocity distribution for silt-like mudstone specimens. The wave velocity, primarily concentrated at 2.20–2.30 km/s, fell within the range found in original rock. Selecting specimens with similar wave velocities aided in minimizing variability, thus the specimens demonstrating wave velocities within the range of 2.20–2.30 km/s were selected for further testing (Figure 2). The comparison between the basic properties of the selected silt-like mudstone and the original rock is presented in Table 1.

Table 1. Comparison between basic properties of silt-like mudstone and original rock.

Parameters	Density (g/cm^3)	Water Absorption Rate (%)	Swelling Rate (%)	Longitudinal Wave Speed (km/s)	Uniaxial Strength (MPa)	Tensile Strength (MPa)	Softening Coefficient
Original rock	2.15–2.34	2.14–8.16	0.14–0.18	1.75–2.85	6.9–23.8	0.3–1.4	0.4–0.66
Specimens	2.26	6.05	0.15	2.20–2.30	8.15	1.38	0.56

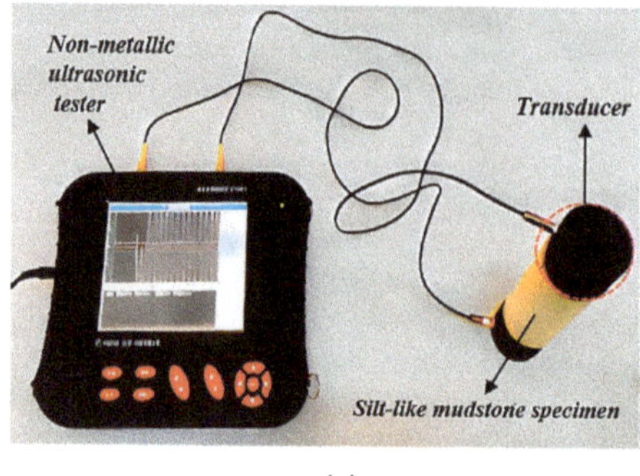

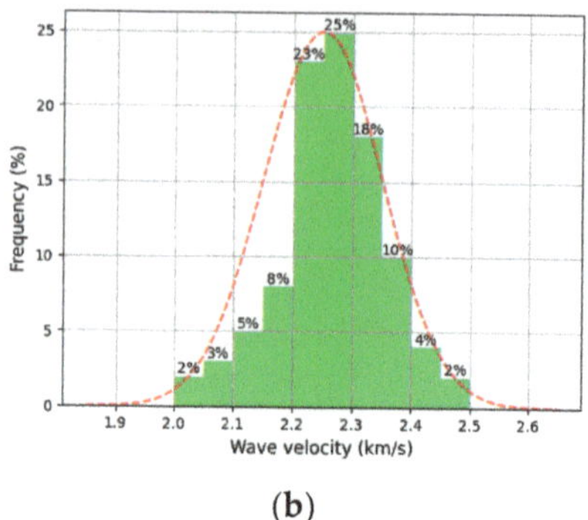

(a)

(b)

Figure 1. Ultrasonic longitudinal wave velocity measurements: (**a**) RSM-SY5(T) non-metallic ultrasonic tester; (**b**) wave velocity distribution for silt-like mudstone specimens.

Figure 2. Selected silt-like mudstone specimens.

2.2. Test Apparatus

The test employed the DZSZ-150 multi-field coupling rock triaxial testing machine, as illustrated in Figure 3.

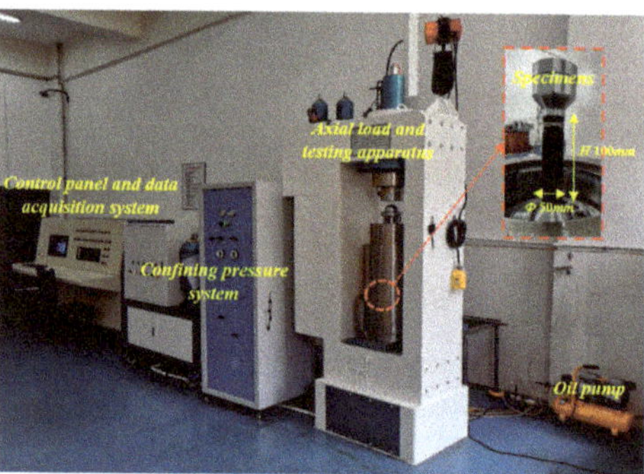

Figure 3. Test apparatus: DZSZ-150 multi-field coupling rock triaxial testing machine.

The machine utilized a full-digital servo controller as its control system, which incorporates high-precision load sensors and displacement sensors to automatically and accurately control and display data such as test force, confining pressure, and specimen deformation.

The machine was equipped with 1500 kN axial actuators, along with a 70 MPa confining pressure boosting system. Additionally, it featured a 70 MPa pore pressure boosting control system. The control panel allowed for the manual adjustment of the loading and unloading rate, pore pressure, back pressure, and axial pressure. The system was designed to collect test data automatically. The confining pressure in this system was generated by the oil pump.

3. Test Methods

This study employed non-standard loading and unloading pathways that mirrored actual excavation scenarios involving silty mudstone cutting slopes. Therefore, two unloading paths are considered in this experiment, as follows. (1) The first path involves the application of an initial deviatoric stress under unloading confining pressure conditions (i.e., the specimen is loaded until a particular level of predetermined deviatoric stress and confining pressure is reached). Subsequently, the deviatoric stress is kept constant while the confining pressure is gradually released at a specific rate until the specimen fails. This typically happens during the early stages of excavation when the horizontal confining pressure is reduced while the deviatoric stress caused by overburdening remains constant. (2) The second path concerns the application of a confining pressure that is left unchanged throughout the whole process. The specimen is then loaded to different levels of deviatoric stress and the confining pressure is unloaded until the specimen fails. This occurs in deep excavation scenarios, wherein the confining pressure remains steady due to the overlying rock layers, but the deviatoric stress varies due to changes in terms of overburdening or other external loading conditions [29,30].

The test procedure was conducted as follows: first, the confining pressure was applied in a stress-controlled manner at a rate of 0.075 MPa/s until it reached a predetermined value (20 MPa, 30 MPa, 40 MPa, and 50 MPa). Subsequently, with the confining pressure held steady, the axial pressure was applied at the same rate of 0.075 MPa/s to a predetermined axial pressure value (35 MPa, 45 MPa, 55 MPa, and 65 MPa). Finally, the confining pressure was incrementally unloaded at a rate of 0.075 MPa/s, while the axial pressure was kept constant until the specimen failed (Table 2).

Table 2. Test plan.

Path Number		Axial Pressure σ_1 (MPa)	Confining Pressure σ_3 (MPa)	Deviatoric Stress σ_1–σ_3 (MPa)	Axial Pressure after Damage (MPa)	Confining Pressure σ_3 (MPa)
Path I	I-1	35	20		35	0.075 MPa/s unloading rate until damage
	I-2	45	30	15	45	
	I-3	55	40		55	
	I-4	65	50		65	
Path II	II-1	50		10	50	0.075 MPa/s unloading rate until damage
	II-2	55	40	15	55	
	II-3	60		20	60	
	II-4	65		25	65	

4. Test Results and Discussions

4.1. Stress Path I

The stress–strain relationship curve of the silt-like mudstone when stress path I was used is depicted in Figure 4. It exhibits certain commonalities under different initial confining pressures when the deviatoric stress level was set to 15 MPa. More specifically, during this axial loading stage, the stress–strain curves demonstrate linear deformation, indicating the specimens' linear elastic behavior. During the unloading stage, the axial strain continues to increase as the deviatoric stress increases, and a significant increase in elastic modulus is observed compared with the axial loading process. Post-failure, the stress continuously decreases, as the strain concurrently amplifies. And residual strain occurs at this stage, indicating the presence of apparent plastic flow features in the specimen.

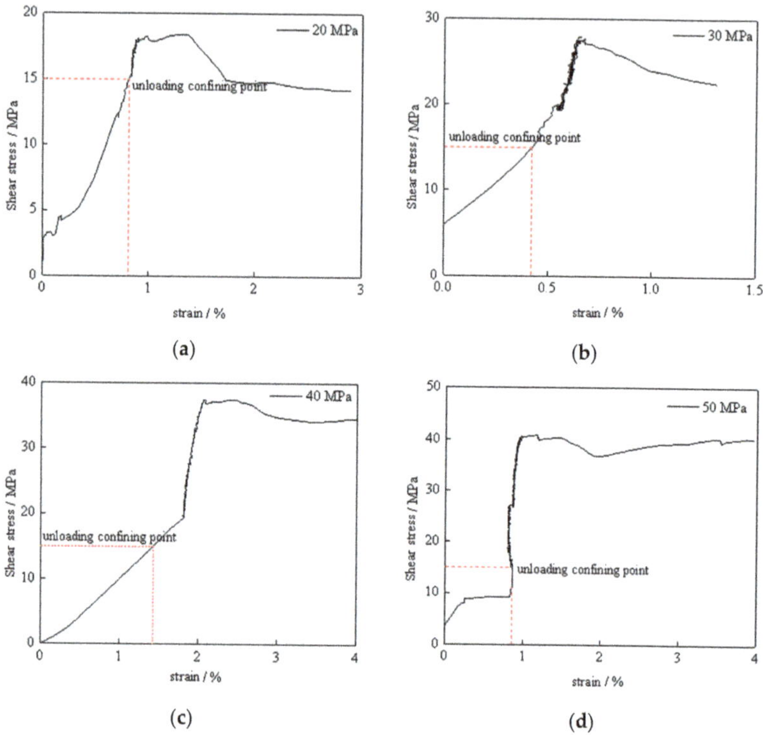

Figure 4. Stress–strain curve of silt-like mudstone specimens (path I): (**a**) initial confining pressure 20 MPa; (**b**) initial confining pressure 30 MPa; (**c**) initial confining pressure 40 MPa; (**d**) initial confining pressure 50 MPa.

The influence of the initial confining pressure on mechanical characteristics when unloading path I was used is depicted in Figures 5 and 6, and the values are presented in the graph. As can be observed in the figure, for initial confining pressures of 20 MPa, 30 MPa, 40 MPa, and 50 MPa, the specimens display strains of 0.84%, 0.44%, 1.59%, and 0.87% until the unloading pressure was reached. The corresponding peak stresses and strains were 18.43 MPa (1.27%), 27.71 MPa (0.675%), 37.46 MPa (2.40%), and 40.41 MPa (0.96%), respectively. Likewise, the results for the elastic modulus are 6.74 GPa, 8.51 GPa, 10.01 GPa, and 14.98 GPa, respectively.

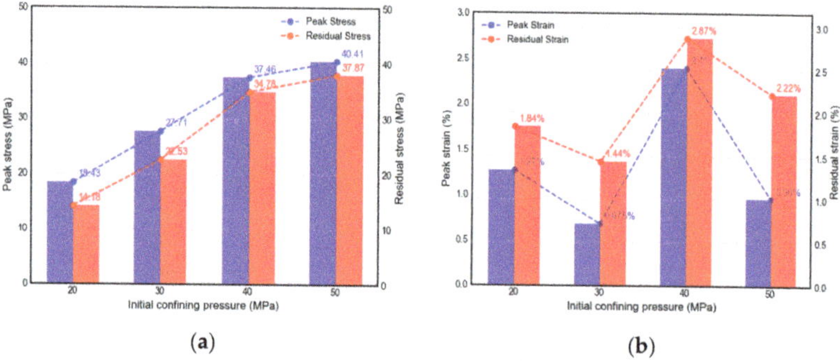

Figure 5. Influence of initial confining stress on mechanical parameters (path I): (**a**) the change law of peak stress and residual stress; (**b**) the change law of peak strain and residual strain.

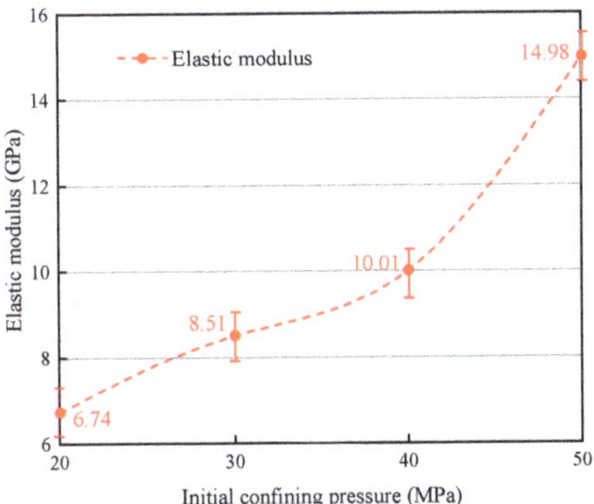

Figure 6. Influence of initial confining stress on elastic modulus (path I).

The results reveal that the deviatoric stress at which failure occurs during unloading increases with higher levels of initial confining pressure. However, increases in deviatoric stress diminish as the confining pressure increases. The strain associated with unloading failure exhibits a trend of initially decreasing slowly and then increasing significantly as the initial confining pressure increases. The elastic modulus shows a notable increase with a higher initial unloading confining pressure, presenting a decelerating, then accelerating trend.

Post-unloading failure, the residual stresses and strains were 14.18 MPa (1.84%), 22.53 MPa (1.44%), 34.78 MPa (2.87%), and 37.87 MPa (2.22%), respectively. Notably, it demonstrated the marked increase in residual strain when the initial confining pressure was 40 MPa. Our interpretation focuses on the increased rock compression at higher confining pressures, which can lead to considerable strain prior to unloading failure. In particular, at this pressure, active micro-crack initiation and propagation are likely to contribute to increased plastic deformation, thus resulting in larger residual strain. These findings suggest a consistent rise in residual stress with increasing confining pressure. Although the specimens display an overall rising trend in residual strain, some fluctuations occur due to the compression density effect caused by the initial confining pressure. As the specimen becomes denser, the residual strain tends to increase. After the specimen fails, there is a certain randomness in the fracture surface and failure morphology, and this characteristic can notably affect the residual stress and strain.

The Mohr stress circle is plotted in accordance with the stress-strain curves of the silt-like mudstone under varying initial deviatoric stress and constant confining pressure conditions, as observed in Figure 7. The shear strength envelope of the silt-like mudstone specimen, as a function of the stress path I unloading confining pressure process, can be obtained as follows:

$$\tau = 0.419\sigma + 0.548, \ R^2 = 0.983 \tag{1}$$

where τ is the shear stress and σ is the positive principal stress. Using Equation (1), the cohesive force c of the silt-like mudstone specimen is 0.548 MPa, and the internal friction angle φ is 26.151°.

From the left to right, Mohr stress circles represent a range from lower to higher initial confining pressure, exhibiting substantial differences. The process of unloading confining pressure shifted the Mohr's circle towards the strength envelope. And Mohr stress circle expand as the confining pressure increases, indicating a decreased likelihood of failure in specimens.

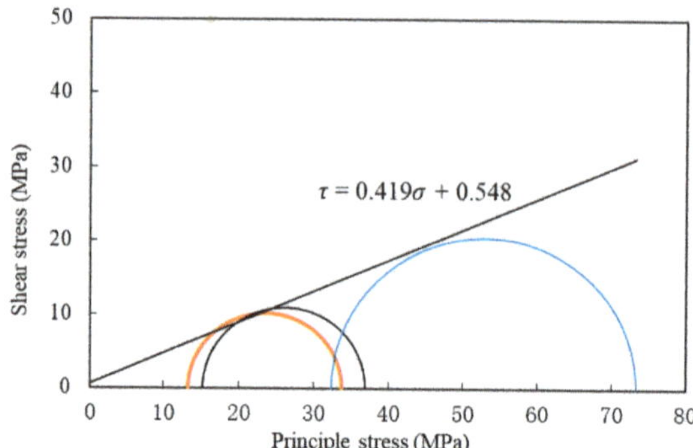

Figure 7. Mohr stress circle and strength envelope diagram of the silt-like mudstone specimen (path I).

For specimens under unloading path I, both the peak and residual stress increase with the initial confining pressure. Peak and residual strain decrease slowly as the initial confining pressure increases, then, they increase significantly and finally drop, both of which follow a similar pattern. The unloading failure time and strain rate are inversely proportional to the confining pressure increase. A higher initial unloading confining pressure corresponds to a substantial increase in elastic modulus.

4.2. Stress Path II

The stress–strain relationship curves for the silt-like mudstone, obtained using stress path II, are depicted in Figure 8. They exhibit consistent trends, demonstrating linear deformation characteristics when the initial confining pressure is maintained at 40 MPa and the deviatoric stress is increased to 10 MPa, 15 MPa, 20 MPa, and 25 MPa, respectively. This indicates linear elastic properties at this stage. As the confining pressure is unloaded, the axial strain continues to increase as the deviatoric stress increases. Upon failure of the silt-like mudstone specimen, the stress diminishes, causing the stress–strain curve to transition into the post-peak stage, which is characterized by a progressive decline in stress and a corresponding increase in strain. This stage, which exhibits significant plastic flow characteristics within the specimen, is where residual strain observed.

The variation in mechanical characteristics is due to initial deviatoric stress, as displayed in Figures 9 and 10, and the mechanical parameters for unloading path II are illustrated in the graph. From the figure, it is evident that under a constant confining pressure of 40 MPa, with initial deviatoric stresses of 10 MPa, 15 MPa, 20 MPa, and 25 MPa, respectively, the peak stresses and strains are 47.07 MPa (1.620%), 37.46 MPa (2.182%), 32.81 MPa (1.740%), 25.80 MPa (1.736%). The results for the elastic modulus are 4.62 GPa, 9.94 GPa, 3.31 GPa, 3.12 GPa, respectively.

The results indicate that peak stress decreases with an increase in initial deviatoric stress, whereas the peak strain exhibits a trend of initially increasing and subsequently decreasing. The elastic modulus demonstrates a substantial increase with higher initial deviatoric stress, followed by a considerable decrease, and finally a slow decrease. The residual stresses and strains were 42.27 MPa (3.228%), 34.78 MPa (3.024%), 28.70 MPa (2.153%), and 22.68 MPa (1.249%) after the failure of the specimen. Both the residual stress and strain exhibited a decreasing trend with increasing initial deviatoric stress.

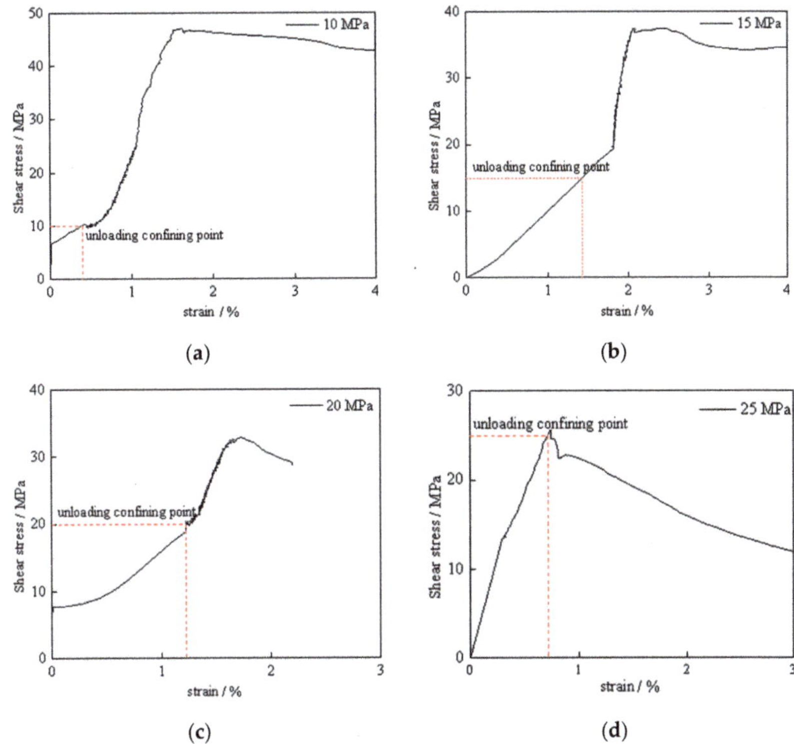

Figure 8. Stress–strain curve of silt-like mudstone specimen (path II): (**a**) deviatoric stress 10 MPa; (**b**) deviatoric stress 15 MPa; (**c**) deviatoric stress 20 MPa; (**d**) deviatoric stress 25 MPa.

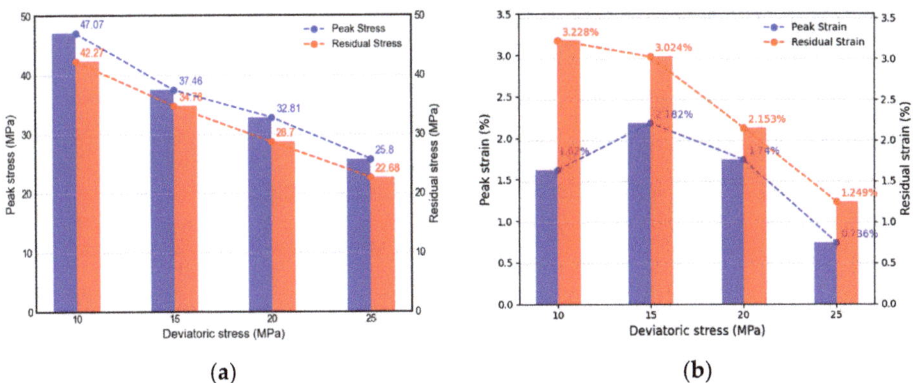

Figure 9. Influence of initial deviatoric stress on mechanical parameters (path II): (**a**) the change law of peak stress and residual stress; (**b**) the change law of peak strain and residual strain.

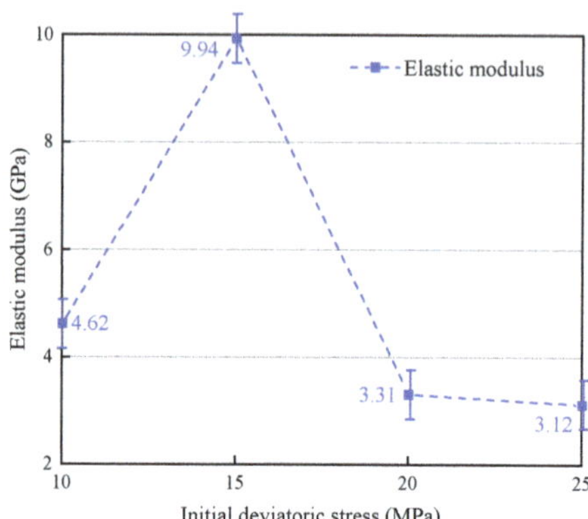

Figure 10. Influence of initial deviatoric stress on the elastic modulus (path II).

Based on the stress–strain curve of silt-like mudstone under varying deviatoric stress conditions at a constant confining pressure, the Mohr stress circle can be plotted as shown in Figure 11. The shear strength envelope function of the silt-like mudstone specimen under stress path II can be obtained, as follows:

$$\tau = 0.459\sigma + 0.477, \ R^2 = 0.968 \tag{2}$$

where τ is the shear stress and σ is the positive principal stress. Using Equation (2), the cohesive force c of the silt-like mudstone specimen is 0.477 MPa, and the internal friction angle φ is 26.335°.

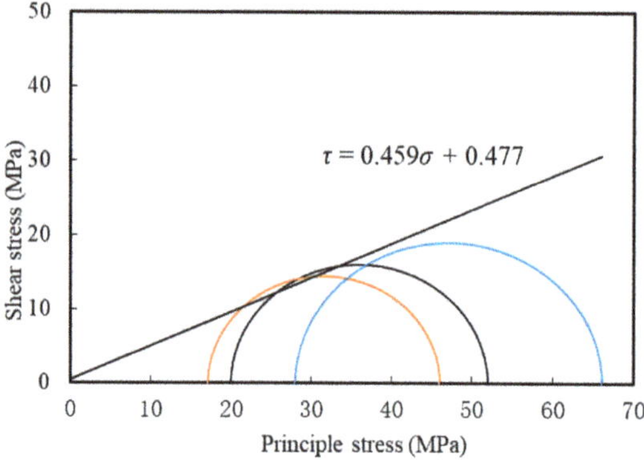

Figure 11. Mohr stress circle and strength envelope diagram of silt-like mudstone specimen (path II).

Arranged from left to right, the Mohr stress circles indicate a progressively larger initial deviatoric stress in the specimens. The widening gap between the principal stresses within each Mohr stress circle suggests an escalating likelihood of failure as the deviatoric stress increases.

For the specimens subjected to the same confining pressure but varying deviatoric stresses when unloading path II was used, both the peak and residual stresses, as well as the residual strain, exhibited a decreasing trend with an increase in initial deviatoric stress; all displayed similar trends. The peak strain increased slowly with the increase in initial deviatoric stress, then, it significantly decreased. A negative correlation was observed between the unloading failure time and unloading strain rate in relation to the initial deviatoric stress. The elastic modulus demonstrated a notable trend of initially increasing and then decreasing with the increase in initial deviatoric stress under unloading confining pressure conditions.

4.3. Effect of Stress Path on Strength

During the triaxial unloading test, the specimen's capacity to withstand the load is indeed influenced by the magnitude of the initial confining pressure and initial deviatoric stress.

By increasing the deviatoric stress by reducing the confining pressure, the ultimate bearing capacity of the specimen is reduced to the ultimate stress level, leading to failure.

Therefore, in this paper, two shear strength parameters, cohesion (c) and the internal friction angle (φ), have been selected to represent the shear strength fluctuations of specimens during unloading failure and to analyze the effect of the unloading path on shear strength [31–33]. The failure parameters of the specimens under different unloading paths are summarized in Table 3. The values of c and φ are comparably close for the specimens undergoing different unloading paths, and the cohesion of path I (0.548 MPa) exceeds that of path II (0.477 MPa). However, the internal friction angle of path I (26.151°) is slightly less than that of path II (26.335°).

Table 3. Failure strength parameters of silt-like mudstone specimens using different stress paths.

Stress Path	Cohesion c (MPa)	Internal Friction Angle φ (°)	Correlation Coefficient R^2
Unloading path I	0.548	26.151	0.983
Unloading path II	0.477	26.335	0.968

5. Analysis of Macroscopic Characteristics of Specimens after Unloading Failure

5.1. Stress Path I

Figure 12 presents a typical image of silt-like mudstone specimens after unloading failure under path I.

(a)   **(b)**   **(c)**   **(d)**

Figure 12. Typical failure morphology of silt-like mudstone specimens (Path I): (**a**) confining pressure 20 MPa; (**b**) confining pressure 30 MPa; (**c**) confining pressure 40 MPa; (**d**) confining pressure 50 MPa.

The image reveals that as the initial confining pressure increases, the failure morphology of the specimens becomes increasingly intricate, and they exhibit a distinct dilation feature, which is closely associated with the unloading confining pressure level of the specimens [9,34,35]. In the case of a low initial confining pressure, unloading failure mainly manifests as brittle shear failure, although, the structural integrity of the failed specimen remains intact. As the initial confining pressure increases, the failure morphology becomes more complex, and it is difficult to maintain the integrity of the failed specimen.

The findings can be summarized as follows: (1) under a low initial confining pressure (20 MPa), shear failure occurs with minimal tensile fractures, maintaining high specimen integrity with a few fine cracks on the fracture surface; (2) as the initial confining pressure increases (30 MPa), the specimen exhibits more tensile cracks in the axial stress direction and a noticeable degree of end fragmentation. Moreover, it demonstrates visible conjugate cracks and numerous microcracks, in which the primary fractured surface appears curved, which extends throughout the entire specimen. (3) as the initial confining pressure (40 MPa) further increases, the microcracks also continue to increase in number after the specimen undergoes shear failure. Penetrating cracks become less apparent, but a significant number of more pronounced microcracks emerge, which are widely distributed throughout the specimen; (4) lastly, after further increasing the initial confining pressure (50 MPa), the specimen under high confining pressures (40 MPa, 50 MPa) undergo significant fragmentation. The entire fractured surface attaches to a multitude of fine particles, resulting in the substantial scattering of particles due to the failure of specimens. The fractured surface exhibits a complex failure morphology, indicating the intact fragmentation of the specimen.

5.2. Stress Path II

Figure 13 illustrates a typical photograph of silt-like mudstone specimens after unloading failure when path II is used.

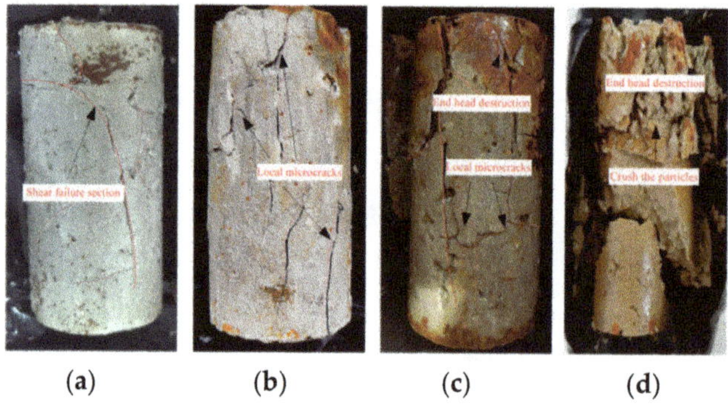

(a) (b) (c) (d)

Figure 13. Typical failure morphology of silt-like mudstone specimens (Path II): (**a**) deviatoric stress 10 MPa; (**b**) deviatoric stress 15 MPa; (**c**) deviatoric stress 20 MPa; (**d**) deviatoric stress 25 MPa.

It indicates that the unloading failure exhibits the pronounced failure characteristics of a fractured surface, which correlates with the confining pressure level of the specimens. Under low deviatoric stress conditions, it takes a relatively long time for the failure of specimen to occur under unloading confining pressure conditions. The failure primarily exhibits itself as brittle shear failure, and substantial integrity is maintained for the failed specimen. However, as the initial deviatoric stress increases, the integrity of the specimen after unloading failure deteriorates, and failure-induced cracks become more complex; in particular, severe end failure coupled with numerous localized micro-cracks may occur. Ultimately, extensive particle fragmentation occurs after failure when the initial deviatoric stress is 25 MPa.

The following can be concluded. (1) At an initial deviatoric stress of 10 MPa, the specimen mainly undergoes penetration fracture failure under constant deviatoric stress during unloading confining pressure, with noticeable tension fractures in the middle and upper sections, maintaining high integrity. (2) When the initial deviatoric stress increases to 15 MPa, the failure degree of the specimen end becomes distinctly visible, along with an observable penetration crack on its surface and the emergence of numerous end cracks. (3) As the initial deviatoric stress further increases to 20 MPa, post-shear failure microcracks proliferate throughout the specimen, with more pronounced local microcracks.

(4) When the initial deviatoric stress is 25 MPa, the specimen as a whole undergoes conspicuous shear failure, at which time the specimen completely fails. In particular, the load-bearing end demonstrates severe failure characteristics. This leads to the generation of a substantial number of broken particles, which is due to the strong frictional effect on the fractured surface of the entire specimen during the process of triaxial unloading failure.

After comparing the stress-strain relationship curves under different unloading conditions (path I and II) and the failure characteristics of the specimens, it is evident that the unloading failure is primarily dictated by the stress path, resulting in significant variations in terms of the macroscopic failure characteristics. In essence, the failure occurs due to the lateral expansion caused by the action of unloading confining pressure [36].

6. Conclusions

In this paper, an analysis was conducted to explore the strength and failure characteristics of silty mudstone that was subjected to various stress paths. The approach capitalized on silt-like mudstone specimens in triaxial unloading tests and it drew the principal conclusions, as follows:

1. Under constant initial deviatoric stress and varying confining pressures, the silt-like mudstone exhibited continuous increases in peak stress, residual stress, and elastic modulus as the initial confining pressure increased during the unloading process. The peak strain and residual strain initially decreased gradually with the increase in initial confining pressure, subsequently displaying a notable increase before eventually decreasing. Additionally, the unloading damage time and unloading strain rate were found to be negatively correlated with the increase in confining pressure.

2. Under different initial deviatoric stress conditions, the peak stress, residual stress, and residual strain under unloading confining pressure conditions exhibited a decreasing trend with an increase in initial deviatoric stress. The peak strain and elastic modulus initially demonstrated an increasing trend, followed by a subsequent decline, with an increase in initial deviatoric stress. Conversely, the unloading failure time and unloading strain rate decreased as the initial deviatoric stress increased.

3. To characterize the shear strength variation of specimens, the cohesion and internal friction angle were obtained in accordance with the Mohr stress circle, the values of which were relatively close when both unloading stress path I and path II were used. Path I exhibited a larger cohesive force than path II, whereas the internal friction angle showed the opposite trend.

4. The failure mechanism observed during the unloading of the specimens was characterized by a collapse-type failure, which may be primarily attributed to lateral expansion induced by the applied unloading confining pressure. The fractured surface predominantly exhibited shear failure. During the unloading process, when subjected to low initial confining pressure and low deviatoric stress conditions, the evolution of tension cracks was not prominently observed, and the specimens exhibited a high level of structural integrity. Conversely, under high initial confining pressure and high deviatoric stress conditions, the macroscopic development of cracks became more pronounced, leading to a greater degree of specimen fragmentation. Additionally, the end failure of the specimen became more evident.

Author Contributions: Conceptualization, J.W. and H.Z.; methodology, J.W. and S.Q.; validation, S.Q., B.L. and X.D.; formal analysis, H.Z.; investigation, X.D. and B.L.; resources, H.Z.; data curation, J.W.; writing—original draft preparation, J.W. and H.B.; writing—review and editing, H.Z. and H.B.; visualization, S.Q.; funding acquisition, J.W. and H.Z. All authors have read and agreed to the published version of the manuscript.

Funding: This research was funded by National Natural Science Foundation of China (No. 52108401), a total of 300,000 RMB, grant received by J.W.; and the China Scholarship Council (202208070060), 1350 Euro per month for 24 months, grant received by H.Z.

Institutional Review Board Statement: Not applicable.

Informed Consent Statement: Not applicable.

Data Availability Statement: Not applicable.

Conflicts of Interest: The authors declare no conflict of interest.

References

1. Omar, H.; Rod, S.; David, R. A study of the effect of strain rate on stiffness and strength in silty mudstone using multistage triaxial testing. *Post-Min.* **2005**, *2005*, 16–17.
2. Corkum, A.G.; Martin, C.D. The Mechanical Behaviour of Weak Mudstone (Opalinus Clay) at Low Stresses. *Int. J. Rock Mech. Min. Sci.* **2007**, *44*, 196–209. [CrossRef]
3. Taheri, A.; Tani, K. Use of Down-Hole Triaxial Apparatus to Estimate the Mechanical Properties of Heterogeneous Mudstone. *Int. J. Rock Mech. Min. Sci.* **2008**, *45*, 1390–1402. [CrossRef]
4. Fu, H.; Jiang, H.; Qiu, X.; Ji, Y.; Chen, W.; Zeng, L. Seepage Characteristics of a Fractured Silty Mudstone under Different Confining Pressures and Temperatures. *J. Cent. South Univ.* **2020**, *27*, 1907–1916. [CrossRef]
5. Huang, X.; Wang, L.; Ye, R.; Yi, W.; Huang, H.; Guo, F.; Huang, G. Study on Deformation Characteristics and Mechanism of Reactivated Ancient Landslides Induced by Engineering Excavation and Rainfall in Three Gorges Reservoir Area. *Nat. Hazards* **2022**, *110*, 1621–1647. [CrossRef]
6. Jian, W.; Wang, Z.; Yin, K. Mechanism of the Anlesi Landslide in the Three Gorges Reservoir, China. *Eng. Geol.* **2009**, *108*, 86–95. [CrossRef]
7. Xie, H.; Liu, Z.; Wen, G.; Chen, H.; Yang, Y. Influencing factors of landslides and rockfalls along the Jinchuan-Xiaojin highway in Sichuan. *Chin. J. Geol. Hazard Control* **2021**, *32*, 10–17. [CrossRef]
8. Wang, J.-J.; Liu, M.-N.; Jian, F.-X.; Chai, H.-J. Mechanical Behaviors of a Sandstone and Mudstone under Loading and Unloading Conditions. *Environ. Earth Sci.* **2019**, *78*, 30. [CrossRef]
9. Wang, Y.; Feng, W.K.; Hu, R.L.; Li, C.H. Fracture Evolution and Energy Characteristics During Marble Failure Under Triaxial Fatigue Cyclic and Confining Pressure Unloading (FC-CPU) Conditions. *Rock Mech. Rock Eng.* **2021**, *54*, 799–818. [CrossRef]
10. Zhang, Y.; Yang, Y.; Ma, D. Mechanical Characteristics of Coal Samples under Triaxial Unloading Pressure with Different Test Paths. *Shock Vib.* **2020**, *2020*, e8870821. [CrossRef]
11. Takeda, M.; Manaka, M. Effects of Confining Stress on the Semipermeability of Siliceous Mudstones: Implications for Identifying Geologic Membrane Behaviors of Argillaceous Formations. *Geophys. Res. Lett.* **2018**, *45*, 5427–5435. [CrossRef]
12. Zhang, Z.; Yu, X.; Deng, M. Damage Evolution of Sandy Mudstone Mechanical Properties Under Mining Unloading Conditions in Gob-Side Entry Retaining. *Geotech. Geol. Eng.* **2019**, *37*, 3535–3545. [CrossRef]
13. Guo, Y.; Wang, L.; Chang, X. Study on the Damage Characteristics of Gas-Bearing Shale under Different Unloading Stress Paths. *PLoS ONE* **2019**, *14*, e0224654. [CrossRef] [PubMed]
14. Hajiabdolmajid, V.; Kaiser, P.K.; Martin, C.D. Modelling Brittle Failure of Rock. *Int. J. Rock Mech. Min. Sci.* **2002**, *39*, 731–741. [CrossRef]
15. Xiao, F.; Jiang, D.; Wu, F.; Zou, Q.; Chen, J.; Chen, B.; Sun, Z. Effects of Prior Cyclic Loading Damage on Failure Characteristics of Sandstone under True-Triaxial Unloading Conditions. *Int. J. Rock Mech. Min. Sci.* **2020**, *132*, 104379. [CrossRef]
16. Huang, D.; Li, Y. Conversion of Strain Energy in Triaxial Unloading Tests on Marble. *Int. J. Rock Mech. Min. Sci.* **2014**, *66*, 160–168. [CrossRef]
17. Zhao, Y.; Cao, P.; Wang, W.; Wan, W.; Liu, Y. Viscoelasto-Plastic Rheological Experiment under Circular Increment Step Load and Unload and Nonlinear Creep Model of Soft Rocks. *J. Cent. South Univ. Technol.* **2009**, *16*, 488–494. [CrossRef]
18. Chen, X.; Tang, C.; Yu, J.; Zhou, J.; Cai, Y. Experimental Investigation on Deformation Characteristics and Permeability Evolution of Rock under Confining Pressure Unloading Conditions. *J. Cent. South Univ.* **2018**, *25*, 1987–2001. [CrossRef]
19. Gupta, N.; Mishra, B. Experimental Investigation of the Influence of Bedding Planes and Differential Stress on Microcrack Propagation in Shale Using X-ray CT Scan. *Geotech. Geol. Eng.* **2021**, *39*, 213–236. [CrossRef]
20. Dai, B.; Zhao, G.; Dong, L.; Yang, C. Mechanical Characteristics for Rocks under Different Paths and Unloading Rates under Confining Pressures. *Shock Vib.* **2015**, *2015*, e578748. [CrossRef]
21. Zhang, L.; Gao, S.; Wang, Z.; Ren, M. Analysis on deformation characteristics and energy dissipation of marble under different unloading rates. *Teh. Vjesn.* **2014**, *21*, 987–993.
22. Cerfontaine, B.; Collin, F. Cyclic and Fatigue Behaviour of Rock Materials: Review, Interpretation and Research Perspectives. *Rock Mech. Rock Eng.* **2018**, *51*, 391–414. [CrossRef]
23. Jiang, C.; Wang, L.; Ding, K.; Wang, S.; Ren, B.; Guo, J. Experimental Study on Mechanical and Damage Evolution Characteristics of Coal during True Triaxial Cyclic Loading and Unloading. *Materials* **2023**, *16*, 2384. [CrossRef] [PubMed]
24. Munoz, H.; Taheri, A. Local Damage and Progressive Localisation in Porous Sandstone During Cyclic Loading. *Rock Mech. Rock Eng.* **2017**, *50*, 3253–3259. [CrossRef]
25. Wang, J.; Shi, Z.; Zeng, L.; Qi, S. The Effects of Different Nanoadditives on the Physical and Mechanical Properties of Similar Silty Mudstone Materials. *Adv. Civ. Eng.* **2020**, *2020*, e8850436. [CrossRef]
26. Fu, H.; Qi, S.; Shi, Z.; Zeng, L.; Zhao, H. Effect of Nano-CaCO3 on the Physical and Mechanical Properties of Analogue to Silty Mudstone Materials. *Arab. J. Geosci.* **2021**, *14*, 2502. [CrossRef]

27. Fu, H.; Qi, S.; Shi, Z.; Zeng, L. Mixing Ratios and Cementing Mechanism of Similar Silty Mudstone Materials for Model Tests. *Adv. Civ. Eng.* **2021**, *2021*, e2426130. [CrossRef]

28. Bustamante, J.G. ASTM D2845. Standard Test Method for Laboratory Determination of Pulse Velocities and Ultra sonic Elastic Constants of Rock. Retrieved July 11, 2023. Available online: https://www.academia.edu/18991867/ASTM_D2845 (accessed on 21 June 2023).

29. Arzúa, J.; Alejano, L.R.; Walton, G. Strength and Dilation of Jointed Granite Specimens in Servo-Controlled Triaxial Tests. *Int. J. Rock Mech. Min. Sci.* **2014**, *69*, 93–104. [CrossRef]

30. Abi, E.; Yuan, H.; Cong, Y.; Wang, Z.; Jiang, M. Experimental Study on the Entropy Change Failure Precursors of Marble under Different Stress Paths. *KSCE J. Civ. Eng.* **2023**, *27*, 356–370. [CrossRef]

31. Brown, E.T.; Trollope, D.H. Strength of a Model of Jointed Rock. *J. Soil Mech. Found. Div.* **1970**, *96*, 685–704. [CrossRef]

32. Lundborg, N. Strength of Rock-like Materials. *Int. J. Rock Mech. Min. Sci. Geomech. Abstr.* **1968**, *5*, 427–454. [CrossRef]

33. Gong, F.; Luo, S.; Lin, G.; Li, X. Evaluation of Shear Strength Parameters of Rocks by Preset Angle Shear, Direct Shear and Triaxial Compression Tests. *Rock Mech. Rock Eng.* **2020**, *53*, 2505–2519. [CrossRef]

34. Taheri, A.; Squires, J.; Meng, Z.; Zhang, Z. Mechanical Properties of Brown Coal under Different Loading Conditions. *Int. J. Geomech.* **2017**, *17*, 06017020. [CrossRef]

35. Yin, G.; Jiang, C.; Wang, J.G.; Xu, J. Geomechanical and Flow Properties of Coal from Loading Axial Stress and Unloading Confining Pressure Tests. *Int. J. Rock Mech. Min. Sci.* **2015**, *76*, 155–161. [CrossRef]

36. Li, J.; Lin, F.; Liu, H.; Zhang, Z. Triaxial Experimental Study on Changes in the Mechanical Properties of Rocks Under Different Rates of Confining Pressures Unloading. *Soil Mech. Found. Eng.* **2019**, *56*, 246–252. [CrossRef]

Article

Dynamic Stability of Tensegrity Structures—Part II: The Periodic External Load

Paulina Obara and **Justyna Tomasik** *

Faculty of Civil Engineering and Architecture, Kielce University of Technology, Al. Tysiąclecia Państwa Polskiego 7, 25-314 Kielce, Poland; paula@tu.kielce.pl
* Correspondence: jtomasik@tu.kielce.pl

Abstract: The paper contains a parametric analysis of tensegrity structures subjected to periodic loads. The analysis focuses on determining the main region of dynamic instability. When load parameters fall within this region, the resulting vibration amplitudes increase, posing a risk to the durability of structures. The study considers structures built using commonly used modules. The influence of the initial prestress on the distribution of the instability regions is examined. Additional prestress can significantly reduce the extent of instability regions, potentially narrowing them by up to 99%. A nondimensional parameter is introduced to accurately assess changes in the extent of the instability region. A geometrically non-linear model is employed to evaluate the behavior of the analyzed structures.

Keywords: tensegrity structures; initial prestress forces; infinitesimal mechanism; parametric resonance; instability region

check for
updates

Citation: Obara, P.; Tomasik, J. Dynamic Stability of Tensegrity Structures—Part II: The Periodic External Load. *Materials* **2023**, *16*, 4564. https://doi.org/10.3390/ma16134564

Academic Editors: Wensheng Wang, Qinglin Guo and Jue Li

Received: 27 May 2023
Revised: 19 June 2023
Accepted: 21 June 2023
Published: 24 June 2023

1. Introduction

Stability refers to the ability of a system to maintain its configuration and resist any tendency to undergo significant changes or deformations under the influence of external loads. This is directly related to the type of equilibrium. The equilibrium of a deformed configuration can be stable or unstable. In the case of periodic dynamic loads, the analysis of the permanent equilibrium can be equated with the study of the stability of motion. Assuming a system with scleronomic constraints, its motion is an oscillation around the static equilibrium position. If a small perturbation induces a new motion with limited solutions, the original motion is static. The instability of the induced motion is the same as the instability of the original motion. It is a case of dynamic instability or parametric resonance. Unstable movements are dangerous in terms of the durability of the structure. The study and assessment of the stability of system movements caused by various types of excitation is called the dynamic stability (Lyapunov stability) [1–3]. Dynamic stability, or what Bolotin referred to as "dynamic instability" [4], involves determining the resonant frequencies of periodic excitations and mapping out stable and unstable regions using Ince–Strutt maps (parametric resonance regions). From the perspective of the physical interpretation, when the load parameters fall within the defined limits of instability, a structure undergoes vibrations characterized by an amplification of amplitude. The Ince–Strutt maps are obtained by solving the equations of motion, which, in the case of structures subjected to periodic load, are generally expressed by a nonhomogeneous Mathieu–Hill equation [5]. This is a second-order linear differential equation that describes the behavior of a parametrically forced system. It is named after the French mathematician Émile Mathieu and the British mathematician George William Hill. The Mathieu–Hill equation captures the essential characteristics of parametric resonance, where the behavior is influenced by the interplay between the natural frequency and the modulation amplitude. The equation allows for the analysis of stability, resonance regions, and the determination of stable and unstable solutions. The stable regions correspond to finite solutions of the Mathieu–Hill equations, while the unstable regions correspond to solutions that increase indefinitely in time [3]. The stability charts of the Mathieu–Hill equations published in the literature are usually calculated using various methods, i.e., harmonic balance method [4,6–11], Galerkin method [11,12], multiscale methods [13], averaging method [14,15], discrete singular convolution [16], homotopy perturbation method [17], and small parameter method (Poincaré method) [3,11]. There is a wealth of literature available on parametric vibrations that comprehensively addresses all fundamental aspects of the subject [3,4,6–8,18,19]. However, the dynamic instability of elastic structures, from analytical, numerical, and experimental perspectives, is an interesting and popular topic [12,13,20,21], and many issues still remain unsolved. A comprehensive overview of current research interests related to the dynamic instability of elastic structures is presented in various studies, including reference [22]. One unresolved issue concerns tensegrity systems that exhibit infinitesimal mechanisms, which are stabilized by a self-balancing system of internal forces known as the self-stress state or initial prestress forces. Based on our current recognition, researchers have carried out many studies on tensegrity systems (see Part 1 [23]). Several recent papers have explored the dynamics of tensegrity structures, including references [24–26]. These studies aim to determine the natural and free frequencies of vibrations in such structures. However, the analysis of dynamic stability, as understood in terms of the Bolotin approach [4,6], has not been conducted. This aspect is often mistakenly conflated with issues related to impulse loads [27]. Examining the behavior of structures under the influence of periodic dynamic loads is crucial for ensuring the durability of tensegrity structures. They are distinguished by an additional parameter, the initial prestress, which has a significant impact on the shape and extent of instability regions. Considering the aforementioned points, it seemed reasonable to explore the subject of dynamic analysis, particularly focusing on the analysis of dynamic stability in tensegrity structures.

This research addresses a gap in the existing literature by examining the response of tensegrity structures to periodic loads. The primary objective is to determine the resonant frequencies of periodic excitations and identify unstable regions as a function of the initial prestress. The study employs the harmonic balance method as a tool for this analysis. The proposed methodology offers the advantage of using a well-established approach to identify instability regions in distinct structures. This study intends to describe the behavior of tensegrity structures considered in the previous part [23], and together with Part 1, it is a complete procedure to investigate the behavior of the tensegrity structure under external loads. To validate the proposed approach, the behavior of a simple two-element truss is considered. Although this structure is not a tensegrity itself, its behavior accurately mirrors that of tensegrity structures. As a result, it enables the explicit determination of the influence of the initial prestress level on unstable regions, as presented in this study. The research focuses on tensegrity structures constructed using modified Simplex and Quartex modules. In this study, a linear connection is examined, encompassing various methods of connecting the modules and considering different support conditions. A nondimensional parameter λ is introduced for the quantitative assessment. This parameter expresses the ratio of the area of the unstable region at the i-th level of initial prestress to the area of the unstable region at the minimum level of initial prestress. It is a measure of changes in the area of the instability region, i.e., the effect of the initial prestress level. The analysis in this study employs a nonlinear approach, assuming the hypothesis of large displacements [23,28,29]. The provided description is adequate for analyzing both planar and spatial lattice structures, including tensegrity structures, under geometrically nonlinear and physically linear conditions.

The paper is organized as follows: Section 2 describes the Mathieu–Hill equations for the tensegrity structures and one of the most popular methods for solving this equation, i.e., the harmonic balance method. In Section 3, the solution procedures are demonstrated using a simple two-element structure that exhibits a self-stress state and an infinitesimal mechanism. Furthermore, the results for tensegrity structures built with the modified Simplex and Quartex modules are given in Section 4. Lastly, in Section 5, several conclusions are drawn based on the findings of the study. The calculation procedure was implemented using the Mathematica environment. This procedure enables the generation of diagrams that illustrate the locations of instability regions as a function of the dynamic force applied and the vibration frequency of the analyzed structures. Moreover, it takes into consideration the impact of the initial prestress on the instability regions.

2. Methods of Analysis

In the analysis of tensegrity structures described in the paper, a discrete formulation based on finite element methods is employed. A tensegrity system is characterized as a spatial, lightweight structure composed of compressed struts and tensioned cables. The structure is an n-element truss with m degrees of freedom, where each element is labeled by $e = 1, 2, \ldots, n$. The displacement of the structure is represented by the vector $\mathbf{q}\left(\in \mathbb{R}^{m \times 1}\right)$. Tensegrity structures are assembled in a self-balanced manner, meaning that in the absence of external loads, there is an equilibrium between the forces in the struts and cables. This equilibrium configuration of internal forces is referred to as the "self-stress state" $(\mathbf{y}_S)$ and its determination is outlined in reference [23]. Once the self-stress state is defined, the geometric stiffness matrix $\mathbf{K}_G(\mathbf{S})\left(\in \mathbb{R}^{m \times m}\right)$ can be constructed, where the longitudinal forces $\mathbf{S}$ depend on the self-stress state $\mathbf{y}_S$ and the initial prestress level $(\mathbf{S} = \mathbf{y}_S S))$. The self-stress state stabilizes one or more infinitesimal mechanisms, which are inherent features of tensegrity structures. Furthermore, tensegrity structures exhibit stiffening behavior when subjected to external loads. The applied load induces displacements according to the form of the infinitesimal mechanism, resulting in additional prestress in the structure. Tensile forces generate further tension in the cables, while compressive forces arise in the struts. The component of the stiffness matrix related to the axial forces $\mathbf{N}$ resulting from the initial load $\mathbf{P}$ is denoted as $\mathbf{K}_{GN}(\mathbf{N})\left(\in \mathbb{R}^{m \times m}\right)$.

Furthermore, like any discrete structure, it is described by the following matrices: $\mathbf{K}_L\left(\in \mathbb{R}^{m \times m}\right)$ is the linear stiffness matrix, $\mathbf{M}\left(\in \mathbb{R}^{m \times m}\right)$ represents the consequent matrix of masses, and $\mathbf{K}_{NL}(\mathbf{q})\left(\in \mathbb{R}^{m \times m}\right)$ is the non-linear displacement stiffness matrix. The explicit forms of these matrices for the tensegrity element can be found, for instance, in reference [28]. To evaluate the behavior of tensegrity structures, a geometrically non-linear model is used [3,28,30]. The non-linear theory of elasticity in the Total Lagrangian (TL, Lagrange's stationary description) approach is adopted as a basis for formulating the tensegrity lattice equations.

The main purpose is the determination of resonant frequencies of periodic extortions and instable regions (Ince–Strutt maps) as a function of initial prestress. The determination of the Ince–Strutt maps with stable and unstable regions (parametric resonance regions) relies on the determination of the resonant frequencies of the periodic load:

$$P(t) = P + P_t \Phi(t), \tag{1}$$

where P is the constant part of the load, P_t is the amplitude of the periodic force, and $\Phi(t)$ is the periodic function that describes the time variation of the force. The unstable regions, or parametric resonance regions, occur at specific frequencies. These frequencies are determined by the equation:

$$\Omega = \frac{\Omega_i}{k} \text{ or } \Omega = \frac{\Omega_i \pm \Omega_j}{2k}; \; k = 1, 2, \ldots; i \neq j, \tag{2}$$

where Ω is the resonant frequency and Ω_i and Ω_j are the free frequencies of the structure under the influence of the constant load P. In the case of periodic resonances, the unstable regions occur at frequencies $\frac{\Omega_i}{k}$, while in combined resonances, the unstable regions occur at frequencies $\frac{\Omega_i \pm \Omega_j}{2k}$. Of particular interest are the main instability regions, which correspond to the first-order periodic resonances ($k = 1$).

2.1. Mathieu–Hill Equations

The equation of motion in the case of structures subjected to periodic load (1) is expressed by a nonhomogeneous Mathieu–Hill equation:

$$\mathbf{M\ddot{q}}(t) + \mathbf{K}_L\mathbf{q}(t) + [1 + v\Phi(t)]\mathbf{K}_G\mathbf{q}(t) = 0; \ v = \frac{P_t}{P}, \tag{3}$$

where $\ddot{\mathbf{q}}\left(\in \mathbb{R}^{m\times1}\right)$ is the acceleration vector and v is the pulsatility index. Equation (3) describes the free vibrations under the influence of the initial load (1), taking into account the geometric stiffness matrix. It should be noted that in the case of tensegrity structures, the geometric stiffness matrix consists of two parts:

$$\mathbf{K}_G = \mathbf{K}_G(\mathbf{S}) + \mathbf{K}_{GN}(\mathbf{N}). \tag{4}$$

The calculation of the second part of the geometric stiffness matrix requires a nonlinear analysis:

$$[\mathbf{K}_L + \mathbf{K}_G(\mathbf{S}) + \mathbf{K}_{NL}(\mathbf{q})]\mathbf{q} = \mathbf{P}, \tag{5}$$

where $\mathbf{P}\left(\in \mathbb{R}^{m\times1}\right)$ is a time-independent external load vector. The non-linear Equation (5), which is solved through an incremental-iterative analysis of large displacement gradients (further details can be found in Part 1 [23]), subsequently results in real normal forces in the elements N^e:

$$N^e = E^e A^e \varepsilon^e \sqrt{1 + 2\varepsilon^e}; \ \varepsilon^e = \frac{1}{2}\frac{\left(l_1^e\right)^2 - \left(l^e\right)^2}{\left(l^e\right)^2}, \tag{6}$$

where E^e is the Young modulus of an element, A^e is the cross-sectional area of an element, l^e is the length of the element in the initial configuration, and l_1^e is the length of the element in the actual configuration:

$$l_1^e = \left(\Delta_{u_2}\right)^2\sqrt{\left(\Delta_{u_2}\right)^2 + \left(\Delta_{u_3}\right)^2 + \left(l^e + \Delta_{u_1}\right)^2}, \tag{7}$$

where $\Delta_{u_i} = q_i^2 - q_i^1 (i = 1, 2, 3)$ represents the displacement increments between nodes two $\left(q_i^2\right)$ and one $\left(q_i^1\right)$.

The nonlinear parametric equation of motion (3) is known as the Hill equation. Considering a harmonic periodic force with frequency θ:

$$\Phi(t) = \cos(\theta t); \ \theta = \frac{2\pi}{T}, \tag{8}$$

the Hill Equation (3) can be expressed as:

$$\mathbf{M\ddot{q}}(t) + [\mathbf{K}_L + (1 + v\cos(\theta t))\mathbf{K}_G]\mathbf{q}(t) = 0 \tag{9}$$

and is referred to as the Mathieu equation. To solve the nonlinear equations of motion (9), the commonly used approach is the harmonic balance method.

2.2. Harmonic Balance Method

The harmonic balance method is a powerful technique used to solve nonlinear equations, particularly in the context of periodic systems. It offers several advantages over other existing methods. The primary benefits of the harmonic balance method encompass its efficiency, accuracy, applicability in frequency response analysis, capability for nonlinear parameter identification, facilitation of system optimization and control, and operation within the frequency domain. To determine the first instability region, the Fourier series with a period of 2T can be employed:

$$q(t) = \sum_{k=1,3,5}^{\infty} \left(a_k \sin \frac{k\theta t}{2} + b_k \cos \frac{k\theta t}{2} \right), \tag{10}$$

where a_k and b_k are constant coefficients. By substituting Equation (10) into Equation (9), a linear combination of trigonometric functions is obtained:

$$A_1 \sin \frac{\theta t}{2} + B_1 \cos \frac{\theta t}{2} + A_3 \sin \frac{3\theta t}{2} + B_3 \cos \frac{3\theta t}{2} + A_5 \sin \frac{5\theta t}{2} + \cdots = 0. \tag{11}$$

The coefficients A_i and B_i are derived by balancing the terms with the appropriate harmonics:

$$
\begin{aligned}
A_1 &= C_1 a_1 - D a_1 + D a_3, & B_1 &= C_1 b_1 + D b_1 + D b_3, \\
A_3 &= D a_1 + C_3 a_3 + D a_5, & B_3 &= D b_1 + C_3 b_3 + D b_5, \\
A_5 &= D a_3 + C_5 a_5 + D a_7, & B_5 &= D b_3 + C_5 b_5 + D b_7,
\end{aligned} \tag{12}
$$

where:

$$D = \frac{1}{2} v \mathbf{K}_G, \ C_k = \mathbf{K}_L + \mathbf{K}_G - \frac{k^2 \theta^2}{4} \mathbf{M}; \ k = 1,3,5,\ldots. \tag{13}$$

The satisfaction of Equation (11) for each t leads to the following condition:

$$A_1 = B_1 = A_3 = B_3 = A_5 = \cdots = 0. \tag{14}$$

This condition, in turn, gives rise to linearly separated homogeneous systems with an infinite number of equations and unknowns, a_k and b_k. Non-zero solutions exist when the following conditional equation is satisfied:

$$\mathrm{Det} \left\{ \begin{array}{cccc} \mathbf{K}_L + \mathbf{K}_G \pm \frac{1}{2} v \mathbf{K}_G - \frac{1}{4} \theta^2 \mathbf{M} & \frac{1}{2} v \mathbf{K}_G & 0 & \cdots \\ \frac{1}{2} v \mathbf{K}_G & \mathbf{K}_L + \mathbf{K}_G - \frac{9}{4} \theta^2 \mathbf{M} & \frac{1}{2} v \mathbf{K}_G & \cdots \\ 0 & \frac{1}{2} v \mathbf{K}_G & \mathbf{K}_L + \mathbf{K}_G - \frac{25}{4} \theta^2 \mathbf{M} & \cdots \\ \cdots & \cdots & \cdots & \cdots \end{array} \right\} = 0. \tag{15}$$

Considering the determinant of the first degree in Equation (15), it is sufficient to obtain the boundaries of the main instability regions:

$$\det \left\{ \mathbf{K}_L + \left(1 \pm \frac{1}{2} v \right) \mathbf{K}_G - \frac{\theta^2}{4} \mathbf{M} \right\} = 0. \tag{16}$$

Solving Equation (16) allows determining the main instability regions (Ince–Strutt maps) in the parameter plane:

$$v = \frac{P_t}{P}, \ \eta = \frac{\theta}{2\pi}, \tag{17}$$

where η is the resonant frequency of external load.

In contrast to conventional cable-strut frameworks, tensegrity structures are characterized by an additional parameter, the initial prestress **S**. This parameter affects the shape and range of instability regions. In order to measure changes in the instability region and assess the effect of the initial prestress level, the following non-dimensional parameter is introduced:

$$\lambda = \frac{A_\eta(S_i)}{A_\eta(S_{min})}, \tag{18}$$

where $A_\eta(S_i)$ is an area of instability region at the i-th initial prestress level, while $A_\eta(S_{min})$ is an area at the minimum initial prestress level.

3. Behavior of Structures Characterized by a Self-Stress State and Infinitesimal Mechanism

To validate the proposed methodology, the behavior of a simple two-element $(n = 2)$ structure is examined (Figure 1a). The elements are characterized by the Young modulus E, the cross-sectional area A, the density ρ, and the length L. The structure has two degrees of freedom $(m = 2)$—$\mathbf{q} = \begin{bmatrix} q_3 & q_4 \end{bmatrix}^T$ and exhibits one self-stress state S characterized by self-stress forces $N^1 = N^2 = S$ and one infinitesimal mechanism (Figure 1b). To analyze the dynamic behavior of the structure, a periodic force (1) is applied. This force consists of the constant part P, the amplitude P_t, and the frequency θ (as defined in (8)). The force is applied to node 2 in the vertical direction:

$$P(t) = P + P_t \cos(\theta t) \tag{19}$$

Due to the symmetry of the structure and the load, the displacement q_3 is equal to zero. Therefore, the structure has only one degree of freedom $(m = 1)$, represented by $q = -q_4$. The characteristics related to mass m_ρ and stiffness $k(k = k_L + k_G(S) + k_{GN}(N) + k_{NL})$ are as follows:

$$m_\rho = \frac{2\rho AL}{3}; \; k_L = 0; \; k_G(S) = \frac{2S}{L}; \; k_{GN}(N) = \frac{2N(P)}{L}; \; k_{NL} = \frac{EA}{L^3}q^2. \tag{20}$$

In the case of structures with one degree of freedom, the first natural frequency of vibration is described by the formula:

$$f_1 = \frac{1}{2\pi}\sqrt{\frac{k}{m_\rho}}; \; k = k_G(S) + k_{GN}(N). \tag{21}$$

It should be noted that on the basis of Equation (21), the natural $f_1(0)$ and the free frequencies $f_1(P)$, which depend on the constant part P of the load, can be calculated as follows:

$$f_1(0) = \frac{1}{2\pi}\sqrt{\frac{3S}{\rho AL^2}}, \; f_1(P) = \frac{1}{2\pi}\sqrt{\frac{3[S + N(P)]}{\rho AL^2}}. \tag{22}$$

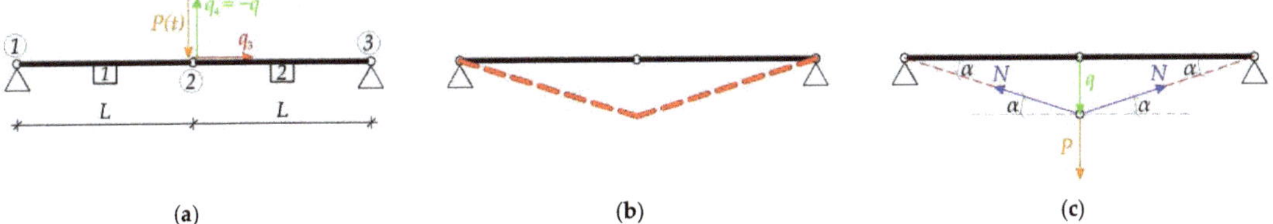

(a)   (b)   (c)

Figure 1. (a) Two-element truss, (b) infinitesimal mechanism, (c) equilibrium of node 2.

The considered structure may not be a traditional tensegrity structure, as it is stable only under tensile forces [23]. However, its behavior serves as a representative model to understand the behavior of tensegrity structures. It allows us to analyze the influence of the initial prestress level S on the unstable regions, which is the focus of this study. In this case, Equation (16) is as follows:

$$k_L + \beta k_G - \frac{\theta^2}{4}m = 0; \; \beta = \left(1 \pm \frac{1}{2}v\right). \tag{23}$$

This equation helps to determine the boundaries of the first instability regions in the parameter plane described by parameters (17):

$$\eta = \frac{1}{\pi}\sqrt{\frac{k}{m_\rho}}; \ k = \beta[k_G(S) + k_{GN}(N)]. \tag{24}$$

For the time-independent load $(P_t = 0 \rightarrow v = 0)$, the resonant frequency (24) depends solely on the constant part P of the load and is twice the free frequency $(22)_2$.

The stiffness $k_{GN}(N)$ depends on the axial force N, which is a result of the load P. It can be determined through the static equilibrium of node 2 in the actual configuration (Figure 1c). The axial force N is calculated as follows:

$$N = \frac{P}{2\sin\alpha} - S; \ \sin\alpha = \frac{q}{\sqrt{L^2 + q^2}}. \tag{25}$$

The displacement q can be obtained using either the quasi-linear (2nd order theory) or non-linear (3rd order theory) theories. In the quasi-linear theory, the displacement is given by:

$$q = \frac{PL}{2S}, \tag{26}$$

which tends to infinity $q(S = 0 \text{ kN}) = \infty$ when there is no initial prestress $(S = 0 \text{ kN})$. In non-linear theory, the displacement is determined by solving the following equation:

$$\frac{1}{L}\left(2S + \frac{EA}{L^2}q^2\right)q = P. \tag{27}$$

In the absence of initial prestress $(S = 0 \text{ kN})$, the displacement is given by the following formula:

$$q(S = 0) = \sqrt[3]{\frac{P}{EA}}L. \tag{28}$$

To illustrate the influence of external loads on the behavior of the structure, specific parameters are given. The cables are assumed to have a length of $L = 1$ m and a diameter of $\phi = 20$ mm ($A = 3.14 \cdot 10^{-4}$ m^2). The material properties of the cables were specified as follows: Young modulus $E = 210$ GPa and density $\rho = 7860$ kg/m^3. Two constant load values P are considered: $P = 1$ kN and $P = 5$ kN. The range of the pulsatility index $(17)_1$ is set between 0 and 0.75. The load capacity is assumed to be $N_{Rd} = 110.2$ kN [31], ensuring that it does not exceed 85% of its maximum value ($S_{max} = 70$ kN) with the range of initial prestressing forces S.

The influence of initial prestress level S on the stiffness k $(24)_2$ is presented in Figure 2 (the stiffness resulting from the load P is marked with a solid line —, while the stiffness generated by the combined effect of the load and prestress forces $P + S$ is marked with a dashed line - - -). Both the 2nd order (Figure 2a,c) and 3rd order (Figure 2b,d) theories are used. As can be seen, the quasi-linear theory is not appropriate to calculate the stiffness for structures characterized by a self-stress state and an infinitesimal mechanism. The graphs in Figure 2a,c coincide, which means that the stiffness does not depend on the load, which is not accurate. This approach neglects an important feature of tensegrity structures, namely, the stiffening effect caused by external loads. The load, which induces displacements according to the form of the infinitesimal mechanism, generates additional prestress in the structure. Tensile forces in the cables lead to additional tensile forces in the cables themselves and compressive forces in the struts. For such regimes, the initial response alone cannot accurately determine the behavior of the structure. Figure 2b,d demonstrate that the influence of nonlinearity is most significant at low initial prestress forces and low load values. Additionally, at lower load values, the initial prestress has a greater impact on the overall rigidity of the structure. The external load induces additional tensile forces in the cables. However, upon introducing the initial prestress, the normal forces resulting from the external load gradually decrease, reducing their influence on the displacement. Figure 3a illustrates the variation in the normal forces arising from the load, $N(P)$, and the normal forces generated by the combined effect of the load and prestress forces, $N(P+S)$.

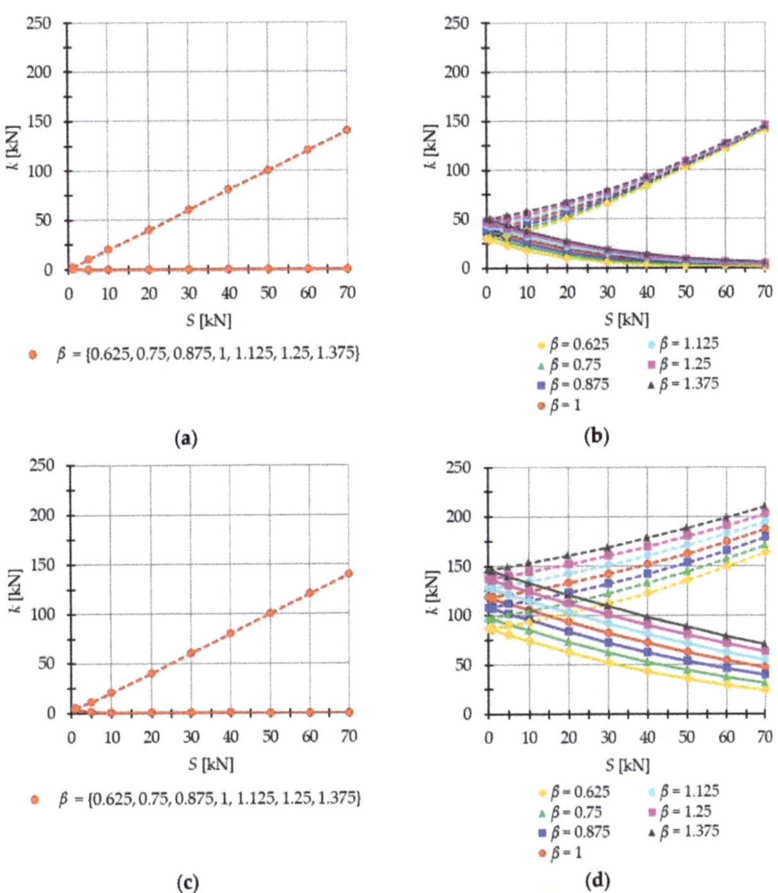

Figure 2. Influence of the initial prestress level S on the stiffness k (23) in the case of $P = 1$ kN (— 1 kN, - - - 1 kN + S): (**a**) 2nd order theory, (**b**) 3rd order theory, and in the case of $P = 5$ kN (— 5 kN, - - - 5 kN + S): (**c**) 2nd order theory and (**d**) 3rd order theory.

The 2nd order theory, which does not consider the stiffening effect of the structure under external loads that induce displacements consistent with the infinitesimal mechanism, is limited in accurately capturing the behavior of the structure. Therefore, in the subsequent analysis, the non-linear theory based on the assumption of large displacements (3rd order theory) is employed to calculate the stiffness of the structure. Initially, the dynamic behavior of the structure is investigated when it is loaded solely by the constant part P of the time-dependent load (t) $(P_t = 0 \rightarrow v = 0)$. The impact of the initial prestress S on the natural frequency $f_1(0)$ and free frequencies $f_1(P)$ is determined using Equation (22) (Figure 3b). The natural frequency varies within the range of $f_1(0) = 0$ to $f_1(0) = 44.6$ Hz. In the case of the free frequency, the external load acts as a prestress on the structure. Figure 3 shows the variation in the normal forces resulting from the load, $N(P)$, and the normal forces generated by the combined effect of the load and prestress forces, $N(P+S)$. The initial dynamic response (at $S = 0$ kN) corresponds to the values of the natural frequency at the following force levels: $N(P = 1$ kN$) = 20.20$ kN and $N(P = 5$ kN$) = 59.12$ kN (Figure 3a—$S = 0$ kN). Although the external load prestresses the structure, the introduction of initial prestress gradually reduces the normal forces caused by the external load, thereby decreasing its influence on the frequency. The analysis reveals that the impact of the initial prestress on the first vibration frequency diminishes as the load value increases.

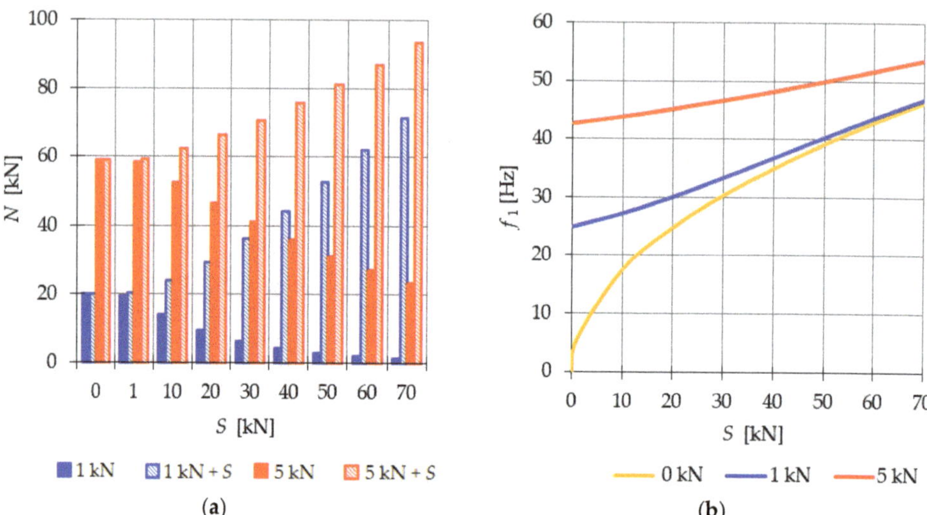

Figure 3. Impact of the initial prestress level S on: (**a**) the axial force N, (**b**) the first natural $f_1(0$ kN$)$, and free frequencies $f_1(1$ kN$)$, $f_1(5$ kN$)$.

In the subsequent analysis, the influence of the initial prestress S on the resonant frequencies η is examined when the periodic force with the constant part P, the amplitude P_t, and the frequency θ is applied. The impact of initial prestress S on the resonant frequencies η is investigated. Figure 4 illustrates the boundaries of the main instability region in the parameter plane of v and η (17) as a function of the prestress level S. The results are presented for three cases of initial prestress: $S = 0$ kN, $S = 30$ kN, and $S = 70$ kN. Since the structure exhibits only one infinitesimal mechanism, a single main instability region is determined. To quantify the changes in the area of the instability region, the nondimensional parameter λ (18) is calculated and depicted in Figure 5. This parameter provides a measure of the size of the instability region. By comparing the values of λ for different prestress levels, the influence of initial prestress on the extent of the instability region can be assessed.

The analysis reveals that as the initial prestress S increases, the resonant frequencies η also increase, and the range of the instability regions decreases. The level of initial prestress has a more significant impact on the extent of the instability regions when lower loads are applied. For example, when considering the load of $P = 1$ kN, the parameter λ, which represents the size of the instability region, takes the values: $\lambda = 0.88$ (at $S = 10$ kN), $\lambda = 0.7$ (at $S = 30$ kN), and $\lambda = 0.12$ (at $S = 70$ kN). This means that the instability region is respectively 12%, 30%, and 88% smaller compared with the case where there is no initial prestress ($S = 0$ kN). Similarly, for the load of $P = 5$ kN, the instability is reduced by 3% ($\lambda = 0.97$), 8% ($\lambda = 0.92$), and 38% ($\lambda = 0.62$), for prestress levels of $S = 10$ kN, $S = 30$ kN, and $S = 70$ kN, respectively.

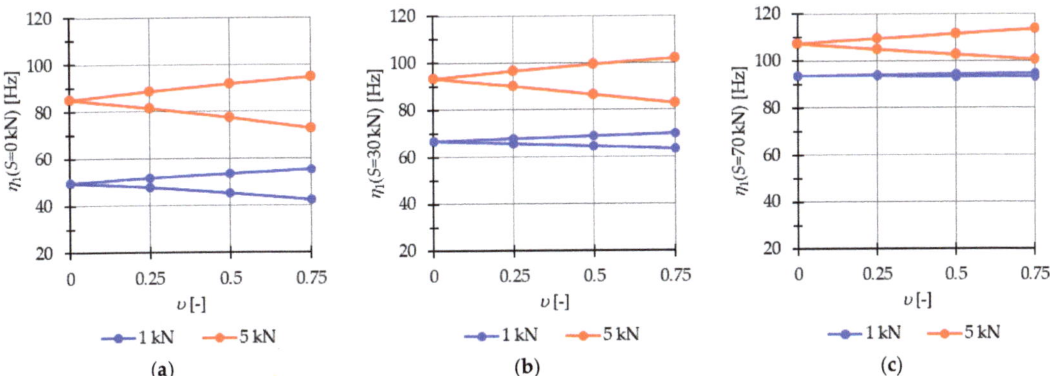

Figure 4. Limits of the main instability region of the two-element structure: (**a**) $S = 0$ kN, (**b**) $S = 30$ kN, (**c**) $S = 70$ kN.

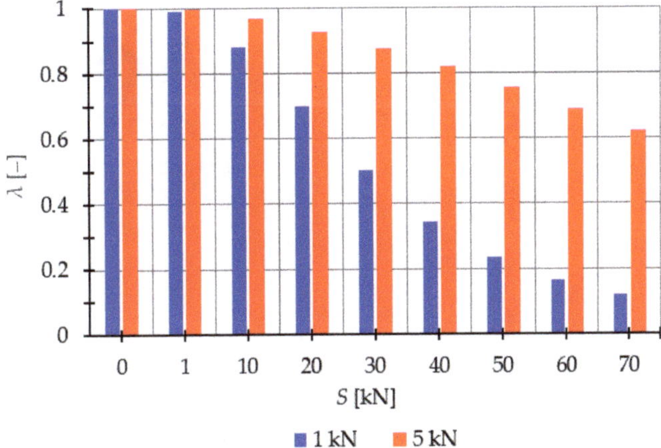

Figure 5. Influence of the initial prestress level S on the range of instability regions of the two-element structure.

Additionally, the extent of the instability regions is influenced by the level of the applied load. If the constant part of the load P increases, the area of the instability region also increases. For instance, in the case of the load $P = 5$ kN the area of the instability regions is 1.09 times larger (at $S = 10$ kN), 1.74 times larger (at $S = 30$ kN), and 5.2 times larger (at $S = 70$ kN) compared with the instability regions obtained for the load of $P = 1$ kN.

The example presented in this study served to validate the proposed approach. The harmonic balance method was employed to derive formulas that enabled the determination of safe regions in the initial prestress function. These formulas provided valuable insights into the behavior and stability of tensegrity structures under varying levels of initial prestress. By utilizing the harmonic balance method, a comprehensive understanding of the safe areas in the prestress function was achieved, contributing to the assessment and design of stable tensegrity structures.

4. Tensegrity Structures

The paper concentrates on dynamic stability analyses of tensegrity structures constructed using widely used tensegrity modules, namely the modified Simplex and Quartex modules. These modules are popular choices in the construction of tensegrity structures due to their versatility and adaptability. The analysis focuses on a linear connection with different ways of connecting modules in the structure. The structures are subjected to a force applied in the z-direction of one of the top nodes. The periodical loads $P(t)$ are taken into account, and the influence of the initial prestress level S ($\mathbf{S} = \mathbf{y}_S S$) on instability regions of the structures is considered. To perform the calculations, the eigenvectors $\mathbf{y}_S$ obtained for a single module in Part 1 [23] are utilized. These eigenvectors represent the normalized self-stress states of the structure. Figure 6 illustrates the values of these normalized self-stress states, with the cables indicated in red, green, and blue and the struts shown in black. The calculations for determining the instability regions were carried out using a geometrically non-linear model implemented in a proprietary program written in the Mathematica environment.

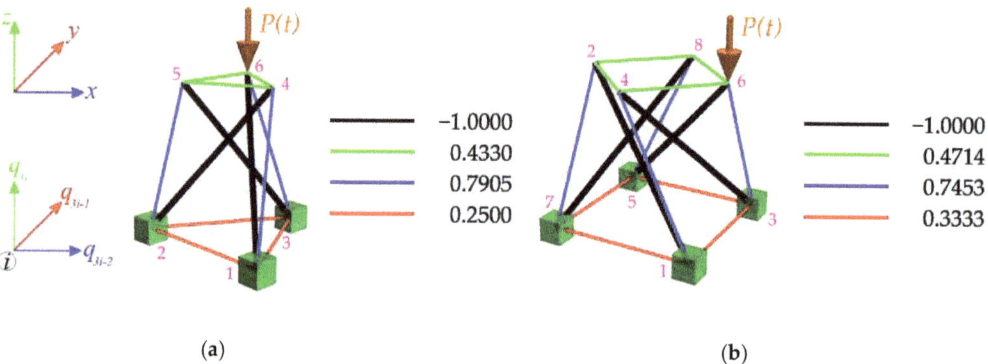

Figure 6. Values of the normalized self-stress states $\mathbf{y}_S$: (**a**) Simplex module, (**b**) Quartex module [23].

The overall size of the modules is such that they can be contained within a cube where each side measures one meter. In the analysis, the structures are constructed using steel as the material, with a density of $\rho = 7860$ kg/m^3. The type A cables are composed of steel grade S460N and have a diameter of $\phi = 20$ mm. They are designed to have a load-bearing capacity of $N_{Rd} = 110.2$ kN [31]. The Young modulus of the cables is $E = 210$ GPa. The struts are composed of hot-finished circular hollow sections with a diameter of $\phi = 76.1$ mm and a thickness of $t = 2.9$ mm. The struts are composed of steel grade S355J2, which has a Young modulus of $E = 210$ GPa. The load-bearing capacity of the struts is $N_{Rd} = 203.5$ kN [32] for the Simplex module and $N_{Rd} = 193.9$ kN for the Quartex module [31]. In the analysis, the loads that cause displacements following the mechanism of the tensegrity structure are taken into consideration. This means that the applied loads are distributed and transmitted through the interconnected members of the structure, resulting in specific displacement patterns governed by the structural configuration. By studying the response of the tensegrity structure to these mechanism-induced loads, the analysis aims to explain how the structure behaves under realistic loading conditions and how it maintains its stability and integrity. This information is crucial for assessing the structural performance and ensuring the safe and efficient operation of tensegrity structures. For single modules, the minimum prestress value is assumed as $S_{min} = 0$ kN, while for structures, it depends on the type of the module. The maximum value is set as $S_{max} = 110$ kN. The main instability regions are determined by the pulsatility index $(17)_1$, ranging from 0 to 0.75.

4.1. Tensegrity Single Modules

The first considered tensegrity structures are the single modified Simplex (Figure 6a) and Quartex (Figure 6b) modules. In these structures, all degrees of freedom (DoF) of the bottom nodes are fixed. For the Simplex module, nine DoFs are fixed, while for the Quartex module, twelve DoFs are fixed. Both modules are characterized by having one infinitesimal mechanism [1]. The modules are subjected to the periodic load $P(t)$, which is applied to the 6th node (Figure 6). Two variants of the constant part P of the periodic force are considered, i.e., $P = 10$ kN and $P = 20$ kN.

The natural $f_1(0)$ and free frequencies $f_1(P)$ of the single-mechanism modules are calculated to analyze their behavior under prestress. In the case of single-mechanism modules, there is only one frequency that depends on the level of prestressing. Figure 7 illustrates the influence of initial prestress S on the first frequency. For the Simplex module (Figure 7a), the natural frequency varies from 0 to 37.8 Hz, while for the Quartex module (Figure 7b), it ranges from 0 to 24.6 Hz. Considering the free frequencies $f_1(10$ kN$)$ and $f_1(20$ kN$)$, the initial dynamic response at $S = 0$ corresponds to the values of the natural frequency at the following force levels:

- For the Simplex module (Figure 8a): $N(P = 10$ kN$) = 25.01$ kN and $N(P = 20$ kN$) = 41.31$ kN;
- For the Quartex module (Figure 8b): $N(P = 10$ kN$) = 35.28$ kN and $N(P = 20$ kN$) = 58.82$ kN.

In both modules, the external load prestresses the structure. However, after introducing the initial prestress, the normal forces resulting from the external load progressively decrease, reducing their influence on the frequency. Figure 8 visualizes the change in the value of normal forces caused by the load $N(P)$ and the normal forces generated by the combined action of the load and prestress $N(P + S)$.

The analysis revealed that the Simplex module is more sensitive to changes in the level of prestress compared with the Quartex module. Both modules exhibit a significant influence of prestressing on the frequency and resonant behavior, especially at lower values of initial prestress forces. Similar to the two-element structure, the effect of prestressing diminishes as the load value increases.

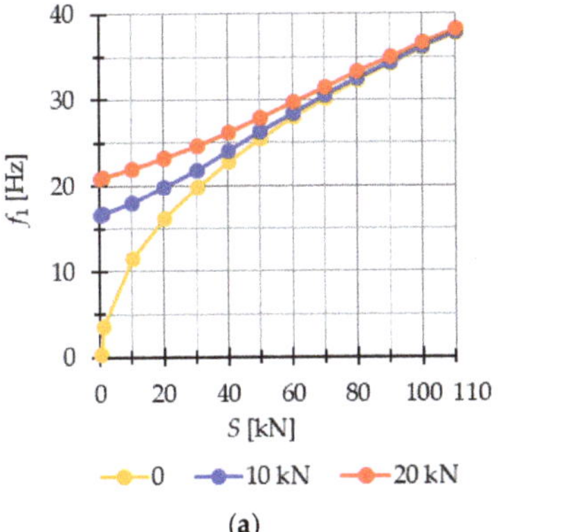

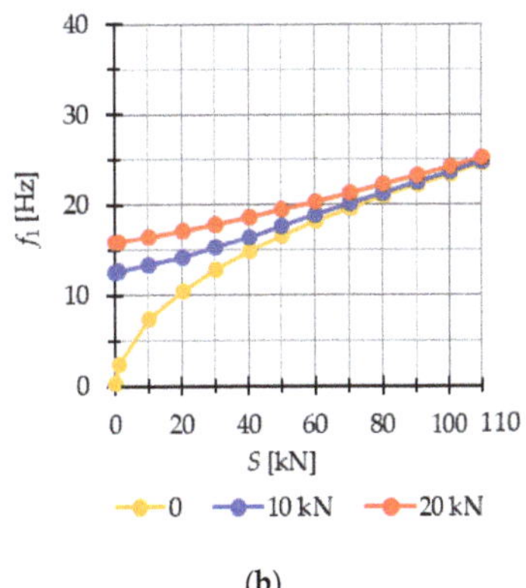

Figure 7. Impact of the initial prestress level S on the axial force N: (**a**) Simplex module, (**b**) Quartex module.

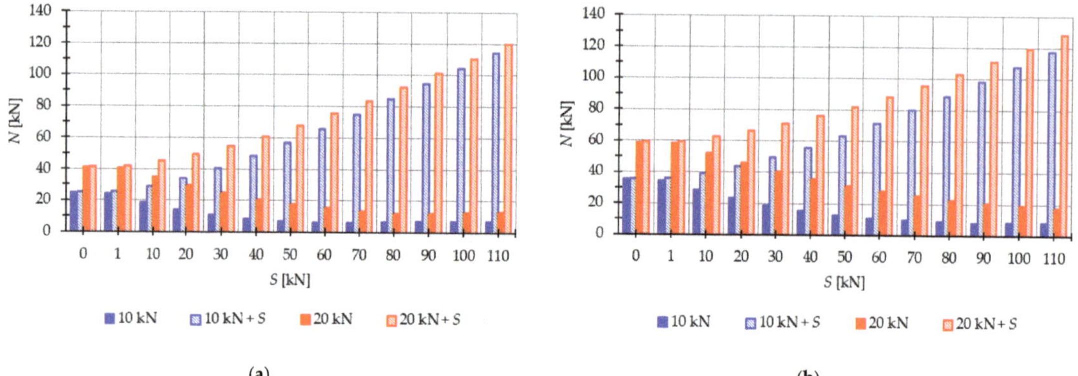

Figure 8. Impact of the initial prestress level S on the first natural $f_1(0)$ and free frequency $f_1(10\,\text{kN})$, $f_1(20\,\text{kN})$: (**a**) Simplex module, (**b**) Quartex module.

Figure 9 presents the limits of the main instability regions for different levels of initial prestress ($S = 0$ kN, $S = 30$ kN, and $S = 60$ kN), while Figure 10 shows the nondimensional parameter λ (18), which measures the changes in the area of the instability regions. Increasing the prestress level leads to higher resonant frequencies and a reduction in the range of the instability regions. The introduction of self-stress significantly improves the stability of the structure.

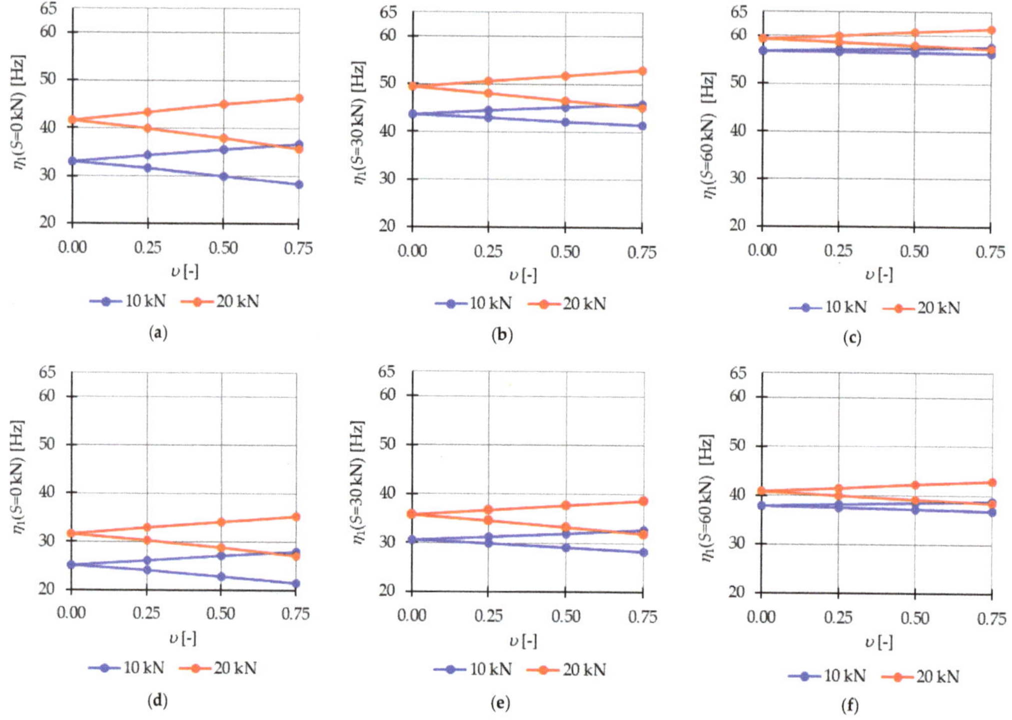

Figure 9. Limits of the main instability region of the Simplex module: (**a**) $S = 0$ kN, (**b**) $S = 30$ kN, (**c**) $S = 60$ kN and of the Quartex module: (**d**) $S = 0$ kN, (**e**) $S = 30$ kN, (**f**) $S = 60$ kN.

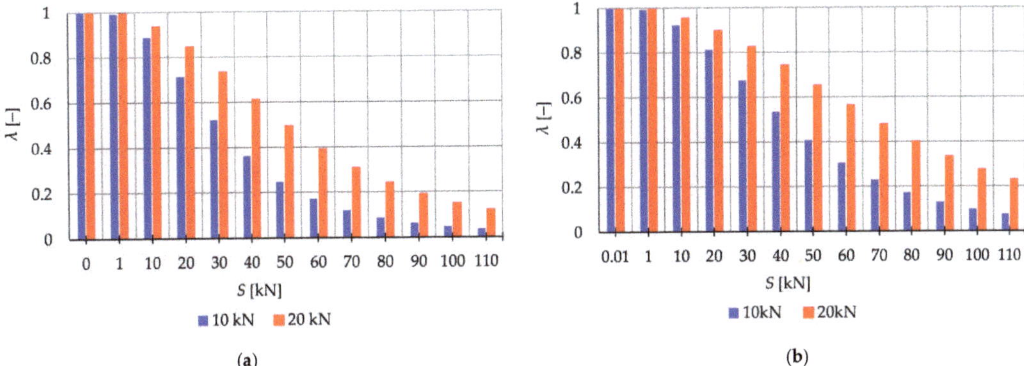

Figure 10. Influence of the initial prestress level S on the range of instability regions: (**a**) Simplex module, (**b**) Quartex module.

For the Simplex module (Figures 9a–c and 10a), at lower values of prestressing forces ($S \in\, < 0$ kN; 40 kN $>$), the parameter λ ranges from 1.0 to 0.36 for $P = 10$ kN and from 1.0 to 0.61 for $P = 20$ kN. This means that at $S = 40$ kN, the instability region is only 64% and 39% smaller, respectively, than the area of the instability region at zero initial prestress. However, further prestressing considerably narrows the instability regions, and at the maximum level, λ ranges from 0.03 to 0.12, resulting in a decrease of the instability region by 97% and 88%.

Similarly, for the Quartex module (Figures 9d–f and 10b), at lower values of prestressing forces ($S \in\, < 0$ kN; 40 kN $>$), the parameter λ ranges from 1.0 to 0.53 for $P = 10$ kN and from 1.0 to 0.74 for $P = 20$ kN. This means that at $S = 40$ kN, the instability region is only 47% and 26% smaller, respectively, than the area of the instability region at zero initial prestress. Further prestressing significantly narrows the instability regions, and at the maximum level, λ ranges from 0.07 to 0.23, resulting in a decrease of the instability region by 97% and 88%, respectively.

In summary, higher values of initial prestress reduce the risk of excitation of motion with increasing amplitudes over time, indicating improved stability of the tensegrity structure.

4.2. Tensegrity Towers

The analysis focuses next on structures built with linearly connected modules, referred to as towers. The Simplex modules can only be connected in one way (Figure 11a), while the Quartex modules can be connected in two ways (Figure 11b,c). The connection method for Quartex modules is important, as confirmed in Part 1. In the analysis of Quartex modules, two types of connections are considered: connection A and connection B. Connection A involves overlapping the struts in a plan view. This means that the struts intersect and overlap each other in the structure, creating a configuration where the struts cross each other at certain points. On the other hand, connection B is characterized by a star-like pattern formed by the struts. In this configuration, the struts radiate from a central point and do not intersect or overlap each other. In the previous part, an abnormality in the dynamic behavior of Quartex towers was observed. The number of natural frequencies, depending on the prestressing, should be equal to the number of infinitesimal mechanisms. This principle holds true for Simplex towers. However, in the case of Quartex towers constructed with four or more modules, there is an additional frequency that depends on the initial prestress. This frequency is non-zero in the absence of initial prestress ($S = 0$ kN) and its value varies with the change in prestress. Therefore, the analysis considers structures built with three (S3, Q3-A, Q3-B) and four (S4, Q4-A, Q4-B) modules. Similar to previous cases, all degrees of freedom of the bottom nodes are fixed, resulting in nine fixed degrees of freedom for Simplex structures and twelve fixed degrees of freedom for Quartex structures.

The structures are characterized by three and four infinitesimal mechanisms for structures built with three and four modules, respectively. The structures are loaded with the periodic load $P(t)$ applied to one of the upper nodes. In this analysis, the constant part of the periodic force is set as $P = 5$ kN.

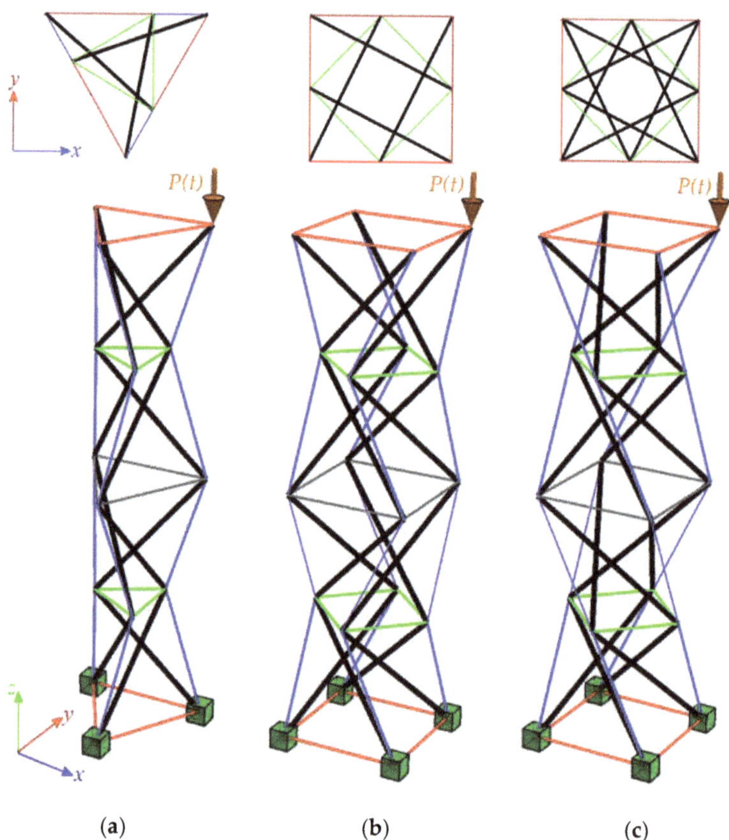

Figure 11. Linear four-module models: (**a**) S4, (**b**) Q4-connection A, (**c**) Q4-connection B. The colors are the same as in Figure 6.

4.2.1. Tensegrity Three-Module Towers

The analysis focuses next on three-module tensegrity towers, specifically the S3, Q3-A, and Q3-B structures. The minimum level of self-stress for all towers is equal to $S_{min} = 1$ kN. These towers are characterized by three infinitesimal mechanisms [1] and behave typically as structures characterized by infinitesimal mechanisms, i.e., the number of natural frequencies depending on the prestressing is equal to the number of infinitesimal mechanisms. In the absence of initial prestress ($S = 0$ kN), the natural frequencies are zero, and their values vary with the change in prestress. The impact of initial prestress S on the natural frequencies $f_i(0)$ is shown in Figure 12. The values of the three natural frequencies vary within the range of

- 0 to 10.96 Hz (f_1), 29.92 Hz (f_2) and 39.81 Hz (f_3)—for the Simplex tower (Figure 12a);
- 0 to 8.91 Hz (f_1), 15.66 Hz (f_2) and 27.19 Hz (f_3)—for the Quartex tower with connection A (Figure 12b);
- 0 to 8.06 Hz (f_1), 25.68 Hz (f_2) and 30.1 Hz (f_3)—for the Quartex tower with connection B (Figure 12c).

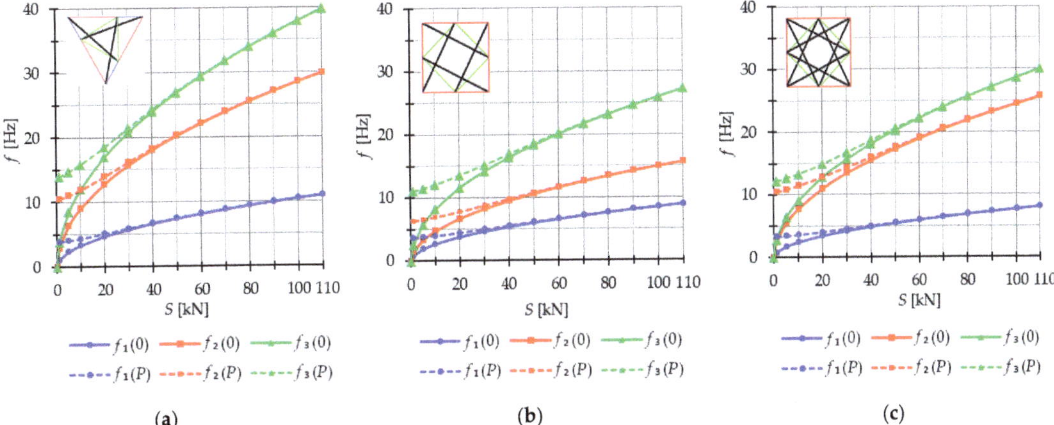

Figure 12. Impact of the initial prestress level S on the natural $f_i(0)$ and free frequencies $f_i(P)$: (**a**) S3, (**b**) Q3-A, (**c**) Q3-B.

In all cases, the frequencies increase with the rise of the initial prestress, and higher frequencies are more sensitive to the change in prestress. Generally, the natural frequencies of the structure built with the Simplex modules (S3) are higher than those of the structure built with the Quartex modules (Q3-A, Q3-B), which is consistent with the behavior observed in the single module case. The impact of initial prestress S on the free frequencies $f_i(P)$ is shown in Figure 12. In the absence of initial prestress ($S = 0$ kN), the free frequencies are equal:

- $f_1(0) = 3.83$ Hz, $f_2(0) = 10.42$ Hz and $f_3(0) = 13.89$ Hz—for Simplex tower (Figure 12a);
- $f_1(0) = 3.63$ Hz, $f_2(0) = 6.31$ Hz and $f_3(0) = 11.05$ Hz—for Quartex tower with connection A (Figure 12b);
- $f_1(0) = 3.28$ Hz, $f_2(0) = 10.41$ Hz and $f_3(0) = 12.16$ Hz—for Quartex tower with connection B (Figure 12c).

The application of the external load increases the vibration frequencies, but at approximately $S = 40$ kN, the free frequencies become equal to the natural frequencies for all towers. The Q3-B model behaves similarly to the S3 model, where the second frequency is closer to the third frequency. On the other hand, for the Q3-A model, the second frequency is closer to the first frequency.

The analysis indicates that the Simplex tower is more sensitive to changes in the level of prestress compared with the Quartex towers. The influence of prestressing on frequency is most significant at low values of initial prestress forces, which are consistent across all towers examined. Similar conclusions can be drawn for the resonant frequencies. Figures 13–15 show the limits of the main instability regions for three cases of initial prestress ($S = 1$ kN, $S = 30$ kN, and $S = 60$ kN). Since the structures have three infinitesimal mechanisms, three main regions of instability are determined, corresponding to three resonant load frequencies: $\eta 1$, $\eta 2$, and $\eta 3$. The area of instability regions is larger at higher frequencies. Regarding free frequencies, the resonant frequencies exhibit a similarity between the Q3-B and S3 models. The locations of the instability regions in these two towers show that the second and third instability regions are close to each other and partially overlap at lower levels of the self-stress state. This behavior is considered dangerous for the structure because it increases the probability of resonance occurring. In contrast, for the Q3-A model, the instability regions do not overlap. To compare the behavior of the towers, the influence of the initial prestress S on the area of instability regions $A_\eta(S)$ is shown in Figure 16. The area of the instability regions is similar for the Q3-B and S3 models, indicating a comparable response to prestress.

The nondimensional parameter λ (18) is calculated to measure the changes in the area of the instability regions, as shown in Figure 17. Similar to the previous example, as the level of initial prestress S increases, the resonant frequencies increase while the range of the instability areas decreases. This trend is consistent across all cases. At low values of initial prestressing forces, the range of the three main instability ranges changes slightly. For example, at $S = 10$ kN, the parameter λ is equal $\lambda = 0.79$ for the Simplex tower and $\lambda = 0.86$ for the Quartex towers. This means that the instability regions are, respectively, 21% and 16% smaller than the regions at the minimum level of initial prestress ($S_{min} = 1$ kN). However, further prestress significantly narrows the instability regions. At $S = 60$ kN, $\lambda = 0.05$ and $\lambda = 0.1$ (the regions decrease by 98% and 90%), and at the maximum level—$\lambda = 0.01$ (the region decreases by 99%). This indicates that the boundaries of instability regions practically coincide at high initial prestress levels.

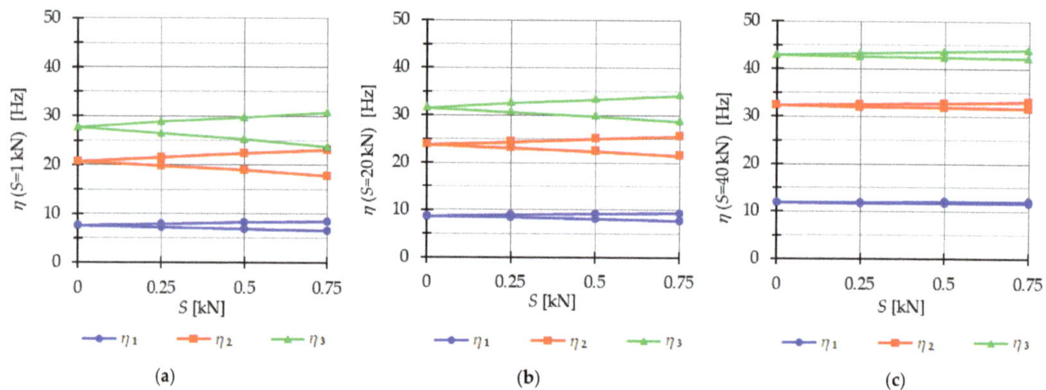

Figure 13. Limits of the main instability region of the Simplex tower (S3): (**a**) $S = 1$ kN, (**b**) $S = 20$ kN, (**c**) $S = 40$ kN.

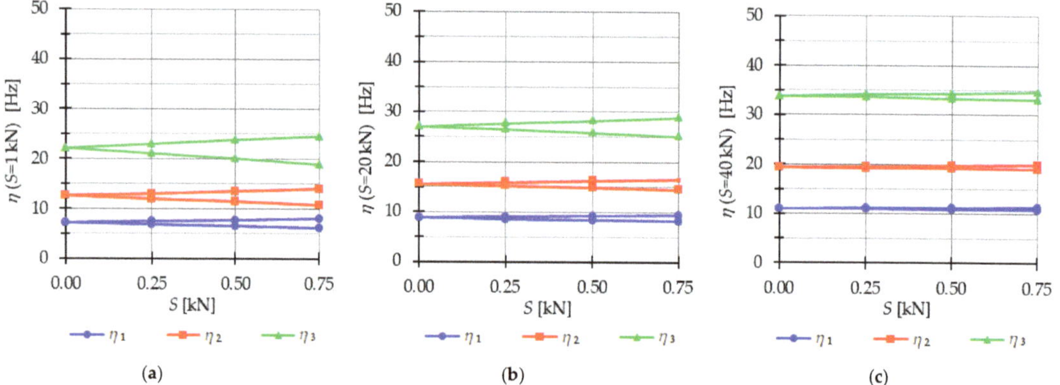

Figure 14. Limits of the main instability region of the Quartex tower with connection A (Q3-A): (**a**) $S = 1$ kN, (**b**) $S = 20$ kN, (**c**) $S = 40$ kN.

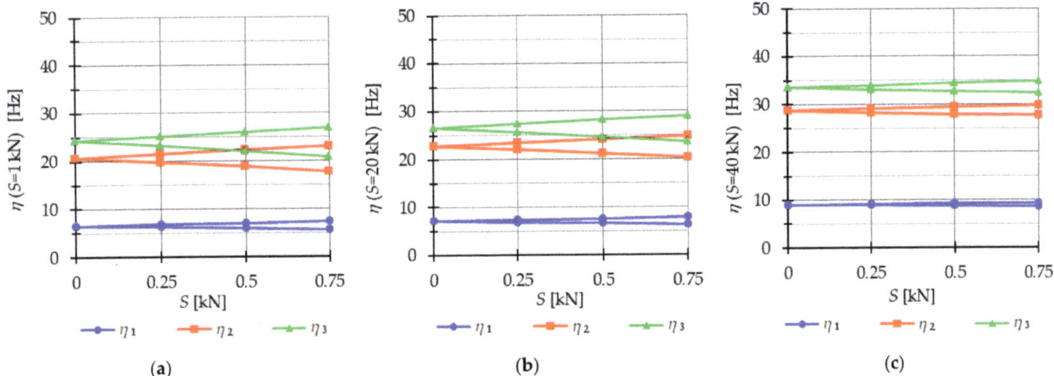

Figure 15. Limits of the main instability region of the Quartex tower with connection B (Q3-B): (**a**) $S = 1$ kN, (**b**) $S = 20$ kN, (**c**) $S = 40$ kN.

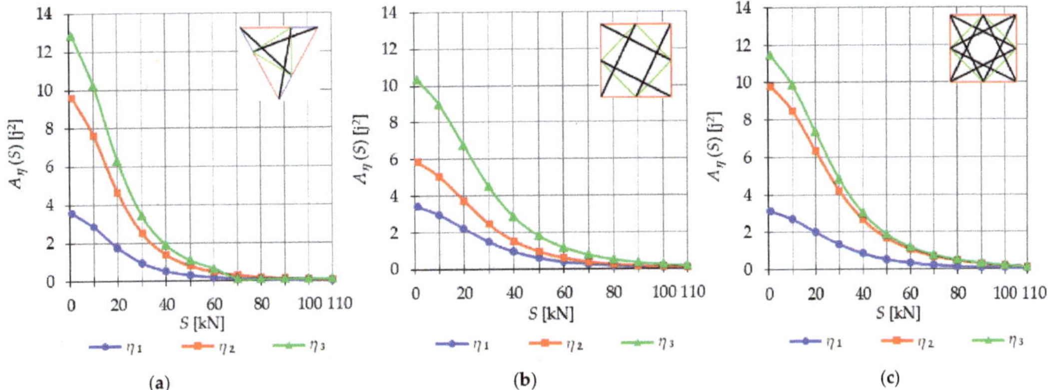

Figure 16. Influence of the initial prestress level S on the area of unstable regions $A_\eta(S)$ of three-module towers: (**a**) S3, (**b**) Q3-A, (**c**) Q3-B.

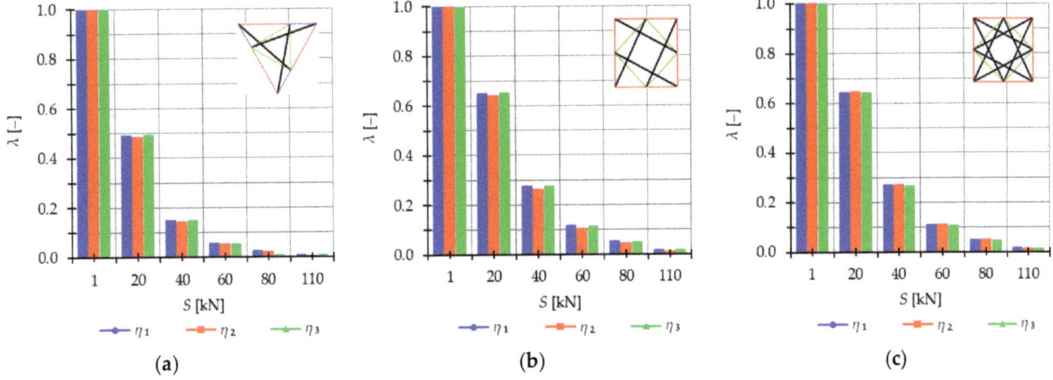

Figure 17. Influence of the initial prestress level S on the range of instability regions of three-module towers: (**a**) S3, (**b**) Q3-A, (**c**) Q3-B.

The results obtained from the dynamic instability analysis confirm the conclusions from the static and dynamic analyses in Part 1 [1]. The impact of the load on the behavior of tensegrity structures is most significant at low initial prestress forces. The area of the main instability regions is closely related to the initial prestress, and the range of these regions decreases with increasing prestress level. This implies that the risk of unstable vibrations is higher at low levels of initial prestress. For all towers, the number of main instability regions depends on the number of infinitesimal mechanisms, which corresponds to the number of natural frequencies related to the initial prestress level.

4.2.2. Tensegrity Four-Module Towers

The analysis focuses next on four-module tensegrity towers, namely S4, Q4-A, and Q4-B. In Figure 18, the natural frequencies $f_i(0)$ and the free frequencies $f_i(P = 5\text{ kN})$ are shown, while in Figures 19–21, the instability regions are visualized. It is important to note that these towers have a minimum level of self-stress that differs for each structure: $S_{min} = 1$ kN for S4, $S_{min} = 13$ kN for Q4-A, and $S_{min} = 15$ kN for Q4-B (for comparison purposes, $S_{min} = 15$ kN is used for both Quartex towers). All considered towers are characterized by four infinitesimal mechanisms [1]. The Simplex tower (S4) behaves typically as a tensegrity structure, where the number of frequencies depending on the prestress is equal to the number of infinitesimal mechanisms (Figure 18a). However, in the case of Quartex towers (Q4-A, Q4-B), there is an additional frequency that is dependent on the initial prestress. This natural frequency is not zero in the absence of initial prestress ($S = 0$ kN), and its value varies with changes in prestress (Figure 18b,c).

Comparing the frequencies of three-module and four-module structures, the number of frequencies related to the initial prestress level changes, but the highest frequency associated with the initial prestress level remains the same for the corresponding structure (compare Figures 12 and 18). For example, comparing the third frequency of the three-module Simplex tower and the fourth frequency of the four-module Simplex tower, they are practically equal. The range of these frequencies changes slightly, such as from 13.9 Hz to 39.8 Hz for the three-module structure and from 13.72 Hz to 39.58 Hz for the four-module structure.

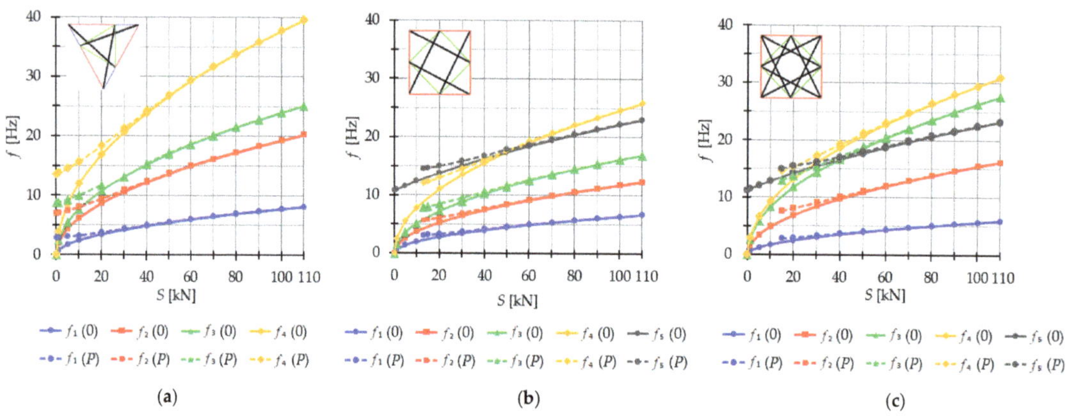

Figure 18. Impact of the initial prestress level S on the natural $f_i(0)$ and free frequencies $f_i(P)$: (a) S4, (b) Q4-A, (c) Q4-B.

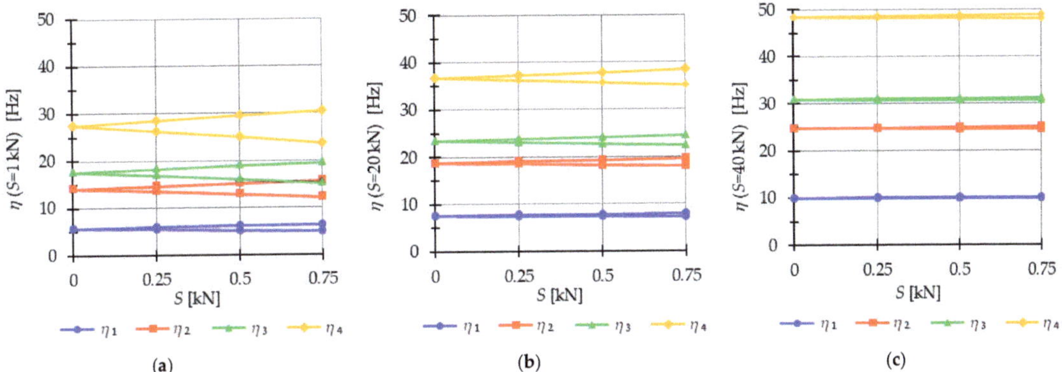

Figure 19. Limits of the main instability region of the Simplex tower (S4): (**a**) $S = 1$ kN, (**b**) $S = 20$ kN, (**c**) $S = 40$ kN.

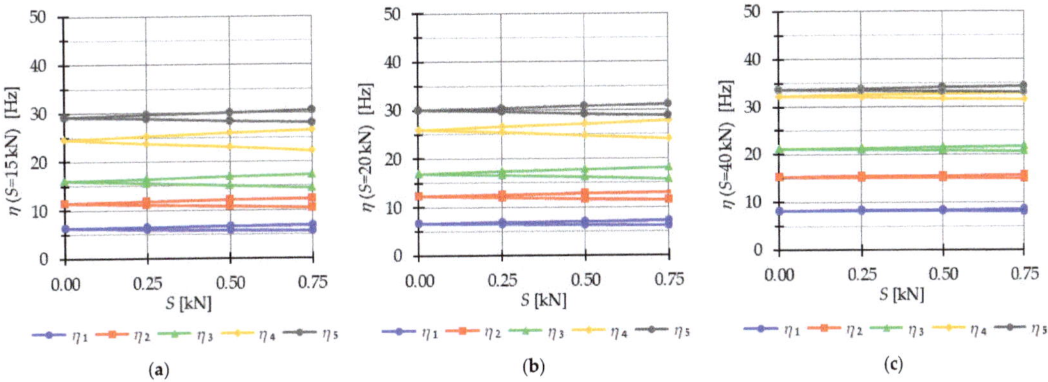

Figure 20. Limits of the main instability region of the Quartex tower with connection A (Q4-A): (**a**) $S = 15$ kN, (**b**) $S = 20$ kN, (**c**) $S = 40$ kN.

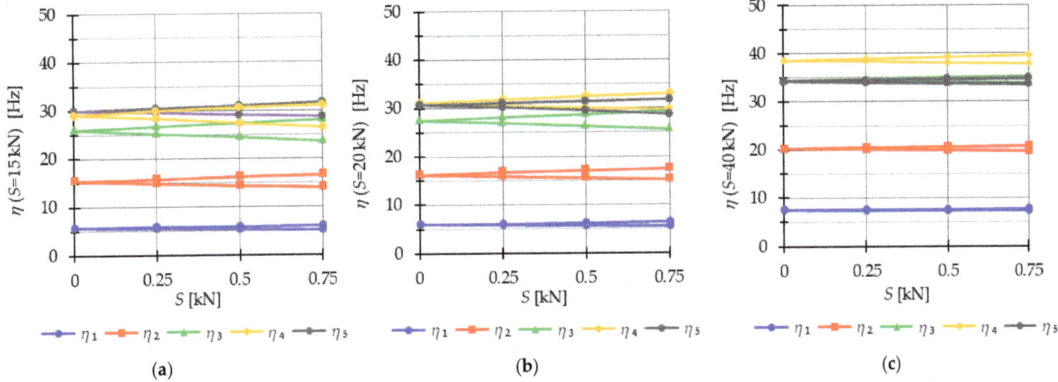

Figure 21. Limits of the main instability region of the Quartex tower with connection B (Q4-B): (**a**) $S = 15$ kN, (**b**) $S = 20$ kN, (**c**) $S = 40$ kN.

The Quartex towers introduce an additional fifth frequency that occurs due to the presence of the fourth module. This frequency is not zero in the absence of initial prestress ($S = 0\,\text{kN}$) and exhibits a more linear variation as the initial prestress level increases. Consequently, for the maximum assumed prestress level, the fifth frequency is lower than the fourth frequency for Q4-A and lower than the third frequency for Q4-B. This behavior is also reflected in the areas of the instability regions, where the fifth region of instability overlaps with other regions (Figures 20 and 21), unlike the "normal" frequencies.

The considerations made for the four-module structures align with the previous deliberations on the three-module towers. The instability regions also overlap for the S4 (Figure 19) and Q4-B (Figure 20) models. For the model Q4-B, the instability regions coincide for the entire prestress range, while for the S4 model, they coincide for lower levels of initial prestress.

5. Conclusions

The work is devoted to the parametric analysis of tensegrity structures subjected to periodic loads. The regions of dynamic instability are initially determined for a simple two-element truss. Subsequently, attention is shifted to the structures constructed using tensegrity modules, i.e., the modified Simplex and Quartex modules. Two methods of connection, labeled A and B, are considered. In connection A, the struts overlap in a plan view, while in connection B, they form a star configuration. The influence of the level of prestress is taken into account in the presented research. The work completes Part 1 [23] of the research, and both papers constitute the complete procedure of the investigation of the dynamic analysis of tensegrity structures.

This research aims to fill a gap in the existing literature by investigating how tensegrity structures respond to periodic loads. The main goal is to identify the resonant frequencies of periodic excitations and determine unstable regions based on the initial prestress level. In terms of physical interpretation, unstable regions indicate that the structure experiences vibrations with amplified amplitudes when subjected to specific load parameters. In this study, the harmonic balance method is employed as a powerful tool for conducting the dynamic stability analysis of tensegrity structures. The harmonic balance method is a well-established approach widely used in structural dynamics to analyze nonlinear systems subjected to periodic excitations. By utilizing this method, the study enables the identification of instability regions within the tensegrity structures being studied.

To validate the proposed approach, the two-element truss is examined as a case study. The harmonic balance method is applied to derive formulas that allow for the determination of safe regions in the initial prestress function. These formulas provide valuable insights into the behavior and stability of tensegrity structures under varying levels of initial prestress. By employing the harmonic balance method, a comprehensive understanding of the safe areas in the prestress function is achieved, which contributes to the assessment and design of stable tensegrity structures. The two-element truss is not a tensegrity structure, but it illustrates the behavior of tensegrity systems and allows deriving the explicit formulas of frequencies in the function of the initial prestress. The considerations serve as an explanation of the use of non-linear analysis in the dynamic analysis of tensegrity systems that stiffen under the application of an external load, and their initial response cannot be used to assess the behavior of the structure.

Regarding the behavior of the single tensegrity modules, the Simplex module is more prone to changes in the level of initial prestress. The introduced initial prestress improves the stability of the structure, and the impact of prestress is more significant at lower external loads.

In the case of towers, four-module structures constructed with Quartex modules exhibit abnormal behavior. For three-module structures, the number of instability regions is the same as the number of frequencies that depend on the level of initial prestress, as expected for tensegrity systems. However, for four-module structures, an additional frequency is identified. In contrast to the "normal" frequencies, this additional frequency is not zero in the absence of self-stress and changes more linearly. Towers built with the Simplex module, the Quartex module, and connection B perform more dangerous behaviors due to the overlapping of the instability regions, increasing the risk of resonance. Towers built with the Quartex module and connection A perform better with dynamic parameters as the instability regions do not overlap. The increase in initial prestress also improves the dynamic behavior of the structures considered.

The dynamic stability analysis reveals that the main instability regions are closely related to the initial prestress. The number of the main instability regions depends on the number of natural frequencies associated with this state, which is not always equal to the number of mechanisms. The range of the instability regions is closely related to the level of prestress, with the largest region observed at the minimum prestress level. As the prestress increases, the extent of the instability regions decreases. This indicates that the risk of exciting unstable vibrations is higher at lower levels of initial prestress.

The results obtained from the dynamic stability analysis confirm the conclusions obtained from the static and dynamic analyses, highlighting that the impact of load on the behavior of tensegrity structures is most significant at low values of initial prestress forces.

Author Contributions: Conceptualization, P.O.; Formal analysis, J.T.; Methodology, P.O.; Software, J.T.; Supervision, P.O.; Visualization, J.T.; Writing—original draft, J.T.; Writing—review and editing, P.O. All authors have read and agreed to the published version of the manuscript.

Funding: Project financed under the program of the Minister of Science and Higher Education under the name "Regional Initiative of Excellence" in the years 2019–2022, project number 025/RID/2018/19, amount of financing 12,000,000 PLN.

Institutional Review Board Statement: Not applicable.

Informed Consent Statement: Not applicable.

Data Availability Statement: The data presented in this study are available within the text of the paper.

Conflicts of Interest: The authors declare no conflict of interest.

References

1. Liapunov, A.M. The General Problem of the Stability of Motion. Ph.D. Thesis, University of Kharkov, Kharkiv, Ukraine, 1892.
2. La Salle, J.; Lefschetz, S.; Alverson, R.C. Stability by Liapunov's Direct Method with Applications. *Phys. Today* **1962**, *15*, 59. [CrossRef]
3. Cunningham, W.J. *Introduction to Nonlinear Analysis*; McGraw-Hill: New York, NY, USA, 1958.
4. Bolotin, V.V. Dynamic Instabilities in Mechanics of Structures. *Appl. Mech. Rev.* **1999**, *52*, R1–R9. [CrossRef]
5. Mathieu, É. Memoir on the Vibratory Movement of an Elliptical Membrane. *J. Math. Pures Appl.* **1868**, *13*, 137–203.
6. Bolotin, V.V.; Weingarten, V.; Greszczuk, L.B.; Trigoroff, K.N.; Gallegos, K.D.; Cranch, E.T. Dynamic Stability of Elastic Systems. *J. Appl. Mech.* **1965**, *32*, 718. [CrossRef]
7. Volmir, A.C. *Stability of Elastic Systems*; Science: Moscow, Russia, 1963.
8. Gomuliński, A.; Witkowski, M. *Mechanics of Buildings, the Advanced Course*; Oficyna Wyd. P.W: Warsaw, Poland, 1993.
9. Briseghella, L.; Majorana, C.E.; Pellegrino, C. Dynamic Stability of Elastic Structures: A Finite Element Approach. *Comput. Struct.* **1998**, *69*, 11–25. [CrossRef]
10. Jani, N.; Chakraborty, G. Parametric Resonance in Cantilever Beam with Feedback-Induced Base Excitation. *J. Vib. Eng. Technol.* **2021**, *9*, 291–301. [CrossRef]
11. Obara, P.; Gilewski, W. Dynamic Stability of Moderately Thick Beams and Frames with the Use of Harmonic Balance and Perturbation Methods. *Bull. Pol. Acad. Sci. Technol. Sci.* **2016**, *64*, 739–750. [CrossRef]
12. Zhang, Q.-C.; Cui, S.-Y.; Fu, Z.; Han, J.-X. Modal Interaction-Induced Parametric Resonance of Stayed Cable: A Combined Theoretical and Experimental Investigation. *Math. Probl. Eng.* **2021**, *2021*, 1–18. [CrossRef]
13. Pomaro, B.; Majorana, C.E. Parametric Resonance of Fractional Multiple-Degree-of-Freedom Damped Beam Systems. *Acta Mech.* **2021**, *232*, 4897–4918. [CrossRef]
14. Tan, T.H.; Lee, H.P.; Leng, G.S.B. Parametric Instability of Spinning Pretwisted Beams Subjected to Sinusoidal Compressive Axial Loads. *Comput. Struct.* **1998**, *66*, 745–764. [CrossRef]
15. Yang, X.-D.; Tang, Y.-Q.; Chen, L.-Q.; Lim, C.W. Dynamic Stability of Axially Accelerating Timoshenko Beam: Averaging Method. *Eur. J. Mech. A Solids* **2010**, *29*, 81–90. [CrossRef]

16. Shen, Y.-J.; Wei, P.; Yang, S.-P. Primary Resonance of Fractional-Order van Der Pol Oscillator. *Nonlinear Dyn.* **2014**, *77*, 1629–1642. [CrossRef]
17. Song, Z.; Chen, Z.; Li, W.; Chai, Y. Dynamic Stability Analysis of Beams with Shear Deformation and Rotary Inertia Subjected to Periodic Axial Forces by Using Discrete Singular Convolution Method. *J. Eng. Mech.* **2016**, *142*, 04015099. [CrossRef]
18. Ghomeshi Bozorg, M.; Keshmiri, M. Stability Analysis of Nonlinear Time Varying System of Beam-Moving Mass Considering Friction Interaction. *Indian J. Sci. Technol.* **2013**, *6*, 1–10. [CrossRef]
19. Życzkowski, M. *Strength of Structural Elements. Part 3: Stability of Bars and Bar Structures*; Polish Scientific Publishers: Warsaw, Poland, 1991.
20. Zahedi, M.; Khatami, I.; Zahedi, A. Parametric Resonance Domain of a Parametric Excited Screen Machine. *Sci. Iran.* **2020**, *28*, 1236–1244. [CrossRef]
21. Liu, W.; Li, Y. Stability Analysis for Parametric Resonances of Frame Structures Using Dynamic Axis-Force Transfer Coefficient. *Structures* **2021**, *34*, 3611–3621. [CrossRef]
22. Mascolo, I. Recent Developments in the Dynamic Stability of Elastic Structures. *Front. Appl. Math. Stat.* **2019**, *5*, 51. [CrossRef]
23. Obara, P.; Tomasik, J. Dynamic Stability of Tensegrity Structures—Part I: The Time-Independent External Load. *Materials* **2023**, *16*, 580. [CrossRef]
24. Ma, S.; Chen, M.; Skelton, R.E. Dynamics and Control of Clustered Tensegrity Systems. *Eng. Struct.* **2022**, *264*, 114391. [CrossRef]
25. Shuo, M.; Chen, M.; Skelton, R. TsgFEM: Tensegrity Finite Element Method. *J. Open Source Softw.* **2022**, *7*, 3390. [CrossRef]
26. Shuo, M.; Chen, M.; Yongcan, D.; Yuan, X.; Skelton, R. The Pulley-Driven Clustered Tensegrity Structure Statics and Dynamics. 2023. Available online: http://dx.doi.org/10.2139/ssrn.4384126 (accessed on 26 May 2023).
27. Shekastehband, B.; Ayoubi, M. Nonlinear Dynamic Instability Behavior of Tensegrity Grids Subjected to Impulsive Loads. *Thin-Walled Struct.* **2019**, *136*, 1–15. [CrossRef]
28. Obara, P.; Tomasik, J. Parametric Analysis of Tensegrity Plate-Like Structures: Part 2—Quantitative Analysis. *Appl. Sci.* **2021**, *11*, 602. [CrossRef]
29. Murakami, H. Static and Dynamic Analyses of Tensegrity Structures. Part 1. Nonlinear Equations of Motion. *Int. J. Solids Struct.* **2001**, *20*, 3599–3613. [CrossRef]
30. Atai, A.A.; Steigmann, D.J. On the Nonlinear Mechanics of Discrete Networks. *Arch. Appl. Mech.* **1997**, *67*, 303–319. [CrossRef]
31. *EN 1993-1-11: 2006*; Eurocode 3: Design of Steel Structures—Part 1-11: Design of Structures with Tension Components. European Union: Strasbourg, France, 2006.
32. *EN 1993-1-1: 2005*; Eurocode 3: Design of Steel Structures—Part 1-1: General Rules and Rules for Buildings. European Union: Strasbourg, France, 2005.

*materials*

Article

Hydration Heat Control of Mass Concrete by Pipe Cooling Method and On-Site Monitoring-Based Influence Analysis of Temperature for a Steel Box Arch Bridge Construction

Tan Zhang [1], Hua Wang [2,*], Yuejing Luo [3,4], Ye Yuan [2] and Wensheng Wang [5,*]

[1]   Yulin Highway Development Center of Guangxi Zhuang Autonomous Region, Yulin 537000, China
[2]   Bridge Engineering Research Institute, Guangxi Transportation Science and Technology Group Co., Ltd., Nanning 530007, China
[3]   Hualan Design and Consulting Group Company Ltd., Nanning 530011, China
[4]   Guangxi Polytechnic of Construction, Nanning 530007, China
[5]   College of Transportation, Jilin University, Changchun 130025, China
*   Correspondence: wanghua15@mails.jlu.edu.cn (H.W.); wangws@jlu.edu.cn (W.W.)

Abstract: The steel box arch bridge in this study will be subjected to various temperature effects from the construction to the operation stage, including the cement hydration heat effect and the sunshine temperature effect caused by an ambient temperature change. Therefore, it is very important to control the temperature effect of steel box arch bridges. In this study, the newly built Dafeng River Bridge is selected as the steel box arch bridge. This study aims to investigate the temperature effect including hydration heat and the sunshine temperature effect of the construction process of a rigid frame-tied steel box arch bridge. The manuscript presents that the heat dissipation performance of concrete decreases with the increase in the thickness of a mass concrete structure. The average maximum temperature values of layer No. 3 are about 1.3, 1.2, and 1.1 times the average maximum temperature value of layer No. 1 for the mass concrete of the cushion cap, main pier and arch abutment, respectively. The higher the molding temperature is, the higher the maximum temperature by the hydration heat effect is. With each 5 °C increase in the molding temperature, the maximum temperature at the core area increases by about 4~5 °C for the mass concrete. The pipe cooling method is conducive to the hydration heat control effect of mass concrete. Based on the monitored temperature change and displacement change, the influences of daily temperature change on the steel lattice beam and arch rib are analyzed. A temperature rise will cause the structure to have a certain camber in the longitudinal direction, and the longitudinal or transverse displacement caused by the sunshine temperature change is no less than the vertical displacement. Due to the symmetrical construction on both sides of the river, the arch rib deformation on both sides presents symmetrical synchronous changes. Based on 84 h of continuous temperature monitoring on-site, the changing trends of the arch back temperature and ambient temperature are consistent and their difference is small during 1:00~4:00 in the morning, which is determined as the appropriate closure time for the newly built Dafeng River Bridge.

Keywords: steel box arch bridge; hydration heat effect; water pipe cooling; sunshine temperature effect; closure temperature

check for updates

Citation: Zhang, T.; Wang, H.; Luo, Y.; Yuan, Y.; Wang, W. Hydration Heat Control of Mass Concrete by Pipe Cooling Method and On-Site Monitoring-Based Influence Analysis of Temperature for a Steel Box Arch Bridge Construction. *Materials* **2023**, *16*, 2925. https://doi.org/10.3390/ma16072925

Academic Editor: Theodore E. Matikas

Received: 8 March 2023
Revised: 30 March 2023
Accepted: 3 April 2023
Published: 6 April 2023

1. Introduction

In recent decades, with the continuous development of highway and bridge construction in China, the number of bridges built has also increased with different spans and types [1]. The steel box arch bridge has been widely used at home and abroad due to its advantages such as light dead weight, large section torsional rigidity, large span capacity, small building height, economic beauty, etc. [2–4]. As an economically and practically important material, concrete is also widely used in infrastructure construction projects such as bridges [5,6]. In order to meet the requirements of bridge crossing capacity, mass concrete cushion caps, abutments, piers and other components are increasingly used in bridge construction. The hydration heat of cement is produced and results in surface tensile stress induced by the temperature difference, leading to temperature cracks if the allowable tensile strength is exceeded [7,8]. In addition, ambient temperature has a significant impact on the stress and alignment of a bridge structure, especially for steel box arch bridges, and temperature has a significant impact on the elevation and axis deviation of the lattice beam and arch rib [9]. Therefore, the problem of the bridge temperature effect is relatively complex, it is very important to control the temperature effect of steel box arch bridges with mass concrete, and it is necessary to conduct in-depth research.

Mass concrete structures first appeared in water conservancy and hydropower engineering, and there are a series of specifications from the temperature field theory to construction methods and temperature and crack control. With the increasing application of mass concrete in engineering, the relevant research is also more comprehensive and in-depth. Lawrence et al. [10] used numerical simulation and experimental verification methods to analyze four groups of mass concrete pouring blocks. The results showed that controlling the maximum temperature difference in mass concrete can ensure that the tensile stress of the concrete is not greater than the tensile strength of the concrete, thus achieving the purpose of controlling cracks. Freitas et al. [11] provided a hybrid finite element technical method that can be used to simulate the basic laws of temperature field changes in mass concrete structures, which can not only improve calculation efficiency, but also improve calculation accuracy. Honorio et al. [12] explored the thermodynamic response of mass concrete structures at the initial stage by numerical simulation analysis. They found that vertical displacement is not only limited by the foundation's deformation, but that ambient temperature directly affects tensile stress, and that ambient temperature fluctuations in a day also have an impact on the structure. Pepe et al. [13] developed a numerical calculation program to simulate the hydrothermal reaction of concrete, fully taking into account the influence of the size of components, concrete strength grade and curing requirements. Cha et al. [14] proposed a thermal stress prediction method using experimental and analytical data, including deformation, elastic modulus and thermal stress experiments, and applied the prediction method to test the thermal stress of mass concrete pouring blocks. Klemczak et al. [15] studied the temperature stress distribution of mass concrete foundations with reinforcement based on theoretical and numerical simulation analysis, and proposed relevant reinforcement measures to control early temperature cracks. Singh et al. [16] used a 3D finite element to simulate the hydration heat of concrete under a forced cooling system with diffusion and dispersion characteristics, and found that there was high-temperature stress in the concrete around the cooling pipe. Tasri et al. [17,18] studied the influence of cooling pipe spacing and water temperature on the temperature stress field of mass concrete, and compared the temperature effect of three different materials of cooling pipe on mass concrete. The results showed that the cooling effect of steel pipe was the best, but the tensile stress was the largest. Han et al. [19] analyzed the original model with a thermal insulation formwork and postcooling system by the layered pouring method based on the temperature stress field of the mass concrete foundation.

A temperature change will make the structure show the deformation trend of "thermal expansion and cold contraction". However, for statically indeterminate structures, because the free deformation is subject to boundary conditions, there will be great temperature stress in the structure, which may bring risks to the construction and use of the bridge structure [20–22]. In order to study the influence of the temperature effect on the various responses of suspension bridges, Xia et al. [23] established the finite element temperature field of the whole bridge and concluded that the most obvious structural response affected by solar radiation is the vertical deflection of stiffened beams through comparison. Xu et al. [24] established a finite element simulation model based on Wuhan Third Yangtze River Bridge to separate the thermal effects in the response of cable-stayed bridges, which provides a reference for further understanding the temperature field of long-span cable-stayed bridges.

Wang et al. [25] took a steel truss bridge with two main trusses as the research object, taking into account the influence of solar radiation and the shielding effect between the members, and found that there were obvious positive and negative temperature differences between the two trusses, resulting in the uneven temperature field on the bridge structure. Based on the long-term measured data of temperature, solar radiation, wind speed and direction, Lei et al. [26] developed the maximum positive transverse temperature gradient prediction model, and obtained the transverse and longitudinal stress caused by temperature through calculation. The results showed that the tensile stress caused by a transverse temperature gradient is large and cannot be ignored. Fang et al. investigated the structural behavior including the axial strength and web crippling strength of cold-formed steel at elevated temperatures by using finite element models, and the proposed design equations could closely predict the structural behavior based on a reliability analysis [27,28]. Therefore, temperature change has a significant impact on the stress and deformation of a structure, and is an important factor to be considered in the design, construction and research of bridge structures. The temperature effect of bridge structures has always been one of the main research objects of scholars at home and abroad.

Most authors of the previous literature have conducted research on mass concrete at a certain location on bridges. Although a large number of temperature effect studies have been conducted, there is a lack of related applications of decision-making in actual bridge construction. Aiming at closing the above gap, this study investigates the cement hydration heat effect on the cushion cap, main pier and arch abutment and the appropriate closure timing based on the sunshine temperature effect in the construction process of the Dafeng River steel box arch bridge. For the mass concrete in the cushion cap, main pier and arch abutment, the pipe cooling method is used to control the hydration heat of cement, and the hydration heat control effect of mass concrete is evaluated based on the arrangement and monitoring results of temperature monitoring points. According to the collected temperature change and displacement change, the influences of daily temperature change on the steel lattice beam and arch rib are analyzed and identified, and the appropriate closure temperature of the arch back is determined for the newly built Dafeng River Bridge.

2. Project Overview

2.1. The Newly Built Dafeng River Bridge

Dafeng River Bridge, located in Qinzhou City, Guangxi, as the main bridge in the Qinzhou–Beihai section of the Lanhai Expressway, crosses the Dafeng River at the bend of the river. The design scheme of the bridge is to demolish and rebuild the upper part and bent cap of the existing bridge without changing its existing navigation, which will be used as the left part of the whole bridge after reconstruction. The newly built Dafeng River Bridge (i.e., the right part) will be built on the downstream side. The appearance of the reconstructed whole bridge is shown in Figure 1.

The newly built Dafeng River Bridge will be a rigid frame-tied steel box arch bridge, with a clear span of 120.0 m, a clear rise of 27.0 m, and a clear rise to span ratio of 1/4.44. As shown in the elevation layout in Figure 1, the arch axis will adopt a secondary parabola, and the arch camber will be 15.0 cm. The main arch rib will be a single box and single room steel box section with an equal section, the section height will be 2.50 m, and the width will be 1.80 m. The main bridge will be a single-span main arch with two arch ribs. A total of five lateral braces will be set between the two arch ribs to ensure the overall stability of the arch bridge. The lateral brace will be a steel box structure, with a horizontal length of 24.30 m, a square section, and a height and width of 1.52 m. The thickness of the top, bottom and web of the cross brace will be 20 mm. Additionally, the suspender will adopt a whole bundle of extruded steel strands (i.e., GJ15-27 with the diameter of each steel strand being 15.20 mm and with 27 strands, Liuzhou Guiqiao Cable Co., Ltd., Liuzhou, China), with the spacing of 26.1 m in the transverse direction and 8.0 m in the longitudinal direction. There will be 14 pairs of suspenders in the whole bridge. A full anticorrosion, full-bundle, replaceable, and adjustable high-strength low-relaxation steel strand-finished cable (i.e., XGK-II 15-31 with the diameter of each steel strand being 15.20 mm and with 31 strands, Liuzhou Guiqiao Cable Co., Ltd., Liuzhou, China) will be used. Additionally, eight tie rod tensioning holes will be arranged under each arch rib, including six permanent tie rods and two reserved cable replacement holes. The bridge deck system will be a steel lattice system, which is composed of longitudinal steel beams, main beams, secondary beams and steel–concrete composite bridge decks. For the bridge deck, an 8 mm thick steel plate will be welded on the beam grid as the bottom formwork, and 15 cm thick C40 steel fiber concrete will be cast in situ. The bridge deck will be paved with 7 cm asphalt concrete.

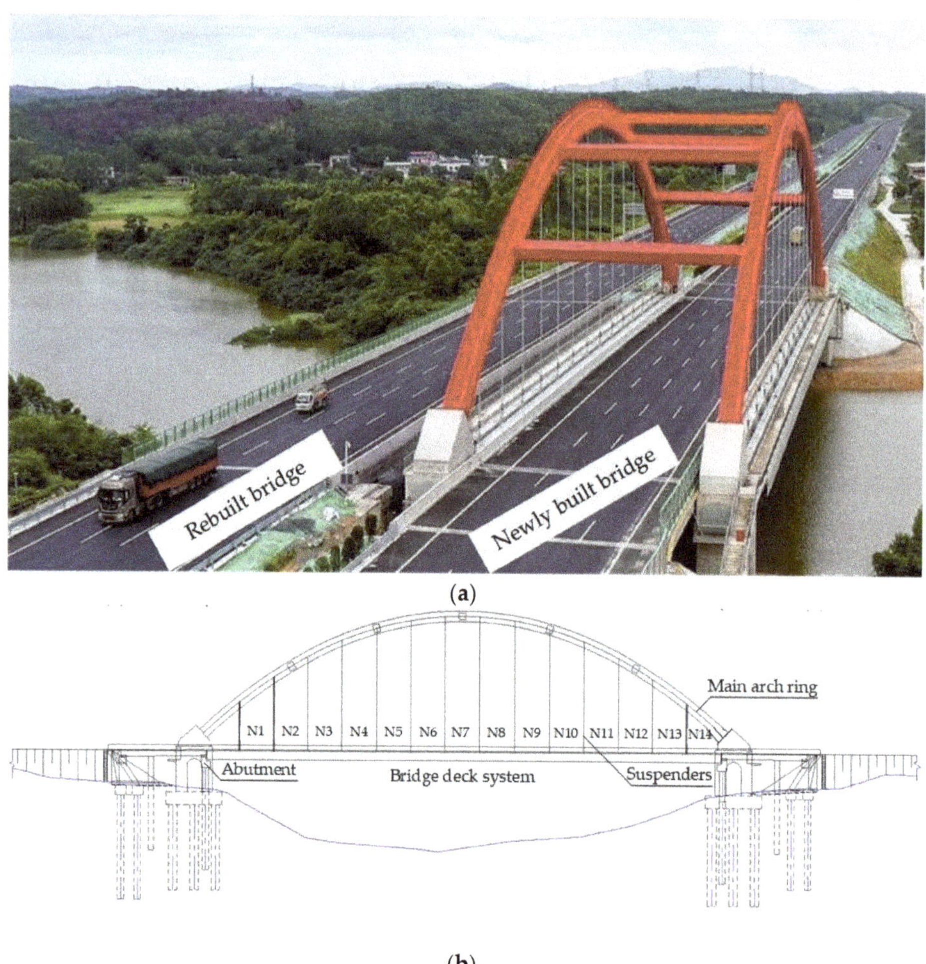

Figure 1. Reconstruction and expansion project of the Dafeng River Bridge in Qinzhou–Beihai section of Lanhai Expressway: (**a**) Dafeng River Bridge; (**b**) structural layout.

2.2. Bridge Construction Based on Segmental Assembling Technique

Before construction, various preparations shall be carried out, such as site cleaning. Then, the approach bridge construction will be carried out, and the main bridge pile foundation, cushion cap, main pier and arch rib arch pedestal will be constructed at the same time. Then, the temporary construction trestle of the main bridge will be constructed. The trestle is a steel pipe pile and has a Bailey beam structure. The arch rib section shall be installed in sections to construct the main arch until the construction of the main arch rib and the transverse brace is completed and the closure of the main arch is completed. The supporting support system of the main arch will be dismantled symmetrically and orderly, and the tie rods N2, N3, N10 and N11 will be tensioned at the same time. Then, the arch rib suspenders N4~N11, and the tie rods N5, N8, N13 and N16 will be tensioned. Then, the arch rib suspenders N1–N3 and N12–N14, and tie rods N6, N7, N14 and N15 will be tensioned. Finally, the temporary trestle will be dismantled orderly and symmetrically, the tie rods N2~N8 and N10~N16 will be tensioned, and the construction of the whole bridge will be completed. The tie rods will be tensioned two times. The first tensioning control force will 1400 kN, the second tensioning control force of tie rods N2~N8 will 2200 kN, and the tensioning control force of tie rods N10~N16 will be 2250 kN. For the main material parameters of the newly built Dafeng River Bridge, Q355C will be used as an arch rib, transverse brace, and lattice beam, and Q235C will be used as a steel deck. C40, C35 and C50 will be used for the main pier, deck, sidewalk and access slab, cushion cap concrete and arch abutment concrete. $\varphi^s 15.24$ will be adopted as a suspender and tie rod.

2.3. Temperature Monitoring

The length, width and height of the cushion cap of the newly built Dafeng River Bridge will be 13.2 m, 8.4 m and 3.0 m, respectively. The length, width and height of the main pier will be 8.5 m, 4.4 m and 11.0 m, respectively, of which the length, width and height of the lower section will be 2.5 m, 4.4 m and 6.0 m, respectively, and the length, width and height of the upper section will be 8.5 m, 4.4 m and 5.0 m, respectively. The length, width and height of the arch abutment will be 7.5 m, 3.4 m and 4.923 m, respectively. Therefore, the cushion cap, main pier including the lower and upper sections, and arch abutment will be made of typical mass concrete.

The temperature problem of the hydration heat of mass concrete is very complex, and the external temperature, construction conditions and raw material changes will cause temperature changes, which could be more accurately understood and grasped only through monitoring. During the performance of mass concrete temperature control, in order to test the construction quality and temperature control effect, the temperature control measures will be adjusted in a timely manner and improved by mastering the temperature control information. During the concrete pouring process, it will be necessary to monitor the concrete pouring temperature. During the curing process, it will also be necessary to monitor the temperature rise and fall of the concrete pouring block, the temperature difference between the inside and outside, the cooling rate and the ambient temperature. In this study, the temperature monitoring system of mass concrete is composed of a temperature sensor, data acquisition system and data transmission system. The monitoring system has the functions of displaying, storing and processing temperature and time parameters, and can draw the temperature change curve of measuring points in real-time. The number of temperature measuring points should not be less than 50. The allowable error of the temperature monitoring system shall not be greater than 0.5 °C. The test range of temperature shall be −30 °C~125 °C. Before the installation of a temperature sensor, it shall be soaked 1 m underwater for 24 h together with the transmission wire without damage. In this study, the pipe cooling method was adopted to control the thermal cracking of the massive concrete structures of the newly built rigid frame-tied steel box arch bridge. The digital temperature sensor and wireless temperature collector (shown in Figure 2) are used in a temperature monitoring system for mass concrete.

In the process of temperature measurement, the following test items need to be defined. At the same time, the layout of measuring points is closely related to that of the test items. The temperature of the concrete mixture when pouring it into the mold (i.e., the molding temperature, $T_{molding}$) shall be measured at least twice per shift. The maximum temperature of concrete in the temperature measuring area is defined as the maximum temperature of concrete (T_{max}). The concrete surface temperature (T_{surf}) refers to the temperature at 50 mm from the concrete outer surface. The concrete bottom temperature (T_{bot}) refers to the temperature at 50 mm from the bottom of the concrete pouring block. The difference between the maximum temperature of concrete and the surface temperature is defined as the temperature difference between the surface and interior of the concrete ($\Delta T_{surf\text{-}int}$). The temperature rise peak (TRP) is the highest temperature rise inside the concrete pouring body. The cooling rate (CR) refers to the temperature reduction value every day or every 4 h after the internal temperature of the concrete reaches the peak temperature under the condition of heat dissipation. The temperature value at the shady ventilation outside the structure is the concrete ambient temperature (T_{air}).

Figure 2. The temperature monitoring system of mass concrete in this study.

3. Results and Discussion

3.1. Monitoring of Hydration Heat of Mass Concrete and Analysis of Early Temperature Control

3.1.1. Hydration Heat of Mass Concrete in Cushion Cap

(a) Arrangement of Temperature Monitoring Points and Cooling Pipe

During the construction process of the project, a detailed temperature monitoring scheme of hydration heat will be developed to determine a comprehensive temperature measuring point arrangement, and the hydration heat temperature of mass concrete will be monitored in real-time to guide the actual construction. Figure 3 shows the arrangement of temperature monitoring points for the mass concrete in the cushion cap. Temperature monitoring points will be arranged at the halfway point of the symmetry axis in the longitudinal and transverse directions of the cushion cap concrete, in which four measuring lines will be arranged in the transverse direction of the bridge, i.e., F, G, H and I, and six measuring lines will be arranged in the longitudinal direction, i.e., A, B, C, D, E and F. Five monitoring points will be arranged along each measuring line in the height direction. A total of 9 measuring lines (45 monitoring points in total) will be arranged for the cushion cap concrete which will be constructed first, and only 5 measuring lines (25 monitoring points in total) will be arranged for the cushion cap concrete which will be constructed later.

Figure 4 shows the cooling pipe layout of the cushion cap. During the construction of each cushion cap, the ambient temperature should be monitored in real-time. According to the cooling pipe layout, 1 water temperature monitoring point will be arranged at the inlet and outlet of the C1, C2 and C3 pipes, respectively, and 9 water temperature monitoring points will be arranged in total. Therefore, there will be 55 temperature monitoring points arranged for the cushion cap constructed first, of which 45 points will be temperature monitoring points of concrete, 9 points will be water temperature monitoring points at the inlet and outlet of the cooling pipe, and 1 point will be an ambient temperature monitoring point. There will be 35 temperature monitoring points arranged for the cushion cap constructed later, of which 25 points will be temperature monitoring points of concrete, 9 points will be water temperature monitoring points at the inlet and outlet of the cooling pipe, and 1 point will be the ambient temperature monitoring point.

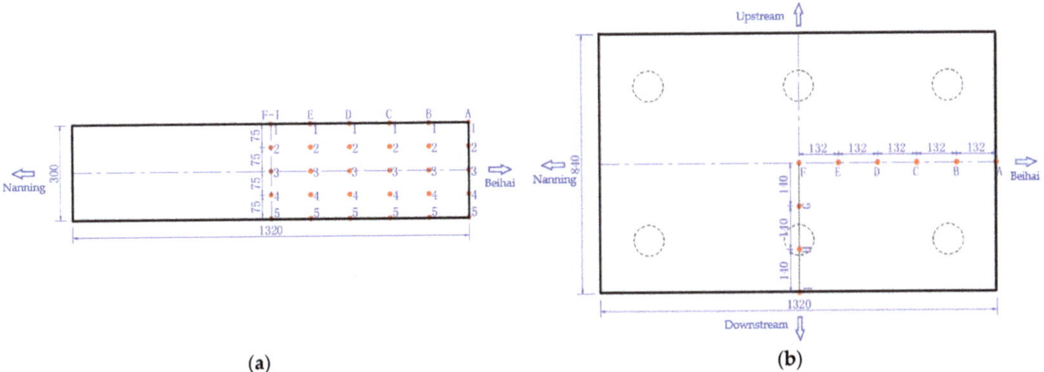

Figure 3. Arrangement of temperature monitoring points for mass concrete in cushion cap: (**a**) elevation; (**b**) plan (unit: cm).

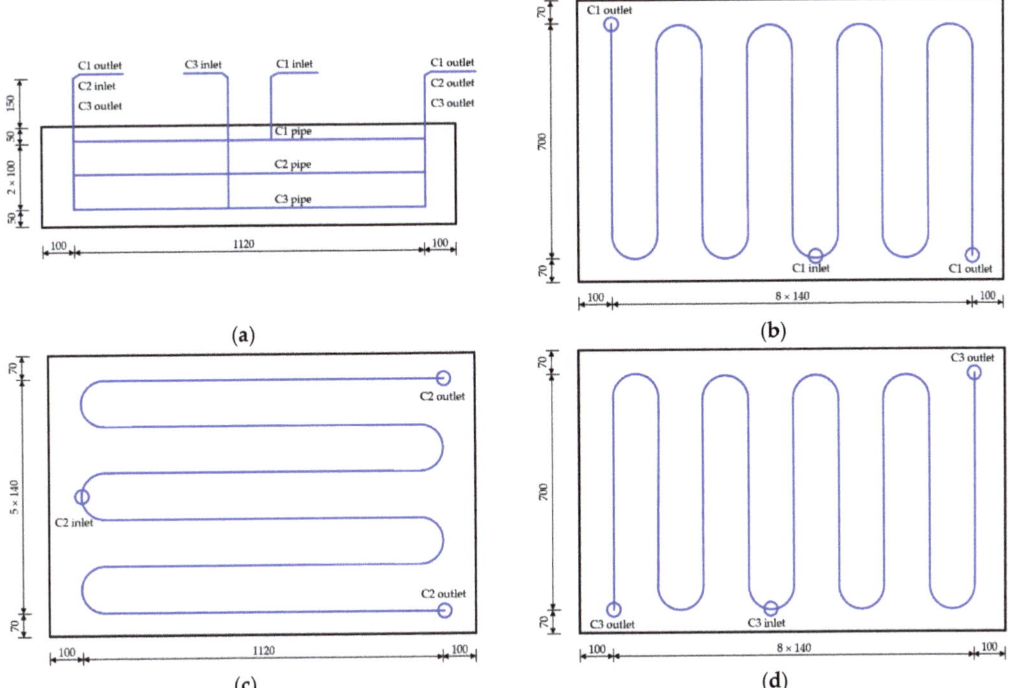

Figure 4. The cooling pipe layout of cushion cap: (**a**) elevation; (**b**) C1 pipe; (**c**) C2 pipe; (**d**) C3 pipe (unit: cm).

(b)   Analysis of Hydration Heat and Cracking Based on Temperature Monitoring

According to the requirements of the Chinese specification "Technical Code for Temperature Measurement and Control of Mass Concrete" (GB/T 51028-2015) [29], the molding temperature ($T_{molding}$), temperature rise peak (TRP), the maximum temperature of concrete (T_{max}), maximum cooling rate (CR_{max}), the maximum temperature difference between the surface and interior ($\Delta T_{surf\text{-}int\text{-}max}$), the maximum temperature difference between the concrete surface and air ($\Delta T_{surf\text{-}air\text{-}max}$), and the maximum temperature when monitoring is stopped ($T_{max\text{-}stop}$) are selected for the statistics of the temperature monitoring of a mass concrete cushion cap.

Figure 5 shows the temperature monitoring results of a mass concrete cushion cap at the upstream and downstream of the Nanning and Beihai sides, respectively. Generally, the higher the molding temperature, the higher the maximum temperature of concrete under the action of hydration heat after pouring. It can be seen that the $T_{molding}$ is strictly controlled at about 35 °C in this study, and the $T_{molding}$ value of each layer is basically stable, in which the relatively low $T_{molding}$ value also avoids an accelerating hydration heat reaction. The TRP value is the difference between the T_{max} and $T_{molding}$ values; therefore, the TRP and T_{max} show a similar change trend, increasing first and then decreasing. This is because the hydration heat reaction of layered concrete is a cumulative process, and the TRP increases with the construction's progress, leading to an increase in the T_{max}. Generally, the temperature of the core area of layer No. 3 (i.e., the middle position) of layered concrete is higher than that of other layers, gradually decreasing to the outward surfaces. The heat dissipation performance of concrete decreases with the increase in the thickness of the mass concrete structure, and the layered pouring of concrete has little impact on the temperature change of the mass concrete. For the mass concrete cushion cap at four different locatitons, the average T_{max} value of layer No. 3 is about 1.3 times the average T_{max} value of layer No. 1. At the same time, the TRP value is less than 45 °C. For the pouring of mass concrete, due to the large volume of concrete, the temperature generated by the hydration heat effect in the concrete increases rapidly, while the heat dissipation is slow, resulting in a high temperature inside the concrete and a large temperature difference between the surface and the inside of the concrete, which makes it very easy for concrete cracking to occur. In this study, it could be found that the temperature difference between the surface and interior ($\Delta T_{surf\text{-}int}$) and the temperature difference between the concrete's surface and air ($\Delta T_{surf\text{-}air}$) will be within the control range. With the help of the cooling pipe layout of the cushion cap, the cooling rate inside the mass concrete will be well controlled. The pipe cooling method is conducive to the hydration heat control effect of mass concrete, and can effectively reduce the central temperature of concrete structures. The cooling rate of mass concrete fluctuates greatly due to many influencing factors. Due to the strong heat dissipation capacity of the surface layer, the cooling rate of layer No. 5 is much higher than that of the other layers. The T_{max} value of layer No. 5 decreases by an average of 9% compared to the lowest T_{max} value of the other layers. When the monitoring is stopped, the maximum temperature ($T_{max\text{-}stop}$) of each layer will be basically near the molding temperature ($T_{molding}$). At the same time, the difference between the maximum temperature of concrete and the minimum temperature of the environment for three consecutive days can be controlled at 20 °C when the monitoring is stopped. By performing temperature control through cooling water, there will be no visible cracks in the mass concrete of the cushion cap at the upstream and downstream of the Nanning and Beihai sides.

3.1.2. Hydration Heat of Mass Concrete in Main Pier

(c)   Arrangement of Temperature Monitoring Points and Cooling Pipe

Temperature monitoring points will be arranged at the halfway point of the symmetry axis in the longitudinal and transverse directions of the main pier concrete, as shown in Figure 6. Three measuring lines will be arranged in the transverse direction of the bridge for the lower main pier constructed first, i.e., C, D and E, and three measuring lines will be arranged in the longitudinal direction of the bridge, i.e., A, B and C. Similarly, five monitoring points will be arranged along each measuring line in the height direction. Thus, a total of five measuring lines (25 monitoring points in total) will be arranged for the concrete of the lower main pier constructed first. Only three measuring lines including A, C and E (15 monitoring points in total) will be arranged for the lower main pier concrete constructed later. On the other hand, three measuring lines will be arranged in the transverse direction of the bridge for the upper main pier constructed first, i.e., J, K and L, and five measuring lines will be arranged in the longitudinal direction of the bridge, i.e., F, G, H, I and J. Thus, a total of seven measuring lines (35 monitoring points in total) will be arranged for concrete of the upper main pier constructed first. Only four measuring lines including F, H, J and L (20 monitoring points in total) will be arranged for the upper main pier concrete constructed later.

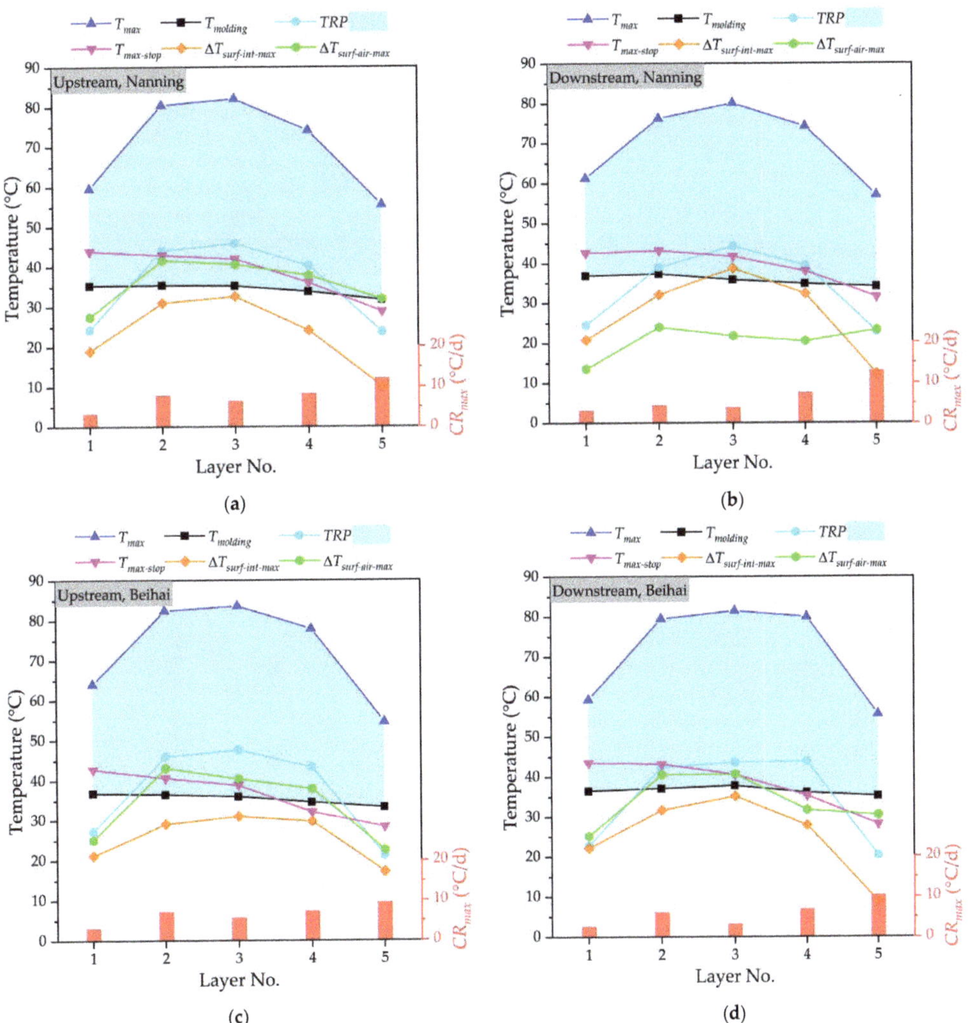

Figure 5. The temperature monitoring of mass concrete cushion cap: (**a**) at the upstream of Nanning side; (**b**) at the downstream of Nanning side; (**c**) at the upstream of Beihai side; (**d**) at the downstream of Beihai side.

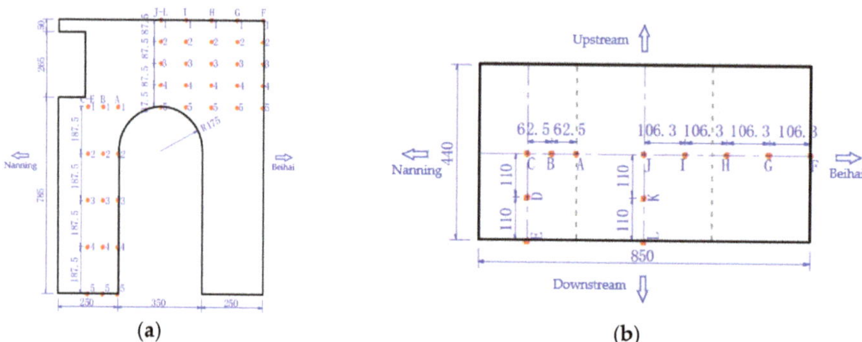

(a)   (b)

Figure 6. The arrangement of temperature monitoring points for mass concrete in main pier: (**a**) elevation; (**b**) plan (unit: cm).

Figure 7 shows the cooling pipe layout of the main pier. According to the cooling pipe layout, one temperature monitoring point will be arranged at the inlet and outlet of the pipe near the electric box for the lower main pier, two temperature monitoring points will be arranged at the cooling pipe, and eight water temperature monitoring points will be arranged at the inlet and outlet of the cooling pipe at the four main piers. In addition, one ambient temperature monitoring point will be arranged near the upper and lower main piers, two temperature monitoring points will be arranged at the pier body, and eight ambient temperature monitoring points will be arranged at the four main piers.

(d)   Analysis of Hydration Heat and Cracking Based on Temperature Monitoring

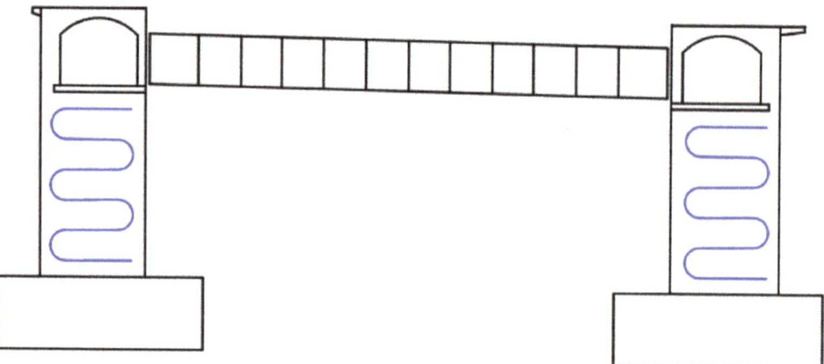

Figure 7. Cooling pipe layout of main pier.

Figures 8 and 9 show the temperature monitoring results of the mass concrete in the lower and upper main piers at the upstream and downstream of the Nanning and Beihai sides, respectively. It can be seen that the $T_{molding}$ of the lower and upper main piers was strictly controlled below 35 °C in this study, and the $T_{molding}$ value of each layer was basically stable, in which the relatively low $T_{molding}$ value also avoided accelerating the action of hydration heat after pouring. According to the comparative temperature analysis of the upstream and downstream with different $T_{molding}$ values, it can be seen that the ambient temperature of concrete has a certain influence on its temperature field. The higher the molding temperature is, the higher the maximum temperature value by the hydration heat effect is. With each 5 °C increase in the $T_{molding}$ value, the T_{max} value at the core area of layer No. 3 increased by about 4.5 °C for the mass concrete in the lower main pier, and the T_{max} value of the core area of layer No. 3 could even be raisedx by about 5.1 °C for the mass concrete in the upper main pier. Considering the stable $T_{molding}$ value and $TRP = (T_{max} - T_{molding})$, the TRP and T_{max} showed basically the same change trend, increasing first and then decreasing. This is because the hydration heat reaction of layered concrete is a cumulative process, and the TRP increases with the construction progress, leading to the increase in T_{max}. The temperature of the core area at layer No. 3 (i.e., the middle position) of layered concrete was higher than that of the other layers, gradually decreasing to the outward surfaces. For the four different locations, the average T_{max} value of layer No. 3 was about 1.2 times the average T_{max} value of layer No. 1 for the lower main pier. Meanwhile, the maximum TRP value was less than 55 °C and its average value was less than 45 °C. For the pouring of mass concrete, due to the large volume of concrete, the temperature generated by the hydration heat effect in the concrete increased rapidly, while the heat dissipation was slow, resulting in a high temperature inside the concrete and a large temperature difference between the surface and the inside of the concrete, which very easily results in concrete cracking. In this study, it could be found that the temperature difference between surface and interior ($\Delta T_{surf-int}$) and temperature difference between concrete surface and air ($\Delta T_{surf-air}$) were within the control range. The average value of the $\Delta T_{surf-int}$ values for all layers was controlled at less than 30 °C. With the help of the cooling pipe layout of the main pier, the cooling rate inside the mass concrete was well-controlled. There were little significant increases in the T_{max} and TRP values of layer No. 3, indicating that the pipe cooling method is conducive to the hydration heat control effect of mass concrete, and can effectively reduce the central temperature of concrete structures. The cooling rate of mass concrete fluctuates greatly due to many influencing factors. Due to the strong heat dissipation capacity of the surface layer, the cooling rate of layer No. 5 was much higher than that of the other layers. For the lower main pier, the T_{max} value of layer No. 5 decreased by an average of 9% compared to the lowest T_{max} value of the other layers. For the upper main pier, the T_{max} value of layer No. 5 decreasesd by an average of about 19% compared to the lowest T_{max} value of the other layers. When the monitoring is stopped, the maximum temperature ($T_{max-stop}$) of each layer woill be basically lower or near the molding temperature ($T_{molding}$). At the same time, the difference between the maximum temperature of concrete and the minimum temperature of the environment for three consecutive days can be controlled at 20 °C when the monitoring is stopped. By performing temperature control through cooling water, there will be no visible cracks in the mass concrete in the lower upper main piers at the upstream and downstream of Nanning and Beihai sides.

3.1.3. Hydration Heat of Mass Concrete in Arch Abutment

(e)   Arrangement of Temperature Monitoring Points and Cooling Pipe

Temperature monitoring points will be arranged at the halfway point of the symmetry axis in the longitudinal and transverse directions of the concrete of the arch abutment constructed first, as shown in Figure 10. Three measuring lines will be arranged in the transverse direction of the bridge for the arch abutment constructed first, i.e., D, E and F, and five measuring lines will be arranged in the longitudinal direction of the bridge, i.e., A, B, C and D. Similarly, five monitoring points will be arranged along each measuring line in the height direction. Thus, a total of six measuring lines (24 monitoring points in total) will be arranged for concrete of the arch abutment constructed first. Only four measuring lines including A, C, D and F (16 monitoring points in total) will be arranged for the concrete of the arch abutment constructed later.

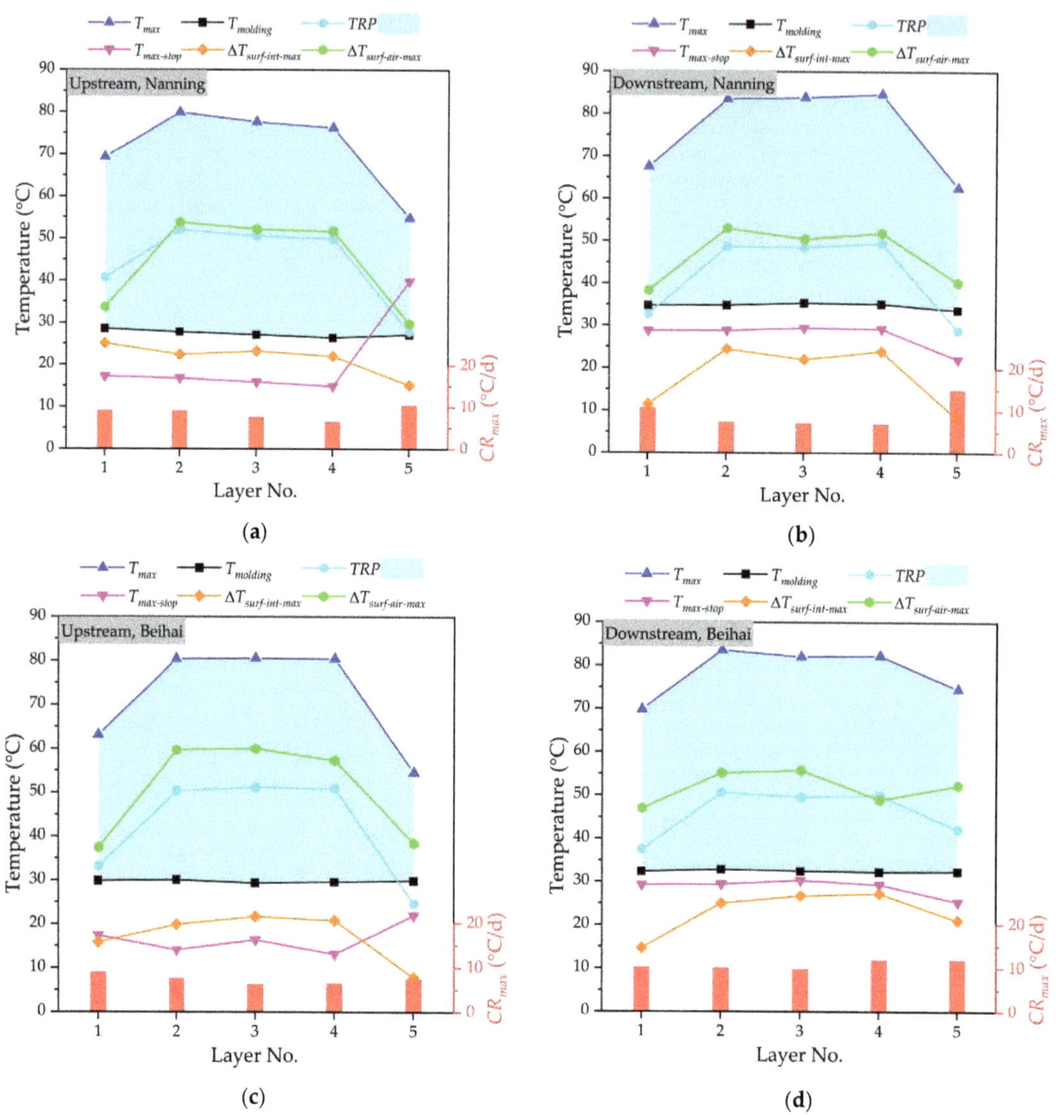

Figure 8. Temperature monitoring of mass concrete in lower main pier: (**a**) at the upstream of Nanning side; (**b**) at the downstream of Nanning side; (**c**) at the upstream of Beihai side; (**d**) at the downstream of Beihai side.

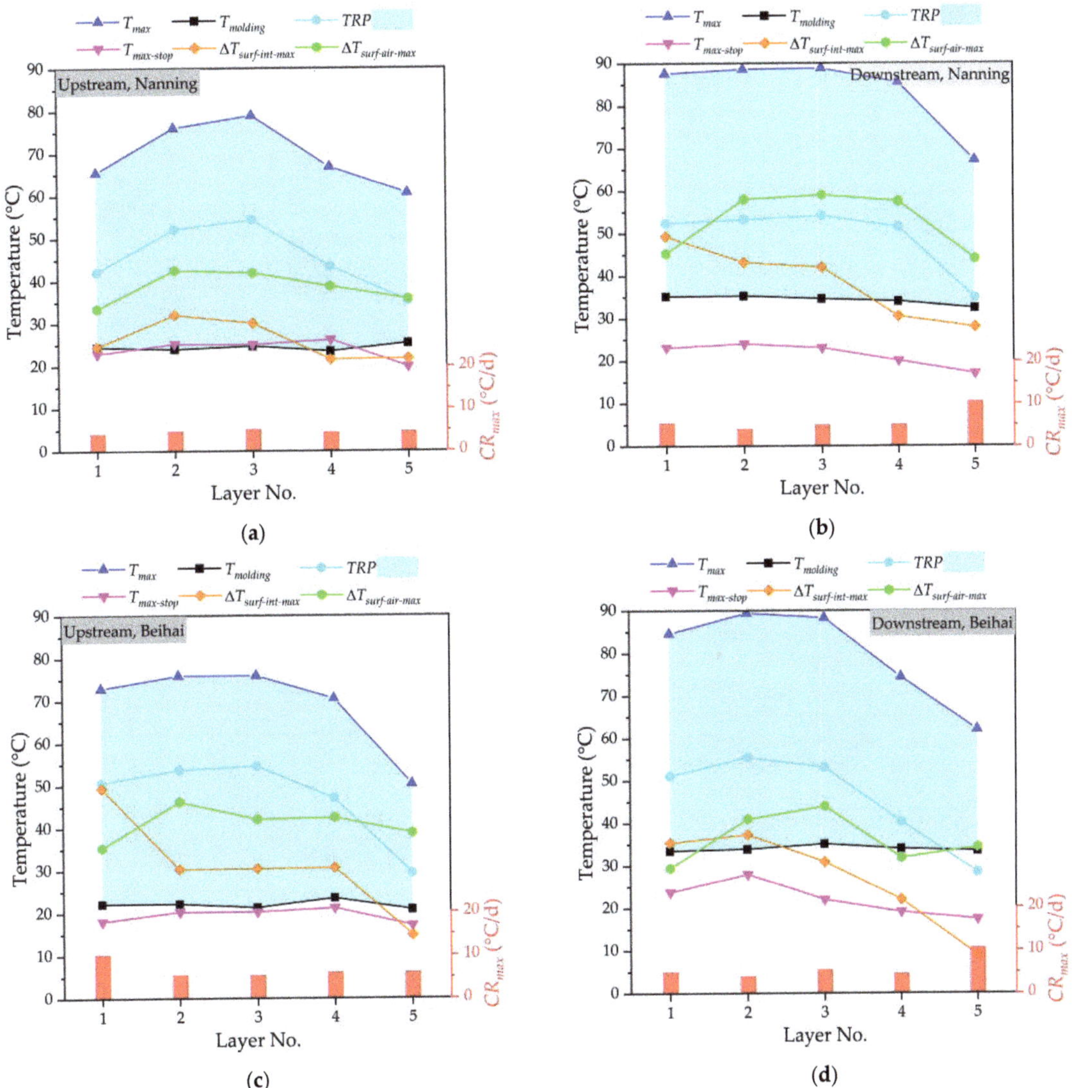

Figure 9. Temperature monitoring of mass concrete in the upper main pier: (**a**) at the upstream of Nanning side; (**b**) at the downstream of Nanning side; (**c**) at the upstream of Beihai side; (**d**) at the downstream of Beihai side.

Figure 11 shows the cooling pipe layout of the arch abutment. According to the layout of the cooling pipe, one temperature monitoring point will be arranged at the inlet and outlet of the pipe, two temperature monitoring points will be arranged at the cooling pipe of a single arch, and eight water temperature monitoring points will be arranged at the inlet and outlet of the cooling pipe for the four arch abutments.

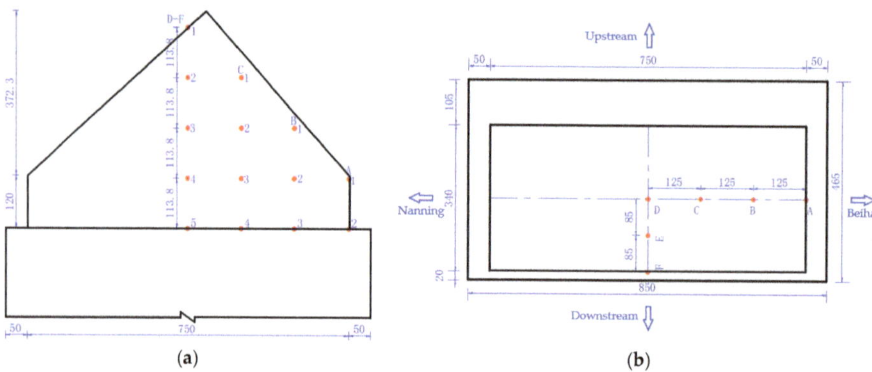

Figure 10. The arrangement of temperature monitoring points for mass concrete in arch abutment: (a) elevation; (b) Plan. (Unit: cm).

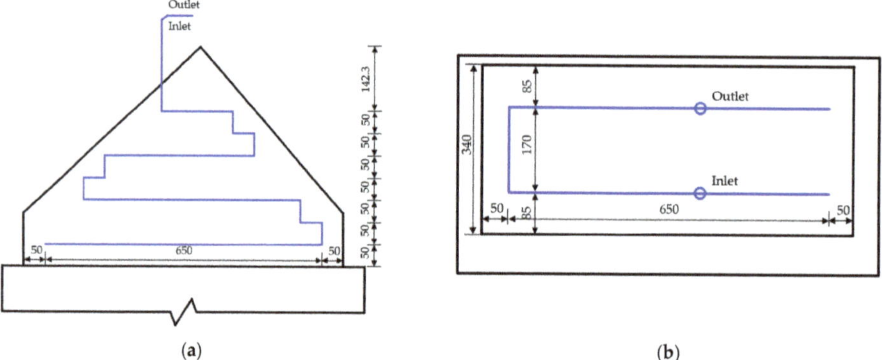

Figure 11. The cooling pipe layout of the arch abutment: (a) elevation; (b) plan (unit: cm).

(f)   Analysis of Hydration Heat and Cracking Based on Temperature Monitoring

Figure 12 shows the temperature monitoring results of the mass concrete in the arch abutment at the upstream and downstream of the Nanning and Beihai sides, respectively. It can be seen that the $T_{molding}$ of the arch abutment was controlled at about 30 °C in this study, and the $T_{molding}$ value of each layer was basically stable, in which the relatively low $T_{molding}$ value also avoided accelerating the action of the hydration heat after pouring. Considering the stable $T_{molding}$ value and $TRP = (T_{max} - T_{molding})$, the TRP and T_{max} showed basically the same change trend, increasing first and then decreasing. This is because the hydration heat reaction of layered concrete is a cumulative process, and the TRP increases with the construction's progress, leading to the increase in the T_{max}. Generally, the temperature of layer No. 3 (i.e., the middle position) of the layered concrete was higher than that of other layers. For the four different locations, the average T_{max} value of layer No. 3 was about 1.1 times the average T_{max} value of layer No. 1. At the same time, the maximum TRP value was less than 55 °C and its average value was less than 45 °C. For the pouring of mass concrete, due to the large volume of concrete, the temperature generated by the hydration heat effect in the concrete increased rapidly, while the heat dissipation was slow, resulting in a high temperature inside the concrete and a large temperature difference between the surface and the inside of the concrete, which very easily causes concrete cracking. In this study, it Was found that the temperature difference between the surface and interior ($\Delta T_{surf\text{-}int}$) and the temperature difference between the concrete's surface and air ($\Delta T_{surf\text{-}air}$) were within the control range. With the help of the cooling pipe layout of the arch abutment, the cooling rate inside the mass concrete was well-controlled. The cooling rate of mass concrete fluctuated greatly due to many influencing factors. When the monitoring is stopped, the maximum temperature ($T_{max\text{-}stop}$) of each layer will be basically lower or near the molding temperature ($T_{molding}$). At the same time, the difference between the maximum temperature of concrete and the minimum temperature of the environment for three consecutive days can be controlled at 20 °C when the monitoring is stopped. By performing temperature control through cooling water, there will be no visible cracks in the mass concrete in the arch abutment at the upstream and downstream of the Nanning and Beihai sides.

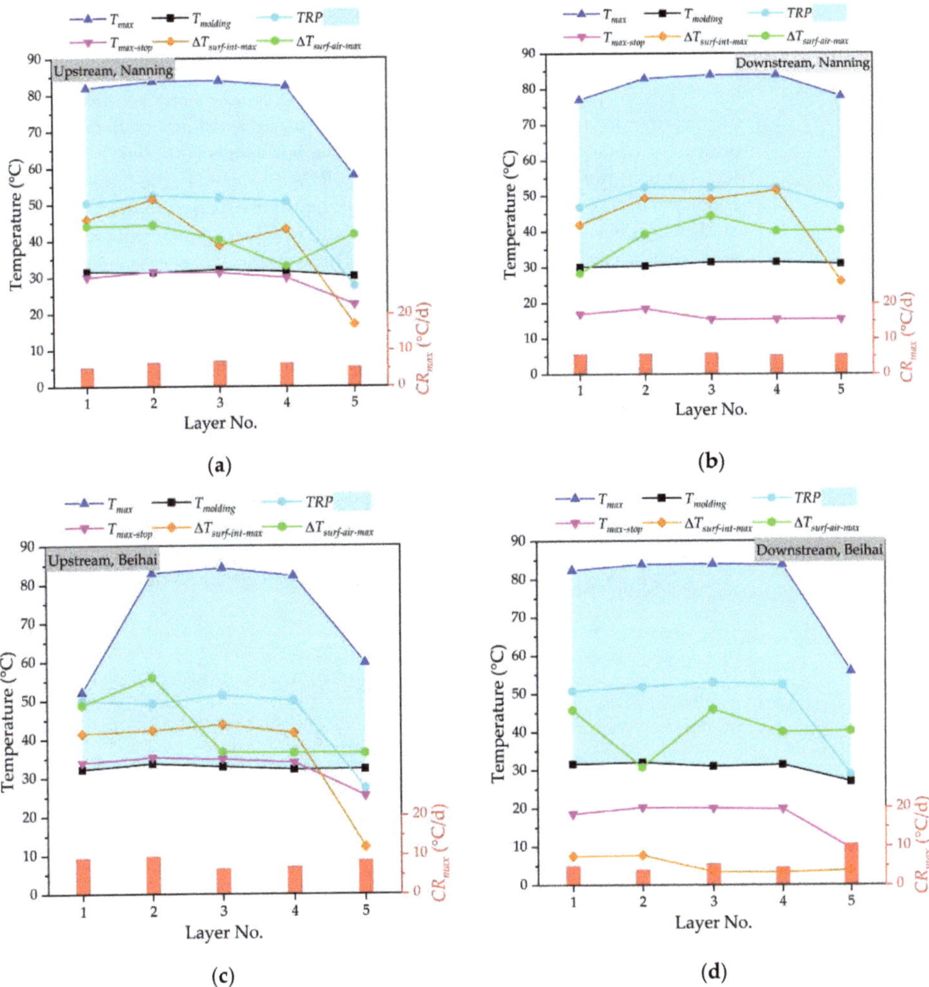

Figure 12. Temperature monitoring of mass concrete in arch abutment: (**a**) at the upstream of Nanning side; (**b**) at the downstream of Nanning side; (**c**) at the upstream of Beihai side; (**d**) at the downstream of Beihai side.

3.2. Analysis of Influence Monitoring-Based Temperature Change on Bridge Deformation

3.2.1. Identification of Temperature Effect on Lattice Beam Deformation

The bridge structure is located in the outdoor environment, and its internal temperature field is affected by many factors such as solar radiation, ambient temperature, etc.

Due to the good thermal conductivity of steel, steel box arch bridges are more sensitive to the ambient temperature compared to ordinary reinforced concrete bridges. Ambient temperature change includes seasonal temperature change and daily temperature change. The influence of seasonal temperature changes on a structure is relatively simple, and the change is uniform. However, daily temperature change is more complex, especially under the effect of sunlight, which causes temperature differences, thus causing the displacement of the structure, and also produces the temperature's secondary internal force for a statically indeterminate structure. The positioning of the lattice beam is the foundation of the arch rib's assembly, and it is important to analyze the influence of temperature on the displacement of the lattice beam in construction monitoring, as it directly affects the assembly accuracy of the arch rib and the verticality of the suspender.

Continuous temperature and deformation monitoring of the lattice beam was carried out to study the relationship between the position change of the suspender beam for the lattice beam and the temperature change in the outdoor environment. On the day of monitoring, it was sunny with a certain temperature difference between morning and noon. The detailed temperature–time curve relationship is shown in Figure 13. It can be seen from Figure 13 that the lattice beam will be displaced in the longitudinal, transverse and vertical directions with the change in temperature. Because the lattice beam is in the nonlinear temperature field, the structure will produce displacement, especially longitudinal displacement as the most obvious. Through the actual measurement of suspender N14 on the left and right for the lattice beam, the maximum temperature change is 10.6 °C, the maximum longitudinal displacement will be 9.3 mm, the maximum transverse displacement is 3.8 mm, and the maximum vertical displacement is 4.1 mm. From the measured displacement curve versus the outside temperature of the suspender for the lattice beam at different times, the position of monitoring points of the suspender also changed constantly with the constant change in temperature. A temperature rise will cause the structure to have a certain camber in the longitudinal direction, and the longitudinal or transverse displacement value caused by the temperature change (temperature rise or drop) under sunshine will be no less than the vertical displacement value. It is necessary to monitor the structure at a time when the temperature is relatively stable to determine the final position of the suspender; thus, the effect of temperature should be considered for the closure of the lattice beam.

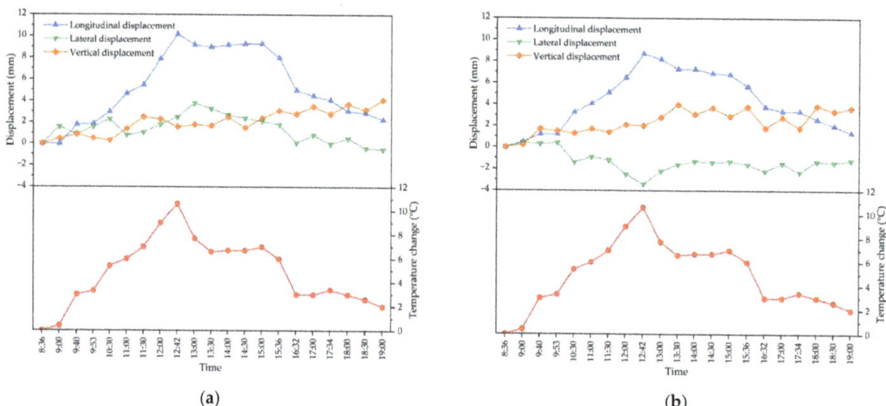

Figure 13. The continuous temperature and deformation monitoring of suspender N14 for the lattice beam: (**a**) on the left; (**b**) on the right.

3.2.2. Identification of Temperature Effect on Arch Rib Deformation

As one of the most important factors affecting the deflection of arch bridges, the ambient temperature change includes the seasonal temperature change and daily temperature change. The daily temperature change is more complex, especially under the effect of sunlight, which will cause the deflection of the arch rib, which will also lead to longitudinal, transverse and vertical deformations. The influence of seasonal temperature change on the structure is relatively simple, and its change is uniform. In order to study the relationship between the position change of segment N4 for the arch rib and temperature change in the outdoor environment, continuous temperature and deformation monitoring of the arch rib was carried out for 24 h. On the day of monitoring, it was sunny with a certain temperature difference between morning and noon, and the maximum temperature change was 9.6 °C. The detailed temperature–time curve relationship is shown in Figure 14. It can be seen from Figure 14 that the arch rib will be displaced in the longitudinal, transverse and vertical directions with the change of temperature. Because the arch rib is in the nonlinear temperature field, the structure will produce displacement, especially longitudinal and transverse displacements as the most obvious. Through the actual measurement of segment N4 on the left and right for the arch rib, the maximum longitudinal displacement is 5.7 mm, the maximum transverse displacement is 5.9 mm, and the maximum vertical displacement is 2.3 mm. From the measured displacement curve versus the outside temperature of the segment for the arch rib at different times, the position of monitoring points of the segment also changed constantly with the constant change in temperature. Additionally, due to the symmetrical construction on both sides of the river, the arch rib deformation on both sides will basically present symmetrically synchronous changes. It is necessary to monitor the structure at a time when the temperature is relatively stable to determine the final position of the segment; thus, the effect of temperature should be considered for the closure of the arch rib.

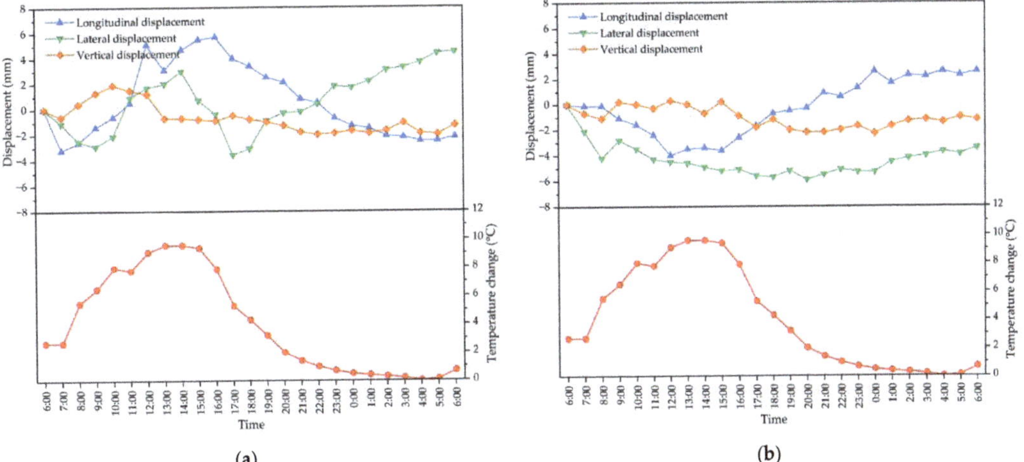

Figure 14. Continuous temperature and deformation monitoring of segment N4 for arch rib: (**a**) at Nanning side; (**b**) at Beihai side.

3.2.3. Determination of the Appropriate Closure Timing

After the whole bridge is closed, the structure will become statically indeterminate, and the temperature stress generated by the bridge will be able to exceed the stress even under a live load. According to the above analysis, the structural temperature effect under the action of sunshine temperature change will be large, which will mainly have a large effect on the deflection during the bridge construction process and the stress state after the transformation of the bridge system, with a small cycle and rapid change. In order to better determine the appropriate time for arch rib closure, this study monitored the arch back temperature at the final closure and ambient temperature for 84 h before the arch rib's closure, and the specific monitoring results are shown in Figure 15. Generally, when the temperature rises, the overall structure will rise upward, and when the temperature drops, the overall structure will deflect downward. The longitudinal displacement value of the structure due to heating or cooling will be no less than the vertical displacement value. Therefore, in the design of the arch rib's closure, the adverse effects of heating or cooling shall be fully considered. From Figure 15, it can be seen that the minimum temperature of the arch back at the final closure before the arch rib closure occurs between 1:00~4:00 in the morning, the changing trends of the arch back temperature and ambient temperature are consistent, and the temperature difference is relatively small, so the appropriate closure time of the arch rib is determined as 1:00~4:00 in the morning.

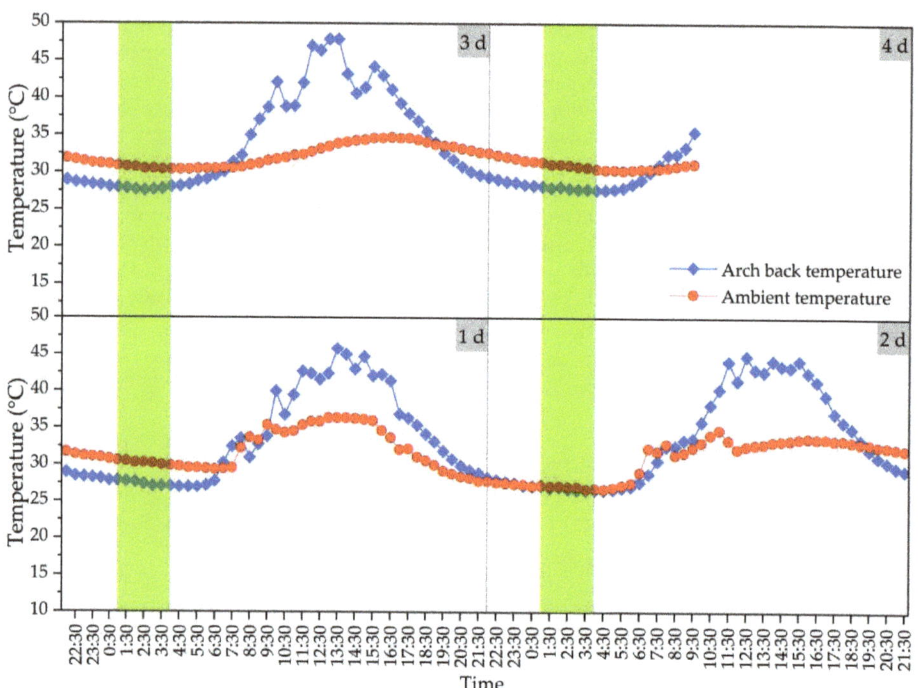

Figure 15. Arch back temperature at final closure and ambient temperature for 84 h (from 1 d to 4 d) before arch rib closure. (note: the green part is the appropriate closure time).

4. Conclusions

This study investigates the temperature effect of the construction process of a rigid frame-tied steel box arch bridge. For the mass concrete in the cushion cap, main pier and arch abutment, the hydration heat control effect of mass concrete by the pipe cooling method is evaluated. Based on the monitored temperature change and displacement change, the influences of daily temperature change on the steel lattice beam and arch rib were analyzed and identified, and the appropriate closure temperature of the arch back was determined. From the experimental and analytical results, the following conclusions can be drawn:

(1) The heat dissipation performance of concrete decreases with the increase in the thickness of a mass concrete structure. The temperature of the core area of the middle layer was the highest for the mass concrete structures, gradually decreasing to the outward surfaces. The average T_{max} value of layer No. 3 was about 1.3, 1.2, and 1.1 times the average T_{max} value of layer No. 1 for the mass concrete in the cushion cap, main pier and arch abutment, respectively.

(2) The ambient temperature of concrete has a certain influence on its temperature field. The higher the molding temperature, the higher the maximum temperature value by the hydration heat effect. With each 5 °C increase in the $T_{molding}$ value, the T_{max} value at the core area of layer No. 3 increased by about 4~5 °C for mass concrete. The pipe cooling method is conducive to the hydration heat control effect of mass concrete, and can effectively reduce the central temperature of concrete structures.

(3) A temperature rise will cause the structure to have a certain camber in the longitudinal direction, and the longitudinal or transverse displacement value caused by the temperature change (temperature rise or drop) under the sunshine will be no less than the vertical displacement value. Due to the symmetrical construction on both sides of the river, the arch rib deformation on both sides will basically present symmetrically synchronous changes.

(4) In view of the large structural temperature effect under the action of sunshine temperature change, the appropriate closure time of the arch rib is determined as 1:00~4:00 in the morning, during which the changing trends of arch back temperature and ambient temperature will be consistent, and the temperature difference is relatively small.

This study investigated the cement hydration heat effect of the cushion cap, main pier and arch abutment and determined the appropriate closure timing based on the sunshine temperature effect in the Dafeng River steel box arch bridge construction. However, the influence of the pipe cooling method on the hydration heat as well as the sunshine temperature effect of steel in bridge construction is worthy of further exploration.

Author Contributions: Conceptualization, T.Z., H.W. and W.W.; methodology, T.Z., H.W. and W.W.; validation, Y.L. and Y.Y.; formal analysis, T.Z., Y.L. and Y.Y.; investigation, H.W., Y.L. and W.W.; writing—original draft preparation, T.Z.; writing—review and editing, H.W., Y.L. and W.W.; project administration, H.W. and W.W.; funding acquisition, H.W. and W.W. All authors have read and agreed to the published version of the manuscript.

Funding: This research was funded by the Science and Technology Key R&D Project of Guangxi (grant number: AB22035074), Scientific Research Project of Department of Education of Jilin Province (grant number: JJKH20221019KJ), and Postdoctoral Researcher Selection Funding Project of Jilin Province.

Institutional Review Board Statement: Not applicable.

Informed Consent Statement: Not applicable.

Data Availability Statement: Not applicable.

References

1. Deng, N.C.; Yu, M.S.; Yao, X.Y. Intelligent Active Correction Technology and Application of Tower Displacement in Arch Bridge Cable Lifting Construction. *Appl. Sci.* **2021**, *11*, 9808. [CrossRef]
2. Gao, H.Y.; Zhang, K.; Wu, X.Y.; Liu, H.J.; Zhang, L.Z. Application of BRB to Seismic Mitigation of Steel Truss Arch Bridge Subjected to Near-Fault Ground Motions. *Buildings* **2022**, *12*, 2147. [CrossRef]
3. Anastasopoulos, D.; Maes, K.; De Roeck, G.; Lombaert, G.; Reynders, E.P.B. Influence of frost and local stiffness variations on the strain mode shapes of a steel arch bridge. *Eng. Struct.* **2022**, *273*, 115097. [CrossRef]
4. Sun, J.P.; Li, J.B.; Jiang, Y.B.; Ma, X.G.; Tan, Z.H.; Zhufu, G.L. Key Construction Technology and Monitoring of Long-Span Steel Box Tied Arch Bridge. *Int. J. Steel Struct.* **2023**, *23*, 191–207. [CrossRef]
5. Bobko, C.P.; Edwards, A.J.; Seracino, R.; Zia, P. Thermal Cracking of Mass Concrete Bridge Footings in Coastal Environments. *J. Perform. Constr. Fac.* **2015**, *29*, 04014171. [CrossRef]
6. Gao, X.L.; Li, C.G.; Qiao, Y.H. Study on On-site Monitoring of Hydration Heat of Mass Concrete for Bridge Slab Based on Measured Data. *Intell. Autom. Soft Comput.* **2019**, *25*, 775–783. [CrossRef]
7. Huang, Y.H.; Liu, G.X.; Huang, S.P.; Rao, R.; Hu, C.F. Experimental and finite element investigations on the temperature field of a massive bridge pier caused by the hydration heat of concrete. *Constr. Build. Mater.* **2018**, *192*, 240–252. [CrossRef]
8. Hamid, H.; Chorzepa, M.G.; Sullivan, M.; Durham, S.; Kim, S.S. Novelties in Material Development for Massive Concrete Structures: Reduction in Heat of Hydration Observed in Ternary Replacement Mixtures. *Infrastructures* **2018**, *3*, 8. [CrossRef]
9. Guo, C.; Lu, Z.R. Effect of temperature on CFST arch bridge ribs in harsh weather environments. *Mech. Adv. Mater. Struc.* **2022**, *29*, 732–747. [CrossRef]

10. Lawrence, A.M.; Tia, M.; Ferraro, C.C.; Bergin, M. Effect of Early Age Strength on Cracking in Mass Concrete Containing Different Supplementary Cementitious Materials: Experimental and Finite-Element Investigation. *J. Mater. Civ. Eng.* **2012**, *24*, 362–372. [CrossRef]

11. de Freitas, J.A.T.; Cuong, P.T.; Faria, R. Modeling of cement hydration in high performance concrete structures with hybrid finite elements. *Int. J. Numer. Meth. Eng.* **2015**, *103*, 364–390. [CrossRef]

12. Honorio, T.; Bary, B.; Benboudjema, F. Factors affecting the thermo-chemo-mechanical behaviour of massive concrete structures at early-age. *Mater. Struct.* **2016**, *49*, 3055–3073. [CrossRef]

13. Pepe, M.; Lima, C.; Martinelli, E. Early-Age Properties of Concrete Based on Numerical Hydration Modelling: A Parametric Analysis. *Materials* **2020**, *13*, 2112. [CrossRef]

14. Cha, S.L.; Jin, S.S. Prediction of thermal stresses in mass concrete structures with experimental and analytical results. *Constr. Build. Mater.* **2020**, *258*, 120367. [CrossRef]

15. Klemczak, B.; Zmij, A. Insight into Thermal Stress Distribution and Required Reinforcement Reducing Early-Age Cracking in Mass Foundation Slabs. *Materials* **2021**, *14*, 477. [CrossRef]

16. Singh, P.R.; Rai, D.C. Effect of Piped Water Cooling on Thermal Stress in Mass Concrete at Early Ages. *J. Eng. Mech.* **2018**, *144*, 04017183. [CrossRef]

17. Tasri, A.; Susilawati, A. Effect of material of post-cooling pipes on temperature and thermal stress in mass concrete. *Structures* **2019**, *20*, 204–212. [CrossRef]

18. Tasri, A.; Susilawati, A. Effect of cooling water temperature and space between cooling pipes of post-cooling system on temperature and thermal stress in mass concrete. *J. Build. Eng.* **2019**, *24*, 100731. [CrossRef]

19. Han, S.Y. Assessment of curing schemes for effectively controlling thermal behavior of mass concrete foundation at early ages. *Constr. Build. Mater.* **2020**, *230*, 117004. [CrossRef]

20. Dong, F.; Li, X.L.; Xie, Q.M.; Ye, R.; Cao, S.C. The influence of weather and temperature on pedestrian walking characteristics on the zigzag bridge. *Int. J. Biometeorol.* **2022**, *66*, 2541–2552. [CrossRef]

21. Fan, J.S.; Li, B.L.; Liu, C.; Liu, Y.F. An efficient model for simulation of temperature field of steel-concrete composite beam bridges. *Structures* **2022**, *43*, 1868–1880. [CrossRef]

22. da Silva, S.; Figueiredo, E.; Moldovan, I. Damage Detection Approach for Bridges under Temperature Effects using Gaussian Process Regression Trained with Hybrid Data. *J. Bridge Eng.* **2022**, *27*, 04022107. [CrossRef]

23. Xia, Y.; Chen, B.; Zhou, X.Q.; Xu, Y.L. Field monitoring and numerical analysis of Tsing Ma Suspension Bridge temperature behavior. *Struct. Control Health Monit.* **2013**, *20*, 560–575. [CrossRef]

24. Xu, X.; Huang, Q.; Ren, Y.; Zhao, D.Y.; Yang, J.; Zhang, D.Y. Modeling and Separation of Thermal Effects from Cable-Stayed Bridge Response. *J. Bridge Eng.* **2019**, *24*, 04019028. [CrossRef]

25. Wang, G.X.; Ding, Y.L.; Liu, X.W. The monitoring of temperature differences between steel truss members in long-span truss bridges compared with bridge design codes. *Adv. Struct. Eng.* **2019**, *22*, 1453–1466. [CrossRef]

26. Lei, X.; Fan, X.T.; Jiang, H.W.; Zhu, K.N.; Zhan, H.Y. Temperature Field Boundary Conditions and Lateral Temperature Gradient Effect on a PC Box-Girder Bridge Based on Real-Time Solar Radiation and Spatial Temperature Monitoring. *Sensors* **2020**, *20*, 5261. [CrossRef]

27. Fang, Z.Y.; Roy, K.; Liang, H.; Poologanathan, K.; Ghosh, K.; Mohamed, A.M.; Lim, J.B.P. Numerical Simulation and Design Recommendations for Web Crippling Strength of Cold-Formed Steel Channels with Web Holes under Interior-One-Flange Loading at Elevated Temperatures. *Buildings* **2021**, *11*, 666. [CrossRef]

28. Fang, Z.; Roy, K.; Lakshmanan, D.; Pranomrum, P.; Li, F.; Lau, H.H.; Lim, J.B.P. Structural behaviour of back-to-back cold-formed steel channel sections with web openings under axial compression at elevated temperatures. *J. Build. Eng.* **2022**, *54*, 104512. [CrossRef]

29. *GB/T 51028-2015*; Technical Code for Temperature Measurement and Controlof Mass Concrete. China Architecture Publishing & Media Co., Ltd.: Beijing, China, 2015.

MDPI

St. Alban-Anlage 66

4052 Basel

Switzerland

www.mdpi.com

Materials Editorial Office

E-mail: materials@mdpi.com

www.mdpi.com/journal/materials